9th International Conference on 3D Radiation Dosimetry (IC3DDose 2016)

Journal of Physics: Conference Series Volume 847

Galveston, Texas, USA
7 – 10 November 2016

ISBN: 978-1-5108-4241-0
ISSN: 1742-6588

Printed from e-media with permission by:

Curran Associates, Inc.
57 Morehouse Lane
Red Hook, NY 12571

Some format issues inherent in the e-media version may also appear in this print version.

Printed by Curran Associates, Inc. (2017)

For permission requests, please contact the Institute of Physics
at the address below.

Institute of Physics
Dirac House, Temple Back
Bristol BS1 6BE UK

Phone: 44 1 17 929 7481
Fax: 44 1 17 920 0979

techtracking@iop.org

Additional copies of this publication are available from:

Curran Associates, Inc.
57 Morehouse Lane
Red Hook, NY 12571 USA
Phone: 845-758-0400
Fax: 845-758-2633
Email: curran@proceedings.com
Web: www.proceedings.com

Table of contents

Volume 847

9th International Conference on 3D Radiation Dosimetry
7–10 November 2016, Texas, USA

Accepted papers received: 11 May 2017
Published online: 5 June 2017

Preface

Papers

IOP Conf. Series: Journal of Physics: Conf. Series **847** (2017) 011001 doi:10.1088/1742-6596/847/1/011001

IC3DDose 2016 - 9th International Conference on 3D Radiation Dosimetry

Preface

It was a great pleasure to welcome participants to IC3DDose 2016, the 9th International Conference on 3D Radiation Dosimetry, held from 7–10 November 2016 in Galveston, Texas.

The series of conferences has evolved considerably during its history. At the first conference, DOSGEL '99, the discussion centered around gel dosimetry. Held in Lexington, Kentucky in 1999, it was timed to coincide with the American Association of Physicists in Medicine (AAPM) Annual Meeting in Nashville, Tennessee. It was my honour to organize that first conference, and it was once again my honour to organize the 9th conference in the series now known as IC3DDose which was held in Galveston, Texas.

As was the case with recent IC3DDose conferences, the topic has broadened considerably beyond gel dosimetry. Not only have newer 3D volumetric dosimeters appeared on the scene, but novel electronic dosimetry systems and software that generate quasi-3D dose information have also. These changes have tracked advances in radiation oncology as techniques such as IMRT, VMAT, and IGRT have become almost ubiquitous. At the same time, dynamic treatments including gating and tracking now enjoy widespread use. Novel treatment technologies have appeared with perhaps the most disruptive being combined MR imaging-treatment units such as the ViewRay MR-cobalt unit and the Elekta/Philips MR-Linac.

The potential benefits offered by 3D dosimetry were explored, compared and evaluated during IC3DDose 2016. Novel and improved readout techniques, some of which take advantage of the contemporary treatment environment and new QA systems and procedures, as well as other aspects of clinical dosimetry were well represented in the program.

Over the past several years, the importance of safety in radiation therapy has been highlighted. The benefits of 3D dosimetry in contributing to safe and accurate treatments cannot be overstated. The implementation of new dosimetry procedures also requires careful attention to safety, accuracy and thorough validation. Therefore, the aims of IC3DDose 2016 were similar to those from previous conferences:

- to enhance the quality and accuracy of radiation therapy treatment through improved clinical dosimetry;
- to investigate and understand the dosimetric challenges of modern radiation treatment techniques;
- to provide a forum to discuss the latest research and developments in 3D and advanced radiation dosimetry; and,
- to energize and diversify dosimetry research and clinical practice by encouraging interaction and synergy among advanced, 3D and semi-3D dosimetry techniques.

IC3DDose 2016 was a stimulating and exciting conference and the program included substantial contributions to scientific progress in the field.

Thanks are offered to those who helped make the conference possible. The Scientific Committee, led by Sam Beddar, conducted thorough peer reviews of submitted papers in a timely manner and made valuable and thoughtful suggestions for improvements. The Committee also contributed to the planning of the meeting and to the organization of the program. The distinguished invited speakers who agreed to present are thanked for the extra work they performed to design and prepare educational and timely talks. My colleagues at MD Anderson Cancer Center are also thanked for their contributions to the planning of the meeting, managing the logistics, and generally helping to make sure that the conference was a success.

Geoff Ibbott
MD Anderson Cancer Center

INTERNATIONAL SCIENTIFIC COMMITTEE

Sam Beddar (Chair) (USA)
Claus Andersen (Sweden)
Sven Back (Sweden)
Clive Baldock (Conference Proceedings Editor) (Australia)
Crister Ceberg (Sweden)
Yves de Deene (Australia)
Simon Doran (United Kingdom)
Anna Karlsson Hauer (Sweden)
Andrew Jirasek (Canada)

IOP Conf. Series: Journal of Physics: Conf. Series **847** (2017) 011001 doi:10.1088/1742-6596/847/1/011001

Kevin Jordan (Canada)
Martin Lepage (Canada)
Daniel Low (USA)
Mark Oldham (USA)
John Schreiner (Canada)
Cheng-Shie Wuu (USA)

LOCAL ORGANIZING COMMITTEE

Geoffrey Ibbott (Chair)
Dana Garrison
Kayla White
Hannah Lee
Yvonne Roed
Mitchell Carroll
Sam Beddar

IOP Conf. Series: Journal of Physics: Conf. Series **847** (2017) 011002 doi:10.1088/1742-6596/847/1/011002

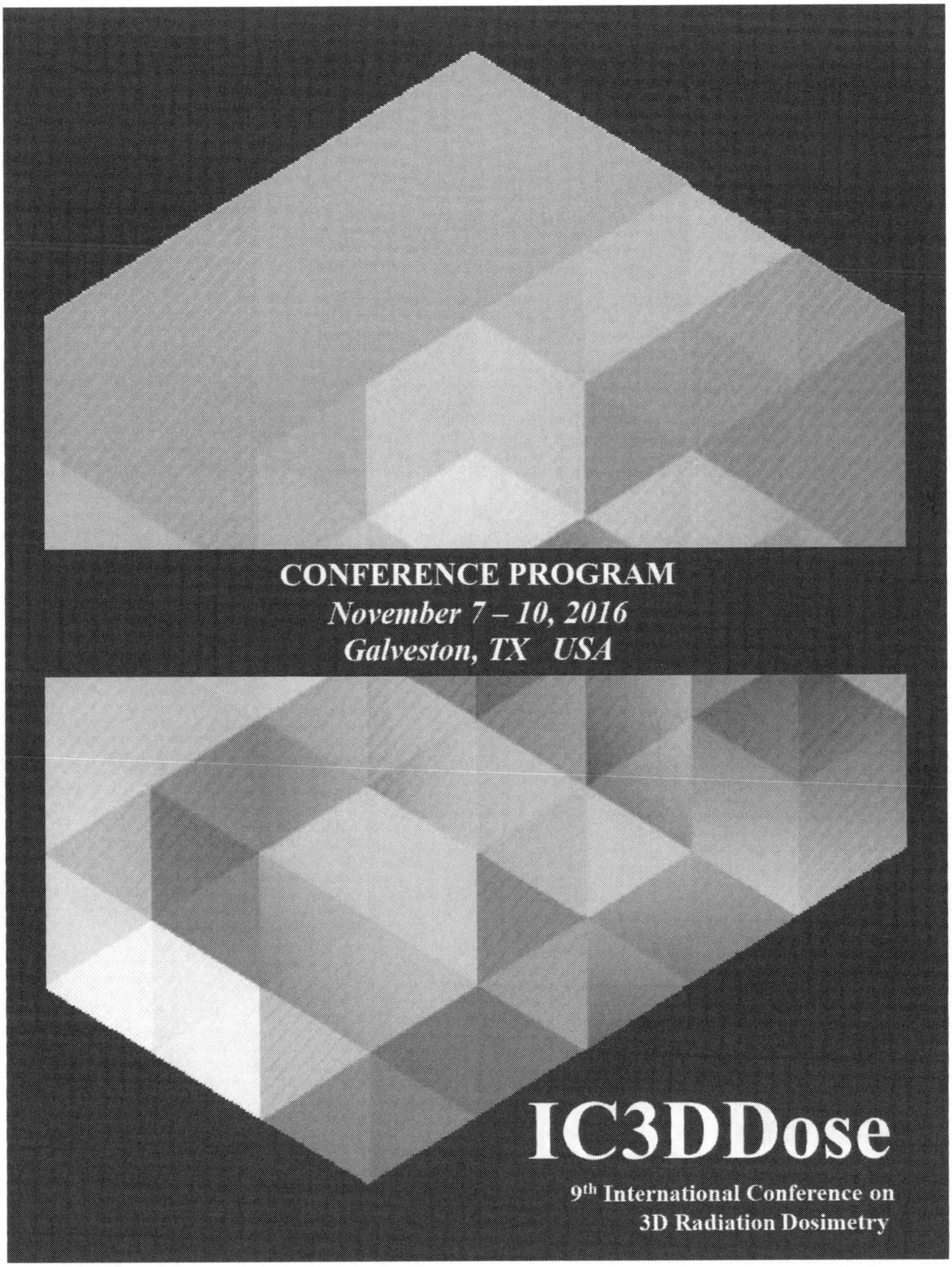

IOP Conf. Series: Journal of Physics: Conf. Series **847** (2017) 011002 doi:10.1088/1742-6596/847/1/011002

Safety. Accuracy.

info@rt-safe.com

www.rt-safe.com

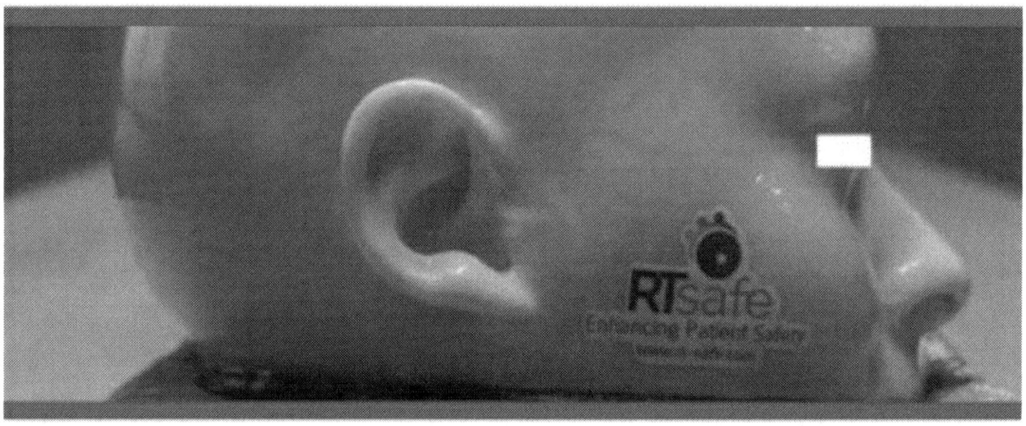

RTsafe is a medical technology company that has developed a unique approach to quality assurance that significantly enhances the safety and accuracy of radiotherapy for cancer and other medical conditions.

We combine proven expertise in medical physics - 3D radiation dosimetry with highly accurate 3D printing technology to create pseudo-in-vivo dosimetry phantoms towards End-to-End commissioning, benchmarking and quality assurance in SRS, IMRT, VMAT and SBRT applications.

Our 3D-printed phantoms offer:

- Commissioning / Periodic QA
- Patient-specific pre-treatment plan verification
- Absolute dosimetry
- High spatial resolution (of ~1mm3) 3D relative dosimetry

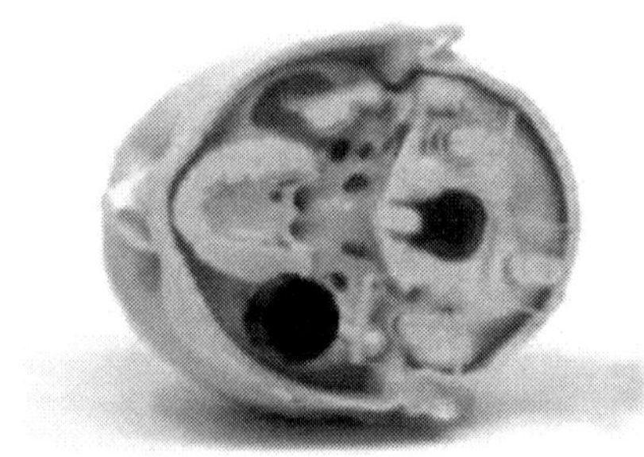

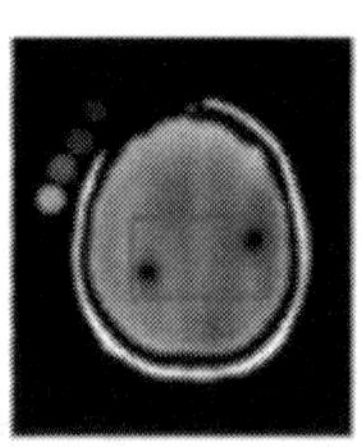

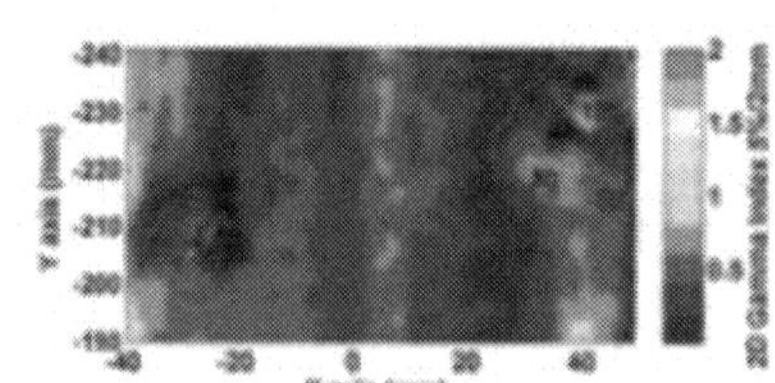

RTsafe

IOP Conf. Series: Journal of Physics: Conf. Series **847** (2017) 011002 doi:10.1088/1742-6596/847/1/011002

IC3DDose

9th International Conference on 3D Radiation Dosimetry

INTRODUCTION AND WELCOME

It is a great pleasure to welcome you to the 9th International Conference on Three-Dimensional Dosimetry: IC3DDose16. The conference has evolved considerably during its history. At the first conference, the discussion centered around gel dosimetry, and in fact, that first conference was called DOSGEL. It was held in Lexington, Kentucky in 1999 and was timed to coincide with the AAPM Annual Meeting in Nashville. It was my honor to organize that first meeting, and it is once again my honor to organize the 9th conference, to be held in Galveston, Texas.

As was the case with recent IC3DDose conferences, the topic has broadened considerably beyond gel dosimetry. Not only have newer 3D volumetric dosimeters appeared on the scene, but also novel electronic dosimetry systems and software that generate quasi-3D dose information. These changes have tracked the advances in radiation oncology as techniques such as IMRT, VMAT, and IGRT have become almost ubiquitous. At the same time, dynamic treatments including gating and tracking enjoy widespread use. Novel treatment technologies have appeared, perhaps the most disruptive being combined MR imaging-treatment units such as the ViewRay MR-cobalt unit and the Elekta/Philips MR-Linac.

The potential benefits offered by 3D dosimetry will be explored, compared and evaluated during the course of the meeting. Novel and improved readout techniques have been developed, some of which take advantage of the contemporary treatment environment. New QA systems and procedures, as well as other aspects of clinical dosimetry are represented in the program.

Over the past several years, the importance of safety in radiation therapy has been highlighted. The benefits of 3D dosimetry in contributing to safe and accurate treatments cannot be overstated. The implementation of new dosimetry procedures also requires careful attention to safety, accuracy and thorough validation. Therefore, the aims of IC3Ddose16 are similar to those from previous conferences, and are:

- To enhance the quality and accuracy of radiation therapy treatment through improved clinical dosimetry
- To investigate and understand the dosimetric challenges of modern radiation treatment techniques
- To provide a forum to discuss the latest research and developments in 3D and advanced radiation dosimetry
- To energize and diversify dosimetry research and clinical practice by encouraging interaction and synergy among advanced, 3D and semi-3D dosimetry techniques.

We sincerely hope that you will find the meeting program to be stimulating and exciting and anticipate that the program will include substantial contributions to scientific progress in the field.

We offer our thanks to those who helped make this meeting possible. The Scientific Committee, led by Sam Beddar, conducted thorough reviews of the abstracts in a timely manner and made valuable and thoughtful suggestions for improvements. The Committee also contributed to the planning of the meeting and to the organization of the program. The distinguished invited speakers who agreed to present are thanked for the extra work they performed to design and prepare educational and timely talks. And my colleagues at MD Anderson for their contributions to the planning of the meeting, managing the logistics, and generally helping to make sure that the conference is a success.

Welcome to North America's Third Coast, and to Galveston, Texas! We hope you will enjoy the scientific interactions, the sharing of knowledge, and all that Galveston Island has to offer!

Dana Garrison, Kayla White, Sam Beddar, and Geoff Ibbott

IC3DDose
9th International Conference on 3D Radiation Dosimetry

IC3DDose16 ACKNOWLEDGES

The scientific and financial support from:

THE UNIVERSITY OF TEXAS
MD Anderson ~~Cancer~~ Center
Making Cancer History®

And the endorsement by:

AAPM

AAPM does not endorse the product in question; rather it endorses the educational component of the program.

The generous support from our sponsors and exhibitors

ScandiDos
modusQA Accuracy. Confidence.™
DoseOptics Visualize radiation therapy
SUN NUCLEAR corporation
MGS
ELEKTA
VARIAN medical systems
GeVero Co.
RTsafe

INTERNATIONAL SCIENTIFIC COMMITTEE

Sam Beddar (Chair) (USA)
Clive Baldock (Conference Proceeding Editor) (Australia)
Claus Andersen (Sweden)
Sven Back (Sweden)
Crister Ceberg (Sweden)
Yves de Deene (Belgium/Australia)
Simon Doran (United Kingdom)
Anna Karlsson Hauer (Sweden)
Andrew Jirasek (Canada)
Kevin Jordan (Canada)
Martin Lepage (Canada)
Daniel Low (USA)
Mark Oldham (USA)
John Schreiner (Canada)
Cheng-Shie Wuu (USA)

LOCAL ORGANIZING COMMITTEE

Geoffrey Ibbott, (chair)
Dana Garrison
Kayla White

IOP Conf. Series: Journal of Physics: Conf. Series **847** (2017) 011002 doi:10.1088/1742-6596/847/1/011002

IC3DDose
9th International Conference on 3D Radiation Dosimetry

PROGRAM

Registration

Online registration on Sunday November 6, 2016
Onsite registration open at 7:00 a.m. on Monday November 7, 2016

Monday, November 7, 2016:

8:00-8:30 **Welcome and Introduction**
Geoff Ibbott and Sam Beddar

8:30-9:30 am **Morning Lecture I**
John Schreiner
Reviewing Three Dimensional Dosimetry: basics and utilization as presented over 17 Years of IC3Ddose

9:30-10:30 am **Session 1 – Role of 3D Dosimetry in QA**
Chair/co-Chair: A K Hauer, D Robertson

A Bäck, F Nordström, M Gustafsson, J Götstedt and A Karlsson Hauer
Complexity metric based on fraction of penumbra dose – initial study
Paul B Ravindran, Ebenezer Suman Babu, Winfred Michael Raj and Amalan S
A Preliminary Study on the use of FX gel and an in-house optical Cone beam CT readout for IMRT and RapidArc Verification
S W Yoon, D Miles, C Cramer, M Reinsvold, D Kirsch, M Oldham
Treatment Planning and 3D Dose Verification of Whole Brain Radiation Therapy with Hippocampal Avoidance in Rats
D. Robertson and S. Beddar
Patient-specific Quality Assurance for Spot Scanning Proton Beams Using a Large-volume Liquid Scintillator Detector

10:30-11:00 am **Break**

11:00-11:30 am **Keynote Presentation**
Mark Oldham
Innovation and the Future of Advanced Dosimetry: 2D to 5D

11:30-12:30 pm **Session 2- New Advances is Detectors based on Optical-CT systems I**
Chair/co-Chair: J Schreiner, S Doran

EE Carpentier, KM Alexander and LJ Schreiner
Quantifying refractive index mismatch effects on cone beam optical CT scanner measurements
K.H. Dekker, J.J. Battista, and K.J. Jordan
Fan-beam scanning laser optical computed tomography for large volume dosimetry

IOP Conf. Series: Journal of Physics: Conf. Series **847** (2017) 011002 doi:10.1088/1742-6596/847/1/011002

IC3DDose
9th International Conference on 3D Radiation Dosimetry

PROGRAM

K.J. Jordan, T. Lindenmaier and K.H. Dekker
Non-diffusing radiochromic leuco-crystal violet hydrogel dosimeter
O Bleuse, R Gschwind, Y Bailly, K Laurent, H Bártová, K. Pilařová, VSpěváčekc, L Makovicka
3D Dosimetry based on a new optical approach for dosimetry gels: Use of the polarisation ratio of the scattering light

12:30-2:00 pm **Lunch – Sponsored Presentation by ModusQA** – *3D Dosimetry products from Modus QA.*

2:00-2:30 pm **Invited Presentation**
Laurence Court
Is there a role for 3D dosimetry in low- and middle-income countries?

2:30-3:15 pm **Session 3 – PAGAT based Gel Dosimeters**
Chair/co-Chair: S Doran, M Oldham

Charles Matrosic, Alan McMillan, James Holmes, Bryan Bednarz, and Wesley Culberson
Dosimetric comparison of DEFGEL and PAGAT formulae paired with an MRI acquisition
Takaoki Takanashi, Hiraku Kawamura, Kazuaki Fukasaku, Daniel Antonio Sahade and Toshimasa Hamada
A comparison of the dose–response behavior of AQUAJOINT®-based polymer gel and PAGAT gel dosimeters measured using Optical CT and MRI
E.JJ Samuel, P.Sathiyaraj, Murat Kurudirek
Discrepancies in Effective atomic number of metal impregnated PAGAT gel dosimeter

3:15-4:00 pm **Break**

4:00-5:00 pm **Session 4 – New Advances in Radiochromic Gel Dosimeters I**
Chair/co-Chair: L Beaulieu, C Huet

C Huet, J Colnot and I Clairand
Preliminary investigation of the dosimetric properties of ClearView™ dosimeter
María del Rocío Bernal Zamorano, Nicolai Højer Sanders, Lars René Lindvold, Claus E. Andersen
Radiochromic and radiofluorogenic 3D solid polymer dosimeter
J Colnot, C Huet and I Clairand
Characterisation of TruView™: a new 3-D reusable radiochromic MethylThymolBlue based gel dosimeter for ionising radiations
M A Silveira, J F Pavoni and O Baffa
Characteristics of modified radiochomic gel Turnbull-Blue

5:00-5:20 pm **Poster Preliminary Highlights (1min/1slide)**
Chair/co-Chair: C Baldock

IOP Conf. Series: Journal of Physics: Conf. Series **847** (2017) 011002 doi:10.1088/1742-6596/847/1/011002

PROGRAM

IC3DDose
9th International Conference on 3D Radiation Dosimetry

Tuesday, November 8, 2016:

8:30-9:30 am **Morning Lecture II**
Kevin Jordan, Michelle Hilts and Andrew Jirasek
Optical and X-ray computed tomography scanning of 3D dosimeters

9:30-10:30 am **Session 1 – Novel 3D detectors I**
Chair/co-Chair: PS Skyt, D Robertson

C D Darne, D Robertson and S Beddar
Development of a liquid scintillator-based 3D detector for range measurements of spot-scanned proton beams
E M Høye , M Sadel, L Kaplan, P S Skyt, L P Muren, J B B Petersen, J Swakoń, G Mierzwińska, M Rydygier, L Malinowski and P Balling
First 3D measurements of proton beams in a deformable silicone-based dosimeter
Fahed Alsanea and Sam Beddar
Evaluating analytical ionization quenching correction models for 3D liquid organic scintillator detector
Louis Archambault, Madison Rilling, Xavier Roy-Pomerleau, Simon Thibault
Dynamic 3D measurement of modulated radiotherapy: a scintillator-based approach

10:30-11:00 am **Break**

11:00-11:30 am **Keynote Presentation**
Ben Mijnheer
EPID-based 3D dosimetry and its relation to other 3D dose measurement techniques

11:30-12:30 pm **Session 2 –New Advances in Detectors based on Optical-CT systems II**
Chair/co-Chair: B Mijnheer, K McAuley
Yi Du, Xincheng Xiang, Xiangang Wang, Yves De Deene
Convergence of SART+OS+TV Iterative Reconstruction Algorithm for Optical CT Imaging of Gel Dosimeters
Devin Miles, Paul Yoon, Jacob Kodra, John Adamovics, Mark Oldham
A prototype cost-effective optical-CT system for PRESAGE 3D dosimeter readout
M. Carroll, M. Alqathami, and G. Ibbott
The quenching effect in PRESAGE® by a proton beam: Investigation of formulation dependence
Simon J Doran, Paolo Pellicioli, John Adamovics and Elke Brauer-Krisch
Enhanced method for determining the low-LET saturation dose of PRESAGE®

12:30-2:00 pm **Lunch – Sponsored Presentation by MGS –** *Functionality of VOLQA software.*

2:00-2:30 pm **Invited Presentation**
David Followill
The IROC Houston QA Program – Potential Benefits of 3D Dosimetry

IOP Conf. Series: Journal of Physics: Conf. Series **847** (2017) 011002 doi:10.1088/1742-6596/847/1/011002

IC3DDose
9th International Conference on 3D Radiation Dosimetry

PROGRAM

2:30-3:15 pm **Session 3 – Clinical Dosimetry Applications I**

Chair/co-Chair: T Olding, Luc Beaulieu

P Sibolt, R O Cronholm, E Heath, C E Andersen and C F Behrens

Automated four-dimensional Monte Carlo workflow using log files and real-time motion monitoring

J C Lizar, L F Santos, F C Brandão, K C Volpato, F S Guimarães and J F Pavoni

Evaluation of the respiratory motion influence in the 3D dose distribution of IMRT breast radiation therapy treatments

T Olding, KM Alexander and LJ Schreiner

Dosimetric Impact of a Change in Breathing Period on VMAT Stereotactic Ablative Body Radiotherapy

3:15-4:00 pm **Break**

4:00-5:00 pm **Session 4 –Clinical Dosimetry Applications II**

Chair/co-Chair: J Schreiner, G Kertzscher

B Mijnheer, R Rozendaal, I Olaciregui-Ruiz, P González, R van Oers, A Mans

New developments in EPID-based 3D dosimetry in The Netherlands Cancer Institute

P Bruza, J M Andreozzi, D J Gladstone, L A Jarvis, J Rottmann and B W Pogue

Real-time 3D dose imaging in water phantoms: reconstruction from simultaneous EPID-Cherenkov 3D imaging (EC3D)

Molham M Eyadeh, Marcin Wierzbicki, Kevin R Diamond

Measurements of superficial dose distributions in radiation therapy using translucent cryogel dosimeters

G Kertzscher and S Beddar

Inorganic scintillator detectors for real-time verification during brachytherapy

IOP Conf. Series: Journal of Physics: Conf. Series **847** (2017) 011002 doi:10.1088/1742-6596/847/1/011002

IC3DDose
9th International Conference on 3D Radiation Dosimetry

PROGRAM

Wednesday, November 9, 2016:

8:30-9:30 am **Morning Lecture III**
Tommy Knoos
3D Dose Computation Algorithms

9:30-10:30 am **Session 1 – Clinical Dosimetry Applications III**
Chair/co-Chair: T Knoos, D Mackin

KM Alexander, J Gooding, LJ Schreiner and T Olding
Clinical Management of Tumour Volume Changes in VMAT Head & Neck Radiation Treatment
A Venning, M Chandroth and B Chick
Clinical experience of the normoxic PAG gel dosimeter using a flattening filter free beam for SABR treatments with MRI readout
P Mann, N Saito, C Lang, A Runz, W Johnen, M Witte, D Schmitt and C P Karger
Validation of 4D dose calculation using an independent motion monitoring by the calypso tracking system and 3D polymer gel dosimetry
J C Xie, E A B Katz, K M Alexander, L J Schreiner and K B McAuley
Leuco-crystal-violet micelle gel dosimeters: Component effects on dose-rate dependence

10:30-11:00 am **Break**

11:00-11:30 am **Keynote Presentation**
Sasa Mutic
ViewRay system: Dosimetry in the presence of low field MRI

11:30-12:30 pm **Session 2 – Novel 3D detectors II**
Chair/co-Chair: P Balling, B Pogue
B W Pogue, V Krishnaswamy, M Jermyn, P Bruza, T Miao, William Ware, S L Saunders, J M Andreozzi, D J Gladstone and L A Jarvis
Real time radiotherapy verification with Cherenkov imaging: development of a system for beamlet verification
Marie-Ève Delage, Marie-Ève Lecavalier, Dominic Larivière, Claudine Ni Allen and Luc Beaulieu
Preliminary investigation of a luminescent colloidal quantum dots-based liquid scintillator
M Sadel, E M Høye, P S Skyt, L P Muren, J B B Petersen and P Balling
Three-dimensional radiation dosimetry based on optically-stimulated luminescence
D Mackin, E Draeger, S Peterson, J Polf, and S Beddar
In Vivo Verification of Particle Therapy: How Compton Camera Configurations Affect 3D Image Quality

12:30-2:00 pm **Lunch – Sponsored Presentation by Sun Nuclear.**

IOP Conf. Series: Journal of Physics: Conf. Series **847** (2017) 011002 doi:10.1088/1742-6596/847/1/011002

IC3DDose
9th International Conference on 3D Radiation Dosimetry

PROGRAM

2:00-2:30 pm **Invited Presentation**

Brian Pogue

Cherenkov imaging in the potential roles of radiotherapy QA and delivery

2:30-3:15 pm **Session 3 – 3D Gel Dosimetry**

Chair/co-Chair: A Jirasek, K McAuley

E Maynard, M Hilts, E Heath and A Jirasek

Accuracy and precision in x-ray computed tomography polymer gel dosimetry

Kalin I Penev, Meng Wang and Kibret Mequanint

Tetrazolium salt monomers for gel dosimetry I: Principles

Yudy Ascención, Jennifer Dietrich, Kibret Mequanint and Kalin I Penev

Tetrazolium salt monomers for gel dosimetry II: Dosimetric characterization of the ClearView™ 3D dosimeter

K.J. Jordan and K.H. Dekker

Active heating of large gel dosimeters

3:15-4:00 pm **Break**

4:00-5:00 pm **Session 4 – Poster Viewing**

(See end of the Program on page 11)

IOP Conf. Series: Journal of Physics: Conf. Series **847** (2017) 011002 doi:10.1088/1742-6596/847/1/011002

IC3DDose
9th International Conference on 3D Radiation Dosimetry

PROGRAM

Thursday, November 10, 2016:

8:00-9:00 am **Morning Lecture IV**
Simon Doran
Dosimetry using MRI: Can it really be that difficult?

9:00-9:30 am **Keynote Presentation**
David Fuller
Clinical Implications of Accurate Spatial Dosimetry: Importance for Patient Care

9:30-10:15 am **Session 1 – New Advances in Radiochromic Gel Dosimeters II**
Chair/co-Chair: K Jordan, C-S Wuu

EE Carpentier, KM Alexander, S Todd and LJ Schreiner
Characterization of a radiochromic silicone dosimeter
Ebenezer Suman Babu S, Timothy Peace Balasingh S, Benedicta Pearlin R, Rabi Raja Singh I and Paul Ravindran B
Cerium nanoparticle effect on sensitivity of Fricke gel dosimeter: Initial investigation
J Kodra, D Miles, S W Yoon, D Kirsch and M Oldham
Characterization of novel preclinical dose distributions for micro irradiator

10:15-10:45 am **Break**

10:45-11:15 pm **Invited Presentation**
Gabriel Sawakuchi
Dosimetry in the presence of strong magnetic fields

11:15-12:15 pm **Session 2 – 3D Detectors in MR Environments**
Chair/co-Chair: J Wang, D Fuller

S Mein, L Rankine, D Miles, T Juang, B Cai, A Curcuru, S Mutic, J Fenoli, J Adamovics, H Li and M Oldham
Remote 3D dosimetry for MR guided Radiation Therapy (MRgRT)
Hannah J. Lee, Gye Won Choi, Mamdooh Alqathami, Mo Kadbi, Geoffrey Ibbott
Using 3D dosimetry to quantify the electron return effect (ERE) for MR-image-guided radiation therapy (MR-IGRT) applications
Filipa Costa, Simon Doran, Simeon Nill, Simon Duane, David Shipley, Ilias Billas, John Adamovics, Uwe Oelfke
Development of a methodology to study the effect of magnetic field on dose distributions in an MR-linac, using PRESAGE® and Monte Carlo calculations

IOP Conf. Series: Journal of Physics: Conf. Series **847** (2017) 011002 doi:10.1088/1742-6596/847/1/011002

IC3DDose
9th International Conference on
3D Radiation Dosimetry

PROGRAM

Y Roed, Y Ding, J Wang, L Pinsky and G Ibbott
The potential of polymer gel dosimeters for 3D MR-IGRT quality assurance

12:15-12:45 pm Summary of Conference - Closing remarks & Goodbyes

12:45-2:00 pm **Lunch**

3D Slicer Workshop – 2:00 pm to 4:30 pm

Workshop Chairs: John Schreiner, Kevin Jordan, and Kevin Alexander
2:00 – 2:10 pm Introduction to workshop – who are we, why are we offering this, introduction to 3D Slicer and SlicerRT

2:10 – 2:20 pm Brief introduction to film analysis, limitations/challenges of current analysis tools

2:20 – 2:45 pm Film dosimetry hands-on example

2:45 – 3:00 pm Break (15 mins max)

3:00 – 3:10 pm Introduce gel dosimetry analysis, previous challenges using CERR, Matlab, Excel

3:10 – 3:45 pm Gel dosimetry hands-on example

3:45 – 4:15 pm Q&A session

4:15 – 4:30 pm Wrap-up and conclude.

Additional information on 3D Slicer and SlicerRT:

1. http://slicerrt.github.io/
2. http://slicerrt.github.io/Download.html
3. https://www.slicer.org/slicerWiki/index.php/Documentation/Nightly/Extensions/SlicerRT

(For those who are planning to attend the workshop, it would be advisable to download SlicerRT beforehand on your laptops or other personal computers)

IOP Conf. Series: Journal of Physics: Conf. Series **847** (2017) 011002 doi:10.1088/1742-6596/847/1/011002

IC3DDose
9th International Conference on 3D Radiation Dosimetry

PROGRAM

Poster presentations

1. L Vazquez-Quino, C Huerta-Hernandez, D Rangaraj
   Clinical experience with Mobius FX software for 3D dose verification for prostate VMAT plans and comparison with physical measurements.
2. KM Alexander, A Robinson, C Pinter, G Fichtinger LJ Schreiner
   Implementation of 3D Slicer Based Film Dosimetry Analysis
3. CJ Watson, JR Supple, AU Yeo, M Geso, RD Franich
   The effect of THPS concentration on nPAG microstructure
4. C Waldenberg, A Karlsson Hauer, C Gustafsson, S Ceberg
   Dose integration and dose rate characteristics of a NiPAM polymer gel MRI dosimeter system
5. E. James Jabaseelan Samuel, K Srinivasan
   Radiological properties of plastics and TLD Materials its application in radiation dosimetry
6. E. James Jabaseelan Samuel, K Srinivasan
   Radiation dose enhancement of gold nanoparticle on different polymer gel dosimeters
7. D Titus, EJJ Samuel K Srinivasan
   Silver nitrate based gel dosimeter
8. E Borguezan Neto, MR Batistuti, JF Pavoni, L Bachmann
   Turnbull Blue Gel (TBG) evaluation as optical dosimter
9. Y Du, X Xiang, X Wang, Y De Deene
   Computational Simulations of the Influence of Background Noise Removal in Optical-CT Imaging of Gel Dosimeters
10. Y Du, X Xiang, X Wang, Y De Deene
    Computational Simulations of the Influence of Angular Rotation Deviations in Optical CT Imaging of Gel Dosimeters
11. A Asena, ST Smith, T Kairn, SB Crowe, RD Franich, JV Trapp
    Reduction of artefacts caused by missing ray-sum data in optical computed tomography imaging of gel dosimeters
12. T Kairn, A Asena, SB Crowe, A Livingstone, D Papworth, S Smith, B Sutherland, S Sylvander, RD Franich, JV Trapp
    Use of a commercial gel dosimetry system to verify stereotactic spinal radiotherapy treatment dose distributions
13. C Baldock
    Trends in gel dosimetry: preliminary analysis of active growth areas, research trends and hot topics

IOP Conf. Series: Journal of Physics: Conf. Series **847** (2017) 011002 doi:10.1088/1742-6596/847/1/011002

NOTES

IC3DDose
9th International Conference on 3D Radiation Dosimetry

IOP Conf. Series: Journal of Physics: Conf. Series **847** (2017) 011002 doi:10.1088/1742-6596/847/1/011002

View of Galveston, TX

Thank you GeVero Co. for providing prizes (free 2-year software licenses) to the best oral presentations of the conference!

IOP Conf. Series: Journal of Physics: Conf. Series **847** (2017) 011002 doi:10.1088/1742-6596/847/1/011002

IOP Conf. Series: Journal of Physics: Conf. Series **847** (2017) 011003 doi:10.1088/1742-6596/847/1/011003

Peer review statement

All papers published in this volume of *Journal of Physics: Conference Series* have been peer reviewed through processes administered by the proceedings Editors. Reviews were conducted by expert referees to the professional and scientific standards expected of a proceedings journal published by IOP Publishing.

IOP Conf. Series: Journal of Physics: Conf. Series **847** (2017) 012001 doi:10.1088/1742-6596/847/1/012001

Reviewing three dimensional dosimetry: basics and utilization as presented over 17 Years of DosGel and IC3Ddose

L J Schreiner
Department of Medical Physics, Cancer Centre of Southeastern Ontario at the Kingston General Hospital, Kingston, Ontario, Canada, K7L 5P9
Departments of Oncology and Physics, Queen's University, Kingston, Ontario, Canada K7L 3N6

E-mail: john.schreiner@krcc.on.ca

Abstract. For seventeen years a community of basic and clinical scientists and researchers has been meeting bi-annually to promote the clinical advance of techniques to measure radiation dose in three dimensions. The interest in this dosimetry was motivated by its promise as an effective methodology for 3D measurement of the complex conformal dose distributions achieved by modern techniques such as Intensity Modulated and Volumetric Arc Radiation Therapy. Each of the International Conferences on 3D Radiation Dosimetry resulted in the publication of informative proceedings [1-8], the majority openly available on the internet. The proceedings included papers that: i) reviewed the basic science of the radiation sensitive materials used to accumulate the dose information, ii) introduced the science and engineering of the imaging systems required to read the information out, iii) described the work flows and systems required for efficient dosimetry, iv) reported the protocols required for reproducible dosimetry, and v) showed examples of clinical use illustrating advantage and limitations of the dosimetry. This paper is intended to use the framework provided by these proceedings to review the current 3D chemical dosimeters available and to discuss the requirements for their use. The paper describes how 3D dosimetry can complement other dose delivery validation approaches available in the clinic. It closes with some personal reflections of how the motivation for, and practice of, 3D dosimetry have changed (or not) over the years.

1. Introduction

Since its early days when megavoltage radiation advances first enabled delivery conforming the shape of the beam to the tumour volume (as described by Trump [9, 10]), conformal therapy has progressed considerably with the development of advanced external beam techniques such as provided by Intensity Modulated Radiation Therapy (IMRT), Volumetric Modulated Arc Therapy (VMAT), Stereotactic Ablative Radiation Therapy (SABR), Fractionated Stereotactic Radiotherapy (FSRT), etc.; and through improved procedures for high-dose rate brachytherapy. The radiation delivery associated by these modern techniques is complex and results in shaped conformal dose distributions that must be correctly registered to the patient's anatomy to achieve the treatment intent. Associated with the increased complexity of the targeting of this radiation delivery is a heightened requirement for quality assurance (QA) for technical components of the delivery, including the treatment unit and planning systems, for validation of the dose delivery planned for individual patients and, perhaps, for end-to-end QA of the multi-staged process associated with the delivery over the full course of treatment [11]. Numerous approaches have been established in the clinic to provide this assurance including, for example, the use of 2D and 3D arrays of detectors for patient specific dose delivery validation [12-14], and electronic portal imaging detector (EPID) based assessment [13, 15-18]. These approaches have their place and are clinically very useful [14]. But they typically provide sparse 3D data and only surrogate validation of 3D dose delivery. Full 3D dosimetry using volumetric chemical dosimeters probed by 3D imaging systems does provide for dose measurements an irradiated volume and hence gives a unique methodology for conformal delivery QA [19, 20] (see Fig. 1).

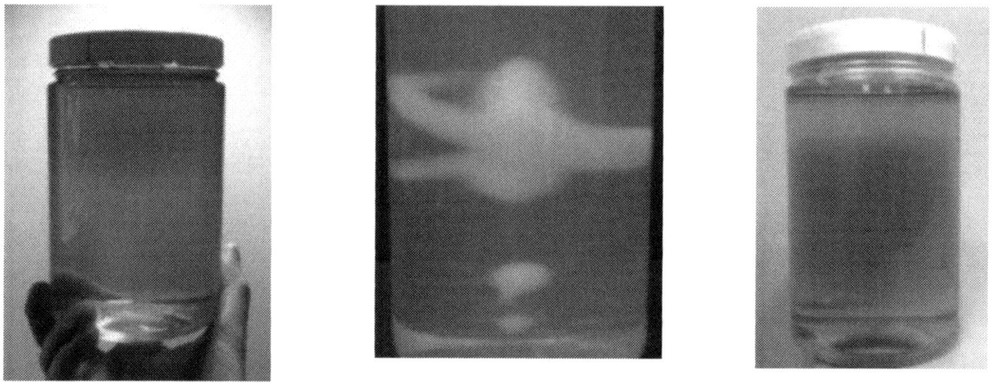

Figure 1. A recent illustration of the 3D dose information captured in three different gel dosimeters from the proceedings of IC3Ddose [21]. (left) The Fricke-xylenol-orange-gelatin dosimeter shows a colour change in the volume irradiated with a12 MeV electron beams. (centre) The polyacrylamide polymer gel dosimeter shows increased scatter in the high dose areas radiated using a Cobalt-60 tomotherapy IMRT delivery. (right) The colour change of a Leuco-crystal-violet micelle gel dosimeter after a VMAT prostate plan irradiation.

The clinical applicability of 3D dosimetry has advanced considerably in the last decade by the development of improved dosimeters [21] (e.g., radiochromic plastics [22, 23], radiochromic gel dosimeters [24,25] and normoxic polymer gel systems [26]) and by improved readout protocols using optical computed tomography or magnetic resonance imaging.

In the discussion above the definition of a 3D dosimeter has been set by elements of the RTAP criteria [27, 28]. Under RTAP an ideal true 3D dosimetry system (dosimeter and associated readout) should be able to deliver dose measurements in a 3D volume with 1 mm isotropic spatial resolution in less than one hour with 3% accuracy and a precision of 1%. As noted in the past [21], criteria for the resolution, accuracy and precision may be relaxed in clinic practice, depending on the specific validation being performed (for example, the criteria in external dose delivery audit under IROC is relaxed considerable for a number of practical reasons [29, 30]). But to date the criterion for high resolution isotropic measurement limits 'true' 3D radiation dosimetry to chemical radiation dosimetry based on quantifying the effects of radiation-induced chemical changes occurring within some volume of material [21, 27, 31]. And this is the condition used in this review to set the "true 3D" designation.

While true 3D dosimetry is, despite its promise, still not widely practiced in the community, clinical adoption beyond research laboratories is now well indicated. In fact, a comprehensive account of how 3D dosimetry has developed and can be used has been presented in 17 years of DosGel and IC3Ddose proceedings of the International Conferences on 3D Radiation Dosimetry [1-8]. The 101 invited review papers and 469 submitted proffered papers therein outline the history of field, report technical improvements and developments as the dosimetry advanced, and describe various roles 3D dosimetry can play in the clinic. They present clear evidence for improvements over time that eliminated the constraints that moderated past clinical acceptance (such as loss of spatial integrity of dose information in Fricke gels [32], response inhibition in oxygen contaminated polymer gels [33], reproducibility and stability challenges as preparation protocols change [34], etc.). And the proceedings have pointed to the benefits which 3D dose measurement could provide for specific clinical problems, often in conjunction with other dosimetry systems.

This paper presents a citation review of 3D dosimetry as collected in these conference proceedings, citing select literature from the proceedings to provide a reader new to 3D dosimetry a convenient and readily accessible introduction to, and presentation of, the field.

2. The Road to IC3Ddose

The International Conferences on 3D Radiation Dosimetry have, from the very first paper [35], presented a strong motivation for the use of true 3D dosimetry. Initially the main applications were thought to be best directed to patient specific dose delivery validation as had been indicated by Hiraoka *et al* in their visionary work using anthropomorphic phantoms and MRI [36, 37]. This promise was quickly supported in proffered reports of clinical applications such as Scheib *et al*'s [38] in-house measurement and analysis system for patient specific dose validation in radiosurgery using polyacrylamide gels, and many more papers that followed.

However, the conference reports over the years have also made clear that the requirements in the clinic [11, 14, 19, 27, 29, 30, 39, 40] do not limit useful dosimetry to the 'true' 3D dosimeters described above [11, 20, 39, 41]; rather it has been noted in the proceedings that other dosimetry techniques may be more efficient and effective in certain roles (see Table 1). For example, patient specific IMRT and VMAT dose delivery validation prior to, or during, treatment is often more efficiently and effectively performed with 2D EPID based techniques [13, 15-18]. The fact that various dosimetry techniques supplement each other and are important in the clinic was acknowledged explicitly in the history of the conferences when they were renamed in 2010 from the series of '*DosGel*' sessions to the '*IC3Ddose*' meetings, establishing a more widespread focus in order to encourage attendance by researchers and clinicians working with point, planar and pseudo 3D dosimetry techniques. This extension to the wider community with expertise in film, EPID based [13, 15-18], scintillation [42, 43] and point array measurements [12, 44-46] enabled a more critical assessment of the role of the different systems and facilitated an improved discussion of how these various dosimetry systems can complement each other [45, 47, 48]. Therefore, the literature reported in the conference proceedings to date provides a broader content not limited only to volumetric chemical 3D dosimeters.

3. The Themes of IC3Ddose as Presented in Invited Papers

There are a number of common themes in each of the proceedings of the previous 8 International Conferences on 3D Radiation Dosimetry (see Table 2). Each conference has included a strong didactic component with invited papers providing attendees the fundamental science and basic mechanisms that guided the development and response of various dosimetry systems. Initial papers in the first conference proceedings review the history and development of 3D dosimeters [33, 35, 49], even providing historical background from before the seminal papers of Gore *et al* [50, 51] that initially established the 3D measurement potential of chemical dosimetry. There have been regular reviews of the fundamentals of Fricke gels [32, 52, 53], polymer gels [34, 54-57], radiochromic gels [23-25] and novel systems throughout the proceedings; each of the review papers providing a comprehensive list of references extending to the broader literature. These descriptions of the basic science of the various dosimeters were typically supplemented by details of how to best prepare and use the various systems to assure consistent and reproducible dosimetry.

This basic didactic component of the proceedings extends also to readout techniques including magnetic resonance imaging (MRI) [58-63], optical computed tomography (optical CT) [64-69], x-ray CT [70-72] and other potential readout systems [72]. These reviews typically describe the radiation properties of the irradiated materials that enable the various imaging modalities to readout the dose, and also provide detailed analyses that inform the imaging protocols required for reproducible and accurate readout. Thus the reviews consistently provide a valuable resource for establishing readout protocols when implementing 3D dosimetry. Additional invited papers further address data analysis

Table 1. A summary of some of the dose measurements and validation experiments required for implementing IMRT as initially summarized in the proceedings from DosGel 2006 [41]. The comments on the utility of volumetric 3D gel or plastic dosimeters in the shaded cells are personal reflections added in this current review based on past DosGel and IC3Ddose proceedings.

Phase and Intent	*Dosimetry / Test Required*	*Typical Tools and Approaches*	*Role for 3D Dosimetry*
1.Commissioning of treatment planning system and benchmarking of performance (both treatment planning and dose delivery)	Acquisition of beam data to dosimetrically characterize beam, and machine data to mechanically characterize linac	Ion chamber, water tank, film, detector arrays	no
		Not strongly indicated for conventional beams but there is likely a role in small field and brachytherapy source commissioning	
	Measurement of test cases planned in phantom under well-defined conditions to ensure correct performance and establish benchmark data for each particular treatment protocol	Regular and anthropomorphic phantoms; film; 2D dose QA systems diode), portal imaging systems	yes
		A strong role for 3D systems especially when used with complementary dosimetry techniques	
2.Periodic QA	Routine testing of the delivery system; QA to ensure continued planning and delivery as at commissioning	Regular and anthropomorphic phantoms; film; 2D dose QA systems (ion chamber and diode), portal imaging systems	yes
	Including end to end QA	*A strong role for 3D systems especially when used with complementary dosimetry techniques*	
3.Routine patient specific treatment QA	QA of Monitor Unit (MU) calculations	Independent validated calculation system or direct measurements in phantom (see cell below) with ion chambers or other point dosimeters	possibly
		Other techniques are much more reliable but, if validated, 3D systems may fill a complementary (not primary) role	
	Testing of delivered dose distributions	Replace patient by standard phantom, expose phantom to same MLC sequence, trajectories and MUs as for patient. 2D dosimeters	yes
		3D systems are powerful tools for initial testing of novel treatments techniques (e.g., small field irradiations, irradiation under motion management, etc.), especially when used to validate other pseudo 3D techniques used regularly for patient specific QA (see below)	
	At treatment measurement of delivered dose and dose distribution.	In-vivo dosimetry or online exit beam dosimetry (using EPID or tomotherapy imaging detector)	no
		In general other techniques may be better indicated, however, new small field techniques (e.g., small field VMAT treatment of multiple brain metastases) may bring an increased clinical role	

procedures required to go from imaging readout to dose, addressing such issues as accuracy and error analysis [73-76], calibration, data processing workflow and 3D dosimetry metrics enabling comparison of measured doses with planned distributions [77, 78]. Over the years the 100 or so invited reviews have presented a detailed set of courses on how to perform 3D dosimetry.

4. The Themes of IC3Ddose as Presented in Proffered Papers

The conference proceedings from the various DosGel and IC3Ddose meetings also review well the development of 3D dosimetry as reported by basic researchers, practical developers and clinical users (see Table 2). (Citing the many individual proffered papers that have advanced the practical and clinical role out of gel dosimetry is beyond the format of this review; but a narrative based on the conference proceedings is possible.)

The initial meeting [1] focussed on Fricke and polymer gel dosimeters as the main systems of interest. But soon reports of new dosimeter systems appeared, often for the first time, in these proceedings. Fong introduced normoxic polymer gels in Brisbane [2]. In Ghent's DosGel04 [3], the first of the proceedings to be reported in the open Journal of Physics: Conference Series, Venning *et al* reported on PAGAT, the normoxic formulation of the more common polyacrylamide gel, while Adamovic and Maryanski described the polyurethane radiochromic systems that developed into PRESAGE. The reduced toxicity N-isopropylacrylamide (NIPAM) system was introduced by Senden *et al* two years later in Sherbrooke [4]. The proceedings from Dosgel08 in Hersonissos [5] had reports of the development and use of the normoxic N-Vinylpyrrolidone polymer (VIP) gel by various Greek researchers and introduced the new family of micelle radiochromic gels through Babic *et al*'s proffered work. The reports of new dosimetry systems has continued in more recent conference series [8] with reports of the development of novel deformable polydimethylsiloxane and silicone based dosimeters by De Deene *et al* and Høye *et al*, respectively. As Juang *et al* showed in the IC3Ddose 2013 [7], the development of these dosimeters enables dose measurements that can validate advanced deformable dose calculation algorithms in modern treatment planning systems. This small sample clearly shows that the proceedings provide a solid account of the historical development of 3D dosimetry systems.

The proceedings do not just report the development of new dosimeters; they also compile a considerable literature on the practical aspects of their use. For example, Keller *et al* showed in the initial DosGel [1] that the then commonly held view that adding mM amounts of benzoic acid during the preparation of the dosimeter would increase sensitivity was false, since the presence of 5% gelatin provided overwhelming alternate reaction pathways already increasing the chemical yield. In Ghent [3], McAuley *et al* confirmed the observation of enhanced dose response at the edge of high dose gradient regions in polyacrylamide polymer gels through mathematical modelling of monomer diffusion in irradiated gel. This modelling was later extended to brachytherapy applications and in Hilton Head [6] the group reported that the perturbation of polymer formation around low dose-rate brachytherapy sources would make dose calibration of polymer gels problematic in LDR brachytherapy. Each of the proceedings over the years has also presented a full set of proffered papers that describe practical issues in sample preparation, particularly highlighting the importance of establishing, and sticking to, careful preparation protocols (including setting timescales for various steps in the preparation of the dosimeters, periods between sample preparation and irradiation, and then time to readout) to ensure consistent and reproducible dosimetry. The proceedings offer a compendium of preparation caveats and practices to avoid; in fact, a search of the proceedings before undertaking the use of a particular dosimeter is well indicated, since this will help the new user set good protocols at the start of their work (see also [31]).

The papers proffered over the last two decades of these international conferences also present a complete review of the potential clinical utility of 3D dosimetry as first promised by Scheib [38]. Even an abridged review of a limited set of conferences makes this clear. In Sherbrooke [4] Oldham reported on IMRT verification with optical CT readout and radiochromic plastic, Pappas reported on small field profile measurement with polymer gels, while Månsson *et al* and Oldham *et al* reported on 4D measurements made with dynamic phantoms simulating respiratory tumour motion. Later, in 2008 in Crete [5], Petrokokkinos *et al*, Moutsatos *et al* and Papas *et al* reported on the use of VIP gels for dosimetry applied to HDR brachytherapy, Gamma Knife and stereotactic radiosurgery commissioning, respectively; Ceberg *et al* discussed RapidArc™ treatment validation using MRI and polymer gels; while Wuu reported on the commissioning of a high definition multileaf collimator using polymer gels and optical readout. Ceberg *et al* showed more on the use of gels for 4D treatment validation in Sydney, reporting the measurement of interplay effects in VMAT delivery on a breathing motion phantom. Over the years our group from Kingston has reported on the use of 3D dosimetry, mainly with optical readout of Fricke gels, for example for stereotactic body radiation therapy with IMRT and VMAT dose delivery validation [7, 8, 48], noting that complete characterization in the clinical setting can benefit from the correlation of measurements from multiple dosimeters such as ion chamber, film and 3D dosimeters [48,47].

Table 2. A summary of the proceedings content (✓indicates significant content, ~ indicates limited or no presentations in a given year, there may still have been some discussion).

	Meeting	DosGel					IC3Ddose		
	Year	1999	2001	2004	2006	2008	2010	2012	2014
	Venue	Lexington USA	Brisbane Australia	Ghent Belgium	Sherbrook Canada	Hersonnisos Greece	Hilton Head USA	Sydney Australia	Ystad Sweden
No of papers	Invited Review/ Refresher	9	9	12	11	10	21	17	12
	Proferred	36	41	39	46	55	80	93	79
Basics of Dosimetry Systems	3D systems (history, background)	✓	✓	✓	✓	✓	✓	✓	✓
	Fricke Gels	✓	✓	✓		✓	✓	✓	
	Polymer Gels	✓	✓	✓	✓	✓	✓	✓	✓
	Other Radio-chromic Gels	~	~	~	~	✓	✓	✓	✓
	Radio-chromic Plastics	~	~	✓	✓	✓	✓	✓	✓
Readout Systems	MRI	✓	✓	✓	✓	✓	✓	✓	~
	Optical-CT	✓	✓	✓	✓	✓	✓	✓	✓
	X-ray CT	~		✓	✓	✓	✓	✓	✓
	other	~	✓	✓	~	✓	✓	✓	✓
Practical/Clinical Issues	Clinical Role	✓	✓	✓	✓	✓	✓	✓	✓
	Data Evaluation	~	~	✓	✓	✓	✓	✓	✓
	Equipment/ source Commissionin g	~	~	✓	✓	✓	✓	✓	✓
	Dosimeter preparation protocols	✓	✓	✓	✓	✓	✓	✓	✓
	End to End QA	~	~	~	~	✓	✓	✓	✓
Complementary Dosimetries	EDID based	~	~	~	~	~	✓	✓	✓
	Point Dosimeter Arrays	~	~	~	✓	~	✓	✓	✓
	Film	~	~	~	✓	✓	✓	✓	~
	Scintillators	~	~	~	~	~	✓	✓	✓
	Others	~	~	~	✓	✓	~	✓	✓

Over the years the groups from Duke and the M.D. Anderson have also investigated the potential for volumetric 3D dosimeters for end-to-end quality assurance in the clinic and in credentialing programs (for example, two papers by Sakhalkar in 2009 [5], work by Newton in 2010 [6], and by Thomas in 2013 [7]). The work outlined in this paragraph is, in large part, the basis for my long held view of a strong role for 3D dosimetry in the clinic as described in the final section (Clinical role - reprise) of the 2015 review article on true 3D dosimetry [21]. A quick read of the submissions for this year's IC3Ddose suggests that proffered papers will again indicate an increased role for 3D dosimetry in the modern radiation therapy clinic.

Over the years attendees at the 3D dosimetry conferences have also reported on one other important factor in 3D dosimetry: data analysis and workflow. Initially the focus was typically on the scientific fundamentals of converting readout to dose, with emphasis on uncertainty and error analysis. Eventually the reports evolved more to description of the analytic tools required for fast efficient data analysis and for the comparison of 3D measurements to the dose distributions from treatment planning as described, for example, by Kozicki in two papers in Crete [5], by Deasy in Hilton Head [6], and by Alexander in Ystad [8].) These papers illustrate the advances in data analysis provided by dedicated systems initially developed in-house and often later commercialized or offered in open source environments.

Finally, the various 3D dose conferences have benefited increasingly over the years by the attendance of commercial vendors offering a variety of services from dosimeter preparation, to sales of dosimeter readout systems (especially optical-CT scanners for radiochromic gels and solids and polymer gels), to actual readout and dose calculation service to clinical users lacking the required facilities in their clinics for in-house preparation and processing. The companies are listed in Table 3 along with some references to proffered papers that touched on the initial research that lead to commercializing, to later reports of the development of protocols for use, or to presentations of clinical applications performed in collaboration with the vendors. The increased availability of these commercial partners is an encouraging development, as it can only help further advance the clinical adoption of 3D dosimetry.

Table 3: A list of commercial service providers and vendors and an open-source system that have contributed to the various IC3Ddose meetings and that offer various 3D dosimetry readout and analysis systems and dosimeters; citations are to select proceedings reports related to their wares.

Provider	Vendor Website	Select DosGel / IC3Ddose papers related to vendor products
GeVero, Poland	www.polygevero.com	[79, 80]
Heuris Pharma, USA	www.presage3d.com	[22, 81-87]
MGS Research Inc., USA	www.mgsresearch.com	[88-90]
Modus QA, Canada	www.modusqa.com	[91-93]
RT Safe, Greece	www.rt-safe.com	(based on [94, 95])
Slicer-RT, Canada	www.slicerrt.org/	[21, 96, 97]

5. Conclusions

This review is intended to provide the reader a comprehensive set of references from past International Conferences on 3D Radiation Dosimetry on which to build their new work. This compendium offers a unique open source library detailing the scientific fundamentals of 3D dosimetry and outlining the development of clinical practice over nearly two decades. It represents a fruitful outcome from meetings born from a perceived need some 20 years ago of the requirement to establish a forum for researchers in a new field [98-100]. It also sets the stage for the remaining 9 invited and 64 proffered papers in these current proceedings. I wish you good reading.

6. References

[1] 1999 *Proc. DosGel 1999: 1st Int. Workshop on Radiation Therapy Gel Dosimetry* Audet C and Schreiner L

J (eds) Edmonton, AB: COMP

[2] 2001 *Proc.DosGel 2001: 2nd Int. Conf. on Radiotherapy Gel Dosimetry* Baldock C and De Deene Y (eds) Brisbane, Australia: Queensland University of Technology

[3] 2004 *Proc. DosGel 2004: 3rd Int. Conf. on Radiotherapy Gel Dosimetry* C Baldock and Y De Deene (eds) *J. Phys.: Conf. Ser.* **3**

[4] 2006 *Proc. DosGel 2006: 4th Int. Conf. on Radiotherapy Gel Dosimetry* Lepage M (ed) *J. Phys.: Conf. Ser.* **56**

[5] 2009 *Proc. of DosGel 2008: 5th Int. Conf. on 3D Radiotherapy Dosimetry* Maris TG and Pappas E (eds) *J. Phys.: Conf. Ser.* **164**

[6] 2010 *Proc. of IC3DDose 2010: 6th Int. Conf. on 3D Radiation Dosimetry* Oldham M and Newton J (eds) *J. Phys.: Conf. Ser.* **250**

[7] 2013 *Proc. of IC3DDose 2012: 7th Int. Conf. on 3D Radiation Dosimetry* Thwaites D and Baldock C (eds) *J. Phys.: Conf. Ser.* **444**

[8] 2015 *Proc. of IC3DDose 2014: 8th Int. Conf. on 3D Radiation Dosimetry* Bäck S and Olsson L E (eds) *J. Phys.: Conf. Ser.* **573**

[9] Wright K A *et al* 1959 *Radiology* **72** 101

[10] Trump J G, *et al* 1961 *Radiology* **76** 275

[11] De Wagter C 2004 *J. Phys.: Conf. Ser.* **3** 4-8

[12] Gutiérrez A N and Calvo O 2010 *J. Phys.: Conf. Ser.* **250** 012049

[13] Bäck A 2015 *J. Phys.: Conf. Ser.* **573** 012012

[14] Low D A 2015 *J. Phys.: Conf. Ser.* **573** 012009

[15] Mijnheer B *et al* 2013 *J. Phys.: Conf. Ser.* **444** 012011

[16] Mijnheer B *et al* 2010 *J. Phys.: Conf. Ser.* **250** 012020

[17] Moran J M 2010 *J. Phys.: Conf. Ser.* **250** 012021

[18] McCurdy B M C 2013 *J. Phys.: Conf. Ser.* **444** 012002

[19] Ibbott G S 2004 *J. Phys.: Conf. Ser.* **3** 58-77

[20] Schreiner L J 2009 *J. Phys.: Conf. Ser.* **164** 012001

[21] Schreiner L J 2015 *J. Phys.: Conf. Ser.* **573** 012003

[22] Guo P *et al* 2006 *J. Phys.: Conf. Ser.* **56** 191-4

[23] Oldham M 2015 *J. Phys.: Conf. Ser.* **573** 012006

[24] Jordan K J 2010 *J. Phys.: Conf. Ser.* **250** 012043

[25] Babic S *et al* 2009 *J. Phys.: Conf. Ser.* **164** 012044

[26] Fong P M *et al* 2001 In *Proc. 2nd Int. Workshop of Radiation Therapy Gel Dosimetry* Baldock C and De Deene Y (eds) Brisbane, Australia: Queensland University of Technology

[27] Oldham M 2014 In *Advances in Medical Physics* Devon J G *et al* (eds) Madison, WI: Medical Physics Publishing

[28] Oldham M *et al* 2001 *Med. Phys.* **28** 1436-45

[29] Ibbott G S and Thwaites D I 2015 *J. Phys.: Conf. Ser.* **573** 012002

[30] Ibbott G S 2010 *J. Phys.: Conf. Ser.* **250** 012001

[31] Schreiner L J and Olding T 2009 In *AAPM Medical Physics Monograph: Clinical Dosimetry Measurements in Radiotherapy* Roger D and Cygler J (ed) Madison, WI: Medical Physics Publishing

[32] Schreiner L J 2001 In *Proc. 2nd Int. Workshop of Radiation Therapy Gel Dosimetry* Baldock C (ed) Brisbane, AU: Queensland University of Technology

[33] Baldock C 2004 *J. Phys.: Conf. Ser.* **3** 1-3

[34] McAuley K B 2004 *J. Phys.: Conf. Ser.* **3** 29-33

[35] Schreiner L J 1999 In *Proc.1st Int. Workshop of Radiation Therapy Gel Dosimetry* Audet C and Schreiner L J (eds) Edmonton, AB: COMP

[36] Hiraoka T *et al* 1992 *Nippon Acta Radiol.* **52** 1039-49

[37] Hiraoka T *et al* 1993 *Med. Dosim.* **18** 73-9

[38] Scheib S G *et al* 2001 In *Proc. 2nd Int. Workshop of Radiation Therapy Gel Dosimetry* Baldock C (ed) Brisbane, AU: Queensland University of Technology

[39] Kron T 2001 In *Proc. 2nd Int. Workshop of Radiation Therapy Gel Dosimetry* Baldock C (ed) Brisbane, AU: Queensland University of Technology

[40] Yin F-F *et al* 2010 *J. Phys.: Conf. Ser.* **250** 012002

[41] Schreiner L J 2006 *J. Phys.: Conf. Ser.* **56** 1–13

[42] Beaulieu L *et al* 2013 *J. Phys. Conf. Ser.* **444** 012013

[43] Guillot M *et al* 2010 *J. Phys. Conf. Ser.* **250** 012006
[44] O'Daniel J *et al* 2010 *J. Phys. Conf. Ser.* **250** 012050
[45] Ceberg S *et al* 2010 *J. Phys.: Conf. Ser.* **250** 012051
[46] Stathakis S 2010 *J. Phys.: Conf. Ser.* **250** 012029
[47] Olding T *et al* 2013 *J. Phys.: Conf. Ser.* **444** 012052
[48] Olding T *et al* 2013 *J. Phys.: Conf. Ser.* **444** 012073
[49] Gore J C and Schulz RJ 2001 In *Proc. 2nd Int. Workshop of Radiation Therapy Gel Dosimetry* Baldock C (ed) Brisbane, AU: Queensland University of Technology
[50] Gore J C *et al* 1984 *Phys. Med. Biol.* **29** 1189-97
[51] Gore J C *et al* 1996 *Phys. Med. Biol.* **41** 2695-704
[52] Bäck S. Olsson L. E. 1999 In *Proc.1st Int. Workshop of Radiation Therapy Gel Dosimetry* Audet C and Schreiner L J (eds) Edmonton, AB: COMP
[53] Schreiner L J 2004 *J. Phys.: Conf. Ser.* **3** 9-21
[54] Maryanski M J 1999 In *Proc.1st Int. Workshop of Radiation Therapy Gel Dosimetry* Audet C and Schreiner L J (eds) Edmonton, AB: COMP
[55] Lepage M 2001 In *Proc. 2nd Int. Workshop of Radiation Therapy Gel Dosimetry* Baldock C (ed) Brisbane, AU: Queensland University of Technology
[56] McAuley K B 2006 *J. Phys.: Conf. Ser.* **56** 35-44
[57] Schacht E H 2004 *J. Phys.: Conf. Ser.* **3** 22-8
[58] De Deene Y 2001 In *Proc. 2nd Int. Workshop of Radiation Therapy Gel Dosimetry* Baldock C (ed) Brisbane, AU: Queensland University of Technology
[59] De Deene Y 2004 *J. Phys.: Conf. Ser.* **3** 87-114
[60] De Deene Y 2010 *J. Phys.: Conf. Ser.* **250** 012015
[61] De Deene Y 2009 *J. Phys.: Conf. Ser.* **164** 012033
[62] Lepage M and Gore J C 2004 *J. Phys.: Conf. Ser.* **3** 78-86
[63] Lepage M 2006 *J. Phys.: Conf. Ser.* **56** 86-94
[64] Jordan K 2004 *J. Phys.: Conf. Ser.* **3** 115-21
[65] Jordan K J 1999 In *Proc.1st Int. Workshop of Radiation Therapy Gel Dosimetry* Audet C and Schreiner L J (eds) Edmonton, AB: COMP
[66] Oldham M 2004 *J. Phys.: Conf. Ser.* **3** 122-35
[67] Oldham M 2006 *J. Phys.: Conf. Ser.* **56** 58-71
[68] Doran S J and Krstajić N 2006 *J. Phys.: Conf. Ser.* **56** 45-57
[69] Doran S J 2009 *J. Phys.: Conf. Ser.* **164** 012020
[70] Hilts M 2006 *J. Phys.: Conf. Ser.* **56** 95-107
[71] Jirasek A and Hilts M 2009 *J. Phys.: Conf. Ser.* **164** 012038
[72] Baldock C 2004 *J. Phys.: Conf. Ser.* **3** 136-41
[73] De Deene Y and Jirasek A 2015 *J. Phys.: Conf. Ser.* **573** 012008
[74] Deene Y De and Vandecasteele J 2013 *J. Phys.: Conf. Ser.* **444** 012015
[75] Deene Y De 2006 *J. Phys.: Conf. Ser.* **56** 72-85
[76] Thwaites D 2013 *J. Phys.: Conf. Ser.* **444** 012006
[77] Low D A 2010 *J. Phys. Conf. Ser.* **250** 012071
[78] Schreiner L J *et al* 2013 *J. Phys.: Conf. Ser.* **444** 012016
[79] Kozicki M *et al* 2009 *J. Phys.: Conf. Ser.* **164** 012062
[80] Kozicki M *et al* 2009 *J. Phys.: Conf. Ser.* **164** 012061
[81] Adamovics J and Maryanski M J 2004 *J. Phys.: Conf. Ser.* **3** 172–5
[82] Sakhalkar H S *et al* 2009 *J. Phys.: Conf. Ser.* **164** 012058
[83] Guo P *et al* 2006 *J. Phys.: Conf. Ser.* **56** 187-90
[84] Doran S *et al* 2006 *J. Phys.: Conf. Ser.* **56** 231-4
[85] Guo P *et al* 2006 *J. Phys.: Conf. Ser.* **56** 207-10
[86] Adamovics J *et al* 2006 *J. Phys.: Conf. Ser.* **56** 176-8
[87] Adamovics J *et al* 2006 *J. Phys.: Conf. Ser.* **56** 172-5
[88] Avery S *et al* 2015 *J. Phys.: Conf. Ser.* **573** 012020
[89] Wuu C *et al* 2004 *J. Phys.: Conf. Ser.* **3** 297-300
[90] Scheib S G *et al* 2004 *J. Phys.: Conf. Ser.* **3** 228-31
[91] DeJean P *et al* 2006 *J. Phys.: Conf. Ser.* **56** 183-6
[92] DeJean P *et al* 2006 *J. Phys.: Conf. Ser.* **56** 179-82

IOP Conf. Series: Journal of Physics: Conf. Series **847** (2017) 012001 doi:10.1088/1742-6596/847/1/012001

[93] Olding T *et al* 2009 *J. Phys.: Conf. Ser.* **164** 012030
[94] Pappas E *et al* 2006 *J. Phys.: Conf. Ser.* **56** 245-8
[95] Pappas E *et al* 2009 *J. Phys.: Conf. Ser.* **164** 012054
[96] Alexander K M *et al* 2015 *J. Phys.: Conf. Ser.* **573** 012042
[97] Olding T *et al* C 2015 *J. Phys.: Conf. Ser.* **573** 012019
[98] https://en.wikipedia.org/wiki/Gel_dosimetry
[99] Baldock C 2006 *J. Phys.: Conf. Ser.* **56** 14-22
[100] Baldock C 2009 J. *Phys.: Conf. Ser.* **164** 012002

IOP Conf. Series: Journal of Physics: Conf. Series **847** (2017) 012002 doi:10.1088/1742-6596/847/1/012002

Complexity metric based on fraction of penumbra dose - initial study

A Bäck[1], F Nordström[1], M Gustafsson[1], J Götstedt[2] and A Karlsson Hauer[1]
[1]Department of Therapeutic Radiation Physics, Sahlgrenska University Hospital, Göteborg, Sweden
[2]Department of Radiation Physics, University of Gothenburg, Göteborg, Sweden

E-mail: anna.back@vgregion.se

Abstract. Volumetric modulated arc therapy improve radiotherapy outcome for many patients compared to conventional three dimensional conformal radiotherapy but require a more extensive, most often measurement based, quality assurance. Multi leaf collimator (MLC) aperture-based complexity metrics have been suggested to be used to distinguish complex treatment plans unsuitable for treatment without time consuming measurements. This study introduce a spatially resolved complexity score that correlate to the fraction of penumbra dose and will give information on the spatial distribution and the clinical relevance of the calculated complexity. The complexity metric is described and an initial study on the correlation between the complexity score and the difference between measured and calculated dose for 30 MLC openings is presented. The result of an analysis of the complexity scores were found to correlate to differences between measurements and calculations with a Pearson's r-value of 0.97.

1. Introduction

Modulated radiation therapy treatment techniques, such as volumetric modulated arc therapy (VMAT) has the potential to increase the dose to the tumor and at the same time decrease the dose to healthy tissue as compared to conventional three dimensional conformal radiotherapy (3DCRT). The fields of VMAT treatment plans are divided in control points with multileaf collimator (MLC) openings of various size and shape. Irregularly shaped MLC openings with small sub-opening components are challenging from a dosimetric point of view in consideration of both calculation, delivery and measurement [1, 2]. For example, small MLC openings increase the sensitivity of small MLC positioning errors during delivery [3] and increase the challenge of measuring as well as calculating dose because of volumes lacking charged particle equilibrium (CPE) [4]. Treatment plans with fields composed of small MLC sub-opening components might therefore cause clinically relevant discrepancies between calculated (planned) and delivered dose distributions, and can be considered as complex.

Metrics to quantify this complexity have been suggested as a complement to measurement-based quality assurance (QA) of VMAT treatment plans and/or to be used within the optimization procedure in the treatment planning system (TPS) to reduce the plan complexity (e.g. [5, 6]). In a previous study, aperture-based complexity metrics were evaluated for static MLC openings [5]. The static MLC openings represented control points in a theoretical VMAT treatment plan and the correlation of the metric scores and the difference between the calculated and measured dose of the MLC openings were investigated. Such metrics have the ability to distinguish complex treatment plans that are unsuitable for patient treatments and need to be reconsidered or need extra attention during the quality assurance. However, they cannot give any information on where in the patient the delivered dose will be different from the calculated dose or the clinical relevance of the differences.

The regions lacking CPE and causing dose calculation errors and regions sensitive to MLC positioning errors during delivery are located in the penumbra of the MLC opening. It is likely that an increased amount of dose in a voxel that originates from a penumbra region will correlate to the probability of increased difference between calculated and delivered dose in that voxel. For example, for VMAT treatments, the center of the target volume will probably have a low fraction of penumbra dose while organs at risk most probably will have a higher fraction originating from the penumbra and thereby a higher probability for differences between calculated and delivered dose. A voxel-based complexity score that correlate to the fraction of penumbra dose have the possibility to also reveal information on the clinical relevance of the difference between calculated and delivered dose.

This study is the initial step in a project with the overall objective to develop a complexity metric that correlates to the fraction of penumbra dose in a three dimensional (3D) volume for VMAT treatments. The hypothesis is that such a complexity metric presented as a distribution in a 3D volume of the patient will correlate to the difference between calculated and delivered dose in the volume. In this initial study, a method to calculate the complexity score for static MLC openings in a 2D plane perpendicular to the beam direction is developed to verify that the fraction of penumbra dose correlate to differences between calculated and delivered dose. The next step will be to study the 2D distribution of complexity scores to establish suitable parameters for the calculation of the complexity scores before moving on to 3D calculations.

2. Materials and methods

2.1. MLC-openings

30 MLC openings, created and described by Götstedt *et al* [5] simulating individual VMAT control points in a treatment field were used in this study (figure 1). They were designed in six series (A-F) including five openings in each series with gradually increasing complexity in their shape and size. The MLC openings were created in Eclipse treatment planning system (TPS) (Eclipse™ version 11, Varian Medical Systems) using a 6 MV photon beam of a Clinac iX (Varian Medical Systems) linear accelerator with the Millennium™ 120 MLC with leaves of 5 mm width in the central 20 cm of the field and 10 mm width further out from the central axis.

2.2. Calculation of complexity scores

A C# software with dynamically linked MatLab® (Mathworks, Natick, MA) libraries was developed to calculate the complexity scores. The input to the software is the DICOM-file of the treatment plan from where the MLC and collimator positions are collected. The MLC and collimator positions define the appearance of the field opening in the beams eye view (BEV) 2D plane. The calculation of complexity scores are performed in the following steps:

1. The pixels of the BEV plane (pixel size in this study was 0.25 mm) is structured binary in open beam (1) or blocked beam (0) according to the MLC and collimator positions.
2. The binary BEV from step 1 is convolved with a Gaussian function normalized to 1. This will increase the complexity score of the penumbra dose higher in regions with higher dose and lower in the low dose region of the penumbra. It will mimic the clinical relevance of the penumbra dose since a dose deviation in the low dose region is of less clinical importance than a deviation of a higher dose. The resulting 2D distribution of this convolution can roughly be interpreted as the relative dose distribution of the field opening. In this study the calculations were done with a sigma for the Gaussian function of 0.5, 2 and 5 mm.
3. The pixels of the BEV plane is structured binary in field edge (1) or no edge (0).

4. The binary BEV from step 3 is convolved with a box function (1 inside box, 0 outside). This will define a region, with the width of the box and symmetrically around the field edge, as the region of interest, i.e. the penumbra region, where the complexity metric will have a score $\neq 0$. In this study the width of the box was varied between 3, 5, 7 and 9 mm.
5. Finally the convolution from step 2 is multiplied with the convolution from step 4. This gives a 2D distribution of complexity scores. The complexity score will mimic a dose distribution from the field truncated so that only the dose from the penumbra region remains.

The calculated complexity scores of the 30 MLC openings were compared to differences between measured and calculated dose of the same openings. In this very first analysis of the complexity score this was accomplished by collapsing the 2D distribution of complexity scores into one score for each MLC opening by calculating the *mean complexity score* of each MLC opening relative to the *estimated mean dose* for that MLC opening. In this case the mean dose is estimated by the mean value of the 2D distribution from the convolution in step 2.

2.3. Dose measurements, dose calculations and evaluations

The fields with the different MLC openings were delivered with gantry angle 0° and measured with Gafchromic™ EBT3 film (Ashland) for a source to surface distance of 90 cm at 10 cm depth in a water equivalent plastic slab phantom of 30x30x30 cm^3 (Gammex and PTW). The number of MU for each MLC opening was adjusted to deliver 2 Gy, within ±2%, centrally in the most open part of the MLC opening at a depth of 10 cm in water based on calculations with the analytical anisotropic algorithm (AAA) in Eclipse™ version 10 (Varian Medical Systems). The film was calibrated pixel by pixel based on a foregoing exposure to a known dose using a double exposure method. The calibration and film measurement procedure have been described earlier [5].

The 2D dose distribution of the film measurements were compared to the corresponding 2D dose distribution calculated with AAA using a calculation grid size of 2.5 mm x 2.5 mm. A MatLab software was developed to facilitate alignment using mutual information (RegularStepGradientDescent). The comparison was done pixel by pixel and the 2D distribution of dose differences between measured and calculated dose in absolute numbers for each MLC opening was registered. The region analyzed was limited by a cutoff value of 0.2 Gy calculated dose. The *mean dose deviation* of each MLC opening relative to the AAA *calculated mean dose* for that opening was registered.

3. Results and discussion

The correlation between *mean complexity score relative to estimated mean dose* and *mean dose deviation between measurement and calculation relative to calculated mean dose* for a sigma of the Gaussian function of 5 mm and a width of the box of 5 mm is shown in figure 2. The correlation expressed as Pearson's r-value is 0.97.

The mean complexity score and the mean absolute deviation between measurement and calculation are not expected to yield the same numbers with a 1 to 1 relationship since the complexity score is not in the same units as the absolute dose deviation which is expressed in Gy. However, the correlation seen in figure 2 is promising for the further development of the complexity score.

In this analysis of mean complexity scores, the sigma of the Gaussian function had minor influence on the complexity score for the 30 MLC openings investigated. An increasing width of the box used in the convolution described in step 4 increased the complexity score by a factor that was constant for all MLC openings meaning that it did not influence the correlation shown in figure 2. The influence of those parameters on an analysis of the uncollapsed 2D distribution is ongoing to find optimal values to be used in the future for analyses of complexity scores in a 3D volume.

4. Conclusion

A new method for calculation of complexity scores based on the amount of penumbra dose was described for calculations of static MLC openings in a 2D plane. The result of an analysis of the

complexity scores were found to correlate to differences between measurements and calculations of 30 MLC openings that could represent control points in a VMAT treatment plan. The results are promising for further development of the calculation of complexity scores in a 3D volume.

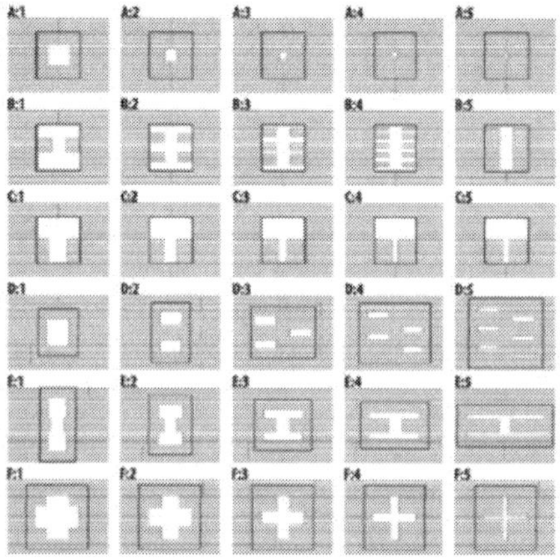

Figure 1. The 30 MLC openings from [5], used in the study. The MLC openings are grouped in Series A-F numbered from 1-5 with increasing number for more complex MLC openings.

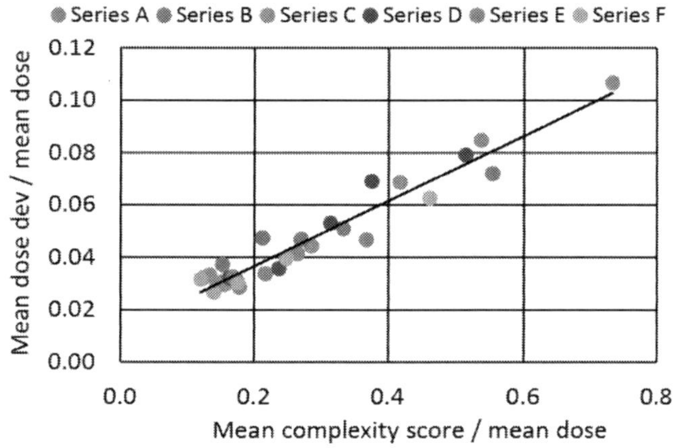

Figure 2. A scatter plot illustrating the correlation between the mean complexity score relative to estimated mean dose and mean dose deviation between measurement and calculation relative to calculated mean dose for the MLC openings (figure 1). Sigma of the Gauss function was 5 mm and the width of the box was 5 mm.

5. Acknowledgements
The authors gratefully acknowledge the financial support by the Swedish Radiation Safety Authority and by the Health & Medical Care Committee of the Regional Executive Board, Region Västra Götaland.

6. References
[1] Vial P *et al* 2006 *Phys. Med. Biol.* **51** 5517-38
[2] Vial P *et al* 2008 *Med. Phys.* **35** 1267-77
[3] LoSasso T *et al* 1998 *Med. Phys.* **25** 1919-27
[4] Das I J *et al* 2008 *Med. Phys.* **35** 206-15
[5] Götstedt J *et al* 2015 *Med. Phys.* **42** 3911-21
[6] Younge K C *et al* 2012 *Med. Phys.* **39** 7160-70

IOP Conf. Series: Journal of Physics: Conf. Series **847** (2017) 012003 doi:10.1088/1742-6596/847/1/012003

A preliminary study on the use of FX-Glycine gel and an in-house optical cone beam CT readout for IMRT and RapidArc verification

Paul B Ravindran, Ebenezer Suman Babu S, Winfred Michael Raj and Amalan S
Medical Physics Division, Department of Radiation Oncology, Christian Medical College, Vellore, India

Email: paul@cmcvellore.ac.in

Abstract. The radiochromic FX gel with Optical CT readout has been investigated by several authors and has shown promising results for 3D dosimetry. One of the applications of the gel dosimeters is their use in 3D dose verification for IMRT and RapidArc quality assurance. Though polymer gel has been used successfully for clinical dose verification, the use of FX gel for clinical dose verification with optical cone beam CT needs further validation. In this work, we have used FX gel and an in- house optical readout system for gamma analysis between the dose matrices of measured dose distribution and a treatment planning system (TPS) calculated dose distribution for a few test cases.

1. Introduction

High precision complex radiotherapy dose delivery techniques have considerably increased the interest in 3D dosimetry. A few dosimetry systems such as the ArcCHECK, Delta4 and Octavius 4D provide dose distribution in 3D, by measuring dose in 2D and reconstructing to 3D geometry. Several investigators have developed gel dosimeters that provide true 3D dose distribution [1-18]. Of the gel dosimeter, the FX gels have the advantages of easy preparation, good spatial resolution, sensitivity, linearity of dose response and tissue equivalence [19, 20]. Moreover, the radio-chromic nature of the FX gel enables the use of less expensive optical method for reading the dose embedded in the gel matrix. However, FX gel suffers from ion diffusion which destroys the spatial dose information rather quickly after the irradiation, thus requiring reading of the gel soon after the formation of ferric ions post irradiation [21].

In this work we have performed a preliminary study on the use of FX-Glycine gel with the in-house optical cone beam readout system for quality assurance of IMRT and RapidArc irradiation.

2. Materials and Methods

In this work the FX gels prepared were irradiated with plans developed using Eclipse planning system for IMRT and RapicArc plans with 6MV beams. The gel was irradiated on Clinac 2100C/D unit that has millennium multi-leaf collimator and on board imager.

IOP Conf. Series: Journal of Physics: Conf. Series **847** (2017) 012003 doi:10.1088/1742-6596/847/1/012003

2.1. The FX Gel

The FX gel used in this study was prepared with 240 bloom gelatine (Sigma), Ferrous Ammonium Sulphate (S D Fine-Chem Limited), xylenol orange (Sigma-Aldrich) and sulfuric acid with glycine (0.3mM) anti-oxidant added to reduce the rate of diffusion as suggested in our earlier paper [22]. The characteristics of this gel including its lower diffusion rate have been studied [22]. Gels were prepared in cylindrical transparent containers suitable for optical scanning. For each batch, two bottles of gels were prepared, one for dosimetry, and the other for calibration

2.2. Optical Cone Beam CT

The optical cone beam CT (OCT) used in this study, is an upgraded version of our in-house optical CT scanner [23]. In the upgraded version, the control and the reconstruction software have been developed in Visual C Sharp. The projection images from the camera are obtained for each degree for 194 degrees. The reconstruction is performed with FDK reconstruction code originally developed for cone beam reconstruction of MV CBCT images [24]. This code has been converted to CUDA so as to work on a Graphic Processor Unit (GPU), thus enabling fast reconstruction. A pre-scan is performed prior to irradiation and the post scan is performed in about 30 minutes after irradiation. The reading of the irradiated gel and the reconstruction takes less than three minutes thus allowing minimal degradation of the gel matrix due to diffusion during scanning. The new version of our reconstruction software also has provisions to apply various filters such as ramp, Butterworth, Gaussian, Hanning, Hamming and Parzen, Shepp-Logan and Metz. In this work, all the reconstructions were performed with Ramp and Hamming filters.

2.3 Verisoft software

The Medphysto software provided by PTW has a module 'Verisoft' that enables calibration of the detector array, acquisition of dose distribution in 2D as well as in 3D and gamma evaluation of the comparison of measured or calculated dose distribution. The software enables 2D and 3D gamma comparison based on the Gamma Index method suggested by Daniel A low *et al* [25]. The Optical Cone beam CT data was converted RT dicom dose file so that it could be read by the Verisoft for gamma evaluation.

2.4 Gel irradiation

The gel phantoms were prepared as per the method described in section 2.1. For gel irradiation, a few test plans were created with the Eclipse planning system on the CT data set of the Gel phantom. The pre-scan of the gel was performed prior to irradiation. For calibration, the gel phantom was irradiated with four 2 cm x 2 cm fields to deliver 100, 200, 300 and 400 cGy to the mid plane of the gel phantom. A plan was created with similar beams on the Eclipse planning system and the dose at the mid plane was determined for calibration against the optical attenuation coefficient.

3. Results and Discussion

3.1 Calibration

The plan generated with the Eclipse planning system on the CT data set of the gel phantom for the calibration of the gel and the optical CT readout of the irradiated gel are shown in figures 1a & b. The optical density to dose relationship was obtained.

3.2 Analysis of gel dosimetry

A couple of RapidArc test plans were generated with the Eclipse planning system with a 'C' shaped and a butterfly shaped PTVs. Verification plans were also generated on the gel phantom for patient plans having PTV smaller than 5cm. The gels were irradiated as per the plan and read with optical CT. Both 2D and 3D gamma analysis between the dose matrices of measured dose distribution and treatment

IOP Conf. Series: Journal of Physics: Conf. Series **847** (2017) 012003 doi:10.1088/1742-6596/847/1/012003

planning system (TPS) calculated dose distribution were carried out using gamma criteria of 3.5%/3.5 mm (% dose difference and mm distance to agreement). The gamma comparison performed with the 'Verisoft' software is shown in figure 2 for the C shaped test plan.

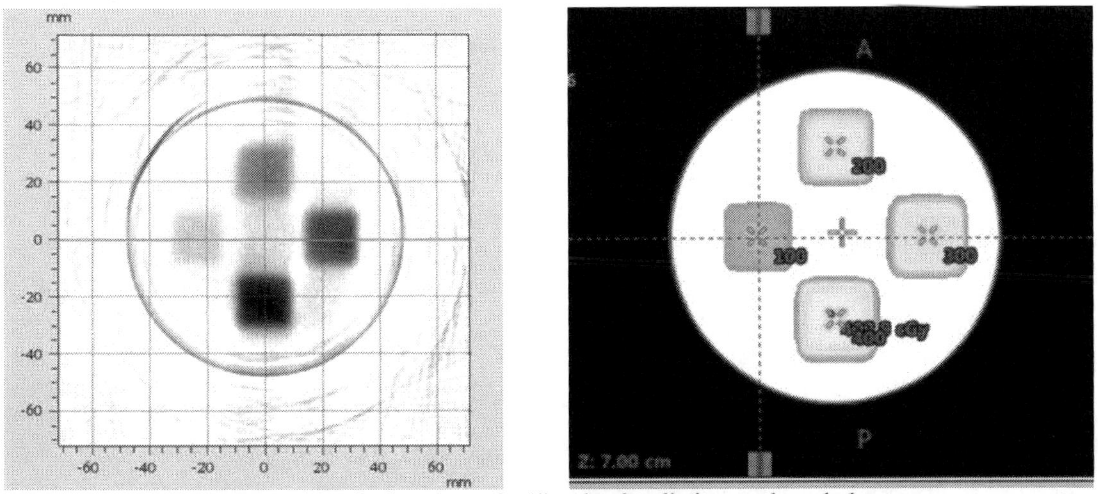

Figure 1. a) Optical readout of calibration irradiation on the gel phantom.
b) Treatment plan generated for calibration

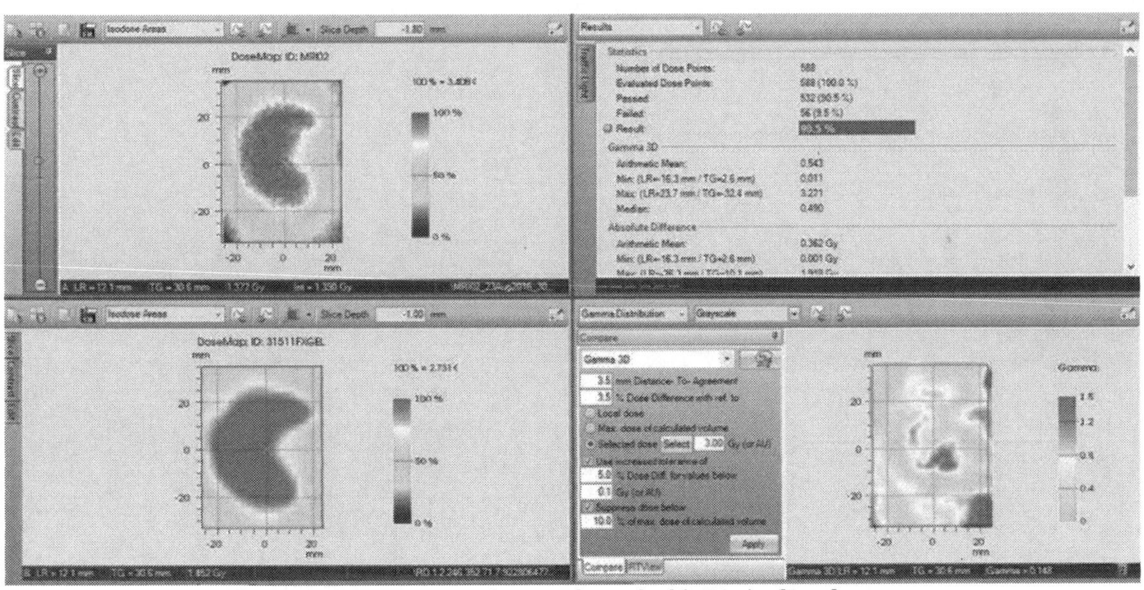

Figure 2. Gamma comparison performed with 'Verisoft' software.

The passing rate for the C shaped plan was above 90% for most points above 50% dose whereas several points below the 50% dose failed the passing criteria. The second test case with the butterfly shaped PTV had only a lower pass percentage of 85%.

4. Conclusion

FX gel with in-house optical cone beam CT readout has been successfully used for 3D dose comparison between the gel matrix and the TPS dose distribution. Though, the pass percentage with gamma criteria of 3.5%/3.5mm was much lower, careful calibration and proper selection of the region for dose comparison could further improve the results.

IOP Conf. Series: Journal of Physics: Conf. Series **847** (2017) 012003 doi:10.1088/1742-6596/847/1/012003

5. References

[1] Hill R *et al* 2014 *Phys. Med. Biol.* **59** R183-231
[2] Hill R *et al* 2009 *Med. Phys.* **36** 3971-81
[3] Baldock C *et al* 2010 *Phys. Med. Biol.* **55** R1-63
[4] Baldock C *et al* 1998 *Phys. Med. Biol.* **43** 695-702
[5] Murry P and Baldock C 2000 *Austral. Phys. Eng. Sci. Med.* **23** 44-5
[6] Venning A J *et al* 2004 *J. Phys.: Conf. Ser.* **3** 155-8
[7] Davies J B and Baldock C 2008 *Radiat. Phys. Chem.* **77** 690-96
[8] Mather M L *et al* 2003 *Ultrasonics* **41** 551-9
[9] Hill B *et al* 2005 *Med. Phys.* **32** 1589-97
[10] Mather M L *et al* 2003 *Phys. Med. Biol.* **48** N269-75
[11] Trapp J V *et al* 2004 *Phys. Med. Biol.* **49** N139-46
[12] Hurley C *et al* 2006 *Nucl. Instrum. Meth. A* **565** 801-11
[13] Baldock C 2006 *J. Phys.: Conf. Ser.* **56** 14-22
[14] Healy B J *et al* 2003 *Med. Phys.* **30** 2282-91
[15] Hill B *et al* 2002 *Phys. Med. Biol.* **47** 4233-46
[16] Baldock C 2009 *J. Phys.: Conf. Ser.* **164** 012002
[17] Brindha S *et al* 2004 *Phys. Med. Biol.* **49** N353-61
[18] Subramanian B *et al* 2006 *J. Med. Phys.* **31** 72-77
[19] Bero M A *et al* 2000 *J. Soc. Radiol. Prot.* **20** 287-94
[20] Schreiner LJ 2004 *J. Phys.: Conf. Ser.* **3** 9-21
[21] Baldock C *et al* 2001 *Austral. Phys. Eng. Sci. Med.* **24** 19-30
[22] Babu S E S *et al* 2015 *J. Cancer Res. Ther.* **11** 580
[23] Ravindran P B and Thomas H M 2013 *J. Phys.: Conf. Ser.* **444** 12064
[24] Thomas T H M *et al* 2009 *Phys. Med. Biol.* **54** 2067-77
[25] Low D A *et al* 1998 *Med. Phys.* **25** 656-61

IOP Conf. Series: Journal of Physics: Conf. Series **847** (2017) 012004 doi:10.1088/1742-6596/847/1/012004

Treatment planning and 3D dose verification of whole brain radiation therapy with hippocampal avoidance in rats

S W Yoon[1], D Miles[1], C Cramer[2], M Reinsvold[1], D Kirsch[1], M Oldham[1]
[1]Department of Radiation Oncology, Duke University Medical Center, Durham, NC, USA
[2]Department of Radiation Oncology, Wake Forest School of Medicine, Winston-Salem, NC, USA

E-mail: mark.oldham@duke.edu

Abstract. Despite increasing use of stereotactic radiosurgery, whole brain radiotherapy (WBRT) continues to have a therapeutic role in a selected subset of patients. Selectively avoiding the hippocampus during such treatment (HA-WBRT) emerged as a strategy to reduce the cognitive morbidity associated with WBRT and gave rise to a recently published the phase II trial (RTOG 0933) and now multiple ongoing clinical trials. While conceptually hippocampal avoidance is supported by pre-clinical evidence showing that the hippocampus plays a vital role in memory, there is minimal pre-clinic data showing that selectively avoiding the hippocampus will reduce radiation-induced cognitive decline. Largely the lack of pre-clinical evidence can be attributed to the technical hurdles associated with delivering precise conformal treatment the rat brain. In this work we develop a novel conformal HA-WBRT technique for Wistar rats, utilizing a 225kVp micro-irradiator with precise 3D-printed radiation blocks designed to spare hippocampus while delivering whole brain dose. The technique was verified on rodent-morphic Presage® 3D dosimeters created from micro-CT scans of Wistar rats with Duke Large Field-of-View Optical Scanner (DLOS) at 1mm isotropic voxel resolution. A 4-field box with parallel opposed AP-PA and two lateral opposed fields was explored with conformal hippocampal sparing aided by 3D-printed radiation blocks. The measured DVH aligned reasonably well with that calculated from SmART Plan Monte Carlo simulations with simulated blocks for 4-field HA-WBRT with both demonstrating hippocampal sparing of 20% volume receiving less than 30% the prescription dose.

1. Introduction

Therapeutic whole brain radiotherapy (WBRT) has been used to primary or metastatic brain tumors since the 1950s [1]. However, many patients experience some degree of neurocognitive decline after treatment [2]. Studies have shown that metastases to the limbic circuit – which includes the hippocampus – are rare, and hence can be spared during WBRT for improved brain functions [3, 4]. The potential value of hippocampal avoidance is supported by pre-clinical and clinical evidence showing that the hippocampus plays a vital role in memory. Yet there is minimal in-vivo pre-clinic data showing that selectively avoiding the hippocampus will reduce radiation-induced cognitive decline. Largely the lack of such evidence can be attributed to the technical hurdles associated with delivering precise conformal treatment to small animal to create an accurate model. Our group published the first rodent model of hippocampal avoidance where treatment planning incorporated MRI-based identification of the hippocampus as an avoidance structure and validated accurate treatment delivery dosimetrically and with tissue immunostaining (Cramer et al. [5]). In brief, in this initial work, we developed lateral-opposed 2-field HA-WBRT plan in 8-week Wistar rats with two conformal lead hand-cut blocks held at a distance in front of the x-ray source shielding the hippocampus. Since then, we have replaced the hand-cut lead blocks with 3D-printed radiation blocks generated from tungsten-doped plastic filaments.

Precise, conformal irradiation of small targets in rodents is exceptionally challenging [5]. High-resolution 3D dosimetry is essential to ensure accurate treatment delivery [6]. Presage® (from John Adamovics, Heuris Inc.) is a 3D radiochromic (i.e. changing color when irradiated), tissue-equivalent solid plastic dosimeter which is ideal for these situations. Presage molding techniques allowed fabrication of rodent-morphic Presage dosimeters [7]. These are dosimeters with exact dimensions of a mice or a rat. In this study, we generated a rodent-morphic Wistar rat dosimeter with the purpose of reconstructing 4-field conformal HA-WBRT dose delivered to a tissue-equivalent, geometrically accurate rat dosimeter on the small animal irradiator with 225kVp x-ray treatment beam.

2. Methods and Materials

In this section, we present methods used to verify that conformal 4-field HA-WBRT has indeed spared the hippocampus with a rodent-morphic dosimeter. All radiation experiments have been performed on Xrad 225Cx MicroCT irradiator from Precision X-ray, Inc. which has image-guided radiotherapy capabilities. All irradiation was performed at 225kVp with a 0.35mm copper filter. Dose read-outs from Presage dosimeter were performed with high-resolution Duke Large Field-of-View Optical Scanner (DLOS), which is an optical CT specifically designed for 3D quantitative reconstruction of Presage [8].

2.1. Conformal radiation therapy on Xrad 225Cx

Conformal radiation therapy was previously performed with Xrad 225Cx unit by attaching a lead radiation block onto a thin piece of paper suspended by 40x40mm^2 collimator and a 3D-printed block holder (Figure 1, left) intended to shield the hippocampus. For further details on our conformal radiation therapy method, the reader is referred to Cramer et al [5].

To improve conformality and remove human error from block fabrication, a metal-composite 3D printing material GMASS (Turner MedTech, Inc., Orem, UT) replaced the lead block. Intended for x-ray shielding, the high-density tungsten-composite ABS filament fit our purposes of performing conformal HA-WBRT. Hippocampal-sparing 3D-printed blocks for 4-field treatment with 1mm margin for each block were attached to our customized collimator attachment for treatment (Figure 1).

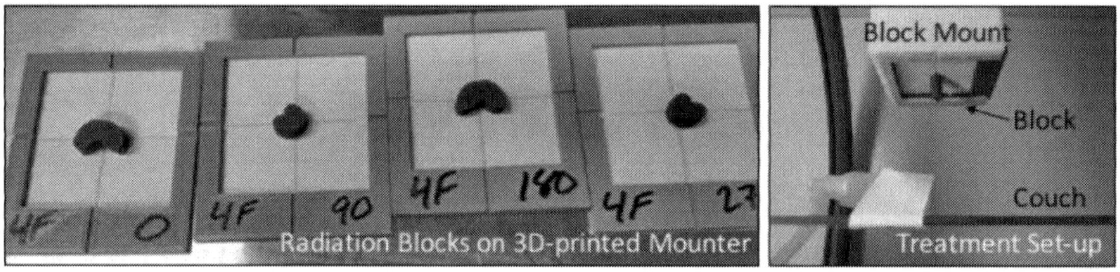

Figure 1. Left: Hippocampal blocks for 4-field HA-WBRT with 1mm margin; Right: Treatment set-up with 3D-printed blocks attached to collimator via block mount.

2.2. Verfication of Radiation Block and 4-field HA-WBRT with Rodent-Morphic Dosimeters

Initial verification of radiation blocking properties of GMASS filament was done using a heart-shaped 3D-printed block with a Gafchromic™ EBT3 film at radiation dose of 6 Gy. Film was sampled at 30 dpi to reduce micro-variations in film optical density.This was to ensure minimal primary radiation is passing through the dosimeter. Once this was verified, rodent-morphic dosimeters were fabricated.

Rodent-morphic dosimeters for this set-up were modelled after the CBCT of an 8-week old Wistar rat (Figure 2). The model of the rat was generated from this CBCT via surface rendering on itk-snap (www.itksnap.org) [9]. The model was then smoothed to exclude extraneous details that could

IOP Conf. Series: Journal of Physics: Conf. Series **847** (2017) 012004 doi:10.1088/1742-6596/847/1/012004

compromise the quality of the resulting Presage mold, then 3D-printed using PLA filament. The 3D-printed rat model was then used to mold and cast two rodent-morphic dosimeters.

The fabricated dosimeters were then subjected to an irradiation with 4-field box with parallel opposed AP-PA and two lateral opposed fields, with conformal hippocampal sparing aided by 3D-printed radiation blocks. The prescription dose for this HA-WBRT was 1 Gy for direct comparison with 1 Gy planning-dose SmART Plan simulation, a full Monte Carlo dose planning software for Xrad 225Cx irradiator [10]. The dose distribution was read out with DLOS at 1mm isotropic voxel resolution. We calculate and compare between dose volume histograms (DVHs) of measured and simulated dose distributions for the hippocampus (plus 1mm margin) and the brain without hippocampus to assess hippocampal sparing and dose agreement.

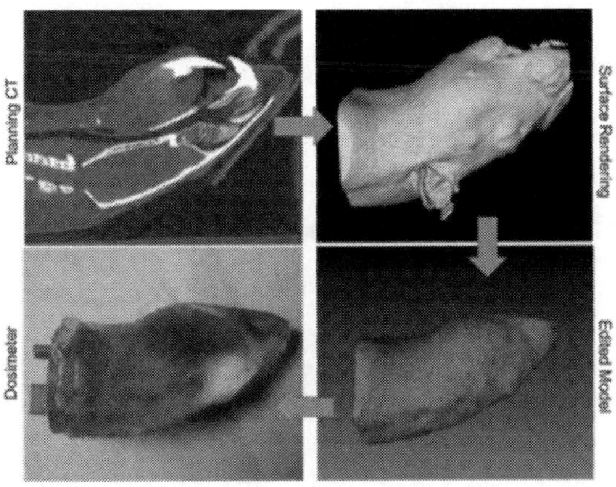

Figure 2. Workflow to generate rodent-morphic dosimeter based on planning CT.

3. Results and Discussion

We present here the result of radiochromic film study and 4-field conformal HA-WBRT delivered to a rodent-morphic Presage dosimeter discussed in section 2.2.

Figure 3 demonstrates that very minimal dose (0.19 Gy for a 6.01 Gy treatment) passes through the 3D-printed blocks. The block edges are well-defined, with 80%-to-20% penumbra spanning less than 1mm.

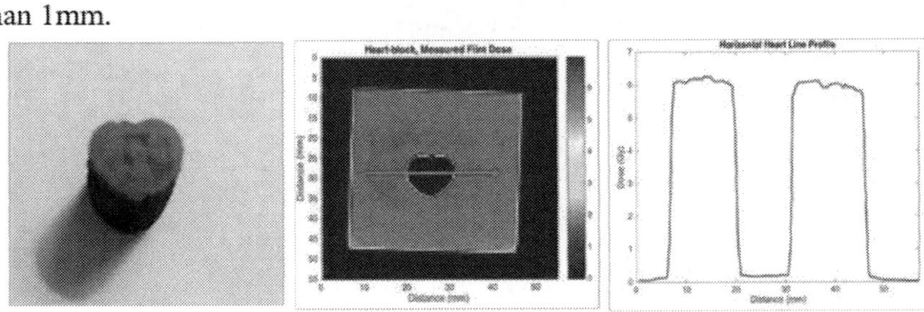

	Blocked Region	Unblocked Region
Mean Dose (Gy)	0.19	6.01
Standard Deviation	0.03	0.15
Minimum Dose (Gy)	0.11	5.52
Maximum Dose (Gy)	0.24	6.51

Figure 3. Heart-shaped block and its radiation properties, verified with dosimetric EBT3 films.

Figure 4 (left) shows the results of the measured rodent-morphic Presage dose distribution for the 4-field HA-WBRT and SmART Plan simulated of HA-WBRT, overlaid on top of rat CBCT. Figure 4 (right) shows a dose volume histogram (DVH) calculated from this dose distribution. Rodent-morphic Presage verification of 4-field HA-WBRT shows significant agreement with the simulated SmART-plan results.

The 4-field DVHs show good agreement between the measured and simulated results. More importantly, both the measurement and the simulation agree that the hippocampus is spared at less than 20% of volume receiving less than 30% of the prescription dose. Slightly more volume of the brain (without hippocampus) received 40% or more of the prescription dose for simulated DVH versus measured DVH. This is probably because of not accounting for divergence of the beam when fabricating the block and should be addressed in subsequent runs.

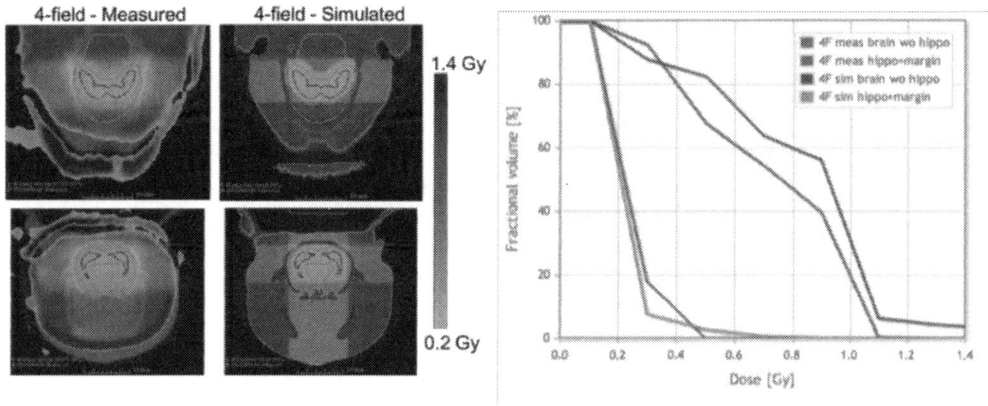

Figure 4. Left: Measured and SmART Plan Monte Carlo simulated dose distribution of 4-field HA-WBRT treatment. Right: Calculated DVHs for brain without hippocampus (brain wo hippo) and the hippocampus with 1mm margin (hippo+margin). Red and green lines were generated using the Presage dosimeter data and blue and green lines were generated via Monte Carlo simulation in SmART Plan.

4. Conclusion
The complex nature of conformal HA-WBRT on a small animal irradiator requires elegant dosimetry to ensure that the delivery is sound. The rodent-morphic Presage dosimeters have fulfilled this role and allowed us to make informed decisions on whether 4-field HA-WBRT is sufficiently hippocampal-sparing. Our planned future studies will further develop the 3D-printed radiation compensators to enable IMRT capabilities in micro-irradiators and verify the treatment once again with rodent-morphic dosimeters.

5. References
[1] Patchell R A 1998 *JAMA* **280** 1485-89
[2] Mehta M *et al* 2012 Radiation Therapy Oncology Group
[3] Gondi V *et al* 2014 *J. Clin. Oncol.* **32** 3810-6
[4] Marsh J C *et al* 2009 *Int. J. Radiat. Oncol. Biol. Phys.* **76** 504-12
[5] Cramer C K *et al* 2015 *PLoS One* **10** 12
[6] Baldock C *et al* 2010 *Phys. Med. Biol.* **55** R1-63
[7] Bache S *et al* 2013 *Med. Phys.* **40** 6
[8] Thomas A and Oldham M 2010 *J. Phys.: Conf. Ser.* **250** 1-5
[9] Yushkevich P A *et al* 2006 *Neuroimage* **31** 1116-28
[10] Van Hoof S J *et al* 2013 *Radiother. Oncol.* **109** 361-6

IOP Conf. Series: Journal of Physics: Conf. Series **847** (2017) 012005 doi:10.1088/1742-6596/847/1/012005

Patient-specific quality assurance for spot scanning proton beams using a large-volume liquid scintillator detector

D Robertson and S Beddar
The Department of Radiation Physics, University of Texas MD Anderson Cancer Center, Houston, TX 77030, USA

Email: danr@byu.net

Abstract. A large-volume liquid scintillator detector was used to measure individual energy layers from a clinical prostate treatment plan on a scanning proton beam system. Lateral and beam's-eye view images of the dose distribution were acquired with two CCD cameras facing adjacent sides of a scintillator tank. The measured dose images were compared with calculated dose distributions from a validated Monte Carlo model. The measured and calculated dose distributions showed good agreement, with the exception of the Bragg peak region of the lateral view, which exhibited ionization quenching. The beam's-eye and lateral views achieved gamma analysis passing rates of 99.7% and 92.5%, respectively, using gamma criteria of 3%, 3 mm. Large-volume scintillator detectors show promise for quick and accurate measurements of patient treatment fields for scanning proton beam systems.

1. Introduction

Proton radiation therapy can produce physical dose distributions that are superior to photon-based treatments because of the proton Bragg peak and the finite range of proton beams in tissue. Most new proton therapy centers employ magnetic beam scanning, which is capable of delivering intensity-modulated proton therapy (IMPT) treatments. However, the complexity of IMPT treatments leads to challenges in treatment plan verification measurements. The steep dose gradients and finite range characteristic of IMPT fields necessitate the use of high-resolution detectors and require measurements at multiple depths in order to verify that the treatment plan is correctly delivered. As a result, patient-specific quality assurance for IMPT is labor-intensive, and it is only practical to measure a subset of locations within each field.

In recent years, interest has grown in the use of large-volume organic scintillator detectors for quality assurance of radiation therapy delivery systems and patient treatment plans. Detectors consisting of plastic or liquid scintillators imaged by CCD cameras have been developed and tested for applications including proton beam range measurement [1-4], proton beam scanning position [5], and measurement of photon-based patient treatment plans [6, 7]. Large-volume scintillator detectors are particularly promising for verification of IMPT patient plan delivery because of their high resolution, their speed, and their ability to image the entire dose distribution in a single measurement. All of these studies have either used a single CCD camera or a relatively small detector volume.

The objective of this study was to evaluate the use of a large-volume scintillator detector employing two orthogonal cameras to measure dose distributions from scanning proton beam treatments.

2. Materials and Methods

A large-volume liquid scintillator detector was constructed, comprising a cubic plastic tank filled with liquid scintillator, two CCD cameras, and a mirror. The inner dimensions of the scintillator tank were 20x20x20 cm^3, and the CCD cameras were placed 75 cm from two adjacent faces of the tank. A mirror, oriented at 45 degrees relative to one face of the tank, was used to re-direct the light towards one of the cameras. The mirror enabled one camera to obtain a beam's eye view of the proton beam without putting the camera's sensitive electronics at risk from radiation damage (Figure 1).

Figure 1. The large-volume liquid scintillator detector, including the tank and two cameras, in position on the patient couch of the Proton Therapy Center - Houston scanning gantry.

The images from the scintillator detector were processed to remove lens distortions, vignetting, blurring, and hot pixels and streaks caused by stray radiation striking the cameras [8]. The images were also corrected for refraction at the tank-air interface [4].

Individual energy layers of a scanning beam prostate treatment were delivered to the scintillator detector and measured with the CCD cameras. The timing structure of scanning beam delivery at the Proton Therapy Center - Houston includes rapid scanning at a given proton energy, followed by a 2-second pause while the synchrotron energy is changed. Consequently, it is convenient to measure one energy layer at a time, allowing the camera data to be read out during the pause in beam delivery. The analysis presented herein is for the 161.6 MeV energy layer of a scanning beam prostate treatment plan. This energy was selected because it was in the center of the field, covering a large area and featuring an unusually-shaped dose distribution. Beam's-eye view and lateral images of the dose distribution were obtained simultaneously during one single delivery of the field.

The dose distribution was calculated using a Monte Carlo proton dose calculation system featuring a validated model of the Proton Therapy Center - Houston scanning beamline [9]. The calculated and measured dose distributions were normalized in the proximal buildup region for the lateral (depth-dose) view, and in a high-dose, low-gradient region in the beam's eye view. The calculated dose distribution was compared with the measured light signal using profile comparisons and gamma analysis [10].

3. Results and Discussion

On visual inspection, the Monte Carlo calculations and measured scintillation light show good agreement, with all features of the calculated dose distribution readily visible in the measured light (Figure 2). Line profiles indicate excellent agreement for the beam's-eye view, while revealing a substantial quenching of the Bragg peak in the scintillator signal (Figure 3). Ionization quenching is caused by linear-energy dependence of the scintillator [11]. Methods have been developed to correct for quenching in scintillator detectors [12]. The beam's-eye view is relatively unaffected by quenching, which decreases the intensity of the distribution without altering its shape. However, it is clear from the lateral camera view that quenching correction is necessary to obtain agreement in the depth direction.

The agreement of the calculated and measured dose distributions was evaluated using gamma analysis. Regions with 10% of the maximum dose or greater were evaluated. Figure 4 shows gamma maps for criteria of 3% dose difference and 3 mm distance to agreement. While the passing rate was above 90%, quenching led to very high gamma values in the Bragg peak in the lateral view. This illustrates the importance of quenching correction when evaluating in the depth direction.

Gamma analysis was repeated using criteria of 2%, 2mm and 1%, 1mm. The passing rates are given in Table 1. At 3%, 3mm, the passing rates were very high, even for the lateral view without quenching correction. This suggests that 3%, 3 mm is too generous a criteria to effectively evaluate the accuracy of IMPT treatment plan delivery. While the individual gamma values in the Bragg peak were very high in the lateral view, they were confined to a small region, leading to passing rates similar to the beam's-eye view. If the entire field, including all energy layers, was evaluated together, passing rates for the lateral view would likely drop due to the presence of multiple Bragg peaks.

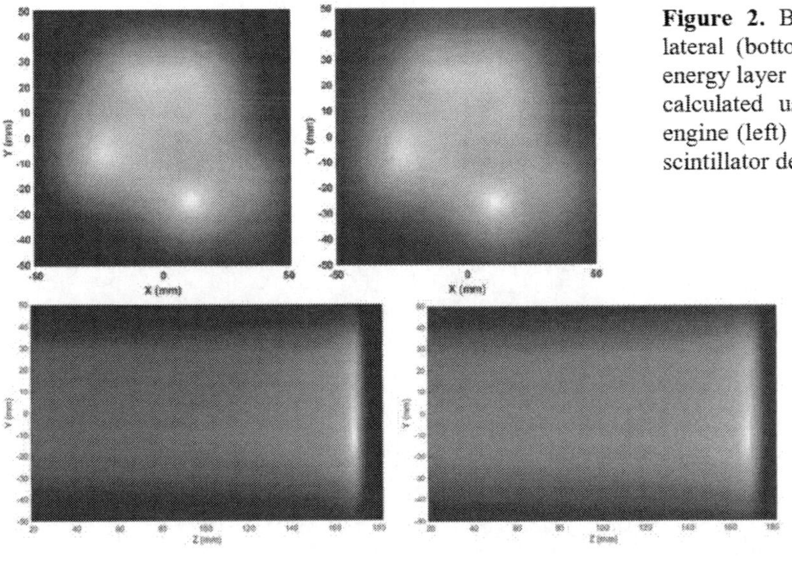

Figure 2. Beam's-eye view (top) and lateral (bottom) projections of a single energy layer of a prostate treatment plan, calculated using a Monte Carlo dose engine (left) and measured with a liquid scintillator detector (right).

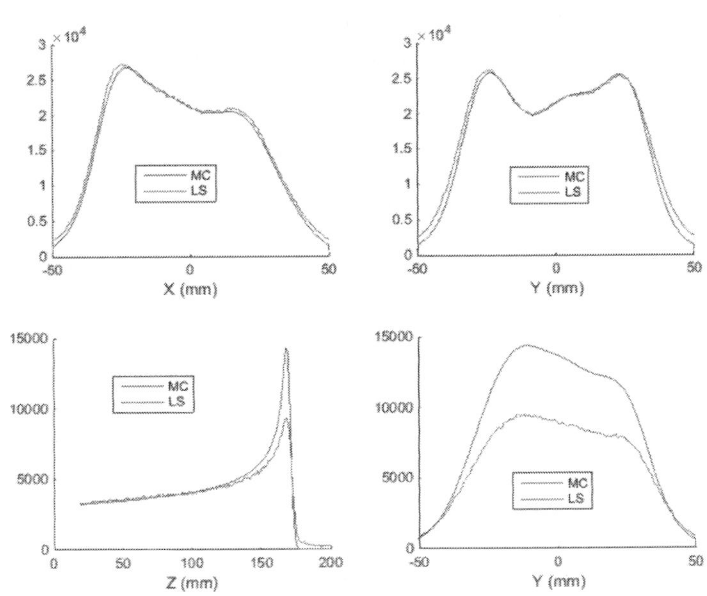

Figure 3. Line profiles of the Monte Carlo-calculated (MC) and liquid scintillator-measured (LS) dose distributions for the beam's-eye view (top) and lateral view (bottom). All profiles pass through the center of the images, with the exception of the bottom right figure, which pases throught he Bragg peak.

IOP Conf. Series: Journal of Physics: Conf. Series **847** (2017) 012005 doi:10.1088/1742-6596/847/1/012005

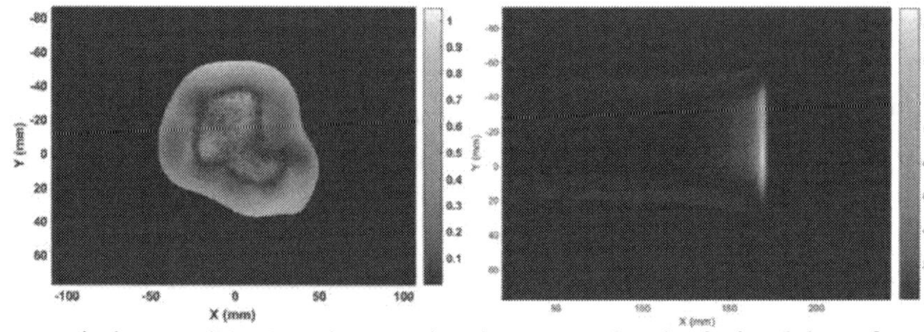

Figure 4. Gamma analysis maps (3%, 3 mm) comparing the measured and calculated doses for a single energy layer, including the beam's-eye view (left) and a lateral view (right).

Table 1. Gamma analysis passing rates for the beam's-eye and lateral views.

Gamma Analysis Criteria	Beam's eye view	Lateral view
3%, 3 mm	99.7%	92.5%
2%, 2 mm	90.4%	88.7%
1%, 1 mm	55.5%	66.5%

4. Conclusions

Patient-specific dose distributions measured with a large-volume liquid scintillator detector agreed well with calculated doses for a single energy layer of a scanned proton beam treatment plan, with the exception of ionization quenching in the Bragg peak. Gamma analysis is useful for comparing measured and calculated dose distributions, but the gamma criteria should be selected carefully to ensure the sensitivity of the test. It is convenient to measure one energy layer at a time using the scintillator detector, and separating the image data into individual energy layers opens the way for application of energy-dependent quenching correction factors. Large-volume scintillator detectors show promise for quick and accurate measurements of patient treatment fields for scanning proton beam systems.

5. Acknowledgements

The authors would like to thank Dr. Dragan Mirkovic and Dr. Narayan Sahoo for their valuable inputs and contributions to this project. The research reported in this publication was supported by the National Cancer Institute of the National Institutes of Health under Award Number R01CA182450. The content is solely the responsibility of the authors and does not necessarily represent the official views of the National Institutes of Health.

6. References

[1] Fukushima Y *et al* 206 *Phys. Med. Biol.* **51** 5927-36
[2] Beddar S *et al* 2009 *Med. Phys.* **36** 1736-43
[3] Kroll F *et al* 2013 *Med. Phys.* **40** 082104
[4] Hui C *et al* 2015 *Biomed. Phys. Eng. Express* **1** 025204
[5] Archambault L *et al* 2012 *Med. Phys.* **39** 1239-46
[6] Ponisch F *et al* 2009 *Med. Phys.* **36** 1478-85
[7] Goulet M *et al* 2014 *Med. Phys.* **41** 449-61
[8] Robertson D *et al* 2014 *Phys. Med. Biol.* **59** 23
[9] Sawakuchi G O *et al* 2010 *Med. Phys.* **37** 4960-70
[10] Low D A *et al* 1998 *Med. Phys.* **25** 656-61
[11] Birks J B *et al* 1951 *Proc. Physical Society. Section A* **64** 874
[12] Robertson D *et al* 2013 *Phys. Med. Biol.* **58** 261-73

IOP Conf. Series: Journal of Physics: Conf. Series **847** (2017) 012006 doi:10.1088/1742-6596/847/1/012006

Innovation and the future of advanced dosimetry: 2D to 5D

Mark Oldham
Dept of Radiation Oncology, Duke University Medical Center, Durham, NC USA

Email: mark.oldham@duke.edu

Abstract. Recent years have witnessed a remarkable evolution in the techniques, capabilities and applications of 3D dosimetry. Initially the goal was simple: to innovate new techniques capable of comprehensively measuring and verifying exquisitely intricate dose distributions from a paradigm changing emerging new therapy, IMRT. Basic questions emerged: how well were treatment planning systems modelling the complex delivery, and how could treatments be verified for safe use on patients? Since that time, equally significant leaps of innovation have continued in the technology of treatment delivery. In addition, clinical practice has been transformed by the addition of on-board imaging capabilities, which tend to hypo-fractionation strategies and margin reduction. The net result is a high stakes treatment setting where the clinical morbidity of any unintended treatment deviation is exacerbated by the combination of highly conformal dose distributions given with reduced margins with fractionation regimens unfriendly to healthy tissue. Not surprisingly this scenario is replete with challenges and opportunities for new and improved dosimetry systems. In particular tremendous interest exists in comprehensive 3D dosimetry systems, and systems that can resolve the dose in moving structures (4D) and even in deforming structures (5D). Despite significant progress in the capability of multi-dimensional dosimetry systems, it is striking that true 3D dosimetry systems are today largely found in academic institutions or specialist clinics. The reasons will be explored. We will highlight innovations occurring both in treatment delivery and in advanced dosimetry methods designed to verify them, and explore current and future opportunities for advanced dosimetry tools in clinical practice and translational research.

1. Introduction

The sophistication capability and complexity of radiation treatments has increased dramatically in recent years. Three main engines of innovation initiated much of the early phase of this transformation. The first originated in the theoretical world, with the discovery and development of mathematical tools to optimize the radiation fluence for a desired dose prescription and normal tissue tolerance [1, 2]. Then came innovations in computer control and electro-mechanical hardware components needed to implement the delivery of fluence-modulated radiation treatments (e.g. IMRT) [3, 4]. The third engine can be attributed to the addition of 'on-board' imaging, in the form of cone-beam-CT capability with the patient in treatment position just prior to treatment [5-7]. Continued innovations in fluence optimization and delivery techniques have led to a range of advanced treatment techniques available at the present time, including IGRT, MRI-guided therapy, VMAT and RapidArc, gating techniques, flattening-filter-free high dose-rate treatments, through to dynamic couch motion treatments [8]. Importantly these techniques have transformed clinical practice, and regularly enable exquisite dose conformation to tumor and substantial dose sparing to healthy structures, even in the most irregularly shaped settings.

The net results of these advances and innovations are substantial gains in clinical capability, flexibility, and efficacy for many patients. However, the price is an elevated potential for errors, and

IOP Conf. Series: Journal of Physics: Conf. Series **847** (2017) 012006 doi:10.1088/1742-6596/847/1/012006

substantially more difficult treatment commissioning and verification procedures. The difficulty of comprehensive verification and commissioning of advanced treatments has now far exceeded the capabilities of conventional dosimetry systems [9]. These topics have received significant recent attention in the literature and at the congressional level in the United States government [10, 11]. Independently, the Imaging and Radiation Oncology Core Houston (IROC Houston, formerly the Radiological Physics Center [RPC]) reported unacceptably high failure rates for dosimetry credentialing in national clinical trials for several IMRT treatments: 18% for head and neck (7%, 4 mm gamma criteria) and 32% for spine (5%, 3 mm criteria) [12]. The high failure rates despite a generous passing criteria are alarming, and a wake-up call, suggesting the quality and accuracy of implementation of advanced treatments like IMRT must be improved. These failures were detected with sparse sampling on a pair of orthogonal 2D film-planes normalized with thermoluminescent dosimeter (TLD) point measurements, which will miss errors in regions that were not sampled, thereby representing an overly optimistic perspective.

In addition to the challenges of treatment commissioning and verification, the advent of on-board imaging created new concerns arising from the tendency to shrink 'safety' margins historically added to allow for patient set-up errors. The ability to shrink margins has the potential to reduce normal tissue toxicity, and hence open the possibility for significantly hypo-fractionated treatments in the curative setting which would have been deemed untenable prior to the on-board imaging era. However, there are real concerns the tendency to shrink margins may lead to the phenomena of 'marginal miss', and a loss of local control. Such concerns have been reported in prostate cancer [13], orbital lymphoma [14], breast and head-and-neck cancer [15-18]. The problem of how to verify accurate dose delivery in moving tumors is a challenge ideally suited to 3D dosimetry, and several studies have been reported [19-22]. Similarly, high resolution 3D dosimetry techniques have potential for a defining role in determining and characterizing the dosimetry of small fields [23-25].

The first 3D dosimetry systems were developed in response to the dramatic increase in complexity represented by IMRT described above. The first commercial IMRT system, The Peacock from Nomos Inc, treated patients tomographically, a 2 or 4cm slice at a time, after which the patient was indexed to treat the next slice. Understanding the dosimetry at inter-slice junctions and the accuracy of computer modeling of the dose from a rotational modulated delivery of many small fields was a huge challenge quite beyond the dosimetry systems of the time. The first attempts to verify these dose distributions in 3D were with polymer gel dosimetry [26, 27], based on pioneering work of Maryanski and Gore [28, 29]. Since these early works, polymer gel dosimetry continues to be refined, and has been used successfully [30, 31] with both MRI [32], x-ray-CT [33, 34], and optical-CT readout [34-36]. Innovations in polymer gel dosimeters continue to this day, with new formulations exhibiting increased dose sensitivity and less sensitivity to oxygen [38]. In 2006 a new radiochromic plastic dosimetry system PRESAGE was reported [38]. The potential advantages included reduced light scatter enabling more accurate optical imaging, and greater versatility arising from oxygen insensitivity. PRESAGE has been extensively characterized and used in many dosimetry applications [8, 40-48]. PRESAGE was not the first radiochromic dosimeter however, with earlier Fricke gels having practical limitations of signal diffusion [49-55]. Today recent innovations in radiochromic dosimeters offer exciting potential for further improvements [8, 56-58].
The innovations described above (fluence modulation and on-board imaging) opened new avenues of clinical potential, including dose accumulation in deforming organs, and on-line adaptive procedures. Responding to these developments, 3D dosimetry researchers developed deformable dosimeters with ability to measure and validate dose in deforming dosimeters [22, 59-61] and deformable-image-registration algorithms [62].

In parallel to 3D dosimetry techniques based on chemical dosimeters (polymer gels, plastics, etc) other researchers investigated transit dosimetry: an approach which reconstructs the actual delivered 3D dose from measurements made by on-board flat-panel-detectors acquiring transmission projections during treatment [63-65]. This approach has the significant advantage of an in-vivo measurement of the treatment dose. Efforts have been made to acquire similar capability with chemical dosimeters [66], but without the directness of the EPID approach.

3D dosimetry requirements have also emerged in other areas of clinical and pre-clinical work. 3D dosimetry has tremendous potential in the challenging world of proton and heavy particle therapy. Several groups have investigated proton dosimetry with chemical dosimeters [67-70], but substantial challenges remain in obtaining accurate readings in the Bragg peak. These challenges have led to innovative alternative approaches. Liquid and solid scintillators [71, 72] have shown very promising potential and results. Brachytherapy is a field where high resolution 3D dosimetry systems have been developed and applied to characterize sources and shields, verify monte-carlo calculations, and other aspects with useful effect [73-80]. The pre-clinical world of micro-radiation-treatments has grown tremendously in the last 5 years, with many research groups investigating aspects of complex treatments in small animals. Substantial innovations have been made incorporating 3D printing technology and ultra-high resolution 3D dosimetry [81-83].

2. Conclusions

In conclusion, this presentation will review highlights from key clinical and pre-clinical advances in treatment capability, and responding innovations in the field of 3D and multi-dimensional dosimetry. A number of 3D dosimetry systems are in use at the present time, based on competing or complimentary approaches including chemical systems (polymer and radiochromic), scintillating systems (liquid and solid) and electronic systems (e.g. transit dosimetry with EPIDS, and diode and ion chamber arrays). It is apparent that much progress has been achieved, but also that the goal of widespread clinical implementation remains elusive, and many challenges have not been comprehensively solved. A number of 3D dosimetry systems still require substantial expertise to achieve clinical grade results in a consistent fashion. Some chemical dosimetry systems exhibit batch-to-batch differences in sensitivity and stability, making consistent dosimetry challenging, and demanding high tolerances for manufacturing practices. One approach to addressing these challenges is the emergence of remote dosimetry [84] services, and there are at last three commercial services at the present time from the main manufacturers of chemical dosimetry systems MGS Research, Modus Medical, and Heuris Inc. It is also apparent that there is substantial room for innovation and improvement in many aspects of 3D-5D dosimetry spanning the gamut of materials, read-out, analysis software, and in the economic and practicality of techniques transferrable to the non-specialised clinic.

3. References

[1] Webb S 1989 *Phys. Med. Biol.* **34** 1349-70
[2] Bortfeld T *et al* 1994 *Int. J. Radiat. Oncol. Biol. Phys.* **30** 899-908
[3] Convery D J and Rosenbloom M E 1995 *Phys. Med. Biol.* **40** 979-99
[4] Stein J *et al* 1994 *Radiother. Oncol.* **32** 163-73
[5] Letourneau D *et al* 2005 *Radiother. Oncol.* **75** 279-86
[6] Jaffray *et al* 2002 *Int. J. Radiat. Oncol. Biol. Phys.* **53** 1337-49
[7] Oldham *et al* 2005 *Radiother. Oncol.* **75** 271-8
[8] Oldham M 2014. In *Advances in Medical Physics* Devon Godfrey *et al* (eds) Madison, WI: Medical Physics Publishing.
[9] Hill R *et al* 2010 *Med. Phys.* **37** 4355-63
[10] Moran J M *et al* 2011 *Med. Phys.* **38** 5067-72
[11] Herman M 2010 *Medical Radiation: an Overview of the Issues.* Hearing before the Subcommittee on Health of the Committee on Energy and Commerce, House of Representatives, 111th congress, 2nd session, February 26 2010, Serial No. 111-100, US Government Printing Office
[12] Molineu *et al* 2013 *Med. Phys.* **40** 022101
[13] Engels B *et al* 2009 *Int. J. Radiat. Oncol. Biol. Phys.* 2009 **74** 388-391
[14] Pfeffer *et al* 2004 I *Int. J. Radiat. Oncol. Biol. Phys.* **60** 527-30
[15] Chan A K *et al* 2013 *Oral Oncol.* **49** 255-60
[16] Chen S A *et al* 2013 *Int. J. Radiat. Oncol. Biol. Phys.* **85** 309-14
[17] Chen W-Y *et al* 2012 *Int. J. Radiat. Oncol. Biol. Phys.* **82** 1485-93

[18] Chen A M *et al* 2011 *Int. J. Radiat. Oncol. Biol. Phys.* **80** 1423-29
[19] Feygelman V *et al* 2013 *Med. Phys.* **40** 021708
[20] Thomas A *et al* 2013 *J. Phys.: Conf. Ser.* **444** 012049
[21] Brady S L *et al* 2010 *Phys. Med. Biol.* **55** 2187-201
[22] Ceberg S *et al* 2008 *Phys. Med. Biol.* **53** N387-96
[23] Cramer C K *et al* 2015 *PLoS One* **10** e0143208
[24] Clift C *et al* 2010 *Phys. Med. Biol.* **55** 1279-93
[25] Babic S *et al* 2009 *Phys. Med. Biol.* **54** 2463-81
[26] Oldham M *et al* 1998 *Phys. Med. Biol.* **43** 1113-32
[27] De Deene Y *et al* 1998 *Radiother. Oncol.* **48** 283-91
[28] Maryanski M J *et al* 1996 *Med. Phys.* **23** 699-705
[29] Maryanski M J *et al* 1993 *Magn. Reson. Imaging.* **11** 253-8
[30] Baldock C 2009 *J. Phys.: Conf. Ser.* **164** 012002
[31] Baldock C 2006 *J. Phys.: Conf. Ser.* **56** 14-22
[32] De Deene Y *et al* 2002 *Med. Phys.* **12** 77-88
[33] Jirasek A *et al* 2010 *Phys. Med. Biol.* **55** 5269-81
[34] Johnston H *et al* 2012 *Phys. Med. Biol.* **57** 3155-75
[35] Olding T O *et al* 2011 *Med. Phys.* **36** 3-14
[36] Qian X *et al* 2013 *Phys. Med. Biol.* **58** N321-31
[37] Krstajic N and Doran S J *et al* 2007 *Phys. Med. Biol.* **52** 3693-713
[38] Baldock C *et al* 2010 *Phys. Med. Biol.* **55** R1-63
[39] Adamovics J *et al* 2006 *Radiat. Prot. Dosim.* **120** 107-12
[40] Thomas A *et al* 2011 *Med. Phys.* **38** 4846-57
[41] Sakhalkar H S *et al* 2009 *Med. Phys.* **36** 71-82
[42] Guo P Y *et al* 2006 *Med. Phys.* **33** 1338-45
[43] Brown S *et al* 2008 *Appl. Rad. Isotop.* **66** 1970-1974
[44] Gorjiara T *et al* 2011 *Med. Phys.* **38** 2265-74
[45] Gorjiara T *et al* 2012 *Med. Phys.* **39** 7071-9
[46] Gorjiara T *et al* 2012 *Austral. Phys. Eng. Sci. Med.* **35** 455-63
[47] Doran S J *et al* 2015 *J. Phys.: Conf. Ser.* **573** 012043
[48] Doran S *et al* 2015 *Phys. Med. Biol.* **60** 709-26
[49] Lepage M and Jordan K 2010 *J. Phys.: Conf. Ser.* **250** 012055
[50] Babic S *et al* 2008 *Int. J. Radiat. Oncol. Biol. Phys.* **70** 1281-91
[51] Kelly R G *et al* 1998 *Med. Phys.* **25** 1741-50
[52] Schreiner L J 2004 *J. Phys.: Conf. Ser.* **3** 9-21
[53] Podgorsak M B and Schreiner L J 1992 *Med. Phys.* **19** 87-95
[54] Baldock C *et al* 2001 *Austral. Phys. Eng. Sci. Med.* **24** 19-30
[55] Harris P J et al 1996 *Phys. Med. Biol.* **41** 1745-53
[56] Vandecasteele J and De Deene Y 2013 *Phys. Med. Biol.* **58** 6241-62
[57] Nasr A T *et al* 2013 *Phys. Med. Biol.* **58** 787-805
[58] Jordan K *et al* 2010 *J. Phys.: Conf. Ser.* **250** 012043
[59] Yeo U J *et al* 2012 *Med. Phys.* **39** 2203-13
[60] De Deene Y *et al* 2015 *Phys. Med. Biol.* **60** **1**543-63
[61] Juang T *et al* 2013 *Int. J. Radiat. Oncol. Biol. Phys.* **87** 414-21
[62] Velec M *et al* 2015 *Pract. Radiat. Oncol.* **5** e401-8
[63] Mans A *et al* 2010 *Med. Phys.* **37** 2638-44
[64] Mijnheer B *et al* 2013 *Med. Phys.* **40** 070903
[65] Mijnheer B J *et al* 2015 *Pract. Radiat. Oncol.* **5** e679-87
[66] Oldham M *et al* 2012 *Int. J. Radiat. Oncol. Biol. Phys.* **84** 540-6
[67] Zhao L *et al* 2012 *Phys. Med. Biol.* 2012. **57** N431-43
[68] Doran S *et al* 2015 *Phys. Med. Biol.* **60** 709-26
[69] Zhu X *et al* 2010 *Med. Phys.* **37** 183-8
[70] Back S A *et al* 1999 *Phys. Med. Biol.* **44** 1983-96
[71] Beddar S *et al* 2009 *Med. Phys.* **36** 1736-43
[72] Beaulieu L and Beddar S 2016 *Phys. Med. Biol.* **61** R305-43
[73] Vidovic A K *et al* 2014 *Phys. Med. Biol.* **59** 3893-905

[74] Adamson J *et al* 2014 *Med. Phys.* **41**071705
[75] Adamson J *et al* 2013 *J. Phys.: Conf. Ser.* **444** 12100
[76] Palmer A L *et al* 2013 *Med. Phys.* **40** 061707
[77] Adamson J *et al* 2012 *Med. Phys.* **39** 1515-23
[78] Wuu C S *et al* 2003 *Med. Phys.* **30** 132-7
[79] Wai P *et al 2009 Appl. Radiat. Isot.* **67** 419-22
[80] Hurley C *et al* 2006 *Nucl. Instrum. Meth. A* 565 801-11
[81] Rankine L J *et al* 2013 *Phys. Med. Biol.* **58** 7791-801
[82] Bache S T *et al* 2015 *Med. Phys.* **42**846-55
[83] Newton J *et al* 2011 *Med. Phys.* **38** 6754-62
[84] Juang T *et al* 2014 *Med. Phys.* **41** 071706

Quantifying refractive index mismatch effects on cone beam optical CT scanner measurements

E E Carpentier[1], K M Alexander[1] and L J Schreiner[1,2]
[1]Department of Physics, Queen's University, Kingston, Ontario, Canada K7L 3N6
[2]Department of Medical Physics, Cancer Centre of Southeastern Ontario at Kingston General Hospital, Kingston, Ontario, Canada, K7L 5P9

E-mail: 11eec6@queensu.ca

Abstract. A series of experiments are conducted to evaluate the effects of refractive index (RI) mismatch in readout with a Vista cone beam optical CT scanner. The changing optical attenuation in cuvettes of different colours and RI are quantified using a spectrophotometer, which shows decreasing attenuation as RI increases. A small jar of a colourless solution inside a large jar of another colourless solution with different combinations of fluids of varying RI is imaged in the Vista scanner. Large RI solutions surrounding small RI solutions create a converging lens, while small RI solutions surrounding large RI solutions create a diverging lens. The same dual-jar system is used with solutions of varying RI and colour (attenuation) to assess the effect of RI on attenuation coefficient determination.

1. Introduction

Optical computed tomography (CT) has become a popular imaging modality for gel dosimeters [1, 2]. Typically the dosimeter is placed in a tank filled with a solution to minimize refraction as light travels from the source to detector through the dosimeter in the tank. A common issue one encounters in optical CT is refractive index (RI) mismatch between a gel dosimeter and the surrounding substance. A large RI mismatch decreases the usable region of interest (ROI) within a dosimeter [3] caused by a lensing effect [4] which distorts attenuation information near the dosimeter's edges. To reduce this effect, some match the surrounding solution to the RI of the dosimeter, but for high RI dosimeters, such as PRESAGE [5], this requires a highly viscous liquid that may produce other complications [6]. Dekker *et al* [4] used a scanning laser optical CT system and large area detector with a rotating fiducial marker to trace refracted rays and perform an accurate reconstruction. However, this requires very precise scanner geometry and a reference ray trace scan, adding to the total scan time. The Vista cone beam CCD-based optical CT scanner (Modus Medical Devices Inc., London, ON, Canada) uses a diffuse light source, rather than a scanning laser, and large area detector [7]. In this paper we present results of experiments to investigate the effects of RI mismatch on measurements with the Vista optical CT scanner. The experiments were designed to examine both the effects at the edge of an RI mismatch and the possible perturbation of attenuation measurements internal to a volume simulating a three dimensional radiochromic dosimeter.

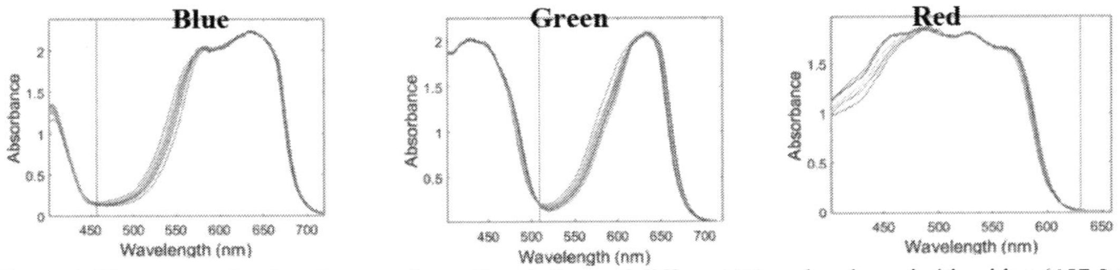

Figure 1. The spectra for the nine cuvettes with solutions of different RI each coloured either blue (457.2 nm, left), green (509.5 nm, centre), and red (630.4 nm, right). A vertical line is drawn to indicate the wavelength of light at which the absorbance values were used for results shown in Figure 3.

2. Materials and Methods

2.1. Fabrication of solutions
Several solutions (in percent by weight unless otherwise stated with an error of ± 1%) were created for these experiments with varying RI: distilled water, 12% propylene glycol (PG), 30% PG, 50% PG, 50% glycerol, 70% PG, 70% glycerol, 100% PG, and 90% glycerol. These solutions have an RI of 1.3324, 1.3463, 1.3659, 1.3880, 1.3976, 1.4105, 1.4266, 1.4320, and 1.4584 respectively, measured with a Reichert r^2 mini refractometer (Reichert Technologies, AMETEK, Inc., Buffalo, NY, USA) with an error of ± 0.0002. The RI of 90% glycerol is from [8] with an error of ± 0.0001 as it exceeds the RI limits of the refractometer. RI values were measured with a 589 nm light source at 20°C ± 1°C.

2.2. Varying colour and index of refraction
Each RI fluid prepared was divided into three small containers. Each container was dyed with 0.99% v/v blue, green, or red food colouring to change optical attenuation values due to absorption. These solutions were poured into 3.5 mL cuvettes and were scanned using a SpectroVis Plus spectrophotometer (Vernier, Beaverton, OR, USA). The spectra determined the maximum absorbance for each solution. Since the maximum absorbance differed slightly for solutions of different RI, a best estimate was used to determine the wavelength of maximum absorbance for green (509.5 nm), blue (457.2 nm), and red (630.4 nm) solutions. The optical attenuation at these peak wavelengths were recorded and plotted against the solutions' RI's to determine a relationship (see figure 3).

2.3. Double jar test 1: colourless solutions
The next set of experiments used a small polyethylene terephthalate (PETE) jar (diameter = 6.8 ± 0.2 cm) suspended in a large (PETE) jar (diameter = 9.5 ± 0.2 cm) by gluing the small jar's lid underneath the centre of the large jar's lid aligning the central axes of the jars. Four non-coloured solutions were initially used: distilled water, 12% PG, 30% PG, and 50% PG. The small inner jar and large outer jar were alternately filled with the various solutions to provide every possible combination of RI. Each combination was scanned with 590 nm diffuse light in the Vista scanner. Three-dimensional (3D) reconstructions of each jar were created with a reference scan of distilled water in both jars. The 3D attenuation information was imported into MATLAB and a horizontal plane midway through the jars was selected for analysis. The attenuation values of a circular region on this plane were averaged to find the mean attenuation in the inner jar for each combination.

2.4. Double jar test 2: coloured solutions
To examine the effects of changing the refractive index on attenuation quantification, a batch of distilled water (RI = 1.3324) and 30% PG (RI = 1.3668) samples were each separated in two; one dyed red and the other green with 0.99% v/v food colouring. Three samples of 12% PG (RI = 1.3461) were prepared: no dye (colourless), dyed red, and dyed green. The red, green, and colourless 12% PG alternated being in the small jar while red and green water and 30% PG alternated surrounding it. Lastly, the colourless

12% PG filled the outer jar, and the inner jar contained red or green water or 30% PG. Each jar combination was imaged in the Vista scanner and reconstructed with a reference scan of colourless distilled water in both jars. The analysis was completed in MATLAB by examining a line profile along a horizontal cross section through the middle of the two jars.

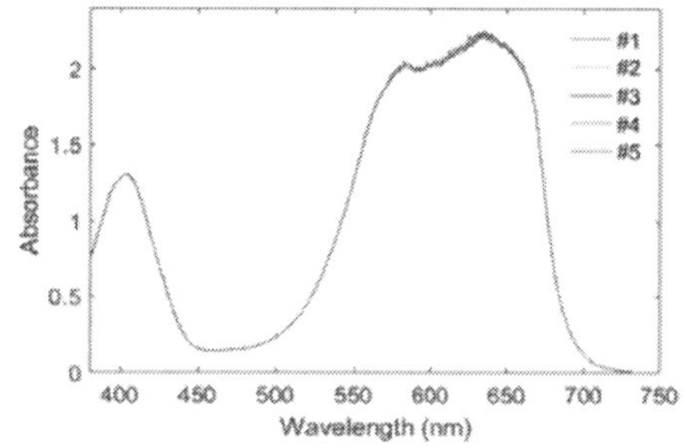

Figure 2. Reproducibility of the spectrophotometer measurements (70% PG sample).

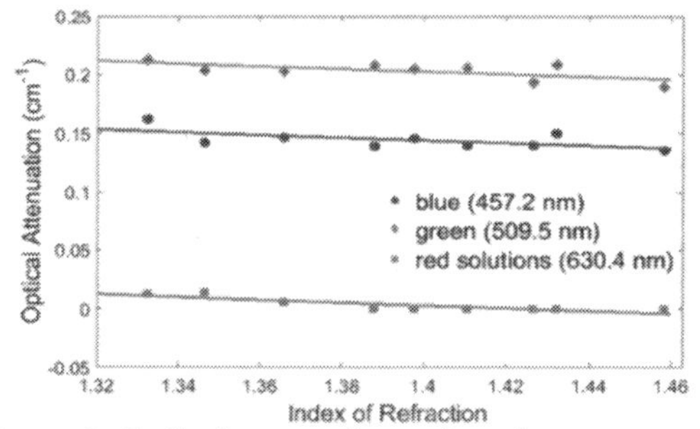

Figure 3. Optical attenuation values for cuvettes of different solutions and colours (error bars are smaller than the symbol size).

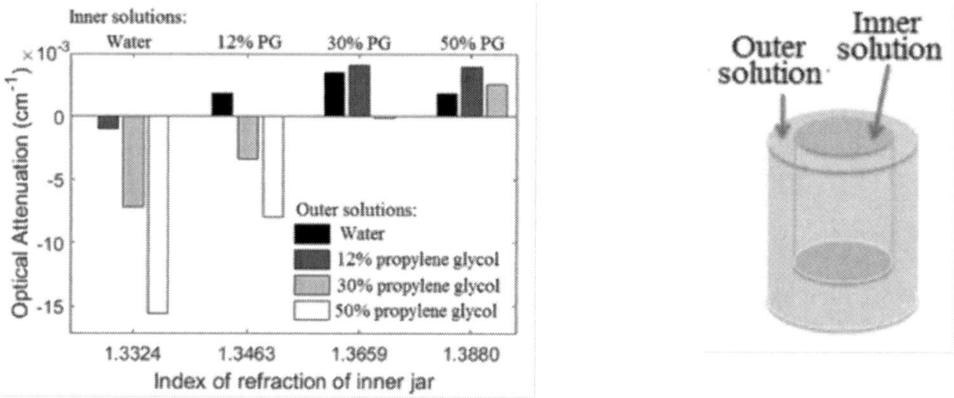

Figure 4. Different combinations of fluids in the dual jars. A schematic of the jar configuration is included on the right.

3. Results and Conclusions

Figure 1 shows the spectra of all nine solutions in different colours in cuvettes as measured in the spectrophotometer; figure 2 demonstrates the reproducibility of the spectra measurements. The graphs show the absorbance value at each colour's wavelength of maximum light transmission (as indicated by a vertical line). Figure 3 shows optical attenuation vs. refractive index for each coloured solution in the cuvettes. The data show a negative slope for each colour, indicating that the attenuation coefficient of each coloured solution decreases slightly with RI.

The Vista scanner measured average attenuation in the inner jar for each jar combination (figure 4). Since all solutions are colourless the expected optical attenuation should be zero, however lensing in the double jars perturbs the measurement. When the solution in the outer jar has a larger RI than the fluid in the smaller jar, the combination becomes a converging lens and the measured optical attenuation in the small jar is negative. Conversely if the solution in the outer jar has a smaller RI than the inner jar's solution, the combination acts as a diverging lens and will produce a positive measured optical attenuation value in the small jar.

The line profiles through both jars containing various coloured liquids are shown in figure 5 with no image averages taken. Figure 5a shows the attenuation profiles for all jar configurations with a red solution in the outer jar. The left and right halves of each figure correspond to a measurement with solutions of different RI surrounding different colours of 12% PG. Observing from left to right in figure

IOP Conf. Series: Journal of Physics: Conf. Series **847** (2017) 012007 doi:10.1088/1742-6596/847/1/012007

5 a and b follows a change in refractive index, while observing from bottom to top tracks a change in attenuation. Perturbations at the jar walls due to RI mismatch are clearly evident; the position of the outer jar walls in Figure 5 are approximately at 35 and 220 pixels and the inner jar walls are near 60 and 195 pixels. This dramatic change in attenuation at the jar walls solidifies the knowledge that a RI mismatch can lead to false attenuation information near the edges of a dosimeter. Since the Vista scanner's light source is 590 nm, one expects green to attenuate more light than red, as can be seen in Figure 5. However, the attenuation through green 12% PG is higher when a red solution surrounds it instead of a green solution, a very unexpected result.

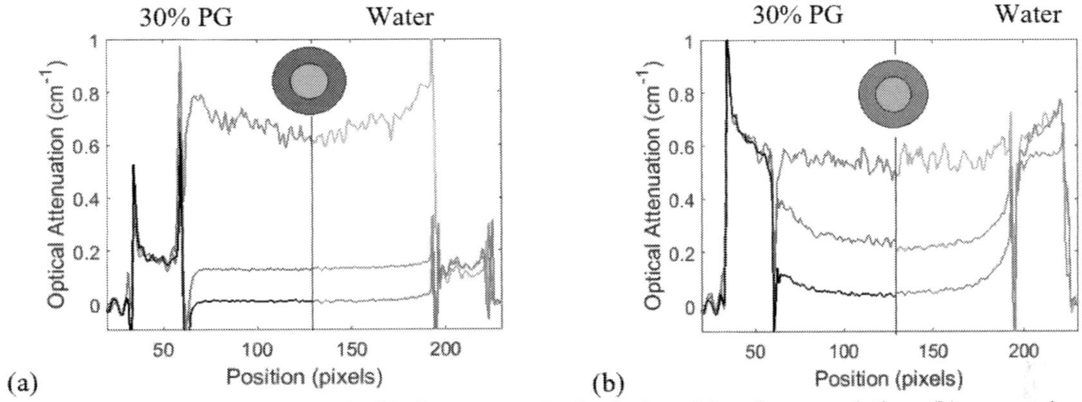

Figure 5. Line profiles through the double jar system. Red solutions (a) and green solutions (b) surround red (light and dark red lines), green (light and dark green lines) and clear (black and grey lines) 12% PG. The left half of (a) and (b) represent 30% PG in the outer jar and the right half of (a) and (b) represents water in the outer jar. A small insert shows a top down view of the two jars, where the grey region represents varying colours of 12% PG.

The results from our experiments indicate that index of refraction has a slight effect on the quantification of the optical attenuation coefficient of the samples in the centre of a jar, and a large effect at the jar walls where the RI mismatch occurs. This supports the measurements from cuvettes seen in figure 3. From figure 4 we can clearly see that a jar will act as a converging or diverging lens and alter the measured attenuation values from the truth, depending on whether a jar's solution has a higher or lower RI than the fluid surrounding it. RI mismatch at the jar walls as shown in figure 5 provides insight into the extent the optical attenuation coefficient can be distorted at the boundary between two materials. This confirms that a RI mismatch will decrease the ROI available for gathering dose information.

4. References

[1] Olding T *et al* 2010 *Phys. Med. Biol.* **55** 2819-40
[2] Baldock C *et al* 2010 *Phys. Med. Biol.* **55** R1-63
[3] Rankine L and Oldham M 2013 *Med. Phys.* **40** 051701
[4] Dekker K *et al* 2016 *Med Phys.* **43** 4585-97
[5] Gorjiara T *et al* 2011 *Med. Phys.* **38** 2265-74
[6] Doran S and Yatigammana D 2012 *Phys. Med. Biol.* **57** 665-83
[7] Bosi S G *et al* 2008 *Phys. Med. Biol.* **54** 275-83
[8] *Physical properties of glycerine and its solutions*. New York: Glycerine Producers' Association; 1963

IOP Conf. Series: Journal of Physics: Conf. Series **847** (2017) 012008 doi:10.1088/1742-6596/847/1/012008

Fan-beam scanning laser optical computed tomography for large volume dosimetry

K H Dekker, J J Battista and K J Jordan
London Regional Cancer Program (London Health Sciences Centre)
Department of Medical Biophysics, Western University
London, ON, Canada

E-mail: kurtis.dekker@lhsc.on.ca

Abstract. A prototype scanning-laser fan beam optical CT scanner is reported which is capable of high resolution, large volume dosimetry with reasonable scan time. An acylindrical, asymmetric aquarium design is presented which serves to 1) generate parallel-beam scan geometry, 2) focus light towards a small acceptance angle detector, and 3) avoid interference fringe-related artifacts. Preliminary experiments with uniform solution phantoms (11 and 15 cm diameter) and finger phantoms (13.5 mm diameter FEP tubing) demonstrate that the design allows accurate optical CT imaging, with optical CT measurements agreeing within 3% of independent Beer-Lambert law calculations.

1. Introduction

Optical computed tomography (CT) has been developed for three-dimensional (3D) radiation dosimetry via scanning of volumetric gels or solids that undergo either radiation-induced colour change (radiochromic materials) or radiation-induced polymerization reactions (polymer gel dosimeters) [1]. Optical CT evolved similarly to x-ray CT, starting from "translate-rotate" scanners, and culminating in broad-beam CCD-based systems operating in either cone-beam [2, 3] or telecentric geometry [4, 5]. The accuracy of broad-beam scanners is perturbed by scatter [6]. It is therefore desirable to make use of single-ray, single-detector geometry, so that transmission measurements within a projection are independent and are only polluted by the scatter associated with a single pencil beam. This motivates a return to scanning-laser systems. Scanning laser geometries have been reported [7-12], but many were too slow for practical applications. Here, we describe a prototype fan-beam scanning laser system which is capable of acquiring 3D imaging of a 15cm diameter, 12cm height cylinder at high resolution (<0.5 mm voxels) in a reasonable scan time (~45 minutes).

2. Methods

Figure 1 shows a schematic of the prototype system. A laser beam is directed to a vertical translation stage, upon which a turning mirror and a fast galvanometer ("galvo") mirror are mounted. The galvo mirror is swept back and forth across the scanned object in a fan-beam pattern. It is driven by a triangular-wave voltage at a rate of 100Hz. The translational stage moves at constant speed during data acquisition, set by the desired slice thickness. The beam falls upon a small area diffuser in front of a vertically translating photomultiplier tube (Hamamatsu). This allows additional collimation to reject stray light, improving upon the previous cone-beam design [12]. The voltage output from the PMT module is sampled by a DAQ device (National Instruments) at a rate of 1 MHz and downsampled as desired. 2D transmission images are generated by assigning the measured voltage at a sampling time to the known ray position at that time.

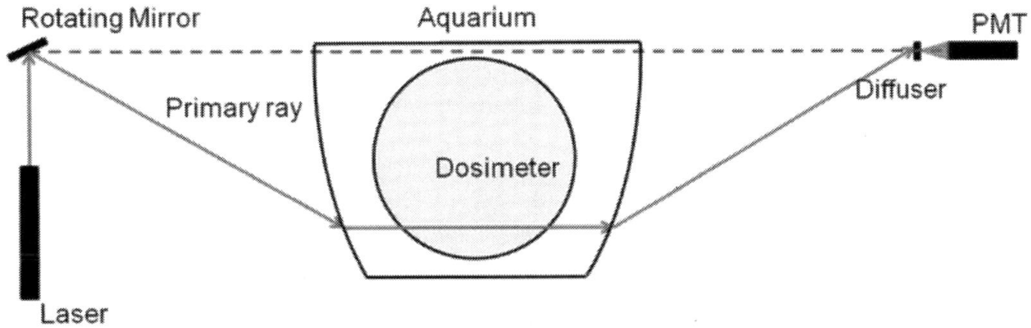

Figure 1. Schematic (top view) of the fan beam scanning laser system.

The shape of the aquarium was generated through ray-tracing simulation to achieve parallel-beam geometry through the scanned volume as well as to focus the light rays down towards a small detector for improved scatter rejection. The asymmetrical design avoids the overlap of regions of important projection data with perpendicular interfaces between the laser beam and the aquarium wall, which result in interference fringes that create rings in reconstructed images (Figure 2).

2.1 Experiments
To illustrate the effect of interference fringes in projection images, optical CT scans of an 11cm diameter uniform solution phantom were obtained using both a symmetric and asymmetric aquarium design. To demonstrate the scanner's ability to image large volumes, a 14.8 cm diameter uniform solution phantom was scanned. Finally, a finger phantom consisting of four FEP tubes (13.5mm outer diameter, 12.5mm inner diameter) [13] was imaged to simulate a small-field dosimetry problem. Carbon black solutions [14] were used as the attenuating liquids in the phantoms. In these preliminary experiments the photomultiplier was stationary. A large area diffuser was used at the detector end with no additional collimation / scatter rejection.

2.2 CT scanning and reconstruction
The data acquisition card samples at a constant temporal frequency, which corresponds to uniform *angular* spacing of rays in this system. However, due to the acylindrical aquarium, this results in *non-uniformly* sampled parallel beam projections. Images of transparent rulers were used to measure this effect. This information was used to re-sample the projections onto a uniformly spaced grid, with spacing equal to the *maximum* inter-sample spacing. In this way, interpolation artifacts from the image re-gridding are avoided. Once this was done, filtered backprojection was performed using a Hamming filter. In all experiments, 1024 projections, spanning 360° of rotation, were acquired (20cm x 12cm FOV, 0.25mm "pixels"). Reconstructions were performed with 0.33 mm isotropic voxels. All reconstruction slices are displayed with 150 mm FOV.

3. Results and Discussion
Figure 2 shows reconstruction slices of a uniformly attenuating solution imaged within the symmetric and the asymmetric aquarium. The ring artifacts near the center of the image in the symmetric case have been avoided by the use of the asymmetric design.

Figure 3 shows the results of the large-volume (15 cm diameter) solution phantom scan as well as the finger phantom. The uniform solution phantom is reconstructed flat (mean value within 1%) over its entire diameter and height. The attenuation coefficient was 0.105 ± 0.003, which agreed within 2%

IOP Conf. Series: Journal of Physics: Conf. Series **847** (2017) 012008 doi:10.1088/1742-6596/847/1/012008

with a Beer-Lambert law calculation (0.107 ± 0.005) made along the central axis of the cylinder with a small aperture (0.5 cm diameter) placed in front of the diffuser.

The finger phantom mean values were measured within 5 mm diameter, 5 mm height ROIs at the center. These were again compared against Beer-Lambert law calculations made by taking the transmission value with the 0.5 cm diameter aperture in place. Excellent linearity between CT and Beer-Lambert law calculations was observed, with a linear regression resulting in $r^2 = 0.9998$ and a slope of 0.983. Note that there is approximately 2-3% underestimation of attenuation in the CT reconstruction compared to the central axis calculation. This may be due to the relatively high scatter of the FEP walls of the phantom, as a large acceptance angle detector was used in this preliminary experiment. However, the fingers were reconstructed circularly and have the correct radius (within 1 pixel), demonstrating the geometric fidelity of the scanner in spite of the complex geometry.

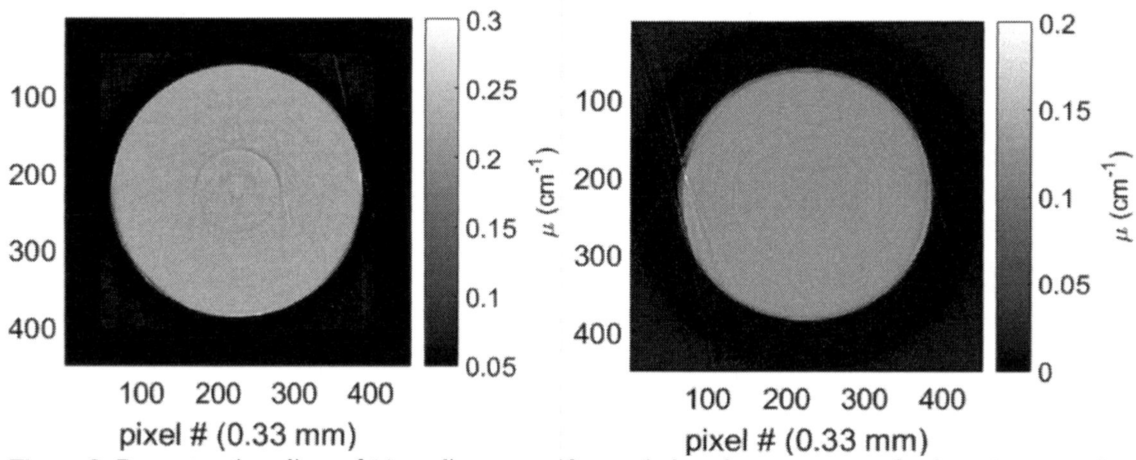

Figure 2. Reconstruction slices of 11cm diameter uniform solution phantoms scanned using: a) symmetric aquarium design. b) asymmetric design. The strong ring in a) is related to interference fringes which have been avoided in the asymmetric design.

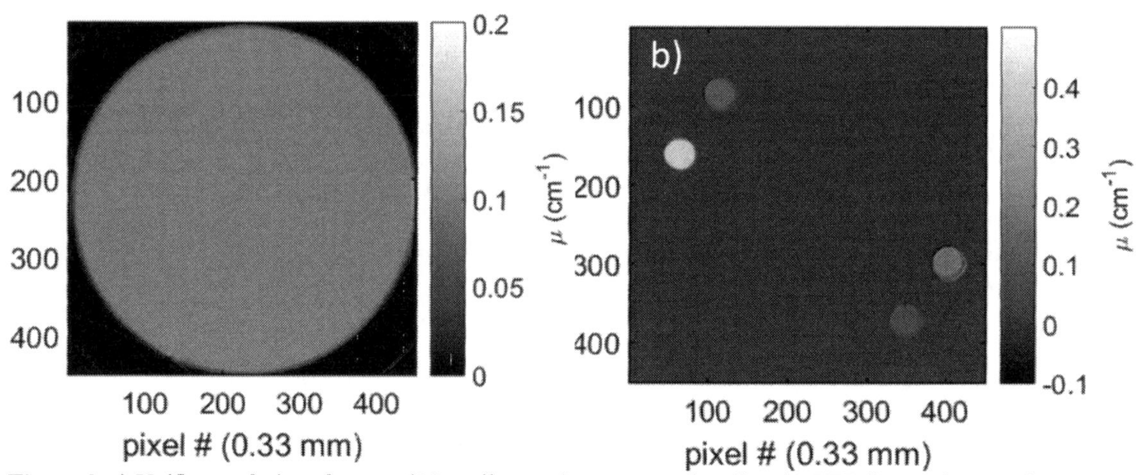

Figure 3. a) Uniform solution phantom (15cm diameter) reconstruction slice. b) FEP finger phantom (13.5mm outer diameter, 12.5mm inner diameter) reconstruction slice.

IOP Conf. Series: Journal of Physics: Conf. Series **847** (2017) 012008 doi:10.1088/1742-6596/847/1/012008

Table 1. CT and Beer-Lambert law attenuation coefficients in finger phantom.

Finger	μ_{CT} (cm^{-1})	$\mu_{Beer-Lambert}$ (cm^{-1})
1	0.064 ± 0.001	0.066 ± 0.001
2	0.115 ± 0.002	0.118 ± 0.001
3	0.175 ± 0.005	0.179 ± 0.004
4	0.400 ± 0.005	0.408 ± 0.008
Linear Regression	$\mu_{CT} = 0.983(\mu_{Beer-Lambert}) - 0.0009$ $r^2 = 0.9998$	

4. Conclusion

We have presented preliminary results from a fan beam scanning laser system making use of a galvo mirror for the fast (horizontal) scan direction and a translational stage for the vertical direction. The system is capable of scanning large volume dosimeters (15cm diameter). The geometric fidelity of the system has been demonstrated here. Currently, it takes about 45 minutes to acquire a 1024 projection dataset (20x20x12 cm^3 imaging volume, 0.25mm voxel resolution). It should be feasible to double the galvo mirror frequency, which will reduce scan time to about 25 minutes. Finally, implementation of detector-side collimation will result in a fast, highly accurate, large volume dosimetry scanner.

5. Acknowledgments

The original version of the ray-tracing code used to generate acylindrical aquarium designs was written by Paul Gallivan. KHD acknowledges scholarship and funding support from NSERC, Western University, and the Ontario Research Fund (OCAIRO project). The authors also acknowledge a grant from the Plunkett Foundation. Disclosure: The authors have a research agreement with Modus Medical Devices Inc. concerning optical CT applications.

6. References

[1] Baldock C *et al* 2010 *Phys. Med. Biol.* **55** R1-63
[2] Wolodzko J G *et al Med. Phys.* **26** 2508
[3] Olding T *et al* 2010 *Phys. Med. Biol.* **55** 2819-40
[4] Sakhalkar H S and Oldham M 2008 *Med. Phys.* **35** 101-11
[5] Krstajić N and Doran S J 2007 *Phys. Med. Biol.* **52** 3693
[6] Granton P V *et al* 2016 *Phys. Med. Biol.* **61** 2893
[7] Gore J C *et al* 1996 *Phys. Med. Biol.* **41** 2695
[8] Kelly R G *et al* 1998 *Med. Phys.* **25** 1741
[9] Maryanski M J and Ranade M K 2001 *SPIE Medical Imaging 2001* pp 764-74
[10] Oldham M *et al* 2003 *Med. Phys.* **30** 623–34
[11] Krstajić N and Doran S J 2007 *Phys. Med. Biol.* **52** N257
[12] Jordan K J *et al* 2013 *J. Phys.: Conf. Ser.* **444** 12062
[13] Jordan K 2013 *J. Phys.: Conf. Ser.* **444** 12076
[14] Jordan K and Battista J 2009 *J. Phys.: Conf. Ser.* **164** 12045

IOP Conf. Series: Journal of Physics: Conf. Series **847** (2017) 012009 doi:10.1088/1742-6596/847/1/012009

Non-diffusing radiochromic leuco-crystal violet hydrogel dosimeter

K J Jordan, T Lindenmaier and K H Dekker
London Regional Cancer Program (London Health Sciences Centre)
Department of Medical Biophysics, Western University
London, ON, Canada

E-mail: kevin.jordan@lhsc.on.ca

Abstract. A systematic study to prepare mixed-micelle, radiochromic hydrogels found that non-diffusing dosimeters can be prepared. Subsequent experiments determined that sodium dodecyl sulfate (SDS) a negatively charged surfactant, binds positively charged crystal violet dye to gelatin below the critical micelle concentration. A typical formulation includes: 4% gelatin by mass, 1 mM hydrogen peroxide, 1 mM SDS, 1 mM leuco crystal violet (LCV) and 25 mM trichloroacetic acid (TCAA). This transparent material has an initial attenuation coefficient of 0.08 cm^{-1} and dose sensitivity of 0.015 cm^{-1} Gy^{-1}. Which is a doubling of the dose sensitivity from the initial formulation with uncharged Triton X100 micelles. Reconstructed beam profiles from 3D optical CT scans performed 1, 14 and 85 hours post irradiation demonstrate no diffusion of the recorded dose distribution.

1. Introduction

Transparent, radiochromic, micelle, gelatin hydrogels are a promising material for 3D water equivalent dosimetry with optical CT readout [1]. The initial transmission and transparency limit which materials can be scanned optically. For dosimeter volumes exceeding one litre the initial optical properties can exclude many radiation sensitive materials. Babic *et al* initially reported that leuco crystal violet dissolved in 4% gelatin hydrogel with trichloracetic acid formed a radochromic gel [2]. The addition of micelles of the uncharged surfactant Triton X100 to this hydrogel dosimeter lowered the diffusion rate of the crystal violet dye tenfold. A previous study had examined positively and negatively charged surfactants as well and found the uncharged micelles had the greatest dose sensitivity [3]. Subsequent studies have reported non-diffusing variations with charged micelles [4, 5]. But these materials have reported larger dose rate affects and were not considered adequate for 3D dosimetry.

This report presents the key result from a study to determine if doping uncharged micelles with charged surfactants to form mixed micelles would lower diffusion rate of charged dye molecules. It was found that micelles were not essential to obtaining low diffusion, and that sodium dodecyl sulfate (SDS) below the critical micelle concentration bound crystal violet to gelatin.

2. Methods

The hydrogel dosimeter (4% gelatin, 1 mM SDS, 1 mM LCV, 25 mM TCAA) was prepared by the following method. For each batch, approximately 90% of the water was used to dissolve the gelatin and 10% to dissolve TCAA and LCV. Gelatin was slowly sprinkled onto water at ~20°C to wet without forming clumps. Hydrogen peroxide, 1 mM was added to increase dose sensitivity and sterilize the gel. The sample was sealed with plastic film and stirred for 5 hours in a 55°C water bath. SDS was then dissolved in the gel solution. Zinc pellets were added for ~30 minutes to scavenge residual oxidizing species and then removed as the sample was cooled to 32°C. The LCV solution was slowly added with stirring and exposure to the air for reoxygenation. Sometimes, the large volume would develop opalescence, however storage at 4°C for several hours resulted in a transparent hydrogel. If the oxidizing species were not scavenged prior to adding LCV the gel would turn light blue over several hours. One centimeter pathlength PMMA plastic cuvettes and 11 cm diameter PETE cylinders ("pop bottle") were filled for irradiation experiments. Transmission measurements were performed at 590 nm with a 10 nm bandwidth with an absorption spectrometer for dose response in cuvettes and a modified Vista10 optical conebeam scanner for 3D measurements [6]. Ultraviolet lights, a ^{60}Co irradiator and a 6 MV photon beams from a Varian medical accelerator were used to write gradients into samples and determine dose sensitivity. The 6MV photon beam was used to perform a small field experiment with the following conditions: jaw defined size at 100 cm (3x3, 2x2, 1x1 and 0.6 x 0.6 cm), gel SSD=90 cm, comparisons at 95 cm, ~20 Gy per beam). Optical CT reconstruction was performed with ModusQA program: projections are ratio of post to pre-irradiation images, 512 projections over 360 degree of rotation, 0.25x0.25x0.25 mm voxels, filtered back projection, Hamming filter. The gel was scanned 1, 14, and 85 hours after irradiation in order to test for diffusion.

3. Results and Discussion

Dose sensitivity was nearly independent of SDS concentration from 0.1 to 2 mM. However, crystal violet diffusion decreased as SDS concentration increased. Dose responses were linear and similar in value for both ^{60}Co at 1 Gy per minute in cuvettes and 12 MeV at 4 Gy per minute at depth of maximum dose in 9.6 cm diameter cylinders. Dose sensitivity was approximately 0.015 $cm^{-1}Gy^{-1}$, nearly double the initially reported value [2]. Most of this sensitivity increase is likely due to the hydrogen peroxide, preconditioning reactions, since SDS had minor sensitivity effect for this range of concentrations. Initial attenuation was 0.08 cm^{-1} which is 50% of the initially reported value [2]. Melting and re-casting previously irradiated gels lowered sensitivity by approximately 15% to a second irradiation dose. This trend is opposite to what was observed in a similar experiment with Triton X100 micelle gels.

Figure 1 shows a white light transmission image from the base of a gel cylinder and a transverse slice from the optical CBCT reconstruction.

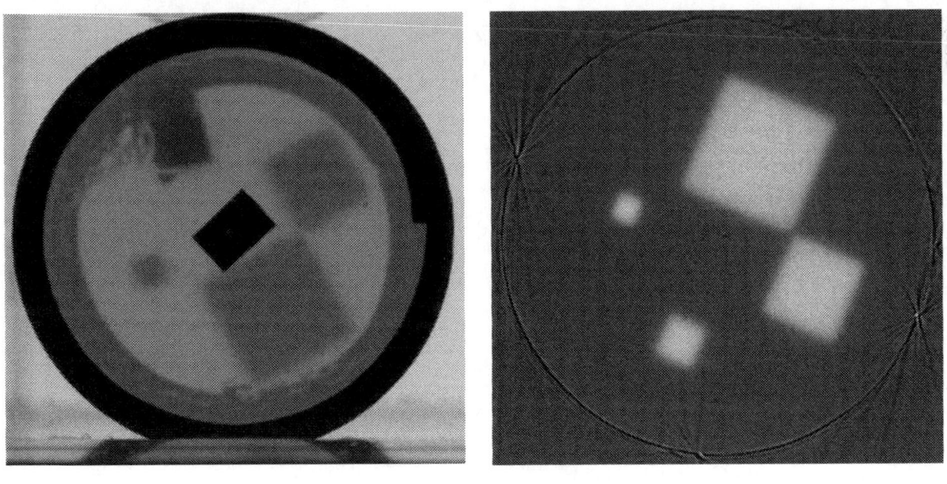

Figure 1. Left: White light transmission image of gel showing small fields, view from bottom to top. Right: Central transverse slice of cone-beam, filtered backprojection reconstruction.

IOP Conf. Series: Journal of Physics: Conf. Series **847** (2017) 012009 doi:10.1088/1742-6596/847/1/012009

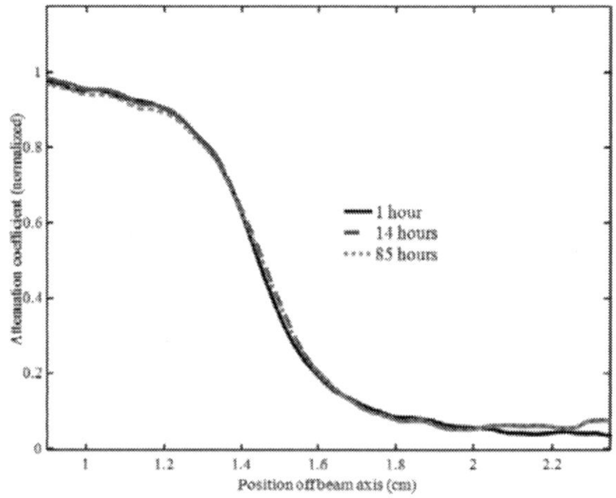

Figure 2. Diffusion test: Attenuation coefficients from reconstructions of optical CT scans 1, 14 and 85 hours post irradiation, background subtracted in low dose area then normalized to central axis. Profiles correspond to 3x3 cm beam at 100 cm.

Figure 2 indicates lack of diffusion since the beam profile is very similar after subtracting an increasing background, approximately 2% of maximum attenuation and normalizing to the central axis of the beam. However, these repeated scans also indicate that a fading of the dose image occurred, ~10% decrease over 14 hours at 4°C. Also the repeat scans indicate a thermal effect related to partially warming the sample and then cooling to 4°C again. The thermal cycling resulted in significant dye formation near the vessel wall providing another example of why thermal histories can be important for accurate 3D dosimetry with hydrogels. Figure 3 demonstrates agreement between Monte Carlo calculation (EGSnrc) and dose measured for same beam conditions.

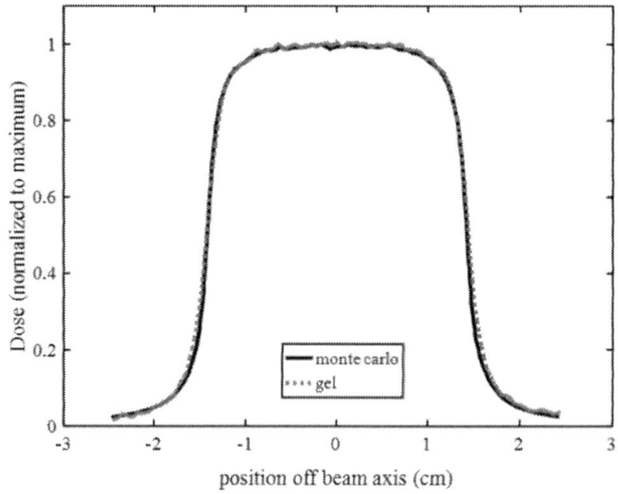

Figure 3. Relative dose profiles comparison: transverse plane, SSD = 90 cm, depth = 5 cm, field size = 3x3 cm at 100 cm, 6 MV, 20 Gy. Solid line: Monte Carlo calculation (EGSnrc). Dashed line: reconstructed gel attenuation coefficient, normalized.

IOP Conf. Series: Journal of Physics: Conf. Series **847** (2017) 012009 doi:10.1088/1742-6596/847/1/012009

4. Conclusion

Non-diffusing radiochromic leuco-crystal violet gel can be achieved below the critical micelle concentration of the negatively charged surfactant SDS [7]. Dose sensitivity has increased due to hydrogen peroxide, preconditioning reaction. Zinc pellets scavenge oxidizing species prior to addition of leuco-crystal violet to the gel. Future work in 15 cm diameter cylinders include measuring: auto-oxidation rate, dose response spatial uniformity, linearity, dose rate effect during irradiation and post irradiation stability. Thermal cycling of dosimeters may provide insights to improve both precision and accuracy of radiochromic hydrogels.

5. Acknowledgments

T L was funded by the Ontario Research Fund (OCAIRO project). KHD acknowledges scholarship and funding support from NSERC, Western University, and the Ontario Research Fund (OCAIRO project). Matt Mulligan performed the Monte Carlo calculation.

6. References

[1] Baldock C *et al* 2010 *Phys. Med. Biol.* **55** R1-63
[2] Babic S et al 2009 *Phys. Med. Biol.* **54** 6791
[2] Jordan K and Avvakumov N 2009 *Phys. Med. Biol.* **54** 6773
[3] Vandecasteele J *et al* 2011 *Phys. Med. Biol.* **56** 6217
[4] Krstajić N and Doran S J 2007 *Phys. Med. Biol.* **52** 3693
[5] Dekker K H *et al* 2016 *Phys. Med. Biol.* **61** 2910
[6] Knox W J and Parshall T O 1970 *J. Colloid and Interface Sci.* **33** 16

IOP Conf. Series: Journal of Physics: Conf. Series **847** (2017) 012010 doi:10.1088/1742-6596/847/1/012010

3D Dosimetry based on a new optical approach for dosimetry gels: Use of the polarisation ratio of the scattering light

O Bleuse[1-3], R Gschwind[1], Y Bailly[2], K Laurent[2], H Bártová[3], K Pilařová[3], V Spěváček[3], L Makovicka[1]
[1]University of Bourgogne Franche Comte, CE UMR 6249 CNRS, 4 place Tharadin, Montbéliard, 25200, France
[2]University of Bourgogne Franche Comte, FEMTO-ST institute UMR 6174 CNRS, Parc Technologique, Belfort, 90000, France
[3]Department of Dosimetry and Application of Ionizing Radiation, Faculty of Nuclear Sciences and Physical Engineering, Czech Technical University in Prague, Břehová 7, Prague, 115 19, Czech Republic

Email: olivier.bleuse@univ-fcomte.fr

Abstract. Several investigations have been carried out by researchers over past two decades to evaluate and perform the reading of gel dosimeters for the three-dimensional measurement of radiation fields. Imaging of the gels has been successfully accomplished with clinical MRI and via laser-based optical scanning using transmission of the light. We report here the methodology and results of a preliminary study carried out to evaluate the utility of a new and simplified approach to make 3D imaging of gel radiation dosimeters based on the scattering light analysis. For the purpose of this initial investigation, nMAG gel has been studied by our method. All pictures were evaluated through a region-of-interest (ROI) analysis to obtain the average change in image density in each sample as a function of the radiation dose. These measured ROI values were subjected to any fit and given a calibration dose and a spatial resolution. This way, we performed a 3D reconstruction of a dosimeter gel.

1. Introduction

One of the most interesting ways to read dosimeter gels is the optical approach because it is not expensive and is more available in comparison with MRI reading [1, 2]. Many studies have focused on its development by the use of the transmission approach due to the irradiated regions in polymer dosimetric gels becoming visibly opaque with an absorbed dose. By this fact, optical computerized tomography (optical-CT) of polymer gel dosimeters has been considered as an alternative to MRI [3]. Several optical scanner types have been developed [4-8] by relying on the principle of filtered back projection to reconstruct sectional images. Much recent optical-CT work has focused on non-scattering dosimeters [9]. Moreover, latterly, some optical devices were upgraded with dual wavelengths [10], and a new reconstruction algorithm was done but it still continues to use still the same approach, namely filtered backprojection (FBP) and algebraic reconstruction technique (ART).

In the optical CT scanning of 3D dosimeters, the dose distribution recorded in a dosimeter is represented by the 3D optical density map caused by the light attenuation and/or absorption from the radiation induced micro-particles in the dosimeter. Therefore, only the portion of the incoming laser beam that survives the multiple Rayleigh-Mie scattering and/or light absorption in the forward direction should be used for image reconstruction. This is why the IRMA research team and the FNSPE at CTU in Prague have previously introduced the new reading method with the scattering light analysis [11, 12] They showed the feasibility of a non-expensive method based on the scattering light analysis to read polymer and radiochromic gels with an acceptable sensibility and a spatial resolution of 2 mm for 3D dosimetry. These measurements were performed on a small and uniform volume and for absorbed doses from 0 to 2 Gy and from 0 to 15 Gy for polymer and Fricke gel respectively. They validated the correspondence of the Beer Lambert law with the scattering intensity and used this calibration curve to read irradiated gel. Latterly, Kristensson suggests the use of more advanced optical diagnostics based on structured illumination and the use of the scattering light to read the gel [13]. In this work, we have created a new optical bench using scattered light, which specifically uses the polarisation ratio to make a link with the absorbed dose. It could also measure size in Particle sizing and velocity measurement of microspheres as shown in on other domains of research [14].

2. Gel Preparation and Irradiation

The nMAG was based on 2% w/w methacrylic acid (MMA) (~99% titration, Sigma Aldrich), gelatin (8% w/w) (Type A, 300; Sigma Aldrich) was used as the matrix substance, and tetrakis (hydroxymethyl) phosphonium chloride (THPC, 80%, H_20 20%, Sigma Aldrich) was used as an oxygen scavenger. The remaining constituent was ultra-pure deionized water which was removed gas from nitrogen during 45 mn before the preparation (Alpha gaz2, Airliquide TM). The gel preparation was carried out according to the standard practice described by [15]. Then, the gel solution was poured into PMMA containers with 4 optical faces of 5 ml (Fisher TM). In addition, we performed a sample which contained gelatin and the polymer gel in order to have an inhomogeneity sample. Then, it was irradiated, and it was used for the 3D imaging. All gels were thereafter stocked in a fridge during 4 hours, and pushed out 1 h before the irradiations. The samples were placed at 10 cm depth in a 500 cm × 500 cm × 500 cm cubic water tank (MP3-M, PTW) at a source-to-surface distance of 100 cm, and irradiated using a 10 cm × 10 cm and 6 MV photons beam, with the linear accelerator Clinac 2100C (Varian Medical system), set to deliver 400 $MU.min^{-1}$ from 0 to 5 Gy. Approximately 12 h after the irradiation, the gels were read by our optical bench.

3. Experimental Measurements and 3D Reconstruction

3.1. Readout system

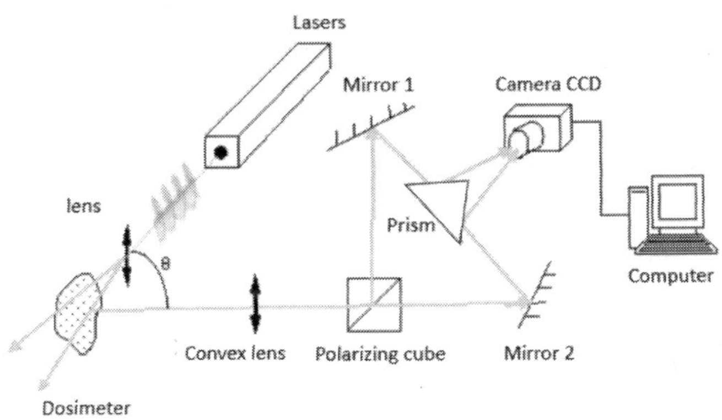

Figure 1. Representation of the optical bench by scattering light.

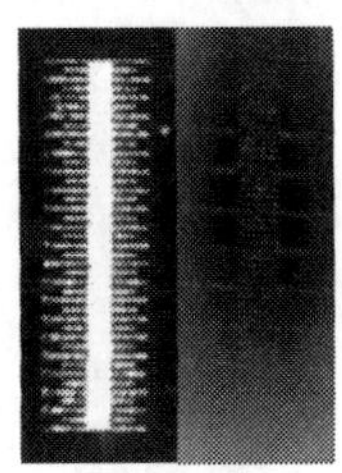

Figure 2. Visualization of the calibration lame (resolution X200 et X800).

Figure 3. Visualization of the scattered light from an irradiated sample (resolution GX200).

The bench contains a laser He-Ne (633 nm, circularly polarised), manual linear stages for the X,Y,Z axis of the sample, and a polarizing cube (Edmunds Optic TM), 2 mirrors (Edmunds Optic TM), a zoom (with a magnification from 200 to 800) (VH Z150, Keyence TM) and a numeric camera 16 bits (ORCA-Flash4.0 LT, Hamamatsu TM).

A circularly polarized incident light is illuminating the sample, creating the scattered light, which is collected by a convex lens at a specific scattering angle. Then the two components of polarisation are separated into two orthogonal directions by a cube polarizer (perpendicular and parallel polarisation). The gel was chosen so that we assumed to measure single scattering light. Both signals are analysed with a homemade software in Python. For each image, a blank image is done with the same exposition parameter of the acquisition, and deleted from the initial picture. This software analyses the average of the image and calculates the polarisation ratio for all zooms (200, 300, 500, and 800) and ROI for 400 to 25 pixels². 20 µm are represented by 10 pixels at a zoom of X200 (1pixel = 2 µm), and 50 pixels for a zoom of X800 (1pixel = 400 nm) (Fig. 2 and 3).

3.2. Scattering Light Analysis

For the first polarisation, the saturation of the ROI started at 3 Gy, which led to a decrease in the camera exposure time and by consequence, the average of the measured signal. On the second polarisation, the signal was linear with the absorbed dose until the saturation of the camera.

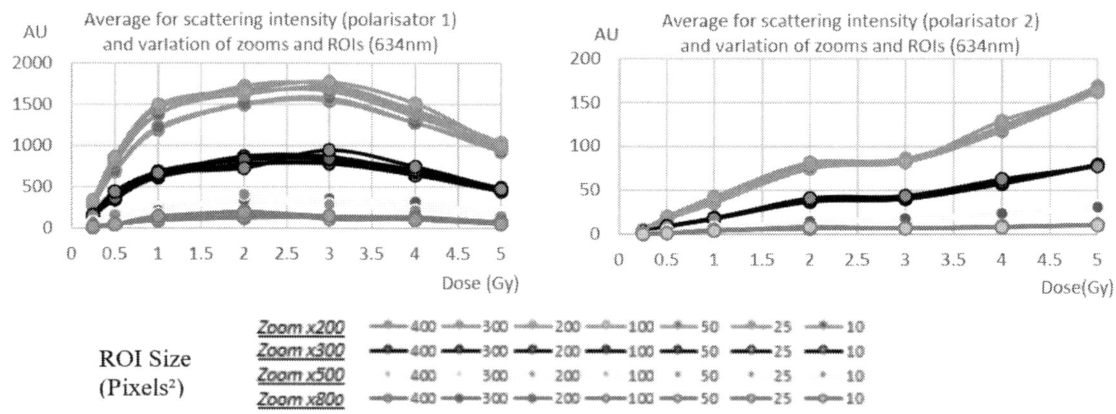

Figure 4. Scattering intensity for polymer gel.

3.3. Calibration Curve and Spatial Resolution

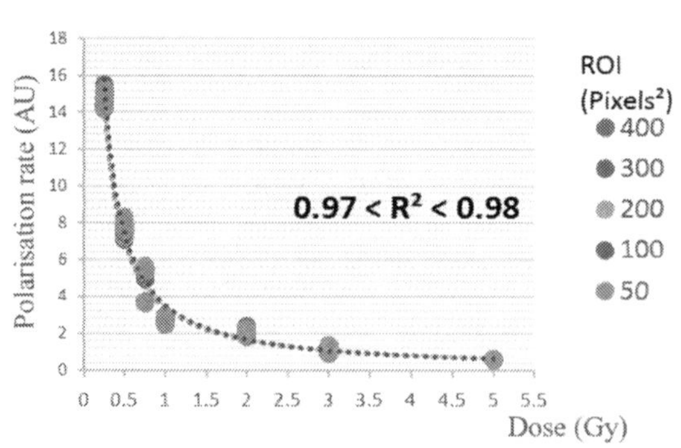

Figure 5. Polarisation rate (I1/I2) measured at 90°as a function of the absorbed dose for 633 nm (all zoom).

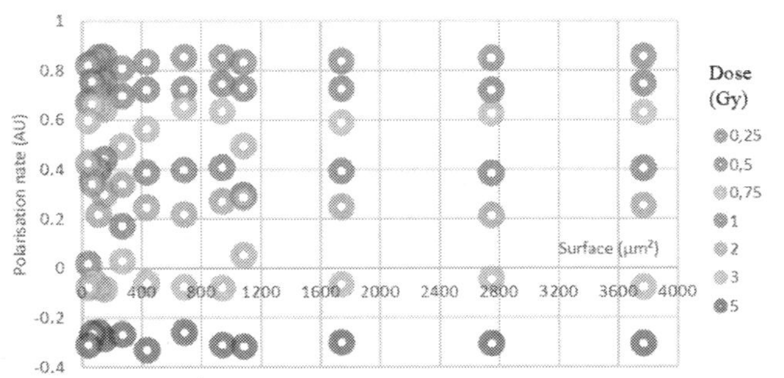

Figure 6. Polarisation rate (I1-I2)/(I1+I2) measured at 90° as a function of the absorbed dose for 633 nm (all zoom).

We performed the calculation of the polarisation ratio for all zooms and ROIs, and obtained the same results, which means, the same area read by different zooms and ROIs gives the same information (fig 5). Then, we applied a spatial resolution to all ROIs, and showed the resolution spatial (in µm^2) as a function of the absorbed dose (fig 6) with a good sensibility, which depends essentially of the gel characteristics. The latter could be adapted to another dose range by modifying gel reagents.

3.4. Results for 3D reconstruction

We performed 66 acquisitions on the Z axis (red square), and made a 3D reconstruction of this volume (400 µm x 400 µm), using the calibration done earlier. We performed voluntary a special sample with gelatin and on the center, the dosimetric reagents. The red area is illustrated by the figure 7. Then, this difference is reconstructed in 3D for a small area.

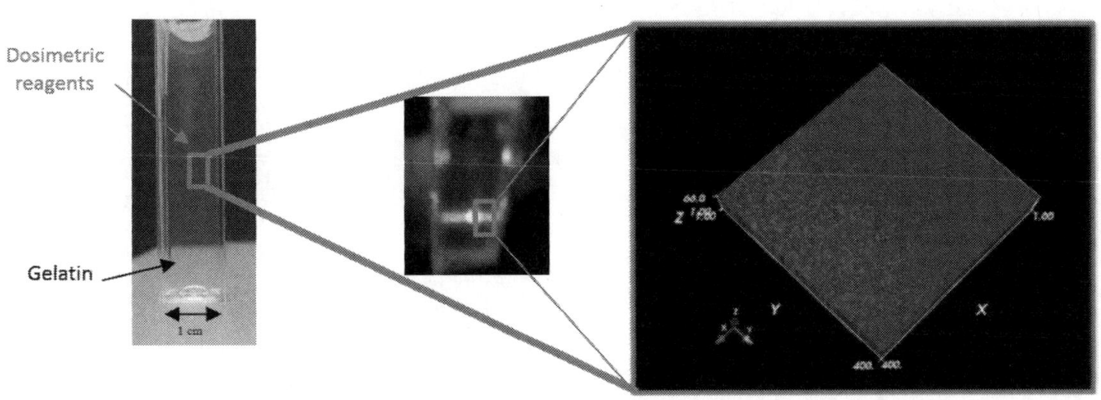

Figure 7. 3D reconstruction of a volume by the scattering method.

4. Conclusion

This preliminary study introduced a new optical bench using the scattering light especially the polarisation ratio. This method is more accurate (µm scale) and more sensitive for low doses than other optical ways, and useful for 3D reconstruction without any algorithmic reconstruction. All data in this work was voluntary not fitted, that could improve results of this approach.

The next steps will consist of testing all kinds of gels, automating the bench, studying all parameters (wavelengths, impact of the volume sample, polarization and depolarization of the scattered light, refractive index, and scattering angulations) and upgrading the zoom and speed of this 3D acquisition. Moreover, we are considering measuring the exact size of microdomains and comparing them with a Mie numerical simulation and finally, to make a link between the absorbed dose and the microdomains' size. To conclude, this approach for dosimetry gel could become a serious and innovative way to measure rays without algebraic reconstruction, and for all applications, especially microdosimetry and nanodosimetry.

5. Acknowledgements

The authors acknowledge the financial support of LCC (Ligue Contre le Cancer région Franche-Comte), the Canceropôle Grand Est, Défi Instrumentation aux limites CNRS and PMA (Pays de Montbéliard Agglomération).

6. References

[1] Baldock C *et al* 2010 *Phys. Med. Biol.* **55** R1-63
[2] Baldock C 2006 *J. Phys.: Conf. Ser.* **56** 14-22
[3] Bosi S G *et al* 2008 *Phys. Med. Biol.* **54** 275-83
[1] Gore J C *et al* 1996 *Phys. Med. Biol.* **41** 2695
[2] Oldham M *et al* 2001 *Med. Phys.* **28** 1436
[3] Doran S J *et al* 2001 *Phys. Med. Biol.* **46** 3191
[4] Xu Y *et al* 2004 *Med. Phys.* **31** 3024-3033
[5] Krstajić N and Doran S J 2006 *Phys. Med. Biol.* **51** 2055-75
[6] Sakhalkar H S and Oldham M 2008 *Med. Phys.* **35** 101
[7] Vandecasteele J and Deene Y D 2013 *J. Phys.: Conf. Ser.* **444** 12053
[8] Alwan R *et al* 2008 *Nucl. Instrum. Meth. B* **266** 834-40
[9] Svoboda J *et al* 2009 *J. Phys.: Conf. Ser.* **164** 12026
[10] Kristensson E *et al* 2015 *J. Phys.: Conf. Ser.* **573** 12010
[11] Huang X Q *et al* 2012 *Opt. Lasers Eng.* **50** 57-63
[12] Karlsson A *et al* 2007 *Phys. Med. Biol.* **52** 4697706

IOP Conf. Series: Journal of Physics: Conf. Series **847** (2017) 012011 doi:10.1088/1742-6596/847/1/012011

Is there a role for 3D dosimetry in low- and middle-income countries?

L E Court
Departments of Radiation Physics and Imaging Physics,
University of Texas MD Anderson Cancer Center, 1400 Pressler Street, Unit 1420, Houston TX 77030, USA

Email: lecourt@mdanderson.org

Abstract. Low- and middle-income countries face significant challenges in developing comprehensive cancer programs. In some cases this involves opening radiation therapy centres in countries where patients have had almost no access to radiotherapy. In additiona to many political and administrative hurdles, staff and equipment shortages are significant. It is these shortages, perhaps, that researchers in high-income countries can have an impact. Perhaps we can adapt our technologies or processes to improve simplicity, robustness and efficiency, or perhaps we can redesign them to reduce cost. The purpose of this paper is to highlight some of the challenges to the 3D dosimetry community, and to promote discussion on whether 3D dosimetry systems can be redesigned or newly developed to help address some of these.

1. Introduction

In the United States and many other countries, much effort is spent researching new technologies that have significant scientific and clinical interest, but often provide only incremental therapeutic benefit [1]. This is likely also true for the last 17 years of the International Conference on 3D Radiation Dosimetry. There are, however, enormous opportunities for the development of technologies for low- and middle-income countries (LMICs) that have the potential to make significant changes to people's lives. More than half the cases of cancer in the world are found in LMICs, but many of these countries have no radiotherapy access at all [2]. In Africa the actual supply of teletherapy machines is around 20% of the actual need. There are, of course, major challenges to improving the situation in these regions, but the potential benefits could be enormous. This is not an area of research/development that will result in incremental therapeutic benefit, but an area that could have a dramatic impact on the lives of the populous in these regions. The purpose of this paper is to highlight some of the challenges and opportunities to the 3D dosimetry community, and promoting discussion on whether 3D dosimetry [3] can be used or adapted to help address some of these.

2. Background to some of the challenges faced in LMICs

Radiation therapy has been shown to be cost-effective for both palliative and curative cancer treatment in LMICs [2, 4]. For example, the cost of an entire course of radiation therapy is $300 in Senegal (independent of the durations or number of fractions) [5]. Reasons for the low cost include the fact that most people undergo treatment as outpatients, equipment and buildings have long lifespans, and equipment can be used for large numbers of patients (high throughput). Also, for many cancers (e.g., locoregionally cervical cancer), radiation is the only effective treatment choice, making for extremely

IOP Conf. Series: Journal of Physics: Conf. Series **847** (2017) 012011 doi:10.1088/1742-6596/847/1/012011

poor prognoses for these cancers if radiation therapy is not available [2]. In high-income countries, 52% of new cancer patients should receive radiation therapy at least once [2]. In LMICs, the epidemiology of cancer cases is different from that in high-income countries, and most patients present with much more advanced disease in the former; in many of these cases, surgical resection is not feasible. For this reason, access to radiation therapy is even more important for these patients. However, access to radiation therapy in LMICs is extremely limited due to deficiencies in equipment, and insufficient skilled personnel.

2.1. Staffing shortages

It is estimated that by 2020, LMICs will have deficits of about 12,000 radiation oncologists, 10,000 medical physicists, and 29,000 radiation therapy technologists [6]. These estimates are based on data from various data sets in the public domain (e.g., IAEA), with staffing levels based on recommendations from the European Society for Radiotherapy & Oncology and IAEA. Hence, there is a crucial need for radiation therapy staff at all levels, in addition to the need for corresponding training for these individuals. For medical physicists, most guidelines recommend a 2- to 3-year internship or residency, often after completion of medical physics graduate school [7-9] - so the training burden is enormous. Various educational initiatives bring young radiation oncologists and medical physicists from LMICs to academic cancer centers in high-income countries, but they are insufficient in addressing the current and future staffing deficits [10]. Furthermore, the failure of professionals who receive training in a high-income country to return home after training in the high-income countries is a historical challenge in radiation oncology and other fields [11-13]. Thus projects that can reduce the number of staff, improve their efficiency, or can accelerate their training, could significantly impact cancer care in LMICs. An example of this from outside the 3D dosimetry world is the automation of treatment planning [14] - this has the potential to make a significant impact because around 50% of physicists time in LMICs is spent doing treatment plans. We should find other solutions, and perhaps there is something in 3D dosimetry?

It should be noted that, in some cases, institutions in LMICs may have the appropriate equipment to treat patients with treatments such as IMRT or VMAT, but lack the staff to perform the additional tasks that this entails, including contouring, treatment planning, machine QA, and patient-specific QA. The author has visited institutions in Africa that have a strong desire to increase their number of VMAT treatments, and have the necessary treatment machines, but do not have all the other resources (particularly staff/time) to make this move. He has also visited an institution where one of the treatment machines was mothballed because of insufficient staff to use it. There are many papers in this and previous conferences on 3D dosimetry which look at patient-specific QA, including potential advantages of 3D dosimetry. Could some of this research be adapted to make patient-specific dose measurements simpler to perform and more efficient? Such a development could help improve the throughput of QA, or all it to be carried out by staff with less training. For example, could a 3D dosimetry system be developed that was relatively insensitive to positioning (and therefore quicker to setup and use), or the control software developed to minimize user interaction?

Similarly, is there a way that 3D dosimetry systems could be developed to facilitate machine QA? The goal would not be to add sophistication to our current techniques, but to allow them to be carried out more efficiently?

Cost is also a significant factor - can current 3D dosimetry systems to be either cheaper, or more robust, than current systems? Often, once equipment breaks, it is difficult to get it serviced - so robustness is of particular importance.

The lack of a significant peer-base can also be a challenge to physicists and other staff in LMICs. They often do not have many experienced peers with whom to discuss their daily challenges. For this reason, links with established institutions is of particular importance in regions where there are limited local experts [2]. One way to establish these links is through research collaborations. Are there opportunities in 3D dosimetry to establish these connections?

Finally, although challenged by insufficient resources, medical physicists in LMICs are involved in research - is there an opportunity for 3D dosimetry here? Perhaps dosimetry systems or readers can be designed at the appropriate price-point to be useful for researchers in LMCs to investigate interesting research topics?

2.2. Equipment shortages

It has been reported that only 4 LMIC countries (Jordan, Lebanon, Suriname and Venezuela) have sufficient teletherapy equipment for their populations, with a median of 37% of patients having access to radiotherapy in the remaining countries (range: 2.3% to 98.8%). In LMICs, there is 0.7 teletherapy units/million population, compared to more than 10 times that number in high-income countries [6]. Patients often have to travel hundreds of kilometres for treatment [15]. The teletherapy equipment available in LMICs varies widely, with much wider use of Cobalt units than we see in the United States. Advantages of Cobalt units include dependability, simplicity and reduced electrical power requirements. This is important, given that in addition to equipment availability, there are many related challenges, such as power shortages and inadequate engineering support [16].

Independent auditing of new treatment equipment is an important part of care in any part of the world. The use of phantoms such as those used extensively by IROC Houston can be a challenge in many LMICs because of the challenge of getting the phantom in and out of the country. Purchasing their own anthropomorphic phantom for end-to-end tests and treatment planning system testing (e.g. as suggested by the IAEA [17, 18]) can also difficult because of the cost. Is there a role for 3D dosimetry here? Perhaps in making the extensive end-to-end testing simpler or more comprehensive? Is there a 3D dosimetry system that could be adapted to make simple measurements of beam characteristics (PDD or profiles) that could then be send to a central location for analysis? Perhaps, if the treatment planning system comes pre-commissioned, the entire acceptance and local commissioning could be performed using a 3D dosimeter - perhaps analysed locally or at a central location?

3. Conclusions

There are many challenges to radiotherapy in LMICs. This paper has highlighted some of them, indicating some of the possible ways in which researchers in the 3D dosimetry community could perhaps change the direction of their research to potentially address these challenges.

4. References

[1] Petereit D G *et al* 2015 *Frontiers in Oncology* **5** 103
[2] Barton M B *et al* 2006 *The Lancet Oncology* **7** 584-95
[3] Baldock C *et al* 2010 *Phys. Med. Biol.* **55** R1-63
[4] Van Der Giessen P-H *et al* 2004 *Radiother. Oncol.* **71** 347-55
[5] Fisher B J *et al* 2014 *Int. J. Radiat. Oncol. Biol. Phys.* **89** 458-61
[6] Datta N R *et al* 2014 *Int. J. Radiat. Oncol. Biol. Phys.* **89** 448-57
[7] Physics IOfM. Policy Statement No. 2: Basic Requirements for Education and Training of Medical Physicists. 2010.
[8] Commission on Accreditation of Medical Physics Educational Programs I. Standards for Accreditation of Residency Educational Programs in Medical Physics. 2014.
[9] Caruana C J *et al* 2014 *Physica Medica* **30** 598-603
[10] Jaffray D A *et al* 2014 *Int. J. Radiat. Oncol. Biol. Phys.* **89** 446-7
[11] Magrath I 2007 *Lancet Oncology* **8** 562-3
[12] Nullis-Kapp C 2005 http://www.who.int/bulletin/volumes/83/2/en/news.pdf
[13] Hagopian A *et al* 2004 *Human Resources for Health* **2** 17
[14] Kisling K *et al* 2016 *Med. Phys.* **43** 3560
[15] Einck J P *et al* 2014 *Int. J. Radiat. Oncol. Biol. Phys.* **89** 462-7
[16] Page B R *et al* 2014 *Int. J. Radiat. Oncol. Biol. Phys.* **89** 476-80
[17] Alfonso R *et al* 2015 *IFMBE Proc.* **51** 543-6
[18] Brunckhorst E 2008 IAEA-TECDOC-1583 1-67

IOP Conf. Series: Journal of Physics: Conf. Series **847** (2017) 012012 doi:10.1088/1742-6596/847/1/012012

Dosimetric comparison of DEFGEL and PAGAT formulae paired with an MRI acquisition

C Matrosic[1], A McMillan[2], J Holmes[2], B Bednarz[1] and W Culberson[1]
[1]The University of Wisconsin School of Medicine and Public Health, University of Wisconsin - Madison, Department of Medical Physics, Madison, WI
[2]The University of Wisconsin School of Medicine and Public Health, University of Wisconsin - Madison, Department of Radiology, Madison, WI

Email: cmatrosic@wisc.edu

Abstract. Normoxic Polyacrylamide Gels, or nPAGs, are 3D gel dosimeters that measure dose through the process of radiation-induced polymerization. Two nPAG formulae are DEFGEL and PAGAT, which are very similar, but differ mainly due to different weight fractions of monomers. The dosimetric resolutions of the two formulae when paired with a Spin-Echo (SE) MRI sequence and a monoexponential fit were compared over a range of 0-15 Gy. It was found that in the dose range 0-6 Gy the PAGAT formula generally showed a much finer dose resolution, while the DEFGEL formula showed a finer resolution from 8-15 Gy.

1. Introduction

Gel dosimeters such as polyacrylamide gels (PAGs) and normoxic polyacrylamide gels (nPAGs) can measure 3D dose distributions, while maintaining tissue equivalence, dose rate independence, and angular independence [1-4]. This family of gel dosimeters operates through the radiation-induced polymerization of the monomer acrylamide and the cross linking monomer N,N'-methylenebis-acrylamide (Bis) [5]. This polymerization changes the optical density (OD) and transverse nuclear magnetic relaxation rate (R_2) as a function of dose, which allows the dose to the gel to be characterized through optical readout or MRI acquisition [6, 7]. Two types of nPAGs are the DEFGEL formula [8] and the PAGAT formula [9], which differ only by the amount of monomers present in each gel type and the form of the antioxidant. Specifically, the DEFGEL formula has a 6% monomer fraction by weight, while the PAGAT formula has 9%. This difference in monomers concentration causes each dosimeter to result in different dose responses and dose response calibrations [10]. The goal of this investigation is to compare the dose response of both gel formulae through an analysis of the dose resolutions of the gels over a range of 0-15 Gy.

2. Method

2.1. Gel Fabrication & Irradiation

Twelve gel dosimeters were fabricated in-house, six using the DEFGEL formula described by Yeo *et al* and six using the PAGAT formula described by Venning *et al* [9, 10]. Both batches of gel were prepared using the methods described by Yeo *et al* [8], but in the case of PAGAT, THPC, the chloride form of the antioxidant THP, was used as opposed to the sulfate form used in DEFGEL. All gel dosimeters were fabricated in 7.0 mL BD Vacutainer® Serum (Becton, Dickinson and Company, Franklin Lakes, NJ) glass vials.

To ensure accurate and precise gel doses, gel irradiations were performed by placing the dosimeters in a Sun Nuclear 1D Scanner™ (Sun Nuclear Corporation, Melbourne, FL). All irradiations were performed using a Varian Clinac 21EX at the University of Wisconsin Medical Radiation Research Center (UWMRRC) with a 6MV beam and a repetition rate of 600 MU per minute. The irradiations were performed at a depth of 5.5 cm in liquid water with a source-to-surface distance (SSD) of 100 cm and a 10 cm x 10 cm field size. The centers of the vials were positioned on the central axis (CAX).

2.2. MR Scanning

Gel R_2 data were acquired using a 3T MRI scanner (SIGNA PET/MR, GE Healthcare, Waukesha, WI) and an 8-channel receive-only head coil. All scans were acquired using a spin-echo (SE) pulse sequence using the parameters shown in Table 1. The TE values used for the scans were based on the optimal SE parameters described in the parameter optimization of De Deene and Baldock [7] The echoes from each scan session were used to create R_2 maps by performing a linear fit of the natural logarithm of the echo signals at each echo time on a voxel-by-voxel basis. Mean R_2 values and their standard deviations were gathered for a central, rectangular, region of interest (ROI) in each of the gels for dose response fits.

Table 1. Spin-Echo scan parameters used to gather the R2 data.

DEFGEL Scan Parameters		**PAGAT Scan Parameters**	
Repetition Time (TR)	5000 ms	Repetition Time (TR)	5000 ms
Echo Time (TE)	20 ms, 512 ms	Echo Time (TE)	20 ms, 330 ms
Echoes (NE)	2	Echoes (NE)	2
Averages (NEX)	1	Averages (NEX)	1
Acquisition Matrix	128 x 128	Acquisition Matrix	128 x128
Field of View (FOV)	128 x 128 mm^2	Field of View (FOV)	128 x 128 mm^2
Spatial Resolution	1 mm^2	Spatial Resolution	1 mm^2
Number of Slices (NS)	1 - Coronal	Number of Slices (NS)	1 - Coronal
Slice Thickness	3 mm	Slice Thickness	3 mm
Scan Time	11.66 min/echo	Scan Time	11.66 min/echo

2.3. Dose Response and Dose Resolution

In order to characterize the dose response of each gel formula the mean R_2 versus dose relationship was fit to the following monoexponential function [7] using MATLAB,

$$R_2(s^{-1}) = R_{2,sat} - \Delta R_2 \cdot \exp(-\alpha * D(Gy)). \qquad (1)$$

In this equation, the fitting parameter $R_{2,sat}$ represents the saturated R_2 value of the gel, the fitting parameter ΔR_2 represents the range of R_2 values possible for the gel, and α is an exponential fitting parameter.

To evaluate the intrinsic dosimetric precision of each of the gels paired with an MRI acquisition, the metric of dose resolution was used. Dose resolution takes into account both the inherent dose sensitivity of the gel, along with the SNR of the MR acquisitions. Dose resolution was calculated using the formalism of Baldock *et al* [11, 12]. In this formalism, the dose resolution of two dose peaks measured from a gel dosimetry system with a confidence level of p is defined as:

$$D_\Delta^p = k_p\sqrt{2}\sigma_D \qquad (2)$$

where k_p is a coverage factor, which is 1.96 for a 95% confidence level, and σ_D is estimated to be the combined dosimetric uncertainty. Given that the Type B uncertainties of the dose are negligible when compared to the Type A uncertainty due to the standard deviation of the R_2 measurements [8], the combined uncertainty simplifies to:

$$u_C^2(D) = \left[\frac{\delta D}{\delta R_2}\right]^2 u^2(R_2), \quad (3)$$

where $u(R_2)$ is the standard uncertainty of the R_2 measurements, which was estimated to be the standard deviations of the mean R_2 values.

Equation 1 can be rearranged to calculate dose as a function of R_2:

$$D(Gy) = \frac{1}{\alpha}\ln\left(\frac{\Delta R_2}{R_{2,sat} - R_2(s^{-1})}\right). \quad (4)$$

Combining equations 3 and 4, the final equation for dose resolution becomes:

$$D_\Delta^p = k_p\sqrt{2}\sqrt{\left(\frac{1}{\alpha(R_{2,sat} - R_2(s^{-1}))}\right)^2 \left(\sigma(R_2)\right)^2}. \quad (5)$$

The value of 1.96 was used for the coverage factor k_p in the following dose resolution calculations to represent a 95% confidence level.

3. Results and Discussion

3.1. Dose Response

The measured mean R_2 values and their standard deviations are shown in Table 2 and the resultant monoexponential fits are shown in Figure 1 and Table 3. Both dose responses showed excellent fits to the monoexponential equation, with R^2 values of 0.9999 and 0.9994 for DEFGEL and PAGAT, respectively. Over the 0-15 Gy dose range investigated, the PAGAT gel resulted in a much larger range of R_2 values due to a much steeper initial slope. The values of $R_{2,sat}$ and ΔR_2 for both gels were quite similar, suggesting that the PAGAT gel response saturated at a much faster rate than the DEFGEL.

Table 2. Mean R_2 values measured within a central ROI of each dosimeter, the standard deviation of each mean, the dose uncertainties calculated from these values, and the resultant dose resolutions with a 95% confidence level.

	DEFGEL					PAGAT				
Dose (Gy)	**R2 (s^{-1})**	**Σ (s^{-1})**	**Σ (%)**	**u_c(D) (Gy)**	**Dose Resolution 95% (Gy)**	**R2 (s^{-1})**	**Σ (s^{-1})**	**σ (%)**	**u_c(D) (Gy)**	**Dose Resolution 95% (Gy)**
15	3.0643	0.0358	1.17	0.48	1.33	5.3044	0.0675	1.27	1.29	3.57
10	2.6567	0.0367	1.38	0.42	1.17	5.0154	0.0718	1.43	0.70	1.95
8	2.4872	0.0321	1.29	0.35	0.96	4.6636	0.0682	1.46	0.42	1.16
6	2.2867	0.0338	1.48	0.34	0.95	4.3060	0.0606	1.41	0.27	0.75
2	1.8791	0.0252	1.34	0.23	0.63	3.0047	0.0382	1.27	0.09	0.24
0	1.6408	0.0232	1.41	0.20	0.54	1.9720	0.0387	1.96	0.06	0.17

Table 3. Fitting parameters for the monoexponential fits of the dose response data.

	$R_{2,sat}$ (s^{-1})	**ΔR_2 (s^{-1})**	**A (Gy^{-1})**
DEFGEL	5.475	3.831	0.0309
PAGAT	5.609	3.648	0.1718

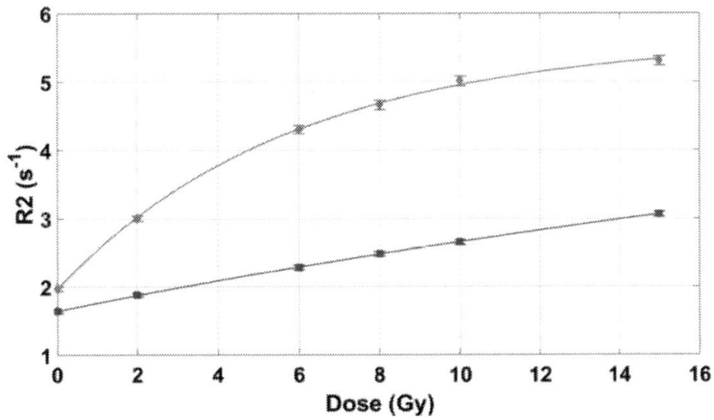

Figure 1. The dose response fits of DEFGEL (blue squares) and PAGAT (red diamonds).

3.2 Dose Resolution

The resulting dose resolution values are provided in Table 2. Over a range from 0-6 Gy, the PAGAT formula showed a much finer resolution than the DEFGEL formula. This is likely due to the larger amount of monomers in PAGAT, which increases the dose sensitivity of the gel. As the dose response of the PAGAT began to saturate, the dose resolution of the DEFGEL formula was improved relative to the PAGAT formula (as seen in the range of 8-15 Gy). This was due to the more linear nature of the DEFGEL formula's dose response, allowing for it to better distinguish higher doses due to no polymerization saturation. The dose resolutions found in this study are generally larger than those found in the literature, except for the values for PAGAT at low doses [9-13]. This is likely because a monoexponential fit, as opposed to a linear fit, was used to calculate dose resolution to account for saturation effects, causing coarser dose resolutions at high doses. Also, the larger Type A uncertainties caused by the small vials and their MRI acquisition may have also worsened dose resolution.

4. Conclusion

In this study the dose responses of the DEFGEL and PAGAT formulae were evaluated and compared using the metric of dose resolution. It can be concluded that the DEFGEL formula showed a better dose resolution at doses ranging from 8-15 Gy, while the PAGAT formula had superior dose resolution at lower doses ranging from 0-6 Gy. This conclusion suggests that when paired with MRI, each gel could be more effective in different applications. For example, the DEFGEL formula would be much more appropriate for a 3D dosimeter to be used as a dose target in cases with high dose fractions, like in phantoms used to measure doses from SBRT, SRS, and other hypofractionated treatment modalities. In contrast, PAGAT would an appropriate dosimeter for measuring the doses from treatment modalities with more standard dose fractions.

5. Acknowledgements

This work was partially funded by NIH grant R01CA190298.

6. References

[1] Baldock C *et al* 2010 *Phys. Med. Biol.* **55** R1-63
[2] Mather M L *et al* 2003 *Ultrasonics* **41** 551-9
[3] Hill B *et al* 2005 *Med. Phys.* **32** 1589-97
[4] Brown S *et al* 2008 *Appl. Rad. Isotop.* **66** 1970-74
[5] Baldock C *et al* 1998 *Phys. Med. Biol.* **43** 695-702
[6] Bosi S *et al* 2007 *Phys. Med. Biol.* **52** 2893-903
[7] De Deene Y and Baldock C 2002 *Phys. Med. Biol.* **47** 3117-41

[8] Yeo U J *et al* 2012 *Med. Phys.* **39** 2203-13
[9] Venning A J *et al* 2004 *J. Phys.: Conf. Ser.* **3** 155-8
[10] Venning A J *et al* 2005 *Phys. Med. Biol.* **50** 3875-88
[11] Baldock C *et al* 2001 *Phys. Med. Biol.* **46** 449-60
[12] Lepage M *et al* 2001 *Phys. Med. Biol.* **46** 2665-80
[13] Vandecasteele J and De Deene Y 2013 *Phys. Med. Biol.* **58** 19-42

IOP Conf. Series: Journal of Physics: Conf. Series **847** (2017) 012013 doi:10.1088/1742-6596/847/1/012013

A comparison of the dose–response behavior of AQUAJOINT®-based polymer gel and PAGAT gel dosimeters measured using Optical CT and MRI

Takaoki Takanashi[1], Hiraku Kawamura[2], Kazuaki Fukasaku[1,3], Daniel Antonio Sahade[1,4] and Toshimasa Hamada[1,5]

[1]3D Gel Dosimeter Research Laboratory, Innovation Center, RIKEN, 2-1, Hirosawa, Wako, Saitama 351-0198, Japan

[2]Department of Radiological Sciences, Ibaraki Prefectural University of Health Sciences, 4669-2, Ami, Ami-machi, Inashiki-gun, Ibaraki, Japan

[3]Department of Neurosurgery, Himon'ya Hospital, 2-9-7 Minami Meguro-ku, Tokyo 152-0013, Japan

[4]Frontier Materials Research Department, Materials Research Laboratories, Nissan Chemical Industries, Ltd., 488-6 Suzumi-cho, Funabashi, Chiba 274-0052, Japan

[5]Planning & Development Department, Performance Materials Division, Nissan Chemical Industries, Ltd., 3-7-1 Kanda-Nishiki-cho, Chiyoda-ku, Tokyo 101-0054, Japan

E-mail: toshimasa.hamada@riken.jp

Abstract Absorbed dose–response characteristics of AQUAJOINT®-based polymer gel and PAGAT gel dosimeters were compared using Optical CT and MRI. AQUAJOINT® gel exhibited a relatively good linear dose–response relationship in the radiation dose range of 0–5 Gy.

1. Introduction

AQUAJOINT® is a stretchable and thermo-irreversible hydrogel formed by mixing two types of water-based liquids (liquid A and B) at room temperature. Water-soluble polymer and inorganic nanoparticles form a hydrogel having a stable three-dimensional network structure after being mixed in water and then left to stand at room temperature, without C–C bond formation reactions. Water-soluble additives can be included into the hydrogel prior to mixing liquids A and B.

Polymer gel dosimetry is a three-dimensional (3D) dosimetric tool for quality assurance. It quantifies radiation-induced polymerization of vinyl monomers as a measure of radiation dose [1, 2]. Spin–spin relaxation rate ($R_2 = 1/T_2$, s^{-1}) can be used to evaluate the degree of polymerization of the irradiated gel and the radiation dose [1-3].

Although various methods, such as X-ray CT [4, 5], ultrasound [6], and Raman spectroscopy [7], have been used for dose evaluation of 3D gel dosimeters, MRI is the most widely used among these [8]. However, the high cost of MRI equipment limits its use for research on gel dosimeter development. Gel dosimeters are based on the property of optical absorption, which varies according to the radiation dose and can be measured through spectrophotometry [9]. An optical CT (OCT) is a device for measuring the absorbance of the gel dosimeter in 3D using this property [10].

In this work, the dose versus relative value curves were assessed using a 2D OCT prototype system and MRI measurements, for a preliminary comparison of AQUAJOINT®-based polymer gel dosimeter and PAGAT gel.

2. Materials and Methods

2.1 AQUAJOINT®-based 3D gel dosimeter formulation

The combination of AQUAJOINT® hydrogel (Nissan Chemical Industries) with suitable radiation sensitive compounds affords good stability to the dosimeter gel matrix even at relatively high temperatures, owing to its heat irreversibility [11]. In this work, we prepared the dosimeter mixture by first dissolving in oxygen-free water, the monomer *N*-vinyl-2-pyrrolidone (NVP), the bifunctional monomer *N,N'*-methylene-bis-acrylamide (Bis), the oxygen scavenger tetrakis (hydroxymethyl) phosphonium chloride (THPC), and AQUAJOINT® B-5 (water content: 86%). Addition of AQUAJOINT® A-5 (water content: 81%) to the aqueous solution yields a homogeneous system comprising of a gel matrix embedded in a monomer mixture solution. The resultant mixture contained 4% NVP, 2% Bis, 25 mM THPC, 11% AQUAJOINT® A-5, and 11% AQUAJOINT® B-5. The gel precursor was poured into 70 mm (height) × 34 mm (diameter) polyethyleneterepthalate (PET) vials. Upon irradiation, a polymer network forms in the gel matrix, which induces a physical change in the irradiated zone that can be quantified using imaging modalities such as MRI or OCT.

2.2 PAGAT gel dosimeter

PAGAT gel dosimeters were prepared using 89% w/w water, 3% acrylamide, 3% Bis, 5% gelatin (300 bloom), and 5 mM THPC. Milli-Q water was heated at 40°C, mixed with gelatin and heated at 50°C on a hot plate/magnetic stirrer [12-14]. After the solution became clear, it was cooled to 45°C, and Bis and acrylamide were added. After complete dissolution of Bis and acrylamide, THPC was added. The gel precursor was then poured into PET vials.

2.3 Sample irradiation

A Radio Flex 320 CG (Rigaku Corporation) was used for X-ray irradiation. At a dose rate of 1 Gy/min, dose control was performed by changing the irradiation time. Irradiation of the sample with X-ray was performed using a simple collimator consisting of lead blocks that was adapted to the apparatus. Radiation dose ranged from 1 Gy to 10 Gy; six of the samples were irradiated (1, 2, 3, 5, 7 and 10 Gy, respectively). A non-irradiated sample was assigned as 0 Gy.

2.4 OCT system

We constructed a low-cost (total material cost: US $400, approximately) OCT to evaluate the 3D gel dosimeter. Educational computer Raspberry Pi was used to control and reconstruct images for our OCT system. A white organic EL panel was used as a light source, and the light receiving unit consisted of a 5 megapixel CMOS image sensor. A chuck mounted to the upper synchronous motor had a structure for fixing the gel dosimeter. The gel dosimeter was placed into a water-filled aquarium in order to carry out the measurements [Fig. 1]. Tomographic image reconstruction was performed using ImageJ [15, 16] software and its plug-ins.

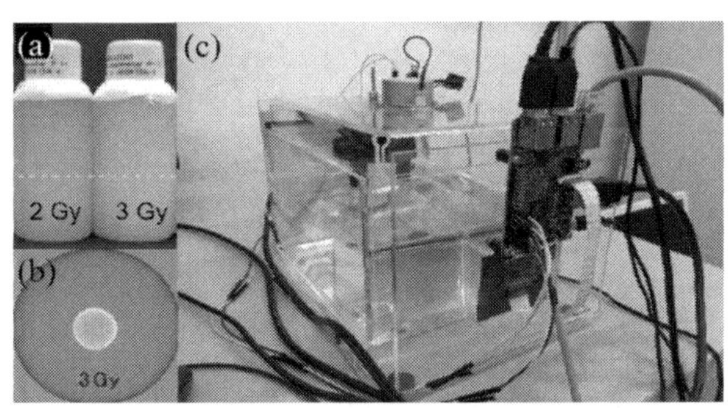

Figure 1. (a) AQUAJOINT® gels irradiated with 2 Gy and 3 Gy of X-rays. (b) Reconstructed cross-section image of the AQUAJOINT® gel (a) irradiated with 3 Gy along the dotted yellow line using OCT. (c) Prototype OCT system using CMOS camera sensor with single board computer Raspberry Pi.

2.5 MRI

We used a 1.5 T MRI scanner (Intera Achieva 1.5 T HP Nova Dual Gradient, Philips Medical Systems, Best, Netherlands). T2 estimation was performed with the turbo mixed sequence [17]. The scan conditions were as follows: for spine echo acquisition, TR = 2800 ms, TE1 = 12, TE2 = 800; for inversion recovery, TR = 3000 ms and TI = 200. Turbo factor was 86. The pixel size was 0.76 mm. Slice thickness was 15 mm. (FOV depended upon the number of samples. The acquisition time also depended upon FOV and the number of acquisitions.)

3. Results and Discussion

The absorbed dose *vs.* relative value curves for AQUAJOINT® and PAGAT gel dosimeters, as measured using OCT and MRI, are shown below [Fig. 2].

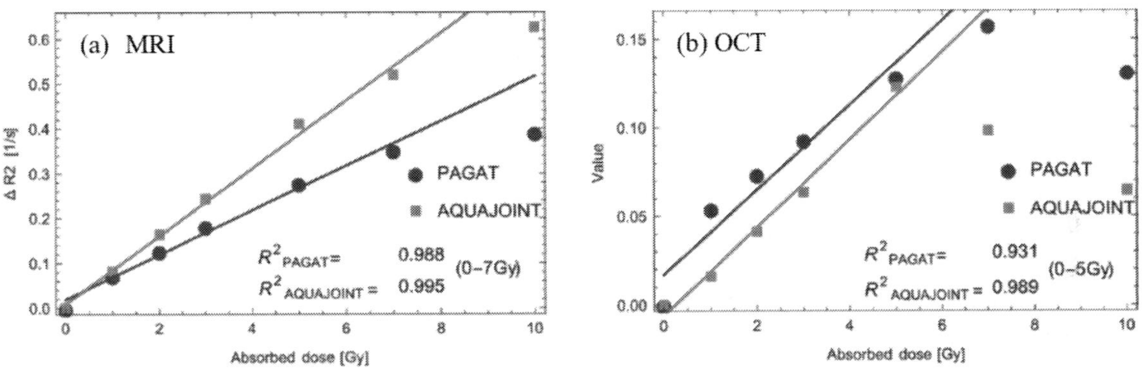

Figure 2. Characteristic curves for absorbed dose *vs.* relative value measured using OCT and MRI. (a) MRI; (b) OCT. For MRI, a fitted straight line was traced using the data points ranging from 0 Gy to 7 Gy to calculate the coefficient of determination R^2 values. For OCT, the data points ranging from 0 Gy to 5 Gy were used to calculate R^2 values.

According to OCT measurements, the relative values of AQUAJOINT® gel at 7 Gy and 10 Gy as well as PAGAT gel at 10 Gy did not fall on a straight line. Absorption coefficient of light in OCT (corresponding to the CT value of the X-ray CT) is estimated from the variation in the intensity of the transmitted light. In the case of samples rendered opaque by higher irradiation, the effect of scattered light could not be omitted. In general, when a sample contains more components causing light scattering than those absorbing light, the algorithms used in image reconstruction (Filtered Back Projection) of OCT are not applicable.

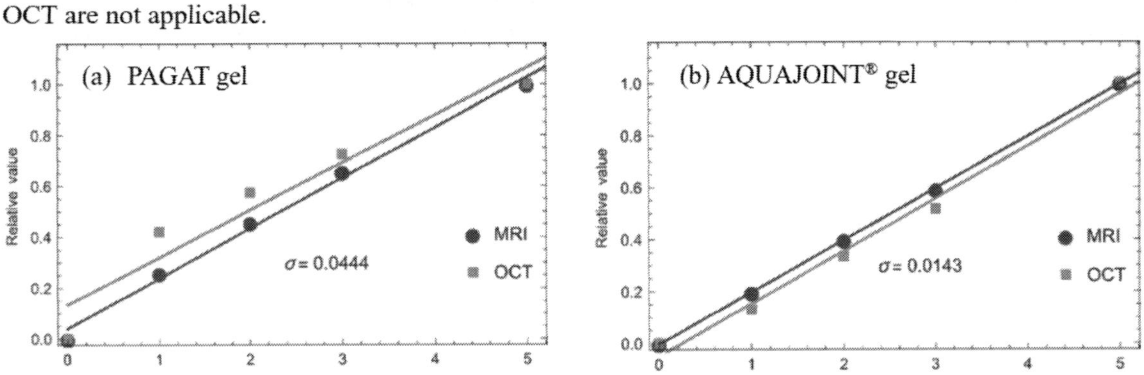

Figure 3. Characteristic curves for absorbed dose *vs.* relative value measured using MRI and OCT. (a) PAGAT gel; (b) AQUAJOINT® gel. Values of OCT and MRI were normalized to 1 relative to the value at 5 Gy and 0 to the value at 0 Gy. The index $\boldsymbol{\sigma = \sum_D |MRI(D) - OCT(D)|^2}$ is of difference between the MRI and OCT values. For both MRI and OCT, a fitted straight line was traced using the data points ranging from 0 Gy to 5 Gy.

In the OCT evaluation using PAGAT gel, the dose response showed relatively small non-linear behavior, whereas the MRI measurements showed a good linearity [Fig. 3 (a)]. This behavior was observed in our previous report on OCT measurement of PAGAT [18]. In contrast, AQUAJOINT® gel dose response showed a linear behavior at a dose range of 0–5 Gy [Fig. 3 (b)].

4. Conclusion

The gel dosimeter using AQUAJOINT® as a hydrogel component was compared with conventional PAGAT gel dosimeter as a reference. Data obtained by OCT correlated well with the MRI data and showed good linearity in the dose range of 0–5 Gy.

The lack of linearity in the dose–response behavior obtained by OCT with AQUAJOINT® gel at doses > 5 Gy is likely to be corrected by optimizing the amount of monomers in the AQUAJOINT® gel.

5. Acknowledgement

Thanks to S. Hara (Yamanashi Gakuin University) who developed the useful plug-ins for ImageJ.

6. References

[1] Baldock C *et al* 2010 *Phys. Med. Biol.* **55** R1-63
[2] Baldock C *et al* 1998 *Phys. Med. Biol.* **43** 695-702
[3] Maryanski M J *et al* 1993 *Magn. Reson. Imaging* **11** 253-8
[4] Hill B *et al* 2005 *Med. Phys.* **32** 1589-97
[5] Hill R F *et al* 2008 *Radiat. Meas.* **43** 1258-64
[6] Mather M L *et al* 2003 *Ultrasonics* **41** 551-9
[7] Baldock C *et al* 1998 *Phys. Med. Biol.* **43** 3617-27
[8] Murry P and Baldock C 2000 *Austral. Phys. Eng. Sci. Med.* **23** 44-51
[9] Davies J B and Baldock C 2008 *Radiat. Phys. Chem.* **77** 690-96
[10] Wuu C S and Xu Y 2006 *Med. Phys.* **33** 1412-9
[11] Maeyama T *et al* Proceedings of 4th 3D Gel Dosimeter Research Meeting at Teikyo University in Tokyo, Japan on November 28, 2015
[12] Venning A J *et al* 2005 *Phys. Med. Biol.* **50** 3875-88
[13] Venning A J *et al* 2004 *J. Phys.: Conf. Ser.* **3** 155-8
[14] Schreiner L J and Olding T 2009 Gel Dosimetry. *AAPM Medical Physics Monograph No. 34: Clinical Dosimetry Measurements in Radiotherapy.* Ed Rogers, D W and Cygler, J E (Madison: Medical Physics Publishing.) pp 979-1025
[15] Rasband W S 1997-2012 ImageJ, U. S. National Institutes of Health, (Bethesda, Maryland, USA) http://imagej.nih.gov/ij/
[16] Schneider C A *et al* 2012 *Nat. Methods* **9** 671-5
[17] Denkleef J and Cuppen J J M 1987 *Magn. Reson. Med.* **5** 513-24
[18] Kawamura H *et al* 2016 Proceedings of 111th Scientific Meeting of the Japan Society of Medical Physics (JSMP) in Yokohama, Japan

IOP Conf. Series: Journal of Physics: Conf. Series **847** (2017) 012014 doi:10.1088/1742-6596/847/1/012014

Water equivalence study of metal impregnated PAGAT gel dosimeter in terms of variation of effective atomic number

P Sathiyaraj[1], E J J Samuel[1], M Kurudirek[2]
[1]Photonics, Nuclear and Medical Physics Division, School of Advanced Sciences, VIT University, Vellore, India
[2]Faculty of Science, Department of Physics, Ataturk University, 25240 Erzurum, Turkey

Email: ejames@vit.ac.in

Abstract. The aim of this study is to evaluate the water equivalency of metal doped PAGAT gel dosimeter in terms of its effective atomic number. Auto-Zeff software and NIST XCOM data base were used to perform this study. Initially Gold (Au) and Silver (Ag) were taken and divided into different concentration (from 0.01 to 100Mm) with standard PAGAT formalism and the elemental composition was determined. Using these elemental compositions Z_{eff} was generated for photon interaction using Auto-Z_{eff} software. Au doped gel was found as a water equivalent in the concentration range of 0.01 to 0.1Mm. Ag doped gel have matched well with water from 0.01 to 1Mm concentration in terms of water equivalence. In these concentrations, it has been concluded that metal doped gel is more water equivalent than plain PAGAT gel based on the Z_{eff} data obtained.

1. Introduction

Effective atomic number (Z_{eff}) is an important parameter to consider a material to be substituted for water for radiation dosimetry [1, 2]. Many researchers reported the method of finding Z_{eff} for a compound or mixture [3, 4]. Z_{eff} of any medium depends on the energy of the incident photon beam so, a single number cannot be a representative of Z_{eff} of a compound material [4]. Hence Z_{eff} depends on the photon energy, it is mandatory to compute the Z_{eff} with respect to different range of energy. There are well reported publications available regarding Z_{eff} of different gel dosimeter [5-7]. As per the best knowledge of the author there is no publication of computing Z_{eff} for metal doped PAGAT gel dosimeter [8-11]. Recently metal nano particles (MNP) are an interesting candidate for dose enhancement effect (DEE) study in gel dosimeter [12, 13]. MNPs increase the probabilities of photon interaction cross sections due to their high atomic number (high Z) and yield greater secondary radiations within the gel causing DEE. Different types of MNPs were reported for gel dosimeter such as Gold (Au, Z=79), Silver (Ag, Z=47), Platinum (Pt, Z=78) and Bismuth (Bi, Z=83) [14-17] for DEE. Though DEE is a great advantage by MNPs, its high Z presence in the gel should be evaluated to consider the MNP doped gel for water equivalency. In this study, Z_{eff} parameter was studied for evaluating the water equivalency of metal doped PAGAT gel dosimeter.

2. Materials and Method

2.1. Computation method

To evaluate the water equivalency of metal doped PAGAT (MPAGAT) Au and Ag dopants with different concentration (0.01, 0.05, 0.1, 1, 5, 10, 50 and 100m M) were considered. In this study, we

IOP Conf. Series: Journal of Physics: Conf. Series **847** (2017) 012014 doi:10.1088/1742-6596/847/1/012014

used Auto-Z_{eff} software for generating the Z_{eff} [18]. In the first step elemental composition analysis was performed by NIST XCOM data base. Elemental compositions were normalized to unity and feed the compositions in the Auto-Z_{eff}. Z_{eff} was generated from 0.01 to 1000 MeV for all the concentrations.

3. Results and Discussion

3.1. Water equivalency of Au doped MPAGAT

The variation of Z_{eff} with respect to energy is shown in figure 1. Initially, Z_{eff}s are increasing from 0.01 to 0.02 MeV. Since photoelectric absorption cross section (PACS) is directly proportional to Z^4 at this low energy region PACS was higher, especially for 10, 50 and 100m M due to the large abundance of Au. The rapid fall was starting at 0.02 to 0.2 MeV by typical Compton scattering interaction. All MPAGAT have almost constant from 0.4 to 2 MeV. The lowest Z_{eff} was found in this energy range for all MPAGAT. Z_{eff} increases with energy starting from 2.044 to 200 MeV due to pair production and above 200 Z_{eff} was constant. Z_{eff} of gel dosimeters was compared with water (figure 2). The maximum percentage of

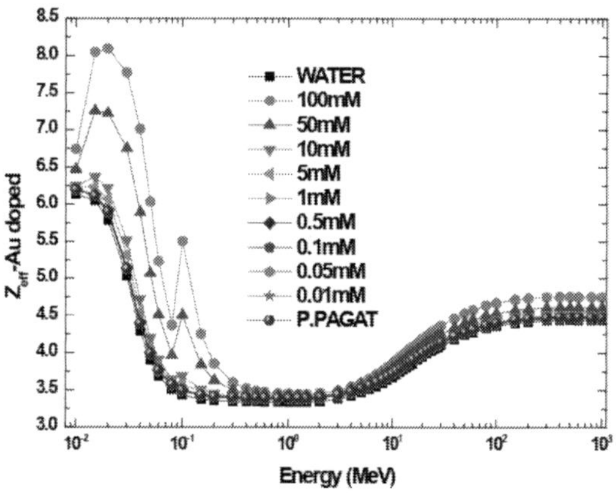

Figure 1. Z_{eff} of MPAGAT for various concentrations of Au.

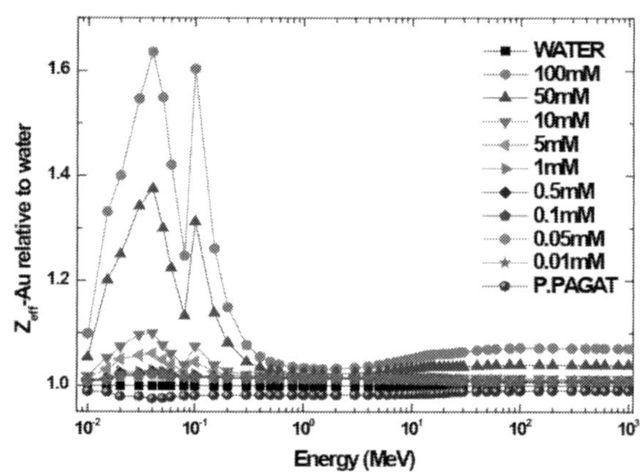

Figure 2. Z_{eff} variation of MPAGAT from water.

Variation obtained at 0.04MeV for all MPAGAT. Highest percentage variation was 63.5% for 100Mm. From 0.01Mm to 0.5Mm the percentage variation was less than 2% over all the energy region. For 1Mm the Z_{eff} variation is less than 2% from 0.15 to 1000MeV. For 5, 10, 50 and 100 Mm the Z_{eff} has greater than 3% from water. In addition to these results we noticed an interesting behavior of MPAGAT that is 0.01 and 0.05Mm concentration of Au make the gel more water equivalent than plain PAGAT (PPAGAT) (figure 3). The arrow mark indicated that the curves are approaching water closer than PPAGAT.

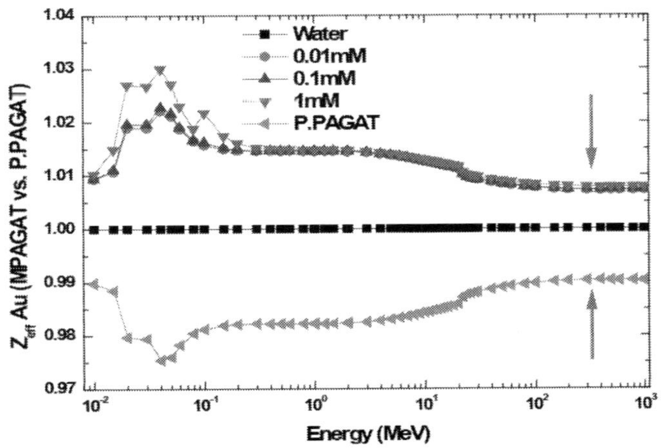

Figure 3. Comparison of PPAGAT and MPAGAT with respect to water.

3.2. Water equivalency of Ag doped MPAGAT

The variation of Z_{eff} with respect to energy is shown in figure 4. In this case all Z_{eff}s started to fall from 0.01MeV. However, 50 and 100Mm MPAGAT rapidly increased from 0.02to0.03MeV. Though the same rapid increment was presented in other concentration, it is very less compared with 50 and 100Mm concentration. The sudden jump in Z_{eff} from 0.02 to 0.03 represents the K-edge of Ag. After K edge all the curves rapidly fell up to 0.2MeV due to Compton scattering. Z_{eff} values are higher in low energy range due to PACS of Ag was higher at this energy, but comparing with Au it was less due to its less atomic number. From 0.3 to 2MeV all the curves were constant (Minimum Z_{eff} was found from these energy region for all concentration values) and slowly start to increase from 2MeV to 300MeV due to the dominance of Pair production. Beyond 300Mev Zeff was constant. Ag doped MPAGAT was compared with water to evaluate the discrepancy in Z_{eff} (figure 5).

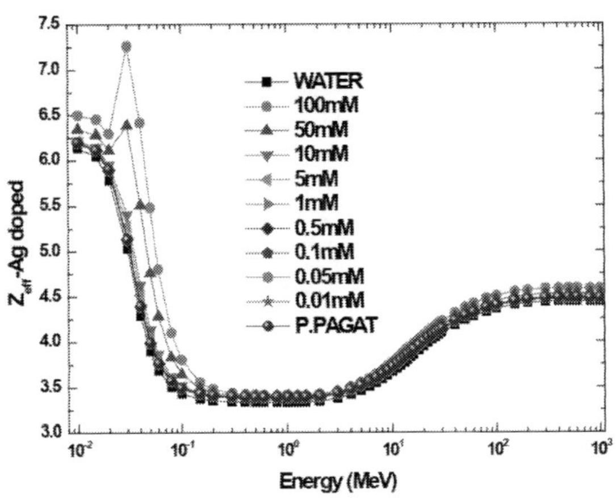

Figure 4. Z_{eff} of MPAGAT for various concentrations of Au.

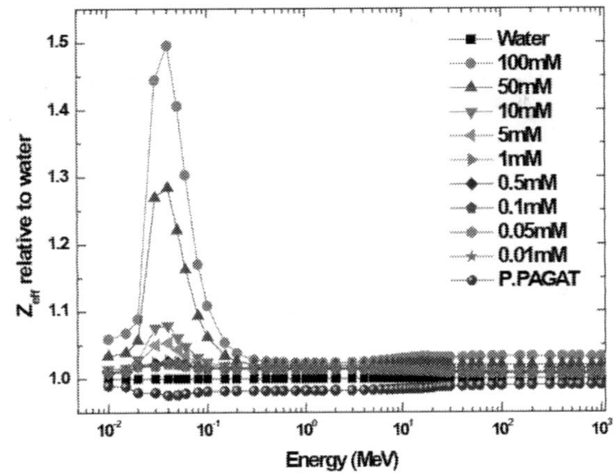

Figure 5. Z_{eff} variation of MPAGAT relative to water.

The highest percentage of variation obtained (49.5%) at 0.04MeV in 100Mm concentration. Up to 1Mm the percentage variation is less than 2%. By doping Ag from 0.01 to 1Mm we can achieve more water equivalency of PAGAT gel (figure 6).

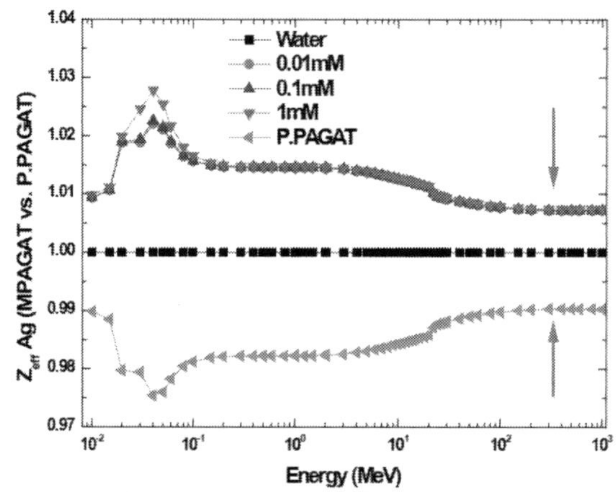

Figure 6. Comparison of PPAGAT and MPAGAT with respect to water.

Both Au and Ag doped gel were more water equivalent than PPAGAT gel because of the increasing effective electron density by addition of high Z material up to certain range of concentration. Over concentrations (1,5,10,50 and 100mM for Au, 5, 10, 50 and 100mM for Ag) make the gel in equivalent to water.

4. Conclusion

Water equivalence of PAGAT gel was evaluated in terms of its Z_{eff} from 0.01 to 1000MeV. By adding Au (0.01 to 0.1mM and Ag (0.01 to 1Mm) they have shown better water equivalence based on Z_{eff} data. At above mentioned concentration MPAGAT was more water equivalent than PPAGAT. One should consider this Z_{eff} variation for any gel while introducing metal nano particle for DEE. More research required in the energy range below 0.01MeV, in addition, different methods should apply and compare the results of Z_{eff} for more reliable results.

5. Acknowledgements

The authors are grateful to the Dr. Suvarna N. Patil, Medical Director of BKLW Hospital, Diagnostic and Research Centre, Maharashtra, India for financial support.

6. References

[1] Hill R *et al* 2009 *Med. Phys.* **36** 3971-81
[2] Hill R F *et al* 2008 *Radiat. Meas.* **43** 1258-64
[3] Manohara S R *et al* 2008 *Phys. Med. Biol.* **53** N377-86
[4] Kurudirek M *et al* 2014 *Radiat. Phy. Chem.* **102** 139-46
[5] Gorjiara T *et al* 2011 *Phys. Med. Biol.* **56** 4685-99
[6] Taylor M L *et al* 2008 *Australas. Phys. Eng. Sci. Med.* **31** 131
[7] Keall P and Baldock C 1999 *Austral. Phys. Eng. Sci. Med.* **22** 85-91
[8] Venning A J *et al* 2005 *Med. Phys.* **32** 1047-53
[9] Venning A J *et al* 2005 *Phys. Med. Biol.* **50** 3875-88
[10] Braun K *et al* 2009 *J. Phys.: Conf. Ser.* **164** 012050
[11] Brown S *et al* 2008 *Appl. Rad. Isotop.* **66** 1970-74
[12] Mahdavi M *et al* 2013 *J. Biomed. Phys. Eng.* **3** 37-44
[13] Rahman W N *et al* 2012 *Australas. Phys. Eng. Sci. Med.* **35** 301-9
[14] Khosravi H *et al* 2015 *J. Biomed. Phys. Eng.* **5** 3
[15] Deyhimihaghighi N *et al* 2014 *J. Phys.: Conf. Ser.* **546** 012013
[16] Hassan M *et al* 2016 *J. Paak. Med. Assoc.* **66** 45-8
[17] Alqathami M *et al* 2013 *J. Phys.: Conf. Ser.* **444** 012025
[18] Taylor M L *et al* 2008 *Australas. Phy.Eng. Sci. Med.* **31** 131-8

IOP Conf. Series: Journal of Physics: Conf. Series **847** (2017) 012015 doi:10.1088/1742-6596/847/1/012015

Preliminary investigation of the dosimetric properties of ClearView™ dosimeter

C Huet, J Colnot and I Clairand
Institut de Radioprotection et de Sûreté Nucléaire (IRSN), Service de Dosimétrie Externe, Laboratoire de Dosimétrie des Rayonnements Ionisants, Fontenay-aux-Roses, France

E-mail: christelle.huet@irsn.fr

Abstract. ClearView™ is a single-use radiochromic 3D dosimeter designed by Modus Medical Devices Inc. Some of the dosimetric characteristics of this gel, such as response as a function of dose and as a function of energy were investigated. Moreover, a 400 mL jar was irradiated with a 1x1 cm^2 field and PDD and cross-profiles were compared to those obtained with active detectors. Small volumes of gels were found to exhibit a linear response to dose in the dose range 10-80 Gy. Promising PDD and cross profiles were obtained. Artefacts obtained in high resolution for the PDD are suspected to come from small inhomogeneities in the gel.

1. Introduction

Gels dosimeters have the ability to retain dosimetry information in three dimensions thus making them good candidate for the dosimetry of advanced radiotherapy techniques which delivered complex dose distributions [1-5]. ClearView™ is a single-use radiochromic 3D dosimeter designed by Modus Medical Devices Inc for remote or on-site read-out by Optical CT. ClearView™ gels are based on a gellan gum matrix which incorporates a tetrazolium salt. The gel also contains propylene glycol, as a radical scavenger. Its melting point is around 70°C. Under ionizing radiations, it turns slightly pink/purple. This color change is due to the chemical formation of a formazan dye within the gel. The dose range is 10-80 Gy while the recommended dose by the manufacturer is 40 Gy. Dosimetric characteristics of the ClearView™ gel, such as batch uniformity, stability, response as a function of dose and as a function of energy, were studied and some are detailed here. Moreover, a 400 mL jar of gel has been irradiated with a 1 x 1 cm^2 photon field and percent depth dose (PDD) and profiles obtained with the gel were compared to diodes and chamber measurements.

2. Materials and Methods

2.1. Storage, irradiation and reading conditions

Gels were provided by Modus Medical Devices Inc in 400 mL jars and in 1 cm x 1 cm x 4 cm spectrophotometer cuvettes. They were stored in a fridge and removed from it at least 8 hours prior to irradiation to allow gel to come to room temperature. During this step, they were protected from light. Gels were irradiated at room temperature (21°C) with an Elekta Synergy linear accelerator. Then they were kept at room temperature without light exposure at least 45 min as recommended by the manufacturer to allow signal to develop. Then, they were read at room temperature.

2.2. Dose response of ClearView™ gel

Spectrophotometer cuvettes filled with ClearView™ gel were irradiated with 4MV photons in the same beam geometry consisting of a 30 x 30 cm^2 field with a source to surface distance of 100 cm. Cuvettes were placed at isocenter and two different batches of gel were irradiated with doses ranging from 10 to 80 Gy in steps of 10 Gy for the first batch and with doses of 10, 25, 40, 55 and 70 Gy for the second batch. The absorbance of irradiated cuvettes was measured 45 minutes after irradiation with a UV1800 (Shimadzu) spectrophotometer. In order to determine the wavelength at which the maximum absorbance is observed after irradiation, spectrum was measured over the range 400-800 nm with intervals of 2 nm. The reference cuvette was a non-irradiated cuvette. In addition, regarding the second batch, cuvettes were also read 90 minutes and 3 hours after irradiation in order to study the evolution of absorbance with time.

2.3. Response as a function of energy

The energy dependency was investigated changing the beam energy. Spectrophotometer cuvettes (3 cuvettes per energy) were irradiated at 40 Gy in a 30 x 30 cm^2 field with an energy of 4, 10 and 18MV, the dose rate being the same for the three energies. Absorbance change was measured 45 minutes after irradiation.

2.4. Relative dosimetry

A 400 mL ClearView™ jar was irradiated vertically without the lid in PMMA slabs with a 1 cm × 1 cm field (4 MV, 1.6 Gy/min 50 Gy at isocenter). Isocenter was placed at 3 cm in depth from the gel surface. Gel was read using the Vista™ Optical-CT (590 nm wavelength, 410 projections) 45 minutes after irradiation. 3-D reconstructions were performed in medium and high resolutions and analyzed using MicroView 2.1.2 software (GE Healthcare). PDD and cross profiles acquired with the gel were compared with PDD and cross profiles acquired with PTW 31014 PinPoint chamber and PTW 60016 and 60017 diodes in a water tank.

3. Results and discussion

3.1. Dose response of ClearView™ gel

The color change of the gel on the 0-80 Gy dose range is shown in Figure 1. The change in absorbance as a function of wavelength for the cuvette irradiated to 40 Gy for the two different batches is shown in Figure 2. The maximum of absorbance occurs at a wavelength of 530 nm for the first batch and of 560 nm for the second batch. As the gel is currently under development, the difference of maximum absorbance wavelength observed may come from a different formulation of the gel. The gel response (attenuation coefficient) as a function of dose for the 530 nm or 560 nm wavelengths as well as for the two wavelengths available in our Vista™ Optical CT system (590 and 632 nm) is displayed in Figure 3. The response is linear for the two batches and for all the wavelengths with a correlation of R^2 around 0.98. Regarding temporal response of ClearView™ (Figure 4), no change is observed when the reading occurred between 45 minutes and three hours after irradiation.

Figure 1. ClearView™ color change on the 0-80 Gy dose range.

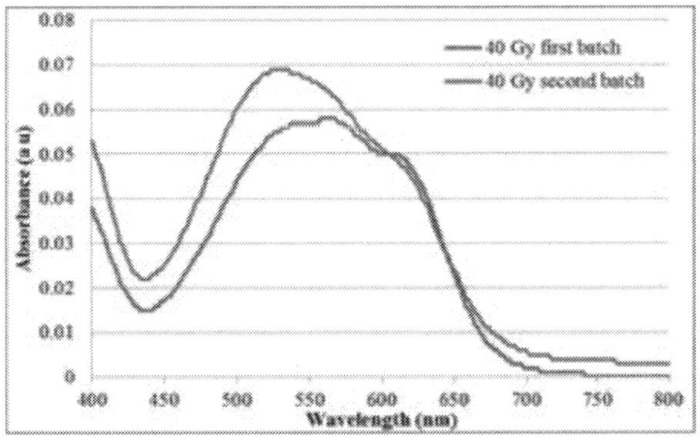

Figure 2. Spectral response of ClearView™ cuvette after irradiation dose of 40 Gy for the two batches.

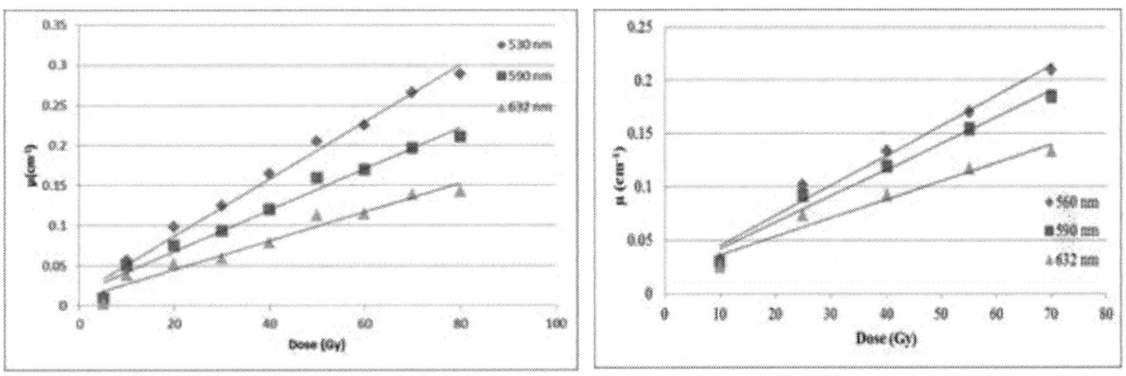

Figure 3. Attenuation coefficient as a function of dose for the first batch (left) and for the second batch (right) for 3 wavelengths.

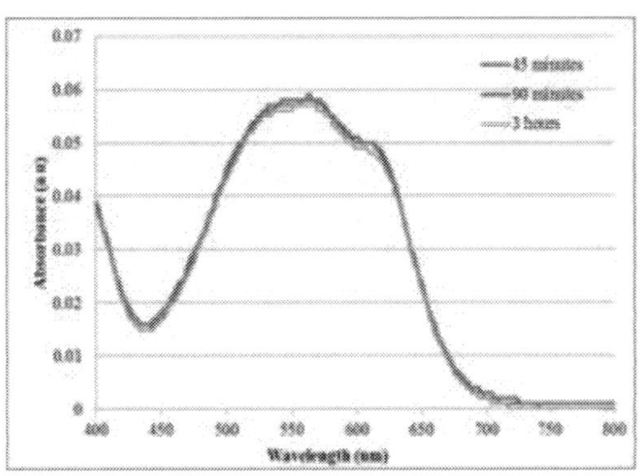

Figure 4. Spectral response of ClearView™ cuvette after irradiation dose of 40 Gy, 45 minutes, 90 minutes and 3 hours after irradiation.

3.2. Response as a function of energy

In the investigated energy range, gel response is independent of energy (Table 1).

Table 1. Absorbance at 560 nm for the three photon beam energies.

	4 MV	10 MV	18 MV
Mean absorbance	0.063	0.063	0.060
Standard deviation	0.006	0.008	0.005

3.3. Relative dosimetry

In Figure 5, PDD and cross profiles acquired with the gel are compared to PDD and cross profiles measured with PTW 31014 chamber and PTW 60016 and 60017 diodes. PPD acquired with the gel in high resolution (HR, 0.5 mm^3 voxel) is very noisy. The artefacts noticed around 30 mm and 65 mm in depth may be due to small gel inhomogeneities; they are no longer present in medium resolution (MR, 1 mm^3 voxel). From 0 mm to nearly 60 mm in depth, a good agreement is observed between MR gel PDD and PDD acquired with the detectors. Above 60 mm, the MR gel PDD is noisier because of the dose range corresponding to a low sensitivity of the gel. A good agreement is found between cross profiles acquired with the gel and the detectors: 90% of points pass the gamma index with criteria of 3%/3 mm. Points which don't pass the gamma index are located in the penumbra region where gel FWHM (9.3 mm) is slightly larger than detectors FWHM (8.6 mm for PTW 31014 chamber).

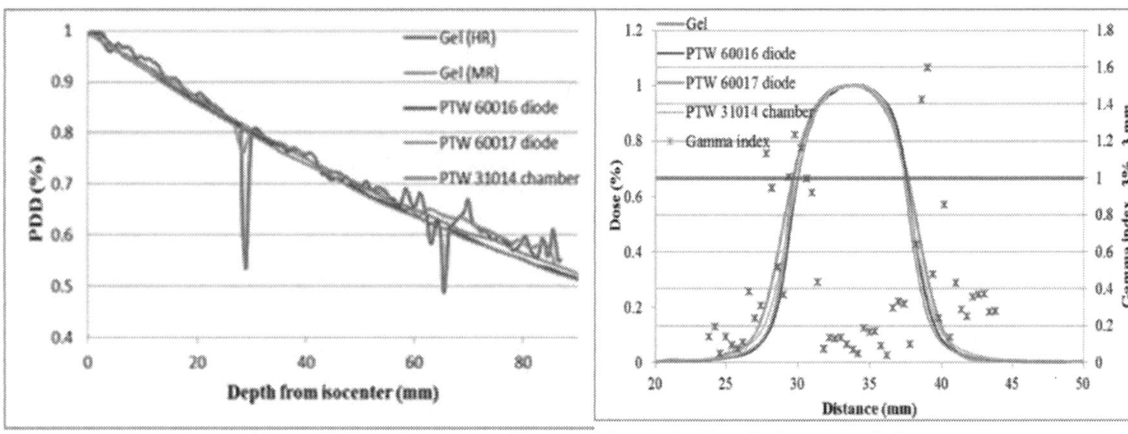

Figure 5. Comparison of PDD (left) and cross profiles (right) acquired with the gel and detectors. HR and MR indicate high resolution and medium resolution scanning.

4. Conclusion

Some of the dosimetric properties of ClearView™ have been investigated. The maximum absorbance is around 550 nm depending on the gel batch. A linear response to dose in the dose range 10-80 Gy has been observed. ClearView™ shows promising features for use in radiotherapy for relative dosimetry. Further work will focus on the identification of the parameters that may impact ClearView™ response and on the quantification of error. An absolute dose comparison is also in progress.

5. Acknowledgements

The authors thank Jennifer Dietrich and Kalin Penev from Modus Medical Devices Inc. for gel shipment and helpful discussions regarding ClearView™ dosimeter.

6. References

[1] Oldham M *et al* 2001 *Med. Phys.* **28** 1436-46
[2] Vergote K *et al* 2004 *Phys. Med. Biol.* **49** 287-305
[3] Baldock C *et al* 2010 *Phys. Med. Biol.* **55** R1-63
[4] Hill B *et al* 2005 *Med. Phys.* **32** 1589-97
[5] Mather M L *et al* 2003 *Ultrasonics* **41** 551-9

Radiochromic and radiofluorogenic 3D solid polymer dosimeter: initial results for high doses

María del Rocío Bernal Zamorano, Nicolai Højer Sanders, Lars René Lindvold, Claus E. Andersen
Center for Nuclear Technologies, Technical University of Denmark, 4000 Roskilde, Denmark

E-mail: mroc@dtu.dk

Abstract. The complexity of dose distributions has increased with the advent of intensity modulated radiation therapy (IMRT). For that reason, experimental measurements using 3D dosimeters with high spatial resolution are required to check the delivered dose. In this study a new 3D solid polymer dosimeter with absorbance and fluorescence responses to radiation is presented. Measuring fluorescence instead of absorbance improves the spatial resolution and eases the read out of the dosimeter. The proposed dosimeter is tissue-equivalent and can be moulded in any shape by a controllable and fast photopolymerization process.

1. Introduction

Modern radiotherapy requires complex dose distributions; therefore, there is an increasing demand for a high spatial resolution system suitable for dose verification to ensure treatment quality [1]. Over the last years, 3D dosimetry systems have been developed for this purpose [2]. They may be classified into polymerizing dosimeters or radiochromic dosimeters, which use magnetic resonance imaging (MRI) [3] or optical computed tomography (CT) [4] as readout systems. These dosimeters change chemical properties upon irradiation [5]. While polymerizing dosimeters consist of a gel matrix that polymerizes with radiation [6], radiochromic dosimeters consist of a gel, plastic or silicone matrix with a radiation sensitive dye that changes color with radiation exposure [7, 8].

It is important to remember that a dosimetry system is not only the dosimeter itself but also the readout technique used for extracting the dose information after irradiation. Oldham *et al* [9] applied both techniques (MRI and CT) to the same gel dosimeter and they obtained a higher resolution for CT. Besides that, MRI scanners have significant disadvantages like its cost and availability, and they may be susceptible to several uncertainty sources like field homogeneity and temperature [10]. These uncertainties are well understood and can be compensated. However, there are still issues concerning the material properties of the gels. It is known that the radiation induced cross-linking of the gel is highly dependent on its oxygen level, pH, and temperature, just to mention a few parameters. For those reasons CT is more commonly used [11].

Some problems of the CT scan are the light scattering and the acquisition time. Gel dosimeters require the use of a container, which adds scatter artifacts in the read out due to reflections [12]. Therefore, the scattering may be reduced by using a solid dosimeter that does not need a holder. Solid dosimeters also avoid the diffusion problem present in gel dosimeters, which causes blurring of the dose distribution image over time. For some systems, the acquisition of the image may be time consuming since current CT readout systems require scanning of several slices while the sample rotates to acquire data at different angles. For example, the 3D scan of the commercially available PRESAGE™/OCTOPUS™ dosimetry system needs 15 slices and takes 8-9 minutes per slice [13]. This gives a total scanning time of 2 hours, followed by a computer intensive image reconstruction. A simple, *in-situ* and fast reading of the dosimeter would facilitate the use of the 3D dosimeter in a clinical basis.

A way to increase the scanning speed is by using scanners based on charge-coupled device (CCD) cameras, since it is possible to obtain a complete 2D image in one go [11]. This technique can be used to measure the fluorescence [14] instead of the attenuation as the CT does. We have developed a 3D readout based on a black and white CCD camera. Detecting small signals is difficult when measuring the absorbance (or attenuation) in the dosimeter; however, measuring fluorescence allows us to use color filters to ensure that the vast majority of the signal comes from the dye in the sample.

In this paper we present a sensitive, soft-tissue equivalent and moldable 3D solid polymer dosimeter that is not only radiochromic but also radiofluorogenic. The underlying mechanism involves the conversion of a non-fluorescent dye molecule into a fluorescent form when incorporated into a rigid polymeric matrix and irradiated. The imaging of the dose distribution is obtained by measuring the fluorescence intensity, which can be recorded using a conventional digital camera. This gives a higher accuracy, a higher spatial resolution and a faster read out, compared to the current commercially available 3D dosimetry systems.

2. Materials and methods

2.1. The dosimeter

We use a poly(ethylene glycol) diacrylate matrix (PEGDA-575 g/mol) containing pararosaniline leuco dye [15]. The radiation chemistry involved in the transformation from leuco dye to dye is presented in figure 1 [16]. This polymer enables the solidification of the material through a photopolymerization process. We use diphenyl(2,4,6-trimethylbenzoyl)phosphine oxide (TPO) as photoinitiator.

The dosimeter can be molded in any shape by light of approximately 400 nm, which does not affect the dye. This is a controllable and fast process that only takes up to 10 minutes. After curing, the dosimeter can be removed from the mold. This property could open up for a significant number of clinical applications relative to making patient-like geometries, or using it as a thin 2D dosimeter (film) for the radiation field verification as a quality control test.

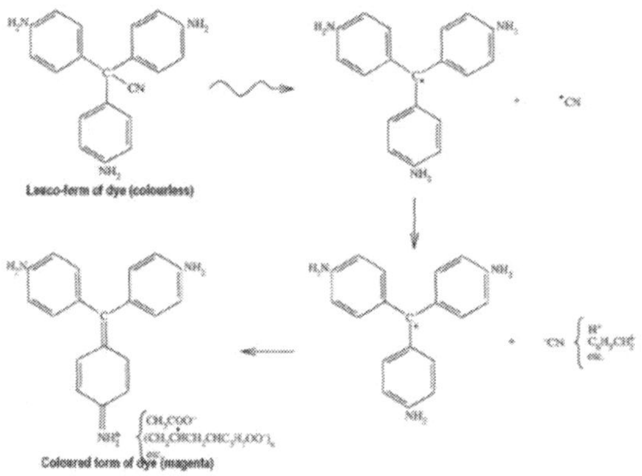

Figure 1. Chemical reaction of the leuco dye due to radiation.

2.2. The readout system

Optical fluorescence tomography is used for read out of the dose distribution in the dosimeter (figure 2) [17]. The dosimeter is submerged in an index matching fluid tank and excited with a green laser. Pictures of the fluorescent emission are taken with a CCD black and white camera to get the 2D dose distribution. 3D information is obtained by moving the sample through the light sheet while taking pictures.

The scanning speed is up to 2 millimeters per second, so the total scanning of the dosimeter is finished within minutes, which avoids problems of chemical changes during scanning. This technique allows having an *in-situ* reading and the device is easy to use. These characteristics would facilitate its use in the clinic.

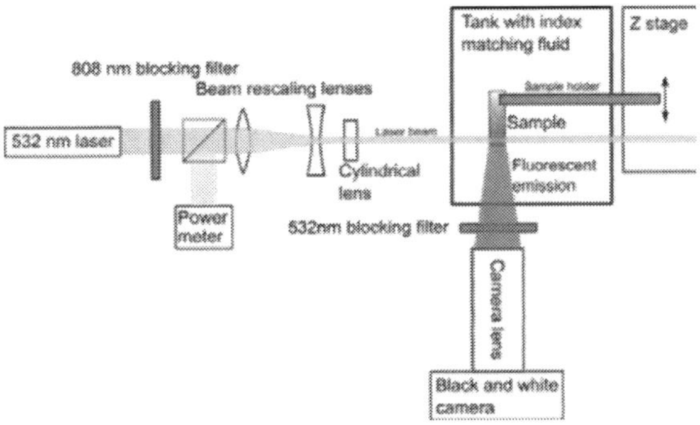

Figure 2. Picture of the dosimeter inside the index matching fluid tank and scheme of the readout system.

2.3. Irradiations and measurements
Irradiations were carried out in a ^{60}Co gamma source with a dose rate of 139.3 Gy/min. The uncertainty on the delivered dose is 1.3%. Absorbance was measured with a Shimazdu UV-2700 spectrophotometer. Fluorescence is excited with a 532 nm Nd:YAG laser, and measured with an Ocean Optics QE6500 spectrometer.

3. Results and discussion
A slide of 1 mm thickness is cured between two glass plates to ensure the best optical quality and irradiated up to 1 kGy in steps of 100 Gy. Normalized absorbance and fluorescence spectra as function of the dose are shown in figure 3. The results show that the fluorescence response is strong enough to be used. In figure 4 we present the fits of the responses with the absorbed dose for different parts of the spectrum. While the absorbance response is linear with the dose, the fluorescence response is exponential in overall and linear for the clinical dose range.

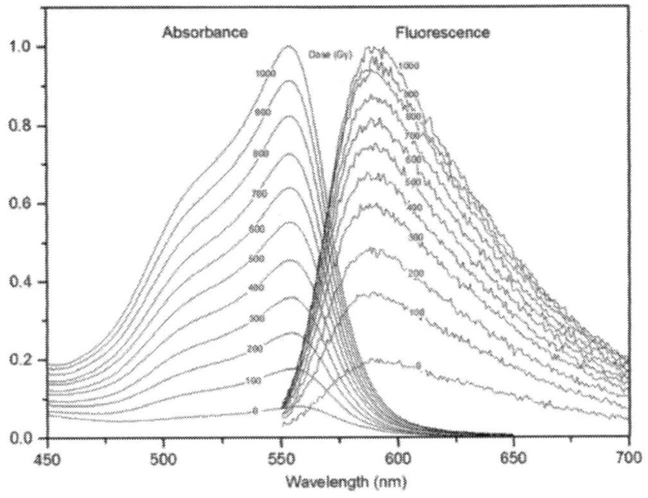

Figure 3. Normalized absorbance and fluorescence spectra.

IOP Conf. Series: Journal of Physics: Conf. Series **847** (2017) 012016 doi:10.1088/1742-6596/847/1/012016

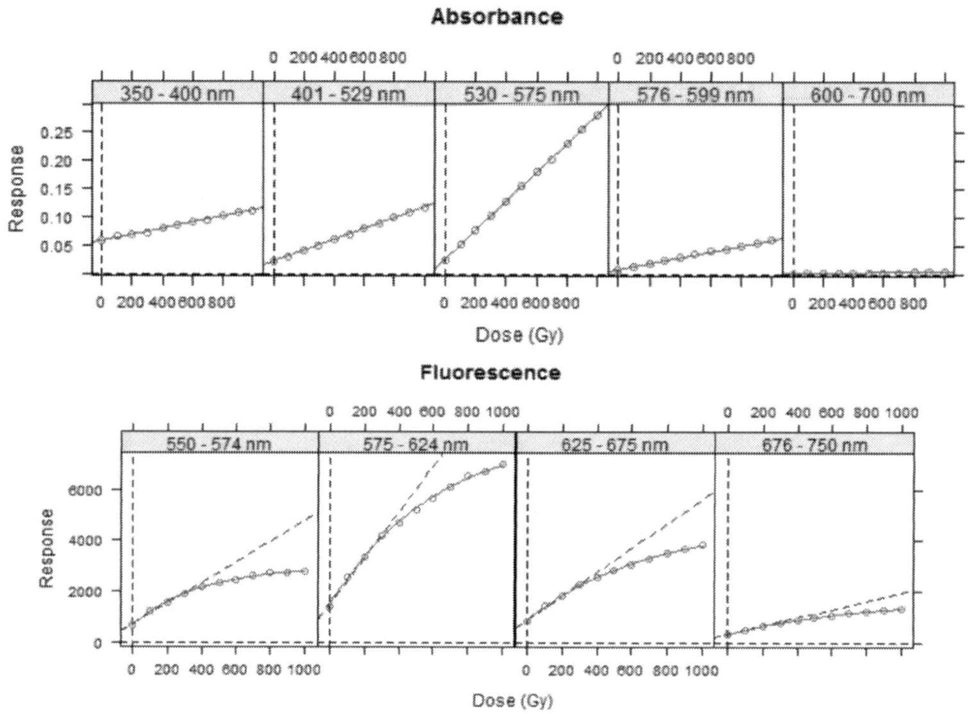

Figure 4. Absorbance and fluorescence fits as function of the absorbed dose.

4. Conlusions

We have developed a passive solid dosimeter that is tissue equivalent, can be molded in the desired form and does not need a container. The dosimeter responds to irradiation in the same way as radiochromic films (oxidation of a leuco dye) but we have established a polymer matrix for this dye that allows us to measure not only the absorbance of the material but also its fluorescence. The main difference with current 3D solid polymer dosimeters is the readout technique. While current 3D dosimetry systems mainly use CT to extract dose information, we can measure the fluorescence, which potentially is a faster and more sensitive method. The response range of this dosimeter is very high (up to 1 kGy), but future studies will focus on optimizing the dosimeter at low doses for its clinical application.

5. References

[1] Baldock C 2006 *J. Phys.: Conf. Ser.* **56** 14-22
[2] Mather M L *et al* 2003 *Ultrasonics* **41** 551-9
[3] Gore J *et al* 1984 *Phys. Med. Biol.* **29** 1189-97
[4] Gore J *et al* 1996 *Phys. Med. Biol.* **41** 2695-704
[5] Hill B *et al* 2005 *Med. Phys.* **32** 1589-97
[6] Baldock C *et al* 2010 *Phys. Med. Biol.* **55** R1-63
[7] Høye E *et al* 2015 *J. Phys.: Conf. Ser.* **573** 012067
[8] Davies J B *et al* 2008 *Radiat. Phys. Chem.* **77** 690-96
[9] Oldham M *et al* 2001 *Med. Phys.* **28** 1436-45
[10] Vandecasteele J *et al* 2013 *Phys. Med. Biol.* **58** 63-85
[11] Doran S *et al* 2006 *J. Phys.: Conf. Ser.* **56** 45-57
[12] Oldham M 2006 *J. Phys.: Conf. Ser.* **56** 58-71
[13] Sakhalkar H *et al* 2009 *Med. Phys.* **36** 71-82
[14] Warman J *et al* 2013 *Rad. Phys.Chem.* **85** 179-8
[15] Bernal Zamorano B *et al* 2016. (Presented at the 18th Solid State Dosimetry conference, Munich)
[16] Journal of the ICRU Vol 8 No 2 (2008) Report 80
[17] Sanders *et al* 2016. (Presented at the 18th Solid State Dosimetry conference, Munich.)

IOP Conf. Series: Journal of Physics: Conf. Series **847** (2017) 012017 doi:10.1088/1742-6596/847/1/012017

Characterisation of TruView™: a new 3-D reusable radiochromic MethylThymolBlue based gel dosimeter for ionising radiations

J Colnot, C Huet and I Clairand
Institut de Radioprotection et de Sûreté Nucléaire (IRSN), Service de Dosimétrie Externe, Laboratoire de Dosimétrie des Rayonnements Ionisants, Fontenay-aux-Roses, France

E-mail: christelle.huet@irsn.fr

Abstract. TruView™ is a new water-equivalent reusable Fricke gel dosimeter based on MethylThymolBlue reactive dye. Details of the characterisation of the TruView™ MTB gel dosimeter by spectrophotometric measurements and of its reading with the Optical-CT Scanner Vista™ are described. In this study, the different parameters influencing TruView™ dose response have been studied and its performances have been compared to chamber and diodes measurements. This gel presents a linear response with dose up to 20 Gy, independent in the investigated range of photon beam energy and dose rate and also a good intra-batch uniformity. Ions diffusion into the matrix homogenizes the gel after a week, losing dosimetric information but allowing a new irradiation to be performed. However, auto-oxidation happens before and after irradiation, degrading the dosimeter response and stability. Storage and reading conditions affect the response as well.

1. Introduction

Cutting-edge radiotherapy techniques deliver customized treatment to patients using both intensity modulation and numerous small beams. Thus, they lead to complex dose distribution with high dose gradients. Accurate and spatially precise dosimeters are needed to verify the match between planned and delivered dose [1-5]. Gel dosimeters are promising technique as they can inform about the whole 3-D dose distribution in one irradiation and without conventional dosimeters issues (electron equilibrium defect, partial volume effect, high dose gradient) [6]. There exist two main types of gel dosimeters: polymer gels [7] and radiochromic ones, each presenting different advantages and limitations. Fricke gel dosimeters [8-10], which are radiochromic, are reference water-equivalent 3-D dosimeters. Ionising radiation induces inside them the Fe^{2+}/Fe^{3+} oxidation defining the irradiation print by a different light absorption [11]. The gel matrix retains ions after irradiation enabling stable spatial information. Optical-CT scanning is known to be the best technique to read this kind of gel combining fast scanning, great spatial resolution and important dynamic range for a low cost compared to MRI, x-ray, laser tomography and ultrasound [12-14]. The TruView™ MTB gel composition, manufactured by Modus Medical Devices Inc., is based on Fricke-MTB composition [15]: water (95%), low diffusive gelatin matrix (5%), sulfuric acid and ferrous ammonium sulfate, methylthymol blue reactive dye generating the color change, and stabilizing additives. This gel needs to be packed in jars and retains structural stability up to 25°C. Low diffusive structure facilitates Optical-CT readout avoiding scatter artifacts. The TruView™ gel presents new characteristics such as a new reactive dye and reusable structure. Dosimetric properties of this gel such as dose response, energy and dose rate dependency have been

IOP Conf. Series: Journal of Physics: Conf. Series **847** (2017) 012017 doi:10.1088/1742-6596/847/1/012017

studied by means of spectrophotometric measurements. Moreover, the influencing parameters during the different steps (container change, storage, irradiation and reading) have been identified and their impact has been studied. Finally, percent depth dose (PDD) as well as dose profiles of a 6 MV 1x1 cm^2 field size have been obtained using the Optical-CT Scanner Vista™ and compared to chamber and diodes measurements.

2. Materials and Methods

2.1. Dose response and uniformity of the TruView™ gel

Gel was packed in 1 cm square spectrophotometer polymethyl metacrylate cuvettes. Gel cuvettes were positioned at field isocenter of a 30 cm × 30 cm 4 MV photon beam with a 2.7 Gy/min dose rate delivered by an Elekta Synergy linear accelerator. Cuvettes were irradiated to doses from 0.25 Gy to 2 Gy in dose increments of 0.25 Gy and from 2 Gy to 20 Gy in dose increments of 2 Gy (one cuvette per dose). Low dose sensitivity was evaluated by irradiating 4 cuvettes per dose from 0.05 Gy to 0.3 Gy in dose increments of 0.05 Gy and at 0.5 Gy. Radiation induced changes in absorbance were measured using an UV1800 (Shimadzu) spectrophotometer 90 minutes after irradiation. Intra-batch uniformity was tested on 20 cuvettes (5 x 4 cuvettes irradiated simultaneously) stored and irradiated in the exact same conditions (10 Gy, 4 MV, 2.7 Gy/min).

2.2. Dose response as a function of energy and dose rate

To evaluate the influence of photon energy, gel dose response from 2 Gy to 20 Gy was performed for 3 energies (4, 10 and 18 MV) with a fixed dose rate (approximatively 2.7 Gy/min). Absorbance change was measured two days after irradiation. Dose rate dependency was studied for 3 different dose rates (1.26 Gy/min, 2.56 Gy/min, 5.02 Gy/min).

2.3. Influence of different parameters on response

Regarding container change, storage, irradiation and reading steps, several parameters can influence gel dose response. For each step, error sources have been identified and quantified. Only the most significant ones are presented here. For gel container change, 10 Gy dose response of a Modus manufactured cuvette was compared to same age gel cuvettes home-conditioned using different preparation techniques: 35°C water bath heating, 50°C and 70°C hot-plate heating. Regarding the storage, to point out gel auto-oxidation, absorbance difference between two cuvettes filled under similar conditions with two gels of different aging (2 months and 1 week after manufacturing and 2 weeks) was measured. Dose response evolution up to 1 month after irradiation was studied using two different storages between readings (at room temperature and in refrigerator). Concerning the irradiation step, gel temperature influence was evaluated irradiating at 0.25 Gy and 2 Gy 4 cuvettes at room temperature and 4 others just got out of the refrigerator. About the reading step, influence of time between irradiation and reading was studied reading cuvettes at different time after irradiation and letting them at room temperature between readings.

2.4. Relative dosimetry

A 400 mL TruView™ jar was irradiated vertically in between PMMA slabs with a 1 cm × 1 cm field (4 MV, 3 Gy/min, 10 Gy at isocenter). Gel phantoms were scanned using the Vista™ Optical-CT, at 632 nm with 410 projections, 90 minutes after irradiation, the day after and 6 days after. Reconstructions were made in high resolution. 3-D reconstructions were analyzed using the MicroView 2.1.2 (GE Healthcare, UK) software. PDD and cross profiles were compared to PTW 31014 PinPoint chamber, PTW 60016 and 60017 diodes measurements.

3. Results

3.1. Dose response and uniformity of the TruView™ gel

In Figure 1, TruView™ dose response is represented as gel attenuation coefficient μ (cm^{-1}) as a function of dose. Gel response with dose is linear from 0.05 Gy up to 20 Gy, this batch sensitivity is around 0,0811 $Gy^{-1}.cm^{-1}$ at 632 nm (wavelength of maximum sensitivity). A good uniformity has been found: 3.56% relative discrepancy between 10 Gy dose response of a same batch.

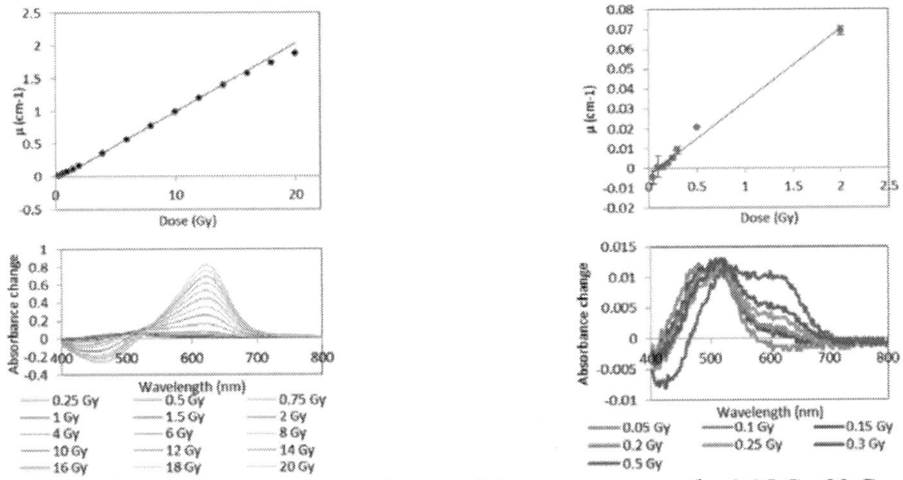

Figure 1. TruView™ dose response and spectral dose response on the 0.05 Gy-20 Gy dose range.

3.2. Dose response as a function of energy and dose rate

In the investigated energy range, dose response is independent of photon beam energy (Figure 2). Moreover, in this study, no observable dose rate dependence was detected (Figure 3).

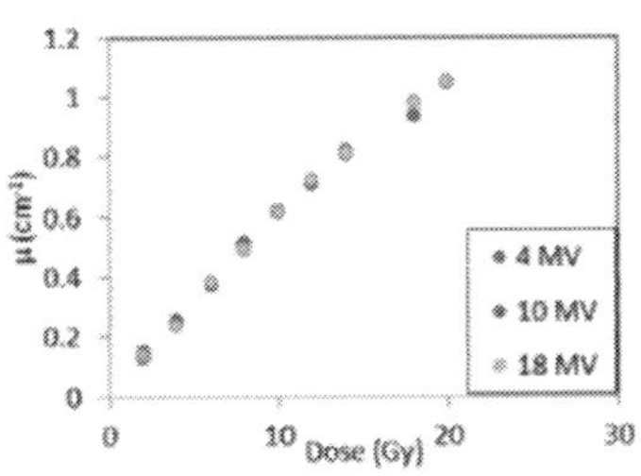

Figure 2. TruView™ dose response for 3 different energies.

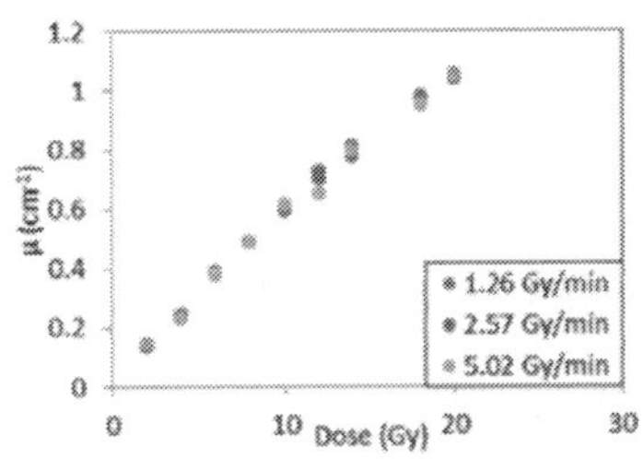

Figure 3. TruView™ dose response for 3 different dose rates (18 MV).

3.3. Influence of different parameters on response

Relative discrepancy between irradiated (10 Gy) Modus manufactured gel cuvettes compared to gel cuvettes made with 35°C water bath heating absorbances is of 2.86% and of 41.5% when compared with 50°C hot-plate heated cuvette. Two gels with different aging (2 months apart) have an absorbance difference of about 0.696 for 448 nm. For 20 Gy, after irradiation, the TruView™ absorbance decreases of 19.02% in 21 days when stored in refrigerator while, it decreases of 23.35% in 5 days when stored at room temperature. Gels irradiated at low temperature are less sensitive than those irradiated at room temperature: 27% of relative discrepancy between responses under 632 nm for 2 Gy and 5% for 0.25 Gy. After irradiation, gel response proves to be stable after 90 minutes.

3.4. Relative dosimetry

In 6 days, maximum dose on cross profiles decreases of 76.9% whereas attenuation coefficient increases in gel non-irradiated areas. A maximum discrepancy of 9.42% and 12% was found between PTW 60016 diode and gel PDD (Figure 4) and cross profiles FWHM respectively (Figure 5). FWHM values obtained for inplane profiles were 8.89 mm, 8.74 mm, 8.75 mm and 9.80 mm for the chamber, the 60017 diode, the 60016 diode and the gel respectively.

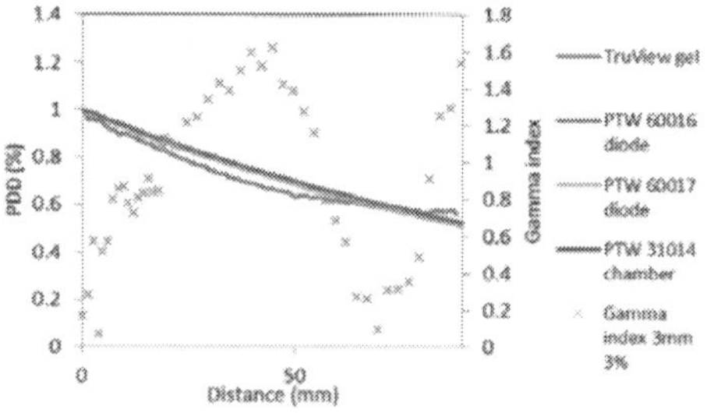

Figure 4. Comparison between PDD obtained with the TruView™ gel and with the active detectors.

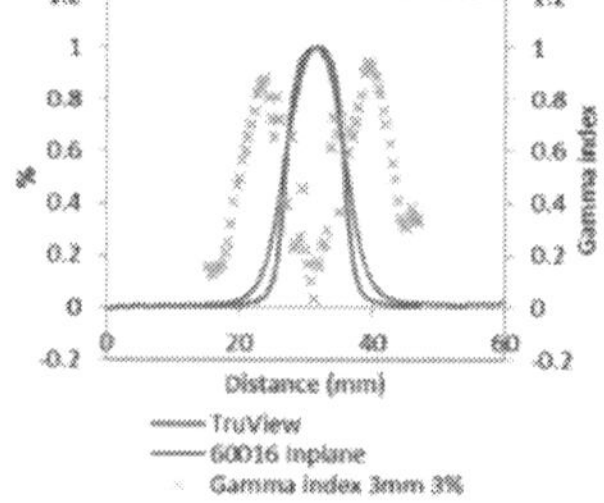

Figure 5. Comparison between gel and diode cross profiles.

4. Discussion

Linear dose response enables convenient calibration for relative dosimetry. Besides, gel independency with energy makes TruView™ an interesting candidate for radiotherapy verification. The dose rate independency study needs to be completed for higher dose rates, but TruView™ seems to be promising. However, this gel is sensitive to numerous packaging, irradiation and reading parameters so it needs to be carefully used to get accurate results. Dose was underestimated in the centre of the jar compared to detectors responses, this might be due to matching errors between profiles and stray light artifacts in the Vista™ system [16]. Then, differences in cross profiles FWHM can be explained by post irradiation ions diffusion inside the gel starting before 90 minutes.

5. Conclusion

3D, accurate and efficient dosimeters are needed to improve quality check in radiotherapy. Recently a new 3D radiochromic gel dosimeter was introduced; it is based on the Fricke dosimeter system with methylthymol blue dye in a gelatin matrix. This new gel presents comparable spectral properties with Fricke gels. It also exhibits great linearity with dose from 0 to 20 Gy, energy and dose rate independency (on the studied ranges). But further studies are required to benchmark this gel for clinical dose verification. In future research, the dose rate dependency will be studied over a larger range of dose rates. Absolute dose comparison is also currently ongoing. Finally, another study to find out how reuse impacts on accuracy and precision is ongoing.

6. Acknowledgements

We thank Jennifer Dietrich and Kalin Penev of Modus Medical Devices Inc. for assistance with gel supply and helpful discussions regarding Vista™ properties.

7. References

[1] Hill R *et al* 2014 *Phys. Med. Biol.* **59** R183-231
[2] Vial P *et al* 2008 *Med. Phys.* **35** 1267-77
[3] Greer P B et al 2007 *Med. Phys.* **34** 4389-98
[4] Mather M L *et al* 2003 *Ultrasonics* **41** 551-9
[5] Hill B *et al* 2005 *Med. Phys.* **32** 1589-97
[6] Babic S *et al* 2008 *Int. J. Radiat. Oncol. Biol. Phys.* **70** 1281-91
[7] Chiu C Y *et al* 2014 *Appl. Radiat. Isoto.* **90** 245-250
[8] Gum F *et al* 2002 *Phys. Med. Biol.* **47** 67-77
[9] Fricke H and Morse S 1927 *Am. J Roent. Radium Ther. Nucl. Med* **18** 430-2
[10] Gore J C *et al* 1984 *Phys. Med. Biol.* **29** 1189-97
[11] Davies J B and Baldock C 2008 *Radiat. Phys. Chem.* **77** 690-6
[12] Baldock C *et al* 2010 *Phys. Med. Biol.* **55** R1-63
[13] Oldham M *et al* 2001 *Med. Phys.* **28** 1436-46

[14] Schreiner J L 2004 *J. Phys.: Conf. Ser.* **3** 9-21
[15] Penev K I and Mequanint K 2015 *J. of Phys.: Conf. Ser.* **573** 012030
[16] Olding T *et al* 2010 *Phys. Med. Biol.* **55** 2819-40

IOP Conf. Series: Journal of Physics: Conf. Series **847** (2017) 012018 doi:10.1088/1742-6596/847/1/012018

Characteristics of modified radiochomic gel Turnbull-Blue

M A Silveira, J F Pavoni and O Baffa
Departamento de Física, FFCLRP, Universidade de São Paulo, Av Bandeirantes 3900,14040-901, Monte Alegre, Ribeirão Preto, SP, Brazil.

E-mail: matsilver@hotmail.com

Abstract. This work presents a dosimetric characterization of the modified radiochromic gel Turnbull-Blue (TB). Formaldehyde a cross-link agent was added to the composition. This modification raised the melting point of agarose gel to 55°C, 22°C above the agarose gel matrix. For this new gel formulation the energy, dose-rate dependence and Diffusion coefficient were verified. The modified TB gel showed higher sensitivity in the 6 MV than 15 MV beam. Dose rate dependence was evaluated from 100 to 500 cGy/min using a 15MV linear accelerator. The results show variations in sensitivity for dose rate dependence. The addition of formaldehyde does not affects the diffusion coefficient for the modified Turnbull-Blue.

1. Introduction

Turnbull-Blue is a radiochromic gel proposed by Solc and Spevacek, it is a gel with low diffusion rate [1] that can be used to perform tridimensional dosimetry [2]. This gel is based on the formation of the Turnbull-Blue (TB) compound when submitted to radiation. The gelatin matrix of TB gel is filled with a ferric salt and potassium ferricyanide. Upon radiation water radiolysis reduces the ferric ions to ferrous ions, which reacts with potassium ferricyanide forming TB complex [3].

Gelling matrices usually have a low melting point what could compromise the use of some gels. Porcine gelatin and agarose matrix have a temperature of 25°C and 33°C of melting point, respectively. In warm environments these kind of gel matrix start to melt and the irradiated gel loses tridimensional dose distribution information. To solve this problem in gel matrix our group proposed the addition of a cross-link agent to the gel matrix. In a previous work the addition of formaldehyde to MAGIC gel [4] formulation increase the dosimeter melting point to 69°C [5, 6]. The new formulation was named MAGIC-f and several studies were performed aiming clinical QA applications [9-11]. Following the same rationale, this work added formaldehyde to the TB's agarose gelling matrix. The cross-linking of agarose and formaldehyde resulted in an increase of the melting point, which allowed us to perform gel dosimetry using the radiochromic TB gel. The aim of this work is the modified TB gel characterization by measuring its, absorbance peak, energy dependence, dose-rate dependence and diffusion coefficient.

2. Materials and Methods

2.1 Gel Preparation and storage

Gel preparation follow the same steps proposed by Solc [7]. It starts with the agarose powder (Uniscience® - electrophoresis agarose) added to water. The solution is heated to 92°C ± 2°C, the fixation point of agarose to dissolution, under continuous stirring. The solution was cooled down to

40°C, when, ferrous chloride solution and potassium ferricyanide were added. After homogenization, formaldehyde was added. Table 1 shows modified turnbull-blue compounds. Gel solution was storage into plastic cuvettes of 1.0 x 1.0 x 5 cm^3 and stored in to refrigerator at 5°C for 12 hours. The original TB formulation was also prepared, but it melted so fast that was not possible to use it for comparison with the modified TB gel.

Table 1. Modified Turnbull-Blue Compounds

Mili-Q water	93.4 w/w%
Agarose	0.6 w/w%
Potassium ferricyanide	1.0 w/w%
Ferrous Chloride solution	1.0 w/w%
Formaldehyde	4.0 w/w%

2.2 Gel Irradiation

Irradiations were performed at Radiotherapy Service of the Hospital das Clinicas de Ribeirão Preto, FMRP - University of São Paulo using a dual energy selection, 6MV and a 15MV, linear accelerator Siemens ONCOR Impression Plus linear accelerator. A SSD of 100 cm was used and the gel dosimeter were irradiated at the depth of the maximum doses. To guarantee the necessary buildup and backscattering for both energies acrylic plates were used. Doses of 3, 6, 9, 12 and 15 Gy were delivered to evaluate the energy dependence with the 6 MV and 15 MV beams. To evaluated dose rate dependence, the same doses were delivered with the 15 MV beam using, 100, 200, 300, 400 and 500 cGy/min.

2.3 Gel scanning.

Optical absorbance measurements were performed few hours after the irradiation process using an Utrospec 2100 pro spectrophotometer. The absorbance of the gel samples scanning band selected were 530 nm to 900 nm for the modified Turnbull-blue samples. Plastic cuvettes of 1cm of optical path were used.

2.4 Diffusion measurements

Gel diffusion measurements were performed in a HP scanner 3707 using an imaging acrylic cuvette, specially constructed to fit the scanner with 1mm of optical path and (5 x 10 cm^2) of area. This cuvette was irradiated with a nominal dose of 18Gy by a 50kv x-ray tube at 5 cm of distance of the tube. The diffusion coefficient was evaluated using the Kron method, [8] for TB gel, described in solc and Spevacek 2009 [1].

3. Results and Discussion

3.1 Absorbance peak

The absorption peak for modified Turnbull-blue gel is around 740 to 750nm (figure 1a), this value differs from the original formulation, that presents the absorption peak at 690 nm [7]. However in this work formulation the molar concentration of potassium ferricyanide and ferrous chloride were increase, both to 1mM and that change the absorption peak.

The gaussian fit analysis, $y = y_0+A.\sqrt{(2/\pi)}/w.e^{(-2.((x-xc)/w)^{\wedge}2)}$, shows a $R^2=0.99$ in all samples plot, y is the absorbance, A is the Area, w the width, x is the wavelength, x_c is the wavelength center. Figure 1b presents the polynomial behavior of the wavelength peak in function of the absorbed dose. The displacement of the wavelength absorption peak as a function of dose, shows absorption peak is dose-dependent with a polynomial function $\lambda = 739.4 + 149.17D - 324.49.D^2$, with Pearson´s correlation coefficient R2 = 0.99, where λ is the wavelength and D is the absorbed dose. This wavelength displacement is according with the Turnbull-blue complex formation. The non-irradiated gel is yellow, upon radiation changes to green and blue, so the when irradiated gel changes the absorption peak in the red visible spectra wavelength region.

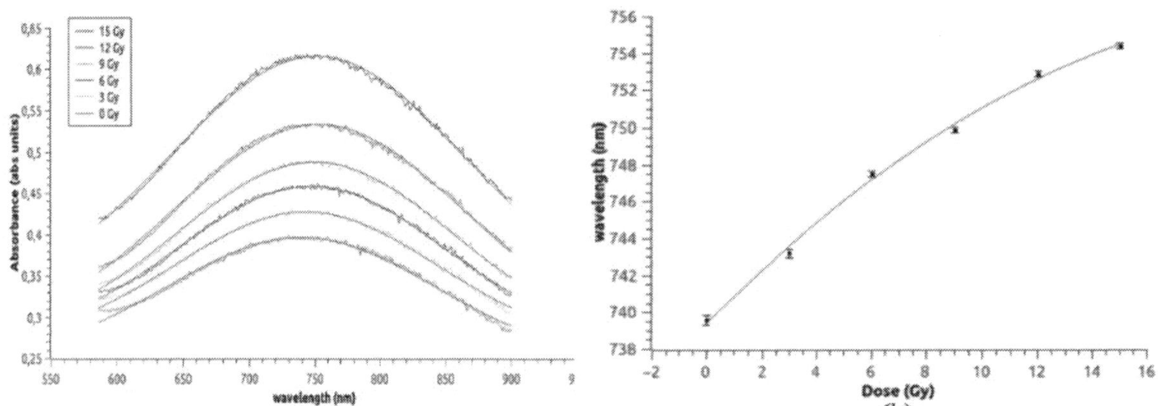

(a) (b)

Figure 1. (a) Absorption spectra of modified TB gel. Irradiated samples 0 to 15 Gy, steps of 3 Gy in a 6 MV beam. (b) Wavelength absorption peak in function of the absorbed dose.

3.2 Energy and dose rate dependence

Figure 2a shows the energy dependence for the two calibration curves achieved when irradiating the dosimeters with 6 MV and 15 MV beams.

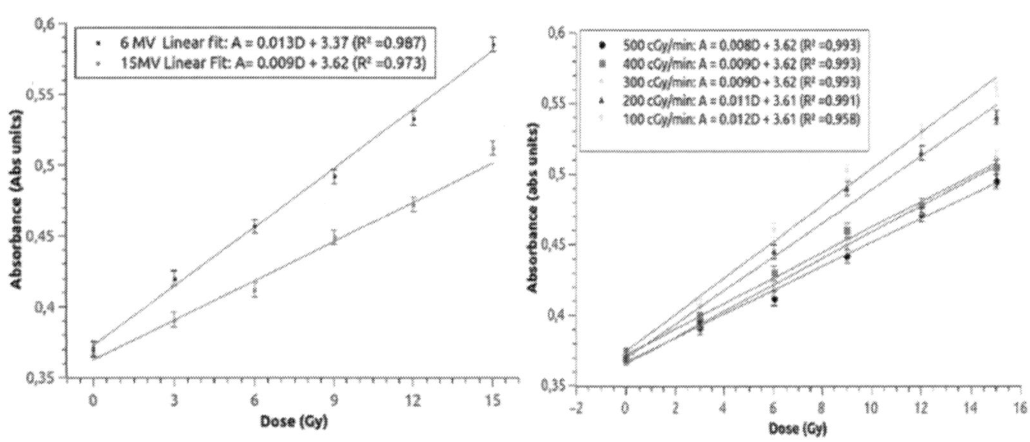

Figure 2. (a) Energy comparison between 6 MV and 15 MV beams. (b) Dose Rate comparison from 15 MV beam, from 100 cGy/min to 500 cGy/min.

The absorbance of the modified TB gel show linear behavior with dose for both energies, but an energy dependence was found with a higher sensitivity for the 6 MV, (S = 0.013 abs.Gy^{-1}) than for the 15 M beam, (0.009 abs.Gy^{-1}), corresponding to an absolute difference of 0.004. Figure 2b shows the dose rate dependence of the modified TB gel. The gel shows dose rate dependence, differing 33% in the sensibility, when varying the dose rate from 100 cGy/min to 500cGy/min, 0.008 and 0.012abs.Gy^{-1}. However, considering only the range of 300 cGy/min to 500 cGy/min, the dose rate dependence is much smaller, varying 12.5% for the sensitivity of the modified TB gel.

3.3 Diffusion measurements

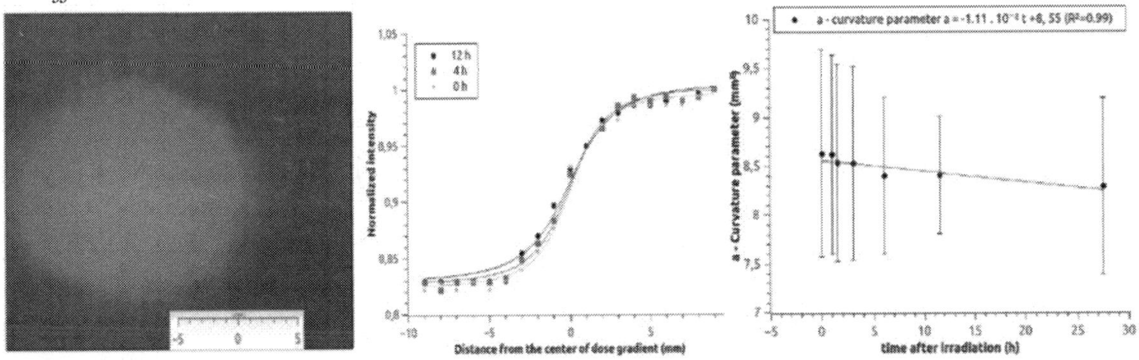

(a) (b) (c)

Figure 3. (a) Image of the cuvette containing the gel irradiated with a dose of 18Gy by a 50kV x-ray source. (b) Normalized intensity profile distance from center of gradient dose. (c) Curvature parameter function of time after irradiation.

Figure 3 show the measurements for the evaluation of diffusion coefficient. Figure 3a show the irradiated gel with a dose of 18 Gy in a 50kv x-ray source. Fitting the normalized intensity (figure 3b) in profile along 10 mm (range of -5mm to 5mm), as shown by a scale in figure 3a, using the inverse of square root function (ISQR) described in [1], it is possible to determinate a(t) parameter. The graph in figure 3c evaluated the slope as a function of parameter a(t), $a = -1.11 \times 10^{-2}\, t + 8.55$ is used to determinate the diffusion coefficient D, by the formula $D = 0.2119.a/t$, resulting in $D = 2.3 \times 10^{-3}\ mm^2.h^{-1}$. This diffusion coefficient corresponds to the same evaluated in the original TB formulation [1, 7], meaning that the addition formaldehyde does not compromise the diffusion effects in the TB gel dosimeter.

4. Conclusions

The present work shows some important dosimetric properties of the modified TB gel dosimeter. The addition of formaldehyde to Turnbull-blue gel dosimeter increased its melting point to 55°C. The displacement of the wavelength absorption peak shows a dependence of the wavelength peak with the absorbed dose. The gel shows a higher sensitivity for a 6 MV beam than 15 MV beam and varying 12.5% in sensitivity of dose rate dependence from when varying from 300cGy/min to 500 cGy/min. The evaluation of diffusion coefficient [12] shows the same result that the original TB formulation, therefore the addition of formaldehyde does not affect the diffusion in the dosimeter. The results in this work shows the dosimetric characteristics of the modified TB gel and gel behavior seems promising to perform tridimensional dosimetry.

5. References

[1] Solc J and Spevacek V 2009 *Phys. Med. Biol.* **54** 5095-107
[2] Solc J *et al* 2010 *Nucl. Instrum. Meth.A* **619** 163-6
[3] Balog J *et al* 1999 *Rad. Phys. Chem.* **55** 483-7
[4] Baldock C *et al* 2010 *Phys. Med. Biol.* **55** R1-63
[5] Pavoni J F and Baffa O 2012 *Radiat. Meas.* **47** 1074-82
[6] Fernandes J P *et al* 2008 *Phys. Med. Biol.* **53** N53-8
[7] Solc J and Spevacek. 2009 *J. Phys.: Conf. Ser.* **164** 012047
[8] Kron T *et al* 1997 *Magn. Reson. Imaging* **15** 211-21
[9] Pavoni J *et al* 2012 *Med. Phys.* **39** 2877-44
[10] Silveira M A *et al* 2014 *Radiat. Meas.* **71** 369-73
[11] Pavoni J F *et al* 2015 *J. Phys.: Conf. Ser.* **573** 012050
[12] Baldock C *et al* 2001 *Austral. Phys. Eng. Sci. Med.* **24** 19-30

Optical and X-ray computed tomography scanning of 3D dosimeters

K J Jordan[1,2], M Hilts[3] and A Jirasek[4,5]
[1]London Regional Cancer Program (London Health Sciences Centre)
[2]Dept. Medical Biophysics, Western University London, ON, Canada
[3]Medical Physics, BC Cancer Agency - Southern Interior, Kelowna BC
[4]Physics, I.K. Barber School of Arts and Science, University of British Columbia, Okanagan Campus, Canada
[5]Dept. Physics and Astronomy, University of Victoria, Canada

E-mail: kevin.jordan@lhsc.on.ca

Abstract. Optical computed tomography (CT) scanners can measure attenuation in large-volume radiochromic dosimeters. Scattered light within the dosimeters is the major factor limiting dynamic range. Scattered light limits the use of large volume polymerization gels for small field dosimetry with broad-beam scanning geometries. Planar and single ray scanning geometries generate and accept less stray light due to scatter in the dosimeter. Transparent, radiochromic dosimeters provide higher dynamic range for accurate transmission imaging. X-ray CT scanning is applicable for large volume polymerization gel dosimeters.

1. Introduction

Optical computed tomography (CT) is being developed for three-dimensional (3D) scanning of radiochromic hydrogels or plastics and radiation-induced polymerization hydrogels. Several scanner geometries have been reported by researchers and "research" versions are available commercially from MGS Research Inc. and Modus QA Inc. In contrast, X-ray CT scanners are present in radiotherapy clinics and can be used without modifications to scan polymerization hydrogels. Polymerization gel formulations are available from MGS Research Inc. (BANG®), radiochromic gels from Modus QA and radiochromic plastic (PRESAGE®) from Heuris Inc. Many researchers have developed both their own CT scanners and dosimeter materials in order to optimize a complete dosimetry system and provide excellent problems for students. This diversity of dosimeter materials, scanner geometries and 3D dosimetry applications has likely impeded commercial development [1]. To further complicate the decision process some of these materials can also be read with magnetic resonance imaging (MRI). But the optimized formulations and dose ranges may be different for optical, X-ray CT and MRI.

Throughout the DOSGEL and IC3DDose series of conferences a large amount of information related to CT scanning of 3D radiation dosimeters has been presented. Until there are edited text books for reference, those interested in efficiently learning more about this topic are encouraged to start with the review talks in the proceedings. Previously optical and X-ray CT have been addressed separately however there are many issues that are common to both. For those unfamiliar with the principles of computed tomography several books are available, for example Kak and Slaney's text is a common reference [2].

Computed tomography is a diverse imaging field and many advances are due to the evolution of increased performance and decreased cost of computers [3]. Graphical processing units and parallel processing algorithms are now making 3D iterative reconstructions practical for clinical dosimetry QA

IOP Conf. Series: Journal of Physics: Conf. Series **847** (2017) 012019 doi:10.1088/1742-6596/847/1/012019

processes. Optical CT evolved in many distinct applications, including: atmospheric [4], chemistry [5], radiation dosimetry [6] and developmental biology [7]. In the focus of this conference, 3D dosimetry, CT scanner development is still active as scan times are reduced and accurate dynamic range increases. Hardware modifications of clinical X-ray CT units has not been actively pursued since the emphasis has been on using available technology.

In order to select which dosimeter and scanner is optimum the dosimetry problem needs to be defined.

Is 3D dosimetry really needed? Could the problem be answered with point measurements with an ion chamber or planar measurements with film of 2D detector arrays? Is the integrated 3D dose distribution required? What is the maximum and minimum doses that must be accurately measured and what is the corresponding spatial resolution? Often, just knowing the maximum dose gradient can limit the system choices. The most challenging problems for external beam 3D dosimetry are small fields in large volumes, for example a 1 cm diameter field in a 15 cm diameter dosimeter. This problem is also challenging for CT readout of dosimeters since scattered visible light or X-rays will have a large impact on the accuracy of the reconstructions and the dose distribution is entirely due to the beam penumbra. In this refresher course the emphasis is on larger dosimeters, greater than 1 litre. Smaller dosimeters are likely to provide greater accuracy due to lower scatter contributions in projection images. Optical and X-ray CT are discussed sequentially but using similar formats.

2. Optical CT

2.1 Scanner geometries

Optical CT scanners can be categorized as broad-beam, planar and single-ray geometries. Stray light due to scatter, fluorescence, reflections and refractions within the hydrogel dosimeter and vessel is maximum for broad beam geometry. In contrast, single-ray geometries minimize "cross-talk" within the dosimeter by sampling different ray paths at different times. Camera based detectors in broad-beam systems can also have up to 5% stray light due to lens glare [8]. As the number of optical interfaces increases the potential for increased stray light also increases. There appears to be a trend of optical CT systems evolving to fewer optical interfaces. This approach should minimize stray light. Simpler systems may have more optical aberrations but computations can sort actual ray paths through dosimeters [9-13]. Lowering acceptance angles is an effective method to reduce stray light. But this increases the scanner's optical path length. Practically, stray light becomes a larger problem as the size of the dosimeter increases. Optical elements such as mirrors, lenses and filters dramatically increase in cost as they become larger and methods to quickly scan larger 15^{+} cm dosimeters is ongoing. In general, full 3D scan speed increases from broad-beam to single-ray. However, if additional scans are required to measure stray light then broad-beam and single-ray scan times may converge. In those geometries where a small detector moves during scanning, the advantages of both small acceptance angle stray light rejection and tolerance of ray deviations of an effectively larger acceptance angle can both be achieved. Based on symmetry, scatter results in a maximum at the optic axis. Depending on system geometry, reflection artifacts may also be maximum at the optic axis due to light travelling perpendicular to optical interfaces. Generally, the centre of the dosimeter is positioned at the optic axis. If there are temperature gradients, they will also be maximized with respect to the dosimeter centre. For all of these reasons, it is prudent not to place the reference dose point at the very centre of the dosimeter. Reference points displaced more than one cm from the optic axis generally agree better with independent point measurements.

There are alternatives to the "transmission CT" geometries. Laser sheet imaging is common for 3D optical microscopy and early results for gel dosimeters has been reported [14]. This geometry has practical advantages if the dosimeter must be rectangular in shape.

2.2 Spectral and source issues

All optical sources require spectral filtering to remove unwanted wavelengths from reaching the detector and increasing the transmission. This is especially important for obtaining high dynamic range. Generally, a narrow bandpass filter is inserted into the optical path at the source. Typically, 10 nm bandpass filters are sufficient, but this choice is coupled to the spectrum of the dosimeter. Preferably, the central probe wavelength corresponds to the peak of the dosimeter absorption spectrum. If the signal spectrum is rapidly changing over the bandwidth of the probe source then spectral changes can be expected and artifact non-linearities in dose responses will be introduced. Alternately, if the signal spectrum is broad the specific wavelength and bandwidth will have minor importance. The narrow linewidths of lasers are not required from a spectral argument and coherence may introduce interference effects which add noise to the measurements. Generally, the intense, narrow, collimated laser beams are nearly ideal for single ray scanning and there is currently no inexpensive X-ray beam of similar properties.

Light emitting diode (LED) sources are rapidly improving in brightness and choice of wavelengths. These sources are very stable when operated with appropriate heat sinks. Intensity drifts of less than 0.1% can be achieved with modest effort. Low power (<1 mW) continuous lasers are sufficient for optical CT scanning, especially for single ray scanning systems. Typical powers are 10 microwatts for dosimeters prior to irradiation. Generally, these lasers may drift up to 3% once they have reached thermal equilibrium and require some method of intensity correction. For example, normalizing profiles to signal in liquid outside the dosimeter is generally sufficient.

Users need to anticipate that all radiation dosimeters are also sensitive to ultraviolet light and minimize exposure. They should, test if photochromic responses are obtained by light levels of ambient lighting and the scanner during readout. In practice, minimize dosimeter exposure to light. Work in dim room lighting if possible and avoid exposure to daylight.

2.3 Polarization factors

While polarization is not an issue at x-ray wavelengths, it is a factor for quantitative imaging with visible light. The Fresnel coefficients describe polarization dependence of reflectivity at an optical interface. In the case of a hydrogel in a cylindrical vessel, incident unpolarized light becomes elliptically polarized due to preferential reflection of light polarized parallel to the rotation axis of the cylinder. The vessel may contain strains that introduces localized optical activity further polarizing the probe light. Gelatin is optically active and can be a further source of polarization artifacts. Polymerization to form plastic dosimeters can also introduce spatially unique optical activity [13]. The effects of polarization need to be evaluated for each dosimeter. In general avoiding rays that are not near perpendicular incidence to interfaces will minimize polarization sensitivity of scanner geometries. Systems with folded optics and high acceptance angles are likely to have polarization artifacts. While polarization sensitivity in 3D dosimeters has not been reported, the radiochromic film product EBT2 is very sensitive to polarization due to a preferentially orientation of the needle-like crystals of the diacetylene monomers. If EBT film is read with optical CT scanners assessment of polarization effects are required.

2.4 Dosimeters

Radichromic, transparent dosimeters are preferred for optical CT scanning. Absorption measurements are robust since ratio methods are used and signal is proportional to the concentration of the molecular product formed due to absorbed dose. Polymerization hydrogels have dose dependent increases in scatter which degrades linearity and spatial resolution as multiple scatter becomes a significant fraction for the attenuation. Changes in refractive index due to polymerization can create another problem for accurate 3D readout [15]. In principle, fluorescent dosimeters may have high dose sensitivity but quantitative 3D emission imaging is a more complicated measurement than transmission imaging. For the clinical dose range, 1 to 60 Gy, likely radiochromic and polymerization dosimeters will be sufficient.

All chemical dosimeters have multiple, temperature dependent reaction rates. Maintaining uniform temperatures is essential for quantitative 3D dosimetry. Reporter molecules such as dyes have high molar extinction coefficients and have much lower temperature sensitivity during absorption measurements than weaker transitions such as ferric xylenol orange. Scanner designs need to minimize thermal gradients during readout. Ideally, a method to measure temperature throughout the dosimeter would be useful for accurate dosimetry.

Isotropic dose response is assumed in 3D dosimetry. However, this must be verified by measuring response to uniform doses. If dosimeters are found to be unique due to their manufacturing then individual, voxel-specific calibrations are required. This is analogous to current situation with radiochromic EBT films where pixel specific corrections are required for non-uniform thicknesses and variations in monomer crystal orientations.

2.5 Vessels

Hydrogel dosimeters are typically made just stiff enough not to flow if tipped horizontally. The vessel provides the mechanical stability. The vessel wall thickness is the dominant factor in how close to the inner wall optical transmission measurements can be obtained. The thinner the wall the more transmission data can be obtained. In the specific case of 15 cm diameter cylinders, 0.2 mm thick PET walls provide a good compromise between mechanical strength and missing projection data. At smaller diameters, 0.1 mm thick walls can also be used. Figure 1, shows reconstructed transverse slice from an 11 cm diameter vessel with 0.2 mm thick PET wall containing a carbon black micelle solution. The scanner was a modified Vista 10 scanner with a Fresnel lens convergent cone beam source [16].

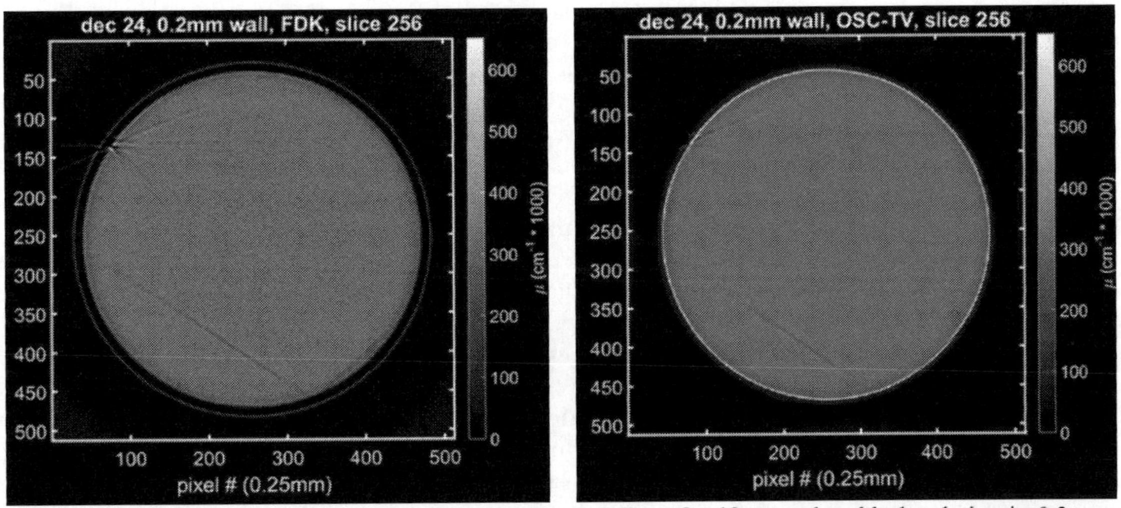

Figure 1. Optical cone-beam CT reconstructed transverse slice of uniform carbon black solution in 0.2 mm thick PET cylinder, 1024 projections, voxel size 0.25x0.25x0.25 mm: left) filtered back projection, right) iterative.

Because the vessel wall has a different refractive index from the hydrogel, no single refractive index liquid will provide minimum deflection of all transmitted rays at increasing radius [6, 17]. This occurs because the fraction of the chord length due to the gel decreases as the radius increases. The net effect of the wall is to bend rays away from the optic axis when the refractive index inside and outside the vessel is same. By lowering the refractive index outside the bending can be compensated and nearly all rays near the inner vessel wall can be sampled, see figure 2. In practice this is an iterative process for each gel formulation and vessel.

Polyethylene terephthalate (PET) is a commonly used polyester plastic in the food industry because of the following features: low oxygen permeability, strength, stiffness, high melting point and transparency. All of these features are preferred for hydrogel dosimeter vessels as well. Teflon formulations such as perfluoroalkoxy (PFA) are nearly transparent and have refractive indexes similar to many hydrogel formulations. From an optical perspective, Teflon is a good match, however it has a high Hounsfield unit for diagnostic energy x-rays, is relatively permeable to oxygen and is a soft plastic.

Also, extruded thin-walled tubes of PFA are optically active which will add uncertainty in polarization sensitive scanners geometries.

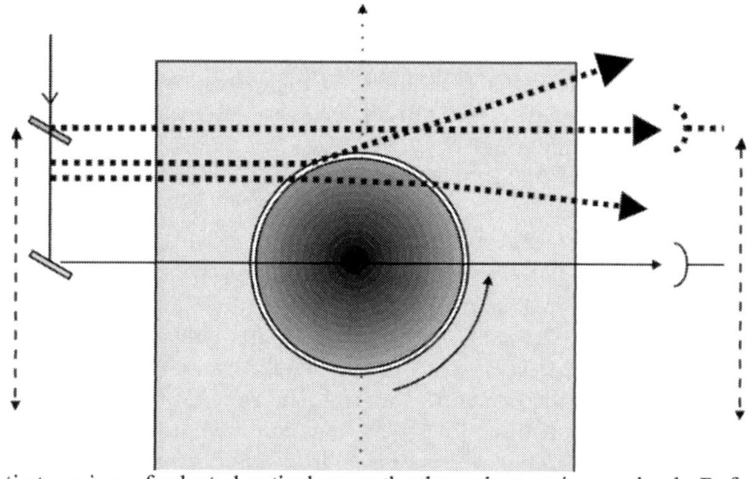

Figure 2. Schematic top view of selected optical ray paths through aquarium and gel. Refractive index liquid is less than gel dosimeter. Vessel wall is thin PET plastic cylinder. Translation of turning mirror and detector is coupled. Ray paths demonstrate: normal incidence (optical axis), refraction near wall (missing data), reflection, reference liquid path.

Generating ratio projection images requires that dosimeters can be reproducibly aligned with pixel resolution between the pre and post-irradiation scans. This is essentially a mechanical problem, how to mount the vessel reproducibly, but it may be possible to use image processing to perform subpixel alignments if required. Fiducial marks such as small dark ink spots of the vessel wall can be used for alignment as well as any sharp feature such as vessel edge or wall seam. Fiducial marks should provide adequate contrast for alignment but low enough contrast not to introduce streaking artifacts in the reconstruction. Optimal placement of fiducial marks may be problem specific.

As the acceptance angle of the scanner decreases, the optical quality of the vessel becomes more important. Stray light in certain scanner geometries essentially masked the effects of scratches, see [16]. By increasing the diameter of the source many optical artifacts such as missing primary rays near the vessel wall can be minimized. Many scanners have some degree of "source defocusing" in order to minimize high spatial frequency features such as scratches of the dosimeter or vessel surface. Laser beam scanners with large effective area detectors are insensitive to small variations in refractive index changes that can deviate the transmitted rays. A projection image of the vessel with a solution with absorption and scatter values similar to the intended hydrogel dosimeter will provide a measure of the optical quality of the entire system.

PRESAGE® is generally scanned with oil based refractive index matching fluid mixtures. These liquids are inconvenient to handle because of high viscosity and ability to interact with plastics. In order to reduce the volume of liquid a hybrid aquarium has been evaluated by [12]. In this approach, a matching solid aquarium is required for each dosimeter size. It is possible to scan PRESAGE® with low viscosity liquids than have much lower refractive indexes with specific scanner geometries that have large effective area detector and calibration of ray paths through the dosimeters [13].

2.6 Scanning strategies

Ratios of projections acquired pre and post irradiation [18], or ratios of dual wavelength post irradiation [19] are helpful for minimizing effects of optical imperfections such as scratches on the dosimeter surface which make each dosimeter optically unique (at this time). Pre and post irradiation scanning can introduce artifacts if the dosimeter is not repositioned with sub pixel accuracy [20]. Dosimeter transfers can introduce dust and debris into the aquarium and Langmuir-Blodgett films get deposited on the dosimeter and aquarium windows due to the changing fluid levels. These sources of noise degrade the SNR of the 3D reconstruction. Post processing of projections is another approach filtering features related to bubbles and debris suspended in the liquid.

Dual wavelength scanning may be an effective alternative to pre and post-irradiation scans since the dosimeter is positioned only once in the scanner. This approach will be more convenient if comparable quality reconstructions can be obtained. Scanner systems that allow rapid wavelength switching may have a practical advantage. However, dual wavelength scanning may compromise the magnitude of the signal depending on the specific spectrum of the dosimeter and wavelengths available for scanning. If spectral scanning proves to be a significant advantage, then tunable sources could be introduced for optical CT scanning.

2.7 Data corrections

Optical aberrations result in non-uniform spacing of pixels in projection images. The most common aberration is due to refraction from non-matched refractive indexes, see vessel wall discussion above.

Examples of ray position correction have been reported by several authors [10, 12, 13, 20]. Scattered light in hydrogels is a common problem especially with broad beam geometries and small fields [17, 21-24]. In general a stray-light image for each transmission image should be measured and subtracted since they are coupled. The character of the stray light will be system dependent. But the major components are scatter within the dosimeter, reflections and refractions from the vessel and each optical interface in the system. This is most important for 'broad-beam' geometries and possibly a minor correction for 'single-ray' scanning geometries. Comparison of scan speeds should include time for additional scatter measurement scans if required. For example, if 3 scans are required for broad beam scanning with a structured illumination grid [25] then a slower single ray scanning approach may be equivalent [26]. Well-designed LED sources should be stable to 0.1% and laser sources to ~3% over typical scan times of one hour. Correction for source stability generally involves normalizing to transmission in liquid outside the dosimeter. Depending on system design there may be a settling time associated with switching wavelengths. Spatial filtering of projection images to remove spurious features such as scratches may be necessary. However, blemish-free optics and clean liquids are essential for low artifact images and maximum SNR of reconstructed images.

Some dosimeters, for example ferrous xylenol orange gelatin hydrogel, will darken with time at room temperature. This drift in absorption must be subtracted in order to calculate the dose dependent change. While other dosimeters may slowly fade following irradiation. Dosimeter specific thermal effects need to be measured and appropriate corrections applied. In general, fast responding dosimeters and fast 3D scanning, for example less than one hour can minimize magnitude of post irradiation corrections.

2.8 Performance tests

Uniform solutions that have refractive indexes, scatter and absorption similar to the dosimeter are important for assessing field of view and linearity. The uniform solutions allow testing of impact of missing data from near the vessel wall. Measuring small fields is another challenging geometry for optical CT. Thin-walled PFA Teflon tubes filled with absorbing aqueous solutions are a substitute for radiochromic hydrogels or custom "finger phantoms" [27]. If these uniform liquids and tubes can be accurately reconstructed then less challenging distributions such as an arc delivery can also be scanned. The scatter and absorption coefficients need to span the range that will be sampled by the irradiated dosimeter to allow interpolation. Spatial resolution QA phantoms have been assessed for 3D optical CT with PRESAGE® [28, 29].

2.9 Reconstruction algorithms

Incomplete projections are the most common problem in optical CT. The missing data generally occurs at grazing angles near the wall of the dosimeter or vessel. Refractive index optimization and thin walled vessels provide the maximum data. Iterative reconstruction algorithms generally result in lower reconstruction artifacts compared to filtered back projection. Typically threefold improvements of signal to noise ratio are observed. Graphical processor units (GPU) allow parallel processing of iterative reconstructions of 512^3 arrays in less than 30 minutes. Researchers can evaluate data sets with GPU enabled open access iterative reconstruction packages such as ASTRA [30] and TIGRE [31]. Implementation of iterative reconstruction has been reported specifically for optical cbct [32] and polymer gel dosimeter [33].

2.10 Dynamic range

Before irradiating the dosimeter, a calculation needs to be performed. The minimum transmission of unknown dose distribution based is estimated from the calculated dose distribution and dosimeter sensitivity. A suggested minimum transmission of 10% for broad beam scanners and 2% for single ray scanners should be adequate for most systems.

Dynamic range can be increased by measuring and subtracting scatter components. In general, this is a 3D problem coupled to the absorbing 3D dose image. This means a scatter image for each projection is required [8]. Structured illumination using a spatially modulated grid and acquiring multiple images is one approach to obtaining scatter subtracted broad beam images [25]. Dual or multiple wavelength scanning is another method to correct for scatter. Finally, single ray scanning will likely provide the lowest scatter projections since only the ray being sampled is illuminated.

3. Basic principles of X-ray CT gel dosimetry

The principle of utilizing x-ray Computed Tomography (CT) for imaging irradiated polymer gel dosimeters was introduced at the first *Dosgel* meeting in 1999 and in a subsequent paper by Hilts *et al* [34]. The fundamental principles revolve around the fact that irradiated polymer gel dosimeters exhibit a small physical density change upon irradiation with ionizing radiation from clinical linear accelerator beams. As the density change occurring in irradiated gel dosimeters is small, on the order of 1%, CT polymer gel dosimetry is a *low contrast* imaging modality and, as such, much effort in CT polymer gel dosimetry have been in the area of contrast resolution improvements. Comprehensive review of the fundamentals of x-ray CT gel dosimetry have been published in the past [35, 36]. In this short overview we highlight some of the main points for consideration in performing CT polymer gel dosimetry.

3.1 Formulations of 3D dosimeters for use with x-ray CT

A primary limitation in CT polymer gel dosimetry is the small density change in irradiated dosimeters. To this end, a significant effort has been expended in developing dosimeters with sufficient density change per Gray (Gy) of irradiation so as to render the technique viable. Early investigations in CT gel dosimetry (e.g. [34, 37]) worked with the standard acrylamide-based dosimeters (termed BANG dosimeters) of Maryanski *et al* [38] and were based on a formulation consisting of acrylamide (3%), N-N'-methelynebisacrylamide (3%), gelatin (5%), and water as remainder. However, three primary factors propelled the further development of dosimeter formulations:

- the dose resolution of the BANG-based system was low (0.78 H/Gy, [33] and ref therein);
- the drive to simpler manufacturing protocols, specifically, the advent of "normoxic" formulations [39], or, dosimeters that could be manufactured on a bench top in normal atmospheric conditions as opposed to within a sealed and nitrogen-filled glove box;
- the toxicity of the BANG systems was moving research interest to consider formulations with reduced toxicity.

Unfortunately, initial moves from BANG dosimeters to normoxic formulations and to formulations with reduced toxicity generally, with few exceptions, lowered the dose resolution in the overall system [37-40-42]. However, Chain *et al* [43] reported on a formulation of normoxic, reduced toxicity dosimeter with reasonable dose resolution (~0.8 - 1 H / Gy) and it is, largely, variants of this formulation that are in use in current CT gel dosimetry. The formulation (termed NIPAM) consists of N-isopropyl-acrylamide (5-7% by weight), N-N'-methelynebisacrylamide (5-7%), gelatin (3-5%), tetrakis-hydroxymethyl-phosphonium chloride (THPC, antioxidant, 5-10mM), and deionized water as remainder. Figure 1 illustrates the progression of NIPAM-based formulation development.

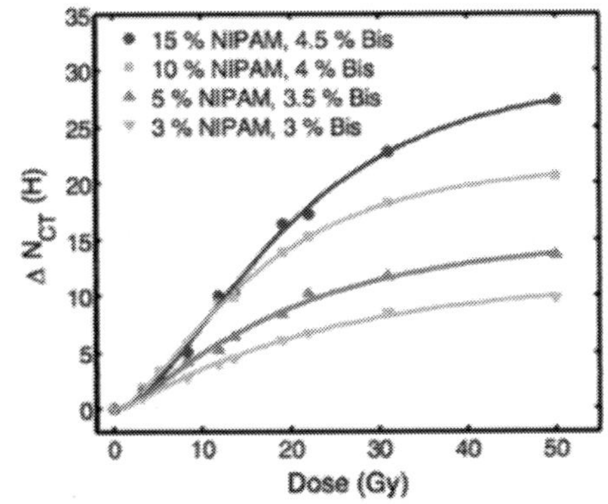

Figure 3. Dose response of NIPAM-based polymer gel dosimeter, as measured with x-ray CT. Shown is the effect of dosimeter composition on resultant dose response. From Chain *et al* [43].

3.2 Dosimeter containers

Not much work has gone into the specific investigation of dosimeter containers for CT polymer gel dosimetry. However, unlike in the case of optical CT imaging, the dosimeters vessel is, in most situations, of secondary importance. Most common dosimeter vessels are largely transparent to CT, however high density materials should be avoided as they can cause artifacts that are difficult to adequately remove [44]. One advantage is the general need for cylindrical symmetry that exists in optical CT is lifted in x-ray CT polymer gel dosimetry and an example of anthropomorphic phantom geometry can be found in the literature [45].

3.3 Scanning strategies

Concomitant with gel formulation development, the parameters used in CT imaging can affect image quality. While this is not a new finding (standard CT imaging theory applies here), the effect of imaging parameters on CT gel dose resolution have been studied [44, 46] and there exist robust recommendations for imaging protocol for CT gel dosimetry. Table 1 outlines the main parameters for consideration in CT imaging of polymer gel dosimeters and they can impact resultant dose measurements.

An optimal choice of imaging parameters is difficult to recommend as this will depend on the specific needs of a given application, considered balancing of noise and spatial resolution needs. For example, to improve contrast resolution it may make sense to scan with a slice thickness that matches the treatment plan you are intending to verify (e.g. 2 mm) rather than choosing the smallest available slice thickness (e.g. 1 mm) which would maximize spatial resolution. That said, typical scan values are listed in Table 1.

Recent work has investigated use of multi-slice CT scanners for CT gel dosimetry [47, 48]. This is a promising advance as multi-slice imaging has proven straightforward to implement for CT gel dosimetry and offers faster imaging time as well as reduced load on x-ray tube. As such, multi-slice imaging, when available, is recommended for CT gel dosimetry. An initial investigation has also looked at cone-beam CT for gel dosimetry [49]. While the dose response observed in this study was significantly poorer than that observed with conventional CT (less than half the sensitivity) this remains an interesting avenue to explore, particularly with any advances in CBCT technology.

Table 1: Image acquisition parameters for consideration in x-ray CT polymer gel dosimetry.

Parameter	Impact on Dose Measurement	Typical Values
Tube current (mAs)	Increasing mAs decreases image noise and improves dose resolution. However tube overheating is a limiting factor.	200 mAs
Tube voltage (kVp)	Increasing kV decreases image noise and improves dose resolution. This outweighs any potential increase in sensitivity at low kV as no significant impact has been observed.	120 or 140kV
Scan time (s)	Increasing scan time decreases image noise (combined with mA, above). Minimize to reduce scan time.	1 s
Slice thickness (mm)	Increasing slice thickness decreases image noise and improves dose resolution, but it reduces spatial resolution in "z".	2 - 5 mm
Reconstruction	Algorithms are designed for various options e.g. to smooth or enhance detail and can affect image noise dramatically.	Standard, Smooth

3.4 Image reconstruction

One area that has received little attention is the effect of image reconstruction technique on resultant gel image quality, beyond a comparison of commercially available reconstructions on a single scanner system [44]. While image reconstruction analysis is a well-documented area of interest (e.g. [50]), little application has been made to the specific case of CT polymer gel dosimeters. In addition to the investigation of image reconstruction filters [51], the primary open area of investigation is in the area of iterative reconstruction techniques. While much effort has gone into the investigation of iterative CT reconstruction, scant literature exists in this area when concentrating on the specific case of polymer gel dosimetry. One such example [52], taken from the optical CT arena, suggests that further investigation in this area may be warranted for x-ray CT. A common issue in iterative CT reconstruction research for CT polymer gel dosimetry is the need for sinogram data output from CT machines. Typically, sinogram data is not provided as standard output on commercial scanner, thus impeding the progress of this line of investigation.

3.5 Image processing

3.5.1 Image averaging and background subtraction

Due to the low contrast images produced in CT polymer gel dosimetry, much work has gone into optimizing post-acquisition image processing. Early efforts [34] established a simple yet effective image averaging and background subtraction technique that is still employed in current research. In summary, recommended imaging protocols include:

- acquire 16 - 25 acquisitions / slice to create averaged images of irradiated dosimeter;
- acquire 16 - 25 acquisitions / slice to create averaged images of background (non-irr) dosimeter;
- subtract averaged background image set from averaged irradiated dosimeter image set.

Figure 2 illustrates the overall efficacy of the protocol.

3.5.2 Image filtering

Significant effort has gone into understanding the need and efficacy of image filtering in CT polymer gel dosimetry. Early efforts [51] involved studying a host of static (e.g. mean filter) and adaptive (e.g. adaptive mean) filtering techniques on resultant image noise reduction and concomitant spatial resolution degradation. Adaptive mean is easy to implement and a detailed investigation showed its efficacy for CT gel dosimetry over a range of dose distributions [53].

As with most instances of image filter analysis, summary recommendations possess a certain qualitative aspect, and hard recommendations are difficult to ascertain, primarily due to the fact that the tradeoff between signal to noise ratio improvements and spatial resolution degradation are not universal across applications. Image signal to noise ratio and gel irradiation pattern, including the presence of steep dose gradients within the irradiation, all affect the choice and degree of filtering. Given the preceding significant caveat, we can, however, make a general conclusion that over most "typical" scenarios, an Adaptive Mean filter of 3x3 or 5x5 kernel size and two iterations has been shown to perform well over a range of dose gradients and irradiation patterns.

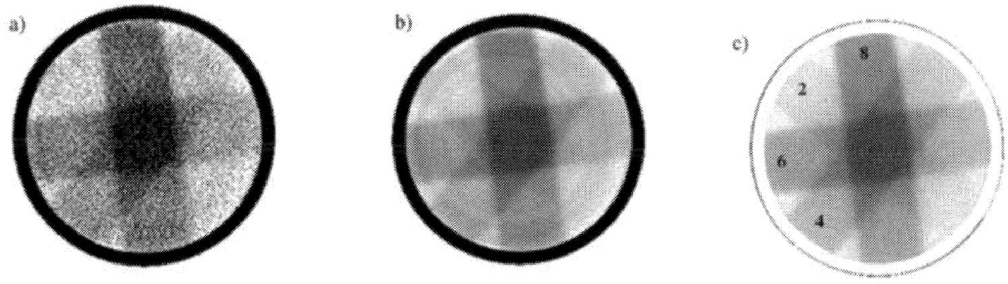

Figure 4. (a) Single CT image of irradiated polymer gel. (b) Image of averaged image of irradiated polymer gel. 25 averages acquired. (c) Averaged and background subtracted image. Numbers within image indicate beam dose (in Gy).

3.5.3 Remnant artefact removal
Even with reasonably heavy filtering it has been shown [54] that persistent, non-stochastic artefacts can remain in filtered and subtracted gel dosimetry images. These "remnant artefact" have been attributed to macroscopic gel structure and subtends local spatial ranges of several image pixels (i.e. mm range). Efforts to mitigate the image degradation from these remnant artefacts have involved the development of a "remnant artefact removal" algorithm. The algorithm is based on a signal removal set of algorithms that involve the selection of three fundamental parameters:

- filter span: the size of the window;
- filter degree: the degree of the fitting polynomial;
- filter iterations: the number of iterations to run the filter over the image.

As with the above comments on filter recommendations, hard recommendations on filter span, degree, and iteration are difficult. In general, the filter span should be on the order of the size of the macroscopic structure to be removed from the image. The polynomial degree is typically low (3-5), and the number of iterations typically run between 10-30. Higher iterations produce diminishing returns at the expense of increased computational time.

3.5.4 Zero scan method
A novel CT imaging method has been proposed for gel dosimetry, known as a "zero scan method", in efforts to account for any dose resulting from CT imaging [55]. In its basic form, the zero-scan method monitors the pixel intensity within each mage pixel as images are acquired on a given slice. A linear function is fit to each set of pixel values, and a "zero scan" pixel intensity is extrapolated from the resultant fit to represent the "true" pixel intensity prior to CT imaging. The method has been shown to improve overall image noise, however, the effects of the zero scan method within high dose gradient regions remains an area of further investigation. A recent work has extended the zero-scan concept to multi-slice imaging [48].

3.6 CT Imaging Dose
Unlike with optical CT or MRI, in x-ray CT read-out of polymer gels ionizing radiation dose will be deposited into the gel as a result of the imaging process [56]. The effect of this imaging dose has been

observed and, in a novel study, an active polymer gel was used to measure the CTDI for a CT scanner [41, 57, 58]. However, the response of normoxic polymer gel dosimeters to kV photon energies, as are emitted in CT scanning, has been shown to be less than half that of MV irradiations and the additional CT contrast resulting from a typical imaging process, incorporating 16 to 32 image averages, is very low, < 0.2H [59]. Further, this effect was measured for fully active gels which were imaged a short time following manufacture. In reality, it is common practice in CT gel dosimetry to expose gels to atmospheric oxygen following polymerization and prior to irradiation which would further reduce the impact of CT dose on read-out by, at least partially, rendering the gel inactive.

3.7 Summary

In summary, the primary considerations that should be evaluated prior to undertaking a CT polymer gel dosimetry experiment include:

- choice of dosimeter;
- smart design of container;
- single or multi-slice CT;
- CT image acquisition parameters;
- image filtering protocol;
- remnant artefact removal protocol, if employed;
- decision on use of zero-scan method.

Suggested areas of further investigation in CT gel dosimetry include: iterative reconstruction techniques, the zero scan method, and development of clinical applications.

4. Conclusion

CT scanning of 3D dosimeters is a practical approaching to achieving accurate dosimetry. Scatter within the dosimeter is often the limiting factor for dynamic range. It is a bigger problem for broad beam geometries, small fields and for larger dosimeters. If the dynamic range is inadequate for the problem of interest then scatter must be measured and subtracted from projection data. Prefiltering of transmission images will reduce reconstruction artefacts associated with the vessel, such as scratches. Iterative reconstruction algorithms will provide higher SNR compared to filtered backprojection.

5. Acknowledgments

Disclosure: K. Jordan has a research agreement with Modus QA Inc. concerning optical CT applications.

6. References

[1] Baldock C *et al* 2010 *Phys. Med. Biol.* **55** R1-63
[2] Kak A C and Slaney M, 1988, *Principles of computerized tomographic imaging*, IEEE Press
[3] Baldock C *et al* 1994 *J. R. Soc. Med.* **87** 806-8
[4] Byer R L and Shepp L A 1979 *Opt. Lett.* **4** 75-7
[5] Burns D H 1994 *Appl. Spect.* **48** 12A-19A
[6] Gore C *et al* 1996 *Phys. Med. Biol.* **41** 2695-704
[7] Wong M D *et al* 2013 *PLOS ONE* **8** e73491
[8] Granton P V *et al* 2016 *Phys. Med. Biol.* **61** 2893-909
[9] Doran S J and Yatigammana D N B 2012 *Phys. Med. Biol.* **57** 665-83
[10] Ramm D *et al* 2012 *Phys. Med. Biol.* **57** 3853-68
[11] Krstajić N and Doran S J 2007 *Phys. Med. Biol.* **52** N257-3
[12] Chisholm K *et al* 2015 *Med. Phys.* **42** 2607-14
[13] Dekker K H *et al* 2016 *Med. Phys.* **43** 4585-97
[14] Alwan R *et al* 2008 *Nucl. Instrum. Meth. B* **266** 834-40
[15] Campbell W G *et al* 2014 *Med. Phys.* **41** 112102
[16] Dekker K H *et al* 2016 *Phys. Med. Biol.* **61** 2910-25
[17] Olding T *et al* 2010 *Med. Biol.* **55** 2819-40

[18] Kelly R G *et al* 1998 *Med. Phys.* **25** 1741-50
[19] De Deene Y 2015 *J. Phys.: Conf. Series* **573** 012058
[20] Campbell W G *et al* 2013 *Med. Phys.* **40** 061712
[21] Bosi S G *et al* 2009 *Phys. Med. Biol.* **54** 275-83
[22] Babic S *et al* 2009 *Phys. Med. Biol.* **54** 2463-81
[23] Olding T *et al* 2011 *J. Med. Phys.* **36** 3-14
[24] Thomas A *et al* 2011 *Phys. Med. Biol.* **56** 4433-51
[25] Babic S and Jordan K 2012 *Phys. Med. Biol.* **57** N377-89
[26] Jordan K J *et al* 2013 *J. Phys.: Conf. Series* **444** 12062
[27] Jordan K 2013 *J. Phys.: Conf. Series* **444** 012076
[28] Rahman A T A *et al* 2011 *Phys. Med. Biol.* **56** 4177-99
[29] Gorjiara T *et al* 2011 *Med. Phys.* **38** 2265-74
[30] Palenstijn W J *et al* 2013 *Proceedings of the 13th International Conference on Computational and Mathematical Methods in Science and Engineering*
[31] Biguri A *et al* 2016 *Biomed. Phys. Eng. Express* **2** 055010
[32] Matenine D *et al* 2015 *Med. Phys.* **42** 6376-86
[33] Shih C-T *et al* 2015 *Physica Medica* **31** 942-47
[34] Hilts M *et al* 2000 *Phys. Med. Biol.* **45** 2559-71
[35] Hilts M 2006 *J. Phys.: Conf. Series* **56** 95-107
[36] Jirasek A and Hilts M 2009 *J. Phys.: Conf. Series* **164** 1-12
[37] Trapp J V *et al* 2001 *Phys. Med. Biol.* **46** 2939-51
[38] Maryanski M J *et al* 1993 *Magn. Reson. Imaging* **11** 253-8
[39] Fong P M *et al* 2001 *Phys. Med. Biol.* **46** 3105-13
[40] Brindha S *et al* 2004 *Phys. Med. Biol.* **49** N353-61
[41] Hill B *et al* 2005 *Br. J. Radiol.* **78** 623-30
[42] Jirasek A *et al* 2006 *Phys. Med. Biol.* **51** 1891-906
[43] Chain J *et al* 2011 *Phys. Med. Biol.* **56** 2091-102
[44] Hilts M *et al* 2005 *Phys. Med. Biol.* **50** 1727-45
[45] Johnson H *et al* 2012 *Phys. Med. Biol.* **57** 3155-75
[46] Subramanian B *et al* 2006 *J. Med. Phys.* **31** 72-7
[47] Johnston H *et al* 2015 *Med. Phys.* **42** 1666-77
[48] Kakakhel M B *et al* 2014 *J. Appl. Clin. Med. Phys.* **15** 367-71
[49] Huang K M *et al* 2010 *Appl. Radiat. Isot.* **68** 607-9
[50] Beister M *et al* 2012 *Physica Medica* **28** 94-108
[51] Hilts M and Duzenli C 2004 *Med. Phys.* **31** 39-49
[52] Yi D *et al* 2015 *J. Phys.: Conf. Series* **573** 012063
[53] Hilts M, Jirasek A 2008 *Med. Phys.* **35** 344-55
[54] Jirasek A *et al* 2012 *Phys. Med. Biol.* **57** 3137-53
[55] Kakakhel M B *et al* 2011 *Med. Phys.* **38** 5130-35
[56] Trapp J V *et al* 2002 *Phys. Med. Biol.* **47** 4247-58
[57] Hill B *et al* 2005 *Med. Phys.* **32** 1589-97
[58] Hill B *et al* 2008 *Phys. Med. Biol.* **24** 149-58
[59] Baxter P *et al* 2007 *Med. Phys.* **34** 1934-43

IOP Conf. Series: Journal of Physics: Conf. Series **847** (2017) 012020 doi:10.1088/1742-6596/847/1/012020

Development of a liquid scintillator-based 3D detector for range measurements of spot-scanned proton beams

C D Darne, D Robertson and S Beddar
Department of Radiation Physics, University of Texas MD Anderson Cancer Center, Houston, TX 77030, USA

Email: cddarne@mdanderson.org

Abstract The goal of the present study is to develop a liquid scintillator-based 3D detector and investigate its performance for beam range measurements in real time. The detector design consists of a tank filled with water-equivalent liquid scintillator that scintillates in response to incident proton beams. Three scientific complementary metal-oxide semiconductor (sCMOS) cameras collect the resulting optical signals. Preliminary measurements have shown that the optical detection system can record light distribution profiles anywhere inside the tank with an average spatial resolution of 0.23 mm. By employing multiple cameras the system is capable of capturing high resolution 2D depth and beams-eye profiles of the delivered beam and can consequently localize the position of beam anywhere inside the tank. It is also shown that it is capable of accurately measuring proton beam range with no more than 0.3 mm difference from the nominal range. The detector system thus demonstrates its ability to perform fast, high-resolution, and precise beam range measurements in 3D.

1. Introduction

Ionization chambers (IC) are the prevailing standard for dosimetry on account of their accuracy and are routinely used for QA measurements of machines and treatment plans [1]. However, drawbacks such as the ability to verify only a limited number of dose points along the beam, volume-averaged readings, and time requirement for dose verification may render them obsolete especially for modern radiation therapy modalities such as intensity modulated radiation therapy (IMRT), or intensity modulated proton therapy (IMPT). Therefore, novel detectors are needed that can measure complex radiation distribution and steep dose profiles in 3D [2]. Our liquid scintillator-based detector has the capability to overcome these difficulties by generating high resolution spatially and temporally resolved 3D dose profiles in clinically relevant time-frames. In this paper, we are examining few important performance characteristics of this 3D detector with scanning proton beam systems.

2. Materials and methods

2.1 Liquid scintillator detector design

The detector consists of a liquid scintillator (LS) filled cubic tank (20 x 20 x 20 cm^3) constructed out of acrylic. Clear tank surfaces allow scintillation light to be captured by the cameras while the opaque surfaces minimize influence of ambient light on the measurements. Optiphase HiSafe 3 (PerkinElmer, Waltham, MA) was our choice of LS because of its water-equivalency (0.986 g/cm^3) and the ability to double up as a phantom as well as an active element thereby ensuring that the radiation field is measured without perturbation. To capture additional beam profiles, two mirrors located on the top and side surfaces of the tank mounted at 45° angles to the tank surfaces redirect the scintillation light to the respective two cameras (Figure 1.a). Multiple cross-hair markers imprinted on the inner tank surfaces allow geometric calibration that converts pixel location in camera frame to physical location within the tank. This geometric calibration takes into account refraction, perspective, and lens distortion, in addition to rotational and translational shift of the camera relative to the tank, and effective distance between the tank and lens center [3]. Finally, the tank along with the three cameras are securely mounted on a breadboard for system portability, rigidity, and robustness.

2.2 Optical system design
We used three identical Zyla 5.5 scientific complementary metal-oxide-semiconductor (sCMOS) cameras for this work (Andor Technology Ltd., Belfast, UK). The camera sensor is equipped with 2560 x 2160 pixels with an individual pixel size of 6.5 µm. A 16-bit analog-to-digital converter digitizes the light signal. The camera is cooled and maintained at 0°C via an on-board thermoelectric cooling system. It is capable of sustaining a full frame transfer rate of 30 frames per second (fps) using USB3.0 interface. Additionally, unlike CCDs, CMOS offers the flexibility to readout smaller region of interests (ROI) that further speeds up data transfer. For example, we use 1100 x 1100 pixel area to image 20 x 20 cm tank. This smaller ROI drastically improves the frame transfer rate to 85 fps, resulting in better time-resolution for the 3D detector.

The cameras are fitted with identical 20.5 mm fixed focal length objective lenses (Schneider Optics, Van Nuys, CA). The lens selection was based on calculations of depth of field (DoF) [4] and field of view (FoV). With 750 mm working distance and f/8 aperture lens setting a 20 cm focal depth around the tank center is obtained. This allows sharp scintillation images to be captured anywhere inside the tank volume. Finally, this work requires all cameras to simultaneously acquire scintillation light from three orthogonal angles. For this study, we connected the three cameras in a daisy chain configuration with the first camera triggering the next with an insertion delay of 2 µsec. Thus, all three cameras fire in tandem capturing instantaneous beam profile from different orientations.

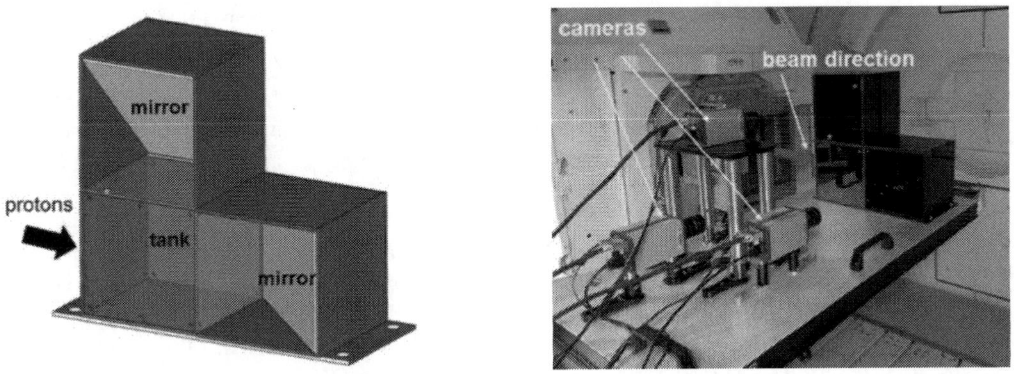

Figure 1. (a) Schematic of liquid scintillator (LS) detector fitted with two mirrors. **(b)** Picture of actual 3D system setup at proton therapy center at MD Anderson Cancer Center at Houston, TX.

All our experiments were conducted within the scanning beam gantry at the Proton Therapy Center at The University of Texas MD Anderson Cancer Center at Houston (PTCH). As seen in Figure 1.b the imaging setup was placed on the patient couch. The beam isocenter was then aligned with the center of the left tank surface. The gantry angle was set to 270°.

2.3 Spatial resolution
In order to accurately localize the beam position inside the tank volume it is critical to determine the effective image pixel size as a function of the distance from the lens. Therefore, using a checkerboard pattern we experimentally recorded the pixel size at three different distances from the lens: (i) the front

wall of the tank (650 mm), tank center (750 mm), and tank back wall (850 mm). We also measured the pixel size as a function of the distance from the image center to the lateral tank edge [5].

2.4 Range measurements of spot scanned proton beams

Proton ranges were estimated using a previously developed geometric calibration technique in our lab [3]. We measured the proton ranges at five different energies: 161.6 MeV, 144.9 MeV, 124 MeV, 100.9 MeV and 85.6 MeV using the liquid scintillator system. They were then compared with measurements from a parallel plate chamber that were collected as part of the beam commissioning, and referred hence forth as nominal measurements. A total of 5 MU dose was delivered for each energy and the images were captured with 100 msec camera integration times. For this study, the beam energies were selected such that the Bragg peaks from the scintillation light profiles fall within the tank. Higher energy beams with ranges greater than the tank dimensions can be measured by placing plastic water buildups in front of the tank side wall.

3. Results and Discussion

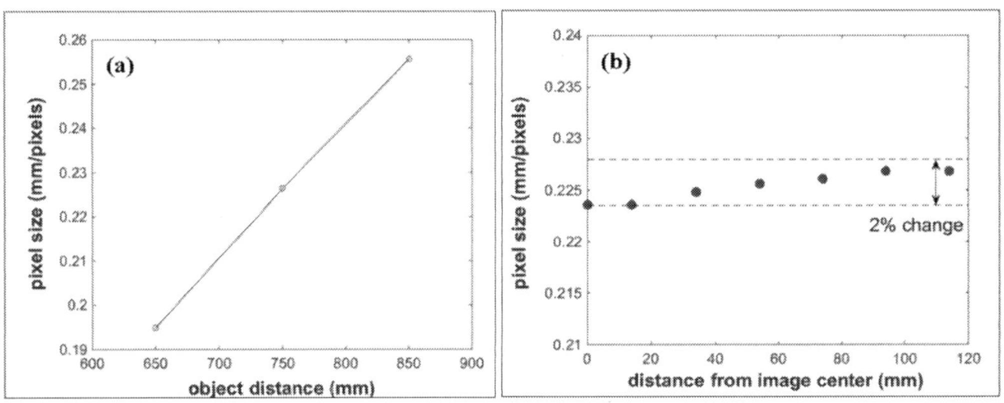

Figure 2. (a) Pixel size as a function of the imaging distance from camera. The solid line shows a linear fit for extracting pixel sizes for any intermediate imaging distances **(b)** Pixel size as a function of the lateral distance from the image center. The dotted line indicates 2% change in pixel size.

Figure 2.a represents change in pixel size (mm/pixel) as a function of the working distance from the camera. A linear fit through the three points can be used to calculate pixel size at any depth inside the tank. A sub-millimeter pixel size suggests that this system is capable of high-resolution measurements. It is therefore useful for comprehensive mapping of areas with steep-dose gradients and for generating 3D plots of dose distributions. Figure 2.b shows the change in pixel size from the image center to the periphery in the central plane of the tank. The dotted line indicates that this change is less than 2% over a distance of 100 mm.

Figure 3.a shows normalized light profiles of five energies delivered to the tank. Measuring beam range involves identifying a typical reference point on dose profile. At PTCH, beam range for scanning beams is defined at 90% of integral depth dose distal to the Bragg peak [6]. Proton studies involving liquid scintillators typically exhibit under-response near the Bragg peak known as ionization quenching. Thus, a point on dose distribution cannot be directly correlated with a point on the light distribution. In order to do so, previous study tested several reference points for minimizing the difference to the nominal range [5]. Our recent study using geometric calibration approach found that distal 90% to light peak provides the most accurate range measurements without having to correct for quenching [3]. Determining quenching is still a challenge and will need to be addressed for successful reconstruction of 3D dose. Table 1 summarizes the nominal and measured proton range showing an excellent agreement to within less than 0.3 mm for all five energies.

4. Conclusion

This preliminary study shows that generating longitudinal and lateral profiles of scanned proton beams with submillimeter resolution (0.23 mm) is now possible with this 3D detector. Post system calibration, proton range is calculated within a duration of few seconds and with a range estimation accuracy of 0.3 mm from the nominal value. The liquid scintillator-based detector can therefore be used for fast, high-resolution, and precise mapping of dose distributions in real-time and potentially for QA of scanning proton beams in 3D.

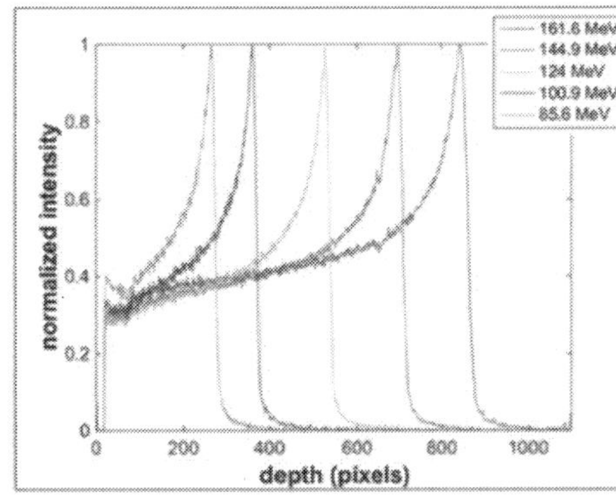

Figure 3. Normalized light profiles for five energies captured by the front camera.

Table 1. Comparison between nominal and measured proton range for five energies.

Beam Energy (MeV)	Nominal Range (mm)	Measured Range (mm)	Difference (ΔR) (mm)
161.6	175.988	175.879	-0.108
144.9	144.940	144.993	0.053
124	109.473	109.709	0.237
100.9	74.976	75.174	0.198
85.6	55.059	55.279	0.219

5. Acknowledgements

We would like to thank Fahed Alsanea and Dr. Narayan Sahoo for their contributions to this project. The research reported in this publication was supported by the National Cancer Institute of the National Institutes of Health under Award Number R01CA182450.

6. References

[1] Hill R *et al* 2009 *Med. Phys.* **36** 3971-81
[2] Baldock C *et al* 2010 *Phys. Med. Biol.* **55** R1-63
[3] Hui C K *et al* 2015 *Biomed. Phys. Eng. Express* **1** 025204
[4] Beddar S *et al* 2009 *Med. Phys.* **36** 1736-43
[5] Archambault L *et al* 2012 *Med. Phys.* **39** 1239-46
[6] Gillin M T *et al* 2010 *Med. Phys.* **37** 154-63

IOP Conf. Series: Journal of Physics: Conf. Series **847** (2017) 012021 doi:10.1088/1742-6596/847/1/012021

First 3D measurements of proton beams in a deformable silicone-based dosimeter

E M Høye[1], M Sadel[2], L Kaplan[2], P S Skyt[1], L P Muren[1], J B B Petersen[1], J Swakoń[3], G Mierzwińska[3], M Rydygier[3], L Malinowski[3] and P Balling[2]
[1]Department of Medical Physics, Aarhus University Hospital, Denmark
[2]Department of Physics and Astronomy, Aarhus University, Denmark
[3]Institute of Nuclear Physics, Polish Academy of Sciences, Poland

E-mail: elhoey@rm.dk

Abstract. 3D dosimetry provides high-resolution dose information of radiation therapy (RT), and is explored to enable and secure high-quality delivery of advanced RT modalities, including proton therapy. We present the first 3D measurements of spot-scanning proton plans in a silicone-based, radiochromic dosimeter with deformation properties. The dose information was read-out by optical CT-scanning. We found that the dosimeter signal was quenched close to the Bragg peak, and that this had a large impact on a measured spread-out Bragg peak. The dose response was linear both in the entrance region and in the Bragg peak, however, the dose response significantly reduced in the Bragg peak. Quenching was attributed to a linear-energy-transfer dependent dose response. Linear energy transfer distributions for each proton treatment plan will provide a means for calibrating the optical measurement to linear energy transfer, as well as dose. This might enable use of the silicone-dosimeter in quality assurance of proton beams.

1. Introduction

Quality assurance for proton therapy (PT) is a current challenge in modern radiotherapy (RT), due to the demanding physical beam properties. A mono-energetic proton depth-dose distribution consists of a Bragg peak; a sharp increase in dose, followed by a steep fall-off, as the protons stop in the irradiated medium. Most new centres employ spot scanning as the irradiation technique. A thin pencil beam of protons is scanned over the target, and by varying the energy, all target depths can be covered with a high dose. In conventional photon-based RT, 3D dosimetry [1, 2] has been used to obtain dose information with sub-millimetre-resolution. This technique would be highly desirable to employ in PT. Many dosimeters have been shown to be sensitive to linear energy transfer (LET), which increases sharply over the proton range [3]. In this study we have for the first time measured 3D proton dose distributions in a deformable silicone dosimeter [4-6], with special attention given to the quenching effects.

2. Material and Methods

Three 3D dosimeters were produced from a single batch, containing, in weight percentages, 5.1 % curing agent (CA) and 93.2 % silicone elastomer from a SYLGARD® 184 kit (Dow Corning) and 0.26 % leuco-malachite green (LMG) dye and 1.5 % chloroform (Sigma-Aldrich). Cone shaped, 11 cm high plastic moulds were used for curing, and could easily be removed after three days storing at room temperature in complete darkness. The dosimeters were fixed in the Vista 10 (Modus Medical Devices Inc.) optical-computed-tomography (CT) scanner and immersed in refractive-index-matching liquid. The highest possible number of 2D projections was used (1024). Read-outs were conducted three days prior to irradiation (pre-scan) and three days after irradiation (post-scan).

Table 1. List of 3D dosimeters irradiated with proton plans.

Name	Irradiation plan	# of deliveries	Total maximum dose
A	Single spot	3	35.22 Gy
B	SOBP	2	30.00 Gy
C	SOBP	4	60.45 Gy

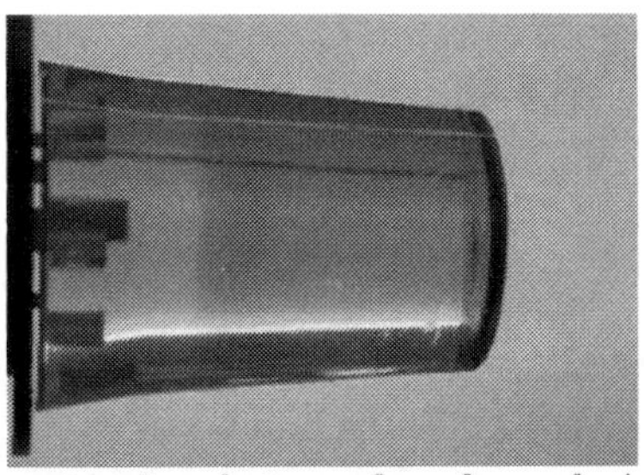

Figure 1. A photograph of a dosimeter, irradiated with an SOBP-field, entering from the right.

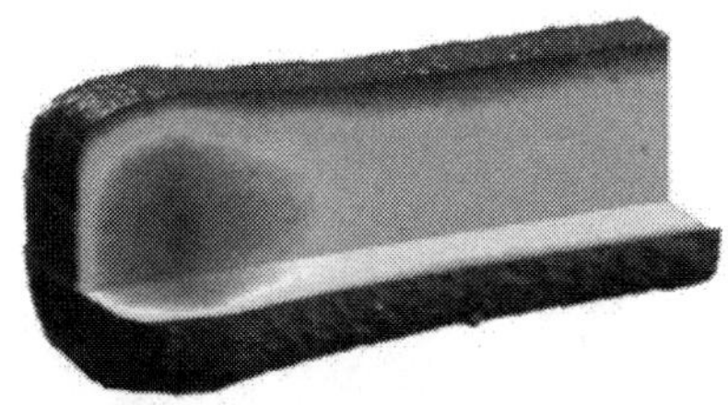

Figure 2. 3D view of the optical measurement of dosimeter C. Protons entered from the right. The optical density is scaled from 0.4 cm-1 (blue) to 0.8 cm-1 (red).

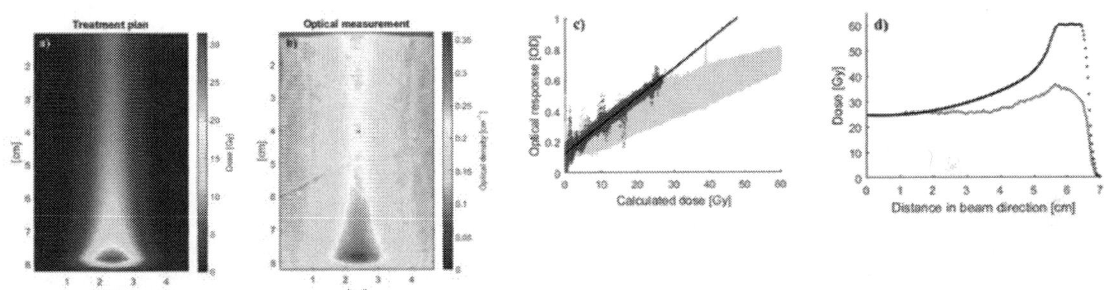

Figure 3. Interpolated planned doses (a) and optical measurement (b) for dosimeter A, irradiated with a single spot, entering from above, and delivering the maximum dose at 7.4 cm is shown. In (c), a correlation plot between individual voxels of the planned dose and the optical response is shown. The blue points correspond to a volume 1.25 cm to 2.2 cm into the entrance region with a planned dose above 1 % of the maximum dose. The black solid line is a linear fit to these points. The green points correspond to the remaining points in the plan. In (d) a projection through the central pixels of the plan (black dotted line) and the optical measurement (red solid line) is shown. The optical measurement was calibrated to the dose response in the plateau.

Dosimeters were irradiated at Institute of Nuclear Physics, Polish Academy of Sciences, six days after production, at a gantry facility with spot scanning (230 MeV cyclotron). Treatment plans were generated in Eclipse™ (Varian Medical Systems, Palo Alto, USA) using a water phantom as target, as CT-scanning was not available. Two plans were made: a single spot with 3 Gy at 2 cm depth and 8 cm range (90 % of distal dose fall-off) and a spread-out Bragg peak (SOBP) shaped field, delivered from one angle, with 15 Gy in the SOBP plateau and 6.56 cm range. Each plan was delivered several times to each dosimeter, with the number of fractions given in table 1. Dosimeters were read out at Aarhus University two days after irradiation, and were placed in a refrigerator for these two days, in order to avoid fading of the signal. The data were reconstructed using the Vista 3-D Reconstruction program (Modus Medical Devices Inc.), with a voxel resolution of 0.25 mm. Following this, an analysis similar to that described in [7] was undertaken. Spatial matching was performed by shifting the overlap between measurement and plan in three dimensions, and minimizing the standard deviation. The planned (i.e. calculated) and measured doses were weighted by their maximum values, in order to minimize sensitivity to quenching of the high-dose volume. Furthermore, the optical measurements were adjusted for a water-equivalent thickness of 0.93 g/cm^3 in the direction of beam entry.

3. Results

The single-spot plan (figure 3) gave a dose response in the entrance region of $(12.4 \pm 0.1) \times 10^{-3}$ cm^{-1} Gy^{-1} and a background of $(132.1 \pm 0.5) \times 10^{-3}$ cm^{-1} (see linear fit in figure 3 (c)). The signal in the measured Bragg peak was 41 % lower than the planned dose, when calibrated with the dose response.

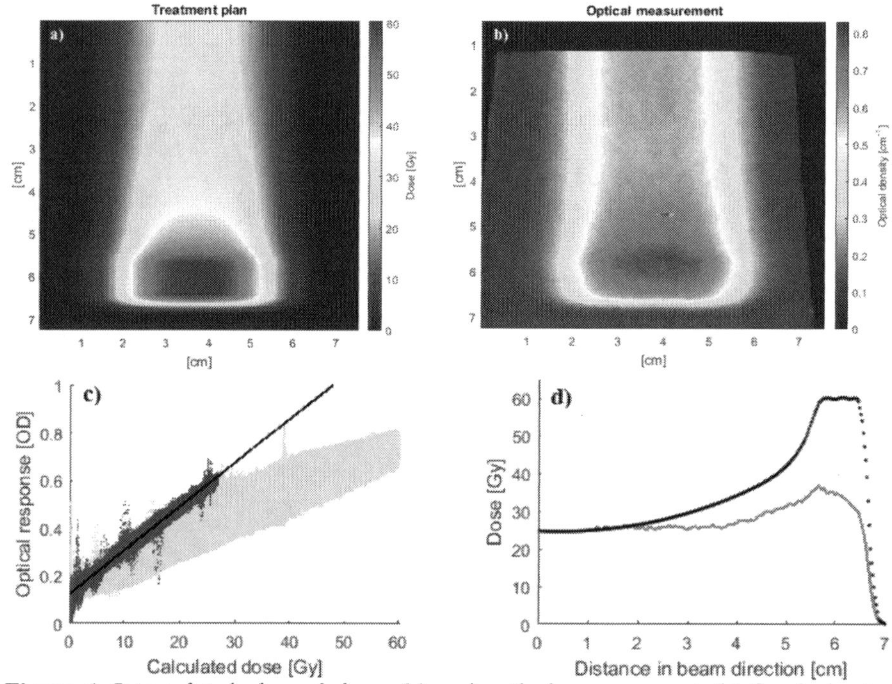

Figure 4. Interpolated planned doses (a) and optical measurement (b) for dosimeter C, irradiated with a SOBP entering from above. Correlation plot between plan and measurement is shown in (c), and (d) shows a projection through the central pixels of the plan and the measurement (see figure 3 for further details).

The SOBP plan delivered to dosimeter C had a dose response in the entrance region of $(18.33 \pm 0.04) \times 10^{-3}$ cm^{-1} Gy^{-1}, with a background of $(123.6 \pm 0.6) \times 10^{-3}$ cm^{-1}. The signal in the beginning of the SOBP plateau was 38 % lower than the planned dose, when calibrated with the dose response (figure 4 (d)), and this fell to 49 % at the distal end of the plateau. The same plan, but with half the dose was delivered to dosimeter B, and this gave a dose response in the entrance region of $(17.94 \pm 0.07) \times 10^{-3}$ cm^{-1} Gy^{-1}, and the background was $(120.3 \pm 0.5) \times 10^{-3}$ cm^{-1}.

4. Discussion

In this study, we have presented the first 3D measurements of a silicone-based, deformable, radiochromic dosimeter, irradiated with spot-scanning proton beams. Measurements of proton beams were clearly feasible, but the dose response was significantly influenced by quenching, i.e. it decreased in the high dose-volumes of the plans, compared to the entrance region. The high-dose regions are associated with a high LET, which is known to inhibit the dose response [8]. Two dosimeters (B and C) were irradiated with the same plan, but to different maximum doses. As expected, these had almost the same dose response in the entrance region. The single-spot dosimeter had a lower dose response, probably due to the different LET-distribution of the beam.

In figure 3(d), the effect of quenching is apparent as a lowering of the signal in the Bragg peak of approximately 41 %. This has a significant effect on the measured SOBP distribution in figure 4(d), where the effects seen for a single Bragg peak accumulate: the plateau region is very shallow, has a maximum value close to the beginning of the plateau, and a tilting plateau region. This is similar to what has been observed for other 3D dosimeters [9].

The correlation plots are much broader for the SOBP measurements than for the single-spot measurement. The larger physical size of the SOBP plan could affect this, but it is likely to arise from the large spread in LET that each dose level represents. The SOBP plan consists of many single spots with different energies, all with an LET that increases steeply towards the end-of-range. Thus several LET levels are found within each dose level.

The correlation plots show that low-dose regions are more sensitive to noise than the high dose regions. In general, noise arises where particles or dust block the light in the optical CT-scan, from imperfect refractive index matching, from suboptimal positioning from pre- to postscan and from artefacts created in the reconstruction.

In a recent study of cuvette size dosimeters, we have shown that the concentration of curing agent greatly affects the quenching in the dosimeter [8]. In the current study, 5 % CA was used, due to the improved temporal stability of the dose response. However, increasing the CA to 9 or even 11 % CA will give a less LET-sensitive dosimeter, albeit at the cost of dose response and temporal stability [5]. Chemical alterations in the Fricke gel dosimeter have also been shown to influence the LET dependence [10]. 5 % CA is furthermore associated with a considerable initial background colour, which limits the dynamical range available in the optical CT-scanner. This study shows that the Vista scanner was successful in reading out these dosimeters.

5. Conclusion

In this study we have performed the first 3D measurements with silicone-based, deformable, radiochromic dosimeters in proton beams. Quenching of the signal, related to an LET-dependent dose response, was observed in all dosimeters. By calculating the LET-distribution in the dosimeter, quenching can possibly be circumvented. Calculations of the expected optical response can be computed, and this can be used in quality assurance to compare with the optically measured 3D distribution.

6. Acknowledgements

This research project was supported by the Danish Cancer Society, award numbers R72-A4486-13-S2.

7. References

[1] Doran S J *et al* 2015 *J. Phys.: Conf. Ser.* **573** 012043
[2] Baldock C *et al* 2010 *Phys. Med. Biol.* **55** R1-63
[3] Doran S *et al* 2015 *Phys. Med. Biol.* **60** 709-26
[4] Høye E M *et al* 2015 *J. Phys.: Conf. Ser.* **573** 012067
[5] Høye E M *et al* 2015 *Phys. Med. Biol.* ***60*** 5557-70
[6] De Deene Y *et al* 2015 *Phys. Med. Biol.* **60** 1543-63
[7] Skyt P S *et al* 2013 *Acta Oncol.* **52** 1445-50
[8] Høye E M *et al* 2016 *Phys. Med. Biol.* **62** N73-89
[9] Heufelder J *et al* 2003*Med. Phys.* **30** 1235
[10] Maeyama T *et al* 2014 *Radiat. Phys. Chem.* **96** 92-96

IOP Conf. Series: Journal of Physics: Conf. Series **847** (2017) 012022 doi:10.1088/1742-6596/847/1/012022

Evaluating analytical ionization quenching correction models for 3D liquid organic scintillator detector

F Alsanea and S Beddar
Department of Radiation Physics, University of Texas MD Anderson Cancer, Houston, TX

E-mail: asbeddar@mdanderson.org

Abstract. Proton therapy offers dosimetric advantage over conventional photon therapy due to the finite range of the proton beam, which improves dose conformity. However, one of the main challenges of proton beam therapy is verification of the complex treatment plans delivered to a patient. Thus, 3D measurements are needed to verify the complex dose distribution. A 3D organic scintillator detector is capable of such measurements. However, organic scintillators exhibit a non-linear relation to the ionization density called ionization quenching. The ionization quenching phenomenon in organic scintillators must be accounted for to obtain accurate dose measurements. We investigated the energy deposition by secondary electrons (EDSE) model to explain ionization quenching in 3D liquid organic scintillator when exposed to proton beams. The EDSE model was applied to volumetric scintillation measurement of proton pencil beam with energies of 85.6, 100.9, 144.9 and 161.9 MeV. The quenching parameter in EDSE model ρ_q was determined by plotting the total light output vs the initial energy of the ion. The results were compared to the Birks semi-empirical formula of scintillation light emission.

1. Introduction

Proton therapy, in particular spot scanning proton therapy, introduced complex treatment plans with conformal dose distribution. This required 3D detectors that are capable of recording the dose deposition in 3D for quality assurance and dose verification [1]. Volumetric organic scintillators are detectors that convert the dose deposited in the scintillator to light, which is captured using CCD cameras [2-4]. Multiple views of the detector are used to reconstruct the 3D dose distribution [5]. However, organic scintillators exhibit a nonlinear response to the ionization density, called the ionization quenching, when exposed to heavy charged particles [6]. Ionization quenching is problematic in proton beams; where the linear energy transfer (LET) changes as a function of depth and increases rapidly at the Bragg peak region. Therefore, a quenching correction factor is needed to obtain accurate dose distribution. The Birks model is a popular and widely used semi-analytical model that can correct for ionization quenching [6]. Others tried to explain the ionization quenching phenomenon by accounting for the difference in the incident charged particle atomic number and not relying on the average LET values [7, 8]. We investigated the EDSE model by Michaelian and Menchaca-Rocha for use in correcting ionization quenching in 3D liquid organic scintillator detectors [7].

2. Methodology

2.1. EDSE Model

The energy deposition by secondary electrons (EDSE) model relates the ionization quenching to the deposition of energy by the secondary electrons along the track of the ion beam. The realization of the EDSE model is based on the difference in the spectrum of secondary electrons produced by ions of different atomic number. The electron energy density as a function of radial distance is given as [7]:

$$\rho(r) = N \frac{e^4}{nm_e} \frac{z^{*2}}{V^2} \frac{1}{r^2} \left[1 - \frac{r}{R_{\max}}\right]^{d+\frac{1}{n}} \quad (1)$$

where N is the number of electrons per unit volume in the medium, m_e and e are the electron mass and charge, n is the exponent for approximating the electron range, z^* and V are the effective ion charge and velocity, R_{max} is maximum range of the secondary electron, and d=0.045Z_{eff}, where Z_{eff} is the effective atomic number of the medium. The energy carrier density per unit path length of the ion is then given by [7]:

$$\frac{dN_e}{dx} = K\left[\int_{b_{min}}^{R_{max}} \frac{\rho(r)}{1+\rho(r)/\rho_q} dr\right] \quad (2)$$

where K is a constant relating the energy deposited to the number of energy carriers. Later, K is absorbed in C, a constant incorporating the scintillators efficiency. The term b_{min} is the minimum impact parameter of the electron, and ρ_q is the quenching parameter that is specific to the scintillator detector. Finally the luminescence in a given region is given by: $\frac{dS}{dx} = C\frac{dN_e}{dx}$.

The EDSE model parameters ρ_q and C are experimentally determined by plotting the total light output vs the energy of the ion.

2.2. Birks Model

The Birks model assumes that the scintillator light emission due to ionizing radiation is dependent on the stopping power (*dE/dx*) of the medium. Therefore, the scintillation light is expressed as [6]:

$$\frac{dS}{dx} = \frac{A\frac{dE}{dx}}{\left(1+kB\frac{dE}{dx}\right)} \quad (3)$$

where *dS/dx* is the luminescence yield per unit length, *A* is the light production efficiency, and *kB* is the quenching factor. The term (*BdE/dx*) is the quantity of damaged molecules and *k* is the probability that a damaged molecule will capture an exciton. The values *A* and *kB* are determined experimentally for the type of the scintillator and ion beam used.

2.3. Experimental data

We used previously published data of a volumetric liquid organic scintillator tank exposed to proton beams [3]. The light scintillation was collected using a CCD camera. The liquid scintillator used was BC-531 (Saint Gobain), which consisted of fluorescing molecules in linear alkyl benzene solvent (density 0.87 g cm^{-3}). We also used the accompanied validated Monte Carlo simulation (MCNPX, version2.7d [9]) of the dose, fluence, and track averaged LET [3].

2.4. Quenching parameter determination

A least square fitting of the EDSE model to the CCD data was used to extract the quenching parameter ρ_q for pristine proton beams with energies of 85.6, 100.9, 144.9 and 161.9 MeV that spans the range of energies used clinically. Pixels along the central axis of the depth dose curve were fitted individually. We used a Monte Carlo tabulated stopping powers of the liquid scintillator to determine the incident ions energy and the energy loss as a function of depth. Then, we plotted the ratio between the calculated light (S_{EDSE}) to measured light (S_{CCD}). We compared the EDSE model fit to the published results that used the Birks model [3].

3. Results and Discussion

In Figure 1, we show the secondary electron deposited energy density as a function of the radial distance (equation 1) for a single 98.6 MeV proton. Also plotted in figure 1, is the electron deposited energy density for the quenched part of the scintillator (equation 2). The principal assumption in the EDSE model is that the ionization quenching in a scintillator happens at regions very close to the ion's track.

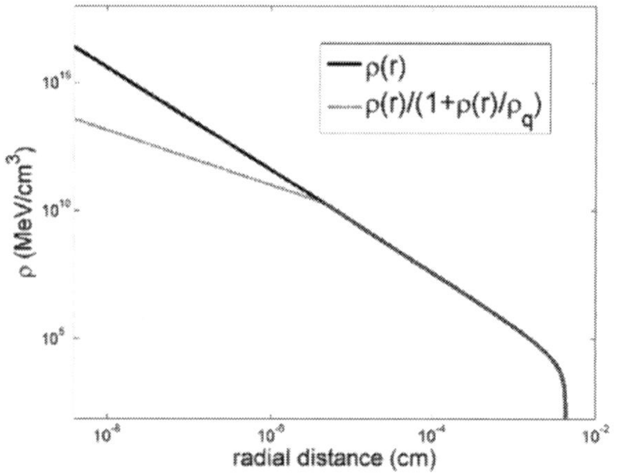

Figure 1. Electron energy deposition density per unit path length of the incident ion beam (proton 98.6 MeV) as a function of the radial distance from the ion's track (solid line). The EDSE model assumes a region close to the ion's track where the scintillator quenches (red dotted line).

We assumed a mono-energetic proton beam to calculate the residual energy of the beam, which caused unrealistic sharp distal edges. Thus, the fit, in particular at the distal edge beyond the Bragg peak, does not match very well. However, the percent difference between calculated and measured light at the Bragg peak was 0.42% as shown in Figure 2. The percent difference between calculated and measured light at the proximal region of the proton beam was within less than ±5%. This is demonstrated in figure 2, which also plots the Monte Carlo depth dose curve relative to the entrance. The EDSE quenching parameter ρ_q was equal to 3.85×10^{13} MeV/cm^3.

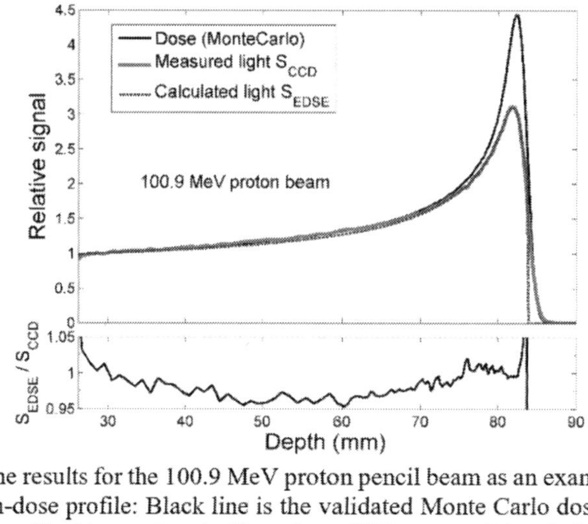

Figure 2. Illustrated is the results for the 100.9 MeV proton pencil beam as an example for one of the 4 energies (Top) Central axis depth-dose profile: Black line is the validated Monte Carlo dose simulation. Red line is the measured light emitted in a liquid organic scintillator by a CCD camera. Blue dashed line is the calculated light by the EDSE model. (Bottom) The ratio between the calculated light and measured light.

IOP Conf. Series: Journal of Physics: Conf. Series **847** (2017) 012022 doi:10.1088/1742-6596/847/1/012022

The usefulness of the EDSE model over the Birks model for calibrating 3D detectors needs further investigations, which are currently being implemented by including other parameters that account for the energy spectrum and range straggling of the proton beam. Therefore, the EDSE model could be applied to single spot proton beams and passive scatter proton beams. Once these parametrizations are determined, a quenching correction factor can be calculated using the ratio of the deposited energy to the emitted light per voxel. Robertson et al used one energy to determine the quenching factor in Birks model [3]. Then they applied it to the four different beam energies. The corrected light compared to the Monte Carlo dose was within ±3% for all energies except for the lowest energy (±10%) [3]. These results could be due the shape of the beam (sharper Bragg peak curve for low proton beam energies). Accounting for the spectrum of energies per unit path length of the proton beam (i.e. not relying on an average LET value) should result in better fits for both Birks and EDSE quenching models.

4. Conclusion

We have explored the ionization quenching phenomenon in organic scintillators by using semi-analytical models like the Birks and EDSE model to correct for ionization quenching. The EDSE model was formulated to calibrate detectors for energy and particle detection. An advantage of the EDSE model is that it will account for the difference of the ion beam's atomic number. The EDSE model can accurately predict the light emission at the Bragg peak, however further parametrization is needed to account for the energy spectrum and generate better fits for the distal edge region.

5. Acknowledgments

The authors would like to acknowledge Dr. Daniel Robertson for sharing and explaining the published scintillation data, Dr. Dragan Mirkovic, and Dr. Gabriel Sawakuchi for help with Monte Carlo simulations. The research reported in this publication was supported by the National Cancer Institute of the National Institutes of Health under Award Number R01CA182450. The content is solely the responsibility of the authors and does not necessarily represent the official views of the National Institutes of Health.

6. References

[1] Baldock C *et al* 2010 *Phys. Med. Biol.* **55** R1-63
[2] Beddar S *et al* 2009 *Med. Phys.* **36** 1736-43
[3] Robertson D *et al* 2013 *Phys. Med. Biol.* **58** 261
[4] Pönisch F *et al* 2009 *Med. Phys.* **36** 1478-85
[5] Hui C *et al* 2014 *Phys. Med. Biol.* **59** 4477
[6] Birks J B 1964 The Theory and Practice of Scintillation Counting (Pergamon)
[7] Michaelian K and Menchaca-Rocha A 1994 *Phys. Rev. B* **49** 15550-62
[8] Luntz M and Heymsfield G M 1972 *Phys. Rev. B* **6** 2530-6
[9] Waters L S *et al* 2002 Monte Carlo N-particle transport code system for multiparticle and high energy applications, Los Alamos National Laboratory, Los Alamos, NM, USA

IOP Conf. Series: Journal of Physics: Conf. Series **847** (2017) 012023 doi:10.1088/1742-6596/847/1/012023

Dynamic 3D measurement of modulated radiotherapy: a scintillator-based approach

Louis Archambault, Madison Rilling, Xavier Roy-Pomerleau, Simon Thibault
Département de Physique, de Génie Physique et d'Optique, Université Laval, Quebec, Canada

E-mail: louis.archambault@phy.ulaval.ca

Abstract. With the rise of high-conformity dynamic radiotherapy, such as volumetric modulated arc therapy and robotic radiosurgery, the temporal dimension of dose measurement is becoming increasingly important. It must be possible to tell both 'where' and 'when' a discrepancy occurs between the plan and its delivery. A 3D scintillation-based dosimetry system could be ideal for such a thorough, end-to-end verification; however, the challenge lies in retrieving the volumetric information of the light-emitting volume. This paper discusses the motivation, from an optics point of view, of using the images acquired with a plenoptic camera, or light field imager, of an irradiated plastic scintillator volume to reconstruct the delivered 3D dose distribution. Current work focuses on the optimization of the optical design as well as the data processing that is involved in the ongoing development of a clinically viable, second generation dosimetry system.

1. Introduction

Three-dimensional mapping of photon-based radiation therapy (RT) dose delivery is increasingly relevant in the era of small tumor targets and high conformity treatments [1]. To assure treatment quality, image guidance can be used to verify that a patient is in the right position and that his or her anatomy has not undergone important changes. However, patient imaging is only a subset of the whole treatment quality monitoring. High conformity RT delivered dynamically can often be strenuous for the treatment machine. It is therefore important to have instruments capable of verifying that a linear accelerator can deliver treatments exactly as planned. Detectors used for this purpose usually share similar properties: a large number of sampling points to map rapidly planar or volumetric doses, a good spatial resolution to track high dose gradients, and a compatibility with some type of phantom to measure dose as it would be received by the patient. An ideal detector would be capable of testing the whole treatment chain from end to end (i.e., from planning to delivery) with high accuracy and high precision. Moreover, the temporal capability of such a detector would enable an identification of the source of any potential deviations. Dosimetric gels have been used with success for this end-to-end task, partly because they are one of the only dosimeters offering complete 3D dose mapping [2]. However, gels can be labor-intensive and they do not yet offer dynamic measurements. Scintillation-based 3D dosimetry has so far shown promising results. Scintillators are water-equivalent and emit light spontaneously (within a few tens of nanoseconds). Feasibility of 3D tomographic reconstruction of scintillation measurements has been demonstrated [3], but it initially required a long acquisition time and a rotating setup. Liquid scintillator-based systems for proton therapy yielded promising results [4], but they exploit the unique physics of proton dose deposition, which cannot be used with

IOP Conf. Series: Journal of Physics: Conf. Series **847** (2017) 012023 doi:10.1088/1742-6596/847/1/012023

photon-based treatments. To image a dynamic treatment, it is preferable to have a static acquisition system to avoid interplay effects. Acquiring a large number of projections required for tomographic reconstruction without rotation is not trivial, but it is possible [5]. The conception of an ideal acquisition system is challenging at best. There is an intrinsic tug-of-war between optimizing the system's spatial resolution, its temporal resolution and its light collection efficiency. Furthermore, the optimal tomographic reconstruction strategy greatly depends on the number of projections and their angles, which are determined by the system's optical design.

While relatively new in medical physics, reconstructing 3D light-emitting objects is actively pursued in combustion research. Several methods for the reconstruction of light patterns from flames and chemiluminescence processes have been proposed [6, 7]. There are several similarities between the reconstruction of a flame and the reconstruction of the light field emitted by a scintillator. Thus, it is possible to use advances in flame/chemoluminescence reconstruction to help optimize a new 3D dosimeter. However, to conceive a system that is clinically viable, both the optical design and data processing (i.e., reconstruction algorithms) must be carefully planned.

2. A 3D scintillator-based prototype using a single plenoptic camera

Over the last decade, plenoptic cameras, or *light field imagers*, have been brought forth as a solution for "3D imaging" of solid objects, namely for digital refocussing, depth measurements and post-acquisition point-of-view adjustment. In essence, the plenoptic camera is simply a conventional digital camera to which a microlens array is added in front of its active sensor (CCD or CMOS). This additional optical component allows a measurement of an imaged light-field, i.e., the 4D function describing the angular distribution of light incident on each pixel of the camera's detector. The microlenses subdivide the rays incident on the camera's main lens, thus generating an array of micro-images on the sensor. Consequently, the signal detected on a single pixel results from the integration of rays incident from within a finite range of angles, which allows for a partial recovery of the angular information lost in the case of a conventional camera. The amount of angular information recovered depends mainly on the placement of the microlens array with respect to the main lens. When coincident with the main lens' focal plane, maximum angular resolution can be attained at the cost of low spatial resolution (plenoptic 1.0). A spatio-angular trade-off is possible when the microlens array is offset from this plane, thus relaying the image formed by the main lens onto the sensor (plenoptic 2.0). A further alteration of the plenoptic 2.0, or focused plenoptic camera, is that of interleaving microlenses of different focal lengths within the microlens array, resulting in what is known as the multifocus plenoptic camera [8, 9]. This type of plenoptic camera serves to increase the usable depth-of-field (DOF), meaning that at least one microlens type is focused for any given imaged point in object space. Usually, increasing the DOF of a camera means reducing the size of the aperture stop of the system, thus involving systems operating at higher f-numbers (f/#). However, the build of the multifocus plenoptic camera allows the system to be used at higher speeds, i.e. lower f/#'s, thus increasing the light collection efficiency of the system and the spatial resolution of the system [8].

In the current 3D scintillation-based dosimetry system, the directionally-weighted information of the scintillation light field acquired by a multifocus plenoptic camera forms the basic input to an iterative tomographic reconstruction algorithm: based on paraxial optics, light rays are propagated from the sensor pixels to the voxelized plastic scintillator volume, first intersecting the microlens array then the main lens. In its current proof-of-concept state [10], the prototype, shown in Fig. 1., uses a plastic scintillator volume of $10\times10\times10$ cm^3 (490 nm emission peak) both as the radiosensitive and phantom material. The cubic scintillator is embedded within cylindrical acrylic plates, allowing for a better angular uniformity during dynamic irradiations (Fig. 1c). The scintillator is imaged with a static, 3-focal length polychromatic multifocus plenoptic camera, used with a 50-mm main lens operated at a f/# of 2.8. The current acquisition rate is of 1 Hz, set as a compromise between the light collection efficiency and temporal resolution of the system. Sampled planes from an absolute 3D dose reconstruction of a lung SBRT treatment are shown in Fig. 2.

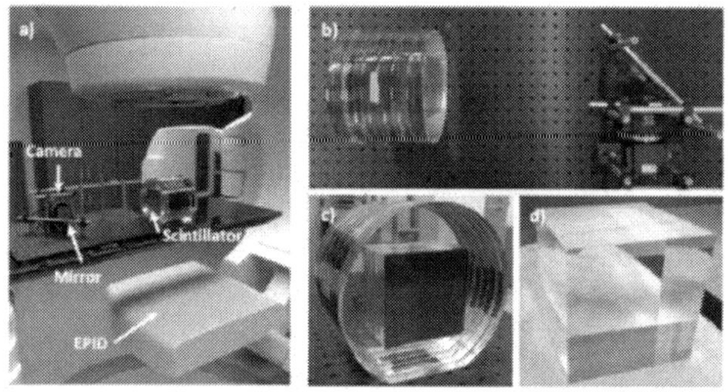

Figure 1. a) 3D detector prototype on a linac's treatment table; b) Top view of the prototype; c) Cubic plastic scintillator embedded in cylindrical acrylic plates; d) Cubic plastic scintillator volume.

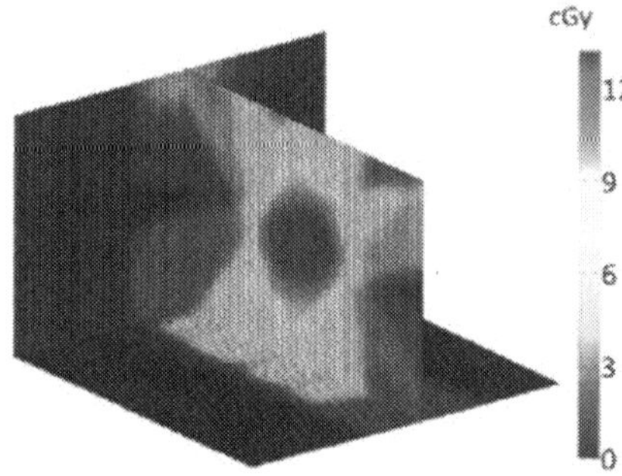

Figure 2. Sampled planes of a measurement (2-mm voxel resolution) reconstructed from of a lung SBRT treatment.

2.1. Towards a second generation prototype: optimizing the system's optics

Improving the dose reconstructions relies on a better knowledge and subsequent optimization of all optical parameters involved in the ray tracing, namely the diameter and focal lengths of the plenoptic camera's main lens and microlenses, as well as the spacing between each component. These parameters influence not only the accuracy of the ray tracing based on the recovered angular information, but they affect the amount of vignetting present in separate micro-images, which limits the number of pixels usable for the reconstruction. Since these optical parameters are difficult to alter and study in the lab, simulations were performed using Zemax (OptiStudio™ 15, Zemax, LLC), a ray tracing and optical design software. Baseline parameters of the current system were first determined by studying the plenoptic camera with an optical microscope and an optical test instrument (OptiSpheric®, TRIOPTICS). These measured values were used as input parameters to reproduce the optical system in Zemax. Simulated images of targets of varying spatial resolution at different camera-to-target distances were compared with plenoptic images acquired in the lab. Using a sequential ray tracing mode with paraxial lenses, and as shown in Fig. 3, it was possible to model appropriately the change in spatial resolution as a function of depth without any lens vignetting effects or noise introduced by the optical system. These simulations form the basis needed for studying the effect of varying different plenoptic parameters.

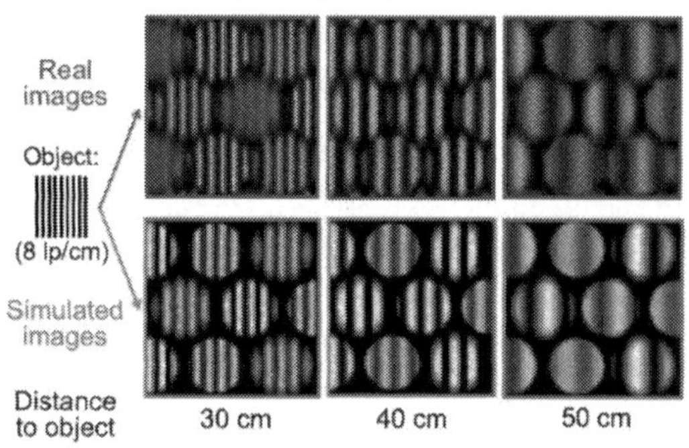

Figure 3. Real plenoptic images vs those simulated in Zemax for an imaged spatial frequency of 8 lp/cm at different camera-to-object distances.

It must be highlighted that the limit to using the plenoptic camera as a solution for 3D imaging is the lack of spatial resolution along the camera's optical axis. From a tomography point of view, it is as if the plenoptic camera acquisition provides a partial sinogram, which needs to be compensated for using some perpendicular acquisition of the radiation beam's eye view (BEV). Images acquired with the electronic portal imaging device (EPID) during dynamic deliveries were first used to provide a constraint on the depth to which the rays were backprojected into the scintillating volume [10]. This solution was initially motivated by the ease-of-use and availability of the EPID, and had the potential to be better exploited in a future prototype. However, in optimizing a 3D dosimetric system for dynamic modulated radiotherapy, the ideal is to provide a fully independent system, making the EPID-based solution imperfect in this perspective. In continuing the optical simulations and optimization of the plenoptic system, it will be important to understand fully the limit of using a plenoptic camera for 3D imaging of translucent media. Knowing this limit will allow for a better consideration of other optical-based solutions aimed at compensating the plenoptic camera's lack of spatial depth resolution.

3. Data processing involved in reconstructing 3D dose distributions

The use of a plenoptic camera to retrieve the 3D light pattern emitted from within a plastic scintillator volume calls for reconstruction algorithms intimately related to those used in single-photon emission computed tomography (SPECT). Various reconstruction methods could have been used, namely analytical algorithms (e.g., filtered backprojection) or iterative algorithms, of which the most popular are algebraic reconstruction techniques (ARTs), maximum-likelihood (ML) and expectation maximization (EM). For the plenoptic image-based reconstruction, an iterative algorithm was selected considering that analytical algorithms, albeit quicker, rely on a large number of projections to attain adequate precision. As for the different iterative algorithms, ARTs were discarded due to the fact that they do not, in their basic form, account for statistical error related to projection measurements. A ML-EM algorithm was retained for the task as it allowed for noise and statistical error modeling as well as incorporation of known physical constraints [11, 12]. The algorithm was implemented in Python because of the language's versatility and its ease of development. However, because of relatively long reconstruction times, other alternatives will need to be explored such as porting the most time-consuming subroutines to a low-level programming language (e.g., C/C++) or to more suitable processing architecture such as graphical processing unit (GPU) using a library such as PyCUDA.

The reconstruction can be divided into four main blocks. (1) All plenoptic images acquired of the scintillating volume during a dynamic beam delivery are imported and filtered. (2) The second block of the algorithm calculates the system projection matrix used in the tomographic reconstruction algorithm. Depending on the scintillator volume, the chosen voxel resolution and the number of pixels sampled from the sensor, the sparse matrix representing the directionally-weighted propagation of rays through the plenoptic system can easily saturate the computer's memory. To prevent memory saturation and to accelerate computation of the matrix, this step is divided into subfunctions of smaller extent, which can then easily be run in parallel using a default multiprocessing package. (3) A BEV constraint is produced from 2D EPID images acquired in cine mode during radiation delivery. For each plenoptic acquisition, a corresponding 3D matrix is generated by scaling the 2D EPID image along the beam's axis. This matrix serves to optimize the shape of the reconstructed 3D dose slice-by-slice based on a Gaussian convolution. (4) Finally, the last block of code takes the outputs of the preceding blocks and iteratively reconstructs the 3D dose distribution corresponding to each plenoptic acquisition. In addition to improving computation speed, future development will include using additional physical constraints to better guide the reconstruction process.

4. Conclusion

With the rise of high-conformity dynamic RT, such as volumetric modulated arc therapy (VMAT) and robotic radiosurgery, the temporal dimension is becoming increasingly important. It must be possible to tell both 'where' and 'when' a discrepancy occurs between the plan and its delivery. A 3D scintillation dosimeter could be ideal for such an end-to-end verification, whereby sources of deviation between planning and delivery could be identified. However, careful optimization of the optical design as well as the data processing is required to have a clinically viable system.

5. References

[1] Baldock C *et al* 2010 *Phys. Med. Biol.* **55** R1-63
[2] Mather M L *et al* 2003 *Ultrasonics* **41** 551-9
[3] Kirov A S *et al* 2005 *Phys. Med. Biol.* **50** 3063-81
[4] Beddar S *et al* 2009 *Med. Phys.* **36** 1736
[5] Kroll F *et al* 2013 *Med. Phys.* **40** 82104
[6] Gilabert G *et al* 2007 *IEEE Trans. Inst. Meas.* **56** 1300-06

[7] Sun, J *et al* 2016 *Opt. Exp.* **24** 1417-25
[8] Georgiev T *et al* 2012 *Proc. SPIE* **8299** 829908
[9] Perwass C and Wietzke L 2012 *Proc. SPIE* **8291** 829108
[10] Goulet *et al* 2014 *Med. Phys.* **41** 082101
[11] Bruyant P 2002 *J. Nuc. Med.* **43** 1343
[12] Shepp L A and Vardi Y 1982 *IEEE Trans. Med. Imaging* **1** 113

IOP Conf. Series: Journal of Physics: Conf. Series **847** (2017) 012024 doi:10.1088/1742-6596/847/1/012024

EPID-based dosimetry and its relation to other 2D and 3D dose measurement techniques in radiation therapy

B Mijnheer
Department of Radiation Oncology, The Netherlands Cancer Institute, Plesmanlaan 121, 1066 CX Amsterdam, The Netherlands

E-mail: b.mijnheer@nki.nl

Abstract. In this paper I will summarize the possibilities and limitations of different 2D and 3D dosimetry techniques used in radiation therapy, and evaluate these features relative to those of EPID-based techniques. After briefly discussing their characteristics, I will review the use of EPIDs for pre-treatment and *in vivo* dosimetry applications by separating them into transit and non-transit approaches, analysed by either forward- or backward-projection methods. I will then review the various types of 3D dosimetry systems by categorizing them into semi-3D, pseudo-3D, to which EPID-based back-projection approaches belong, and full-3D systems. All methods can in principle be used for pre-treatment 3D dose verification; the choice of a specific system depends on the aim of the measurement and the properties of the specific hard- and software. At this moment EPIDs are the only tools available for 3D *in vivo* dosimetry. I will conclude with revealing some trends and future developments in 3D pre-treatment and *in vivo* dosimetry.

1. Introduction

The success of a radiotherapeutic treatment is dependent on the accuracy of the planning and delivery of the 3D dose distribution to a target volume and the surrounding critical organs in a specific patient. 3D dosimetry therefore plays a vital role in radiation therapy [1]. Although it is difficult to generalize, an accuracy in the dose delivery, i.e., a difference between the actual and "true" dose, of the order of a few percent, is often required. For instance, in ICRU Report 83 [2] it is stated that the difference between measured and planned dose should be within 3.5% (1SD) and the distance to agreement within 3.5 mm for low and high dose gradient regions, respectively. As a consequence, the accuracy of all components involved in a radiotherapeutic procedure need a very high accuracy as elucidated and discussed in detail in an upcoming IAEA report [3].

Starting with the acceptance testing of a treatment unit and a treatment planning system (TPS), specific measurements are required for the commissioning and quality assurance of that unit and TPS, as well as for the verification of special treatment techniques. These dose determinations were, and still are, often performed using point dose or 1D measurements, e.g., for the determination of output factors, beam profiles and depth dose distributions [4]. The introduction of advanced treatment techniques such as intensity modulated radiation therapy (IMRT) and volumetric modulated arc therapy (VMAT), both often using small fields, did not change the beam modelling process of the 3D dose calculation algorithms performed by a TPS for these advanced irradiation techniques. It is still based on data obtained by 0D and 1D measurements under reference conditions, even for small ($\leq$ 3 cm x 3 cm) fields. The characterization of detectors to be used for measurements in small fields is, however, complicated. Data concerning their use are scarce and until recently [5] no recommendations were available. Because the TPS is the gold standard in the design of the specific 3D dose distribution for a patient, TPS commissioning verification is needed by means of independent methods. The validation of newly developed software to design novel irradiation techniques, as well as the verification of the delivery of specific patient plans, is therefore a prerequisite for the safe introduction of advanced treatment techniques, requiring new 3D dosimetric approaches.

In this presentation various methods of 2D and 3D dose determination employed in radiation therapy will be discussed. The starting point will be the use of electronic portal imaging devices (EPIDs), their dosimetric characteristics, and their use for specific applications. Possibilities and limitations of 3D EPID dosimetry will then be elucidated, as well as how other 3D dosimetry techniques might complement the work performed with EPIDs. This presentation will partly be similar to that given by Korreman at the IC3DDose 2012 meeting [6] when reviewing the various 3D dosimetry approaches, but in my presentation the emphasis will be on the comparison of these methods with EPID dosimetry.

2. EPID-based dosimetry

2.1. Detector characteristics

In addition to their original and still major use for patient setup verification during external photon beam therapy, EPIDs are also used for many other quality assurance purposes including the verification of MLC leaf positions [7]. Previously, before 2000, various types of technology were used for EPIDs. Current amorphous silicon (a-Si) type EPIDs not only generate better quality images, but also possess useful dosimetric characteristics such as the (almost) linearity of the response with dose and dose rate, good long term stability, high spatial resolution, and real-time readout. These characteristics, and their applications in radiation therapy, have been discussed in detail by McCurdy at the IC3DDose 2012 meeting [8].

In his overview McCurdy also discussed a considerable number of challenges to using EPID dosimetry in clinical practice of which I will summarize the most important ones. First of all, an a-Si EPID over-responds to low energy photons compared to a water equivalent detector due to the increased photoelectric effect in the copper/phosphor screen. As a consequence the EPID response will be modified when the photon spectrum is different compared to the calibration situation, for instance when different field sizes and phantom/patient thicknesses are used. Also the lower energy photons scattered from the phantom/patient to the EPID cause an additional signal, which complicates the analysis. Furthermore, image lag and image ghosting occur due to changes in the trapped charge in the photodiodes, and changes in individual pixel gain, respectively. These effects may result in an under-response up to 10% but are restricted to short irradiation times. Finally, photon scatter generated in the EPID, and sometimes in the EPID-arm, will contribute to the image signal and has to be taken into account. Consequently, correction-or model-based approaches have been employed to mitigate these challenges when using a-Si EPIDs for clinical dosimetry. It is obvious that when performing dosimetric measurements with an EPID, validation of the accuracy of the 3D EPID-based dose determination of that specific irradiation technique, is a prerequisite.

2.2. Methods

The various EPID dosimetry approaches have been discussed in a comprehensive way in a review article by van Elmpt et al. [9] and are schematically presented in Figure 1. If the treatment beams have passed through a patient or phantom before being measured by the EPID, the method is referred to as transit or transmission dosimetry, otherwise, the method is described as non-transit dosimetry. Dose or image comparisons can be made at the EPID level or in the patient/phantom using a forward- or back-projection approach, respectively. Software for comparing a measured EPID image to an expected image in the forward non-transit dosimetry approach (Figure 1a) has been developed by many groups and is commercially available as discussed for instance by Greer [10]. In the forward transit dosimetry approach (Figure 1b) measured grayscale distributions are compared directly or after conversion to dose values, with predicted grayscale or dose distributions [11-13]. In the backward-projection non-transit approach (Figure 1c) the photon fluence measured with the EPID is back-projected to the focus of the linac and then used as input for a 3D dose calculation in the planning CT data of a patient using in-house

[14] or commercially developed [15] software. The resulting 3D dose distribution is then compared with the planned dose distribution for that particular patient, providing information about possible differences between the two calculated dose distributions.

EPID-based 3D transit dose reconstruction methods have been developed by many groups as discussed in the review article (published in 2008) by van Elmpt et al [9]. Two approaches of dose reconstruction can be distinguished. In the first method, the photon fluence measured with the EPID is back-projected through the phantom or patient to the target of the linac, and then used to calculate the 3D dose distribution in the phantom or patient geometry [16, 17]. In the second approach the dose distribution measured with the EPID is directly back-projected in the phantom or patient CT data set [18]. The dose calculation algorithms used in these models are a pencil-beam type [18], a collapsed-cone convolution method [17], or a Monte Carlo approach [16]. Dose calculation algorithms based on a pencil beam model are fast and relatively easy to implement, but the accuracy is inferior to other dose calculation algorithms particularly with respect to tissue heterogeneity corrections. In order to by-pass that problem, a modification of the pencil beam type of dose calculation has been developed providing the same sensitivity for detecting errors as *in vivo* dosimetry of sites without large tissue heterogeneities [19]. In addition to these in-house developed methods, commercial software has been released for 3D dose reconstruction using the transit dosimetry approach [15].

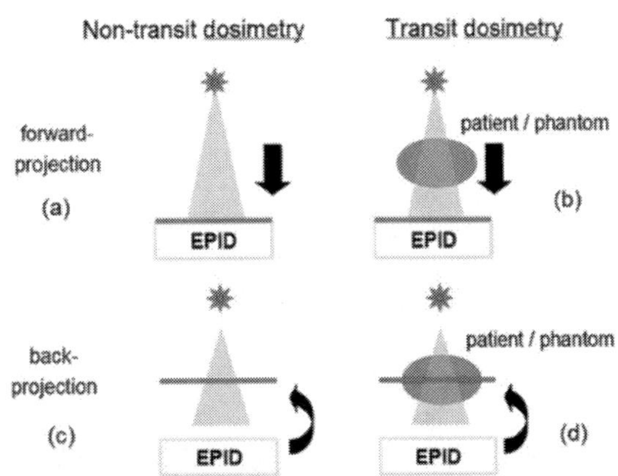

Figure 1. Schematic representation of the various dosimetric applications of EPIDs. Forward projection methods compare measured 2D images or dose distributions with predicted images or dose distributions at the EPID level. Back-projection methods, both transit and non-transit, can provide 3D dose distributions in a phantom or patient, while the transit dosimetry approach with a patient in the beam is the only method to perform 3D *in vivo* patient dosimetry.

2.3. *Pre-treatment dose verification*

Before starting clinically with a new treatment technique, a thorough commissioning of the planning and delivery of that technique should be performed. Such a program generally includes an end-to-end test in which the actual patient treatment procedure is mimicked as much as possible using one of the 2D or 3D systems, often in combination with a suitable phantom, as will be discussed in the next sections. After finishing such a series of systematic pre-treatment measurements, a patient specific pre-treatment verification measurement is performed before the start of each patient treatment with that technique, frequently in combination with an independent dose or MU check. The purpose of such a patient specific pre-treatment measurement, commonly rather carelessly called "patient specific QA", is to ensure that the dose delivered to the patient is identical to what is calculated by the TPS.

EPIDs are often used for pre-treatment verification applying the forward non-transit dosimetry approach (Figure 1a). Basically the EPID images are used to verify if the fluence intensity distribution calculated by the TPS is transferred to and delivered by the linac correctly. This method does, however, not verify the dose calculation for that particular patient/phantom irradiation. Some EPID-based systems using the back-projection non-transit approach (Figure 1c) are able to perform a 3D dose calculation. In this way errors in the dose calculation of the clinically used TPS can also be detected along with the plan deliverability. However, both non-transit verification approaches will miss errors due to changes in patient anatomy and patient set-up, or due to malfunctioning of the treatment machine during the

IOP Conf. Series: Journal of Physics: Conf. Series **847** (2017) 012024 doi:10.1088/1742-6596/847/1/012024

actual patient treatment. In the forward transit dosimetry approach (Figure 1b) not only errors in the data transfer and delivery of the treatment can be determined, but also patient-related errors such as anatomy changes [13]. Both forward approaches perform 2D image comparisons, and observed deviations are not in a simple way related to variations in the dose distribution in a patient.

The main purpose of pre-treatment verification of IMRT and VMAT patient plans is to identify *dosimetrically unacceptable errors*, i.e. deviations from the "true dose" considered from the physical/technical point of view. Pre-treatment verification by means of an EPID, as well as by many of the other dosimetry systems discussed in the next sections, may identify such errors.

2.4. *In vivo dose verification*

Pre-treatment dosimetry using EPIDs checks the dose delivery under well-defined conditions while *in vivo* dosimetry is performed under actual daily clinical conditions. EPID-based systems allow day-to-day monitoring and can be used to highlight any major change in the patient anatomy that may have a clinical impact (see Figure 2). The aim of 3D EPID-based *in vivo* dosimetry is often not to measure the "true" dose delivered to a patient with the highest possible accuracy, but to detect in a simple and reliable way a deviation between the planned and reconstructed 3D dose distribution, which should be within well-specified criteria. The question then arises if a difference in dose distribution, observed when analyzing *in vivo* treatment verification data, can be considered as a *clinically unacceptable error*, i.e. would the treatment outcome in terms of local control or toxicity be affected by this difference. Generally the alerts are first reviewed by an experienced medical physicist. When the error is understood by the physicist, and it was estimated that there might be negative clinical consequences, the case is discussed with a radiation oncologist. A decision for corrective action is then generally made by the physicist and radiation oncologist together.

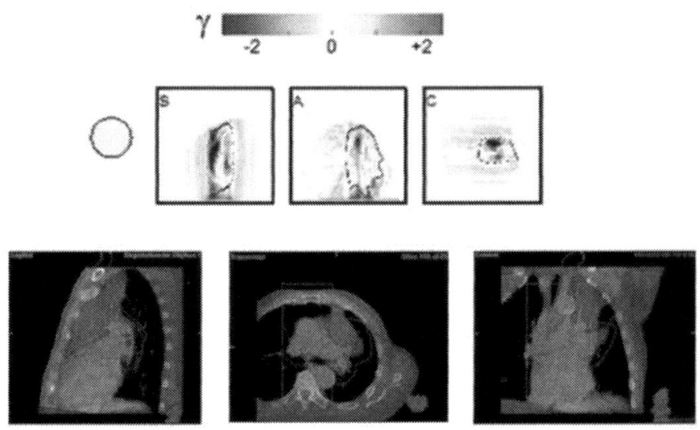

Figure 2. The upper part shows the outcome of a 3D EPID-based *in vivo* dose verification of an IMRT treatment of a lung cancer patient at NKI. Indicated are the results of the 3D gamma evaluation in a sagittal, axial and coronal plane through the isocenter. A signed gamma display is used: the yellow and red color indicate regions where the EPID dose is higher than the planned dose, whereas the green and blue color indicate regions where the EPID dose is lower than the planned dose. The 50% isodose line is shown in black. The yellow dot means that an alert criterion is outside the action level. The bottom part shows a CBCT scan (green) made that day compared with the planning CT scan (purple) in the three orthogonal planes, showing an increase in fluid in the lung.

3. Semi-3D, pseudo-3D and full-3D dosimetry systems

3.1. *Semi-3D detectors*

Multi-dimensional dosimetry systems have largely replaced the use of radiographic film, often in combination with an ion chamber, for many applications in radiation therapy such as IMRT and VMAT pre-treatment verification. Film can be used in the same way as these 2D ion chamber or diode arrays by inserting the film in phantoms, i.e., as a *semi-3D* detector. It has the advantage over these systems of having a much higher spatial resolution, which may be important in verifying IMRT or VMAT plans. These plans often have steep penumbras or a high degree of modulation, which require high resolution

IOP Conf. Series: Journal of Physics: Conf. Series **847** (2017) 012024 doi:10.1088/1742-6596/847/1/012024

equipment for verification. Film provides a relative 2D measurement of the photon fluence while an ion chamber measurement gives an absolute point dose. Such a method is, however, time consuming, and requires extensive resources for calibration, processing and analysis, as well as the costs of single-use films. Also film cannot provide online information, as opposed to the other *semi-3D* equipment. Film dosimetry is therefore mainly used for special applications.

Several multi-dimensional detector systems have been developed, consisting of 2D arrays of diodes or ion chambers that can be positioned in a phantom, which can then be referred to as a *semi-3D* system. Figures 3b and 3c are examples of such an approach. Other types of *semi-3D* dosimetry systems include arrays which make measurements in two orthogonal planes (Figure 3a) or at various points on a helical curve at approximately 2.9 cm below the surface of cylindrical phantom (Figure 3d). All these systems construct a 3D dose distribution by interpolation of the measured data using their own software. The possibilities and limitations of each of these approaches have been reviewed by Korreman [6] and are related to the size of the individual detectors, spatial resolution, angular dependence to beam delivery direction, absolute or relative dose measurement and frequency of calibration. Since that review companies have worked on improvements of their systems. For instance, in addition to the PTW seven29 ion chamber array shown in Figure 3c, there now exists also a the Octavius 1500 system having 1405 vented ion chambers on a large field, and the Octavius 1000 SRS having 977 liquid-filled ion chambers on a 10 cm x 10 cm field. Furthermore a rotating phantom, Octavius 4D, has been developed in which these arrays can be positioned, allowing measurements perpendicular to the beam direction during VMAT delivery.

3.2. Pseudo-3D detectors

Recently a number of transmission chamber devices that have to be attached to the head of the treatment machine became commercially available. These devices measure the photon fluence exiting the linac head. The measured fluence can be used to calculate a 3D dose distribution in the patient anatomy in a similar way as done with an EPID-based in–air measurement shown in Figure 1c. Both approaches, as well as a dose calculation using a linac log file, can be considered as *pseudo-3D* techniques. They are designed to detect transfer errors from the TPS to the linac, and delivery errors by the linac when using them either pre-treatment or during treatment. If the dose calculation algorithm is different from that used in the clinical TPS, then these methods also check the 3D dose calculation for a specific patient treatment in an independent way. These systems do, however, not detect variations in patient anatomy or setup, i.e., cannot be used for *in vivo* dosimetry. It would, however, be interesting to assess their relation to EPID-based 3D *in vivo* transit dosimetry systems.

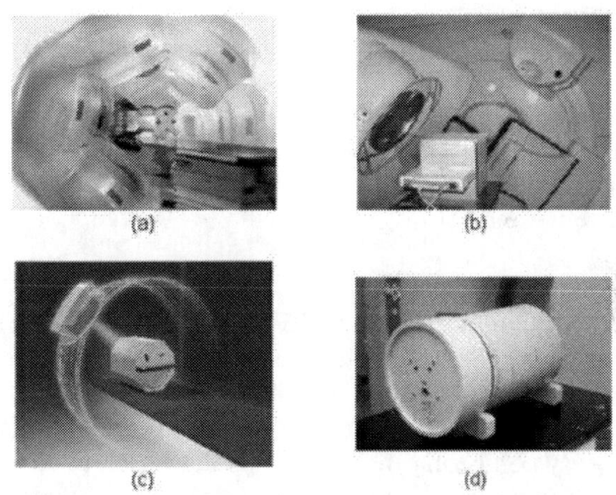

Figure 3. Four commercially available phantoms for semi-3D dosimetry with embedded detector arrays. (a) Delta4® with two orthogonal diode arrays (courtesy of ScandiDos AB), (b) MatriXXEvolution two-dimensional ion chamber array with MULTICube (courtesy of IBA Dosimetry), (c) OCTAVIUS with 2D-ARRAY seven29 ion chamber array (courtesy of PTWFreiburg GmbH), and (d) ArcCHECK with a helical diode array cylindrical around the central axis (courtesy of Sun Nuclear Corporation). (Reproduced from [6])

Both EPID-based back-projection approaches (Figures 1c and 1d), are able to reconstruct the 3D dose distribution in the patient anatomy from 2D EPID images, and can therefore be considered as

IOP Conf. Series: Journal of Physics: Conf. Series **847** (2017) 012024 doi:10.1088/1742-6596/847/1/012024

pseudo-3D methods, a term which I prefer instead of "*virtual 3D*" systems used by Korreman [6]. One of the reasons is that the transit back-projection approach has the possibility to detect errors during real, and not virtual, patient treatments.

3.3. Full-3D detectors
To the best of my knowledge, the only dosimeters currently available that are able to measure a *full-3D*, sometimes called "*true-3D*", dose distribution are polymer gel dosimeters and radiochromic 3D detectors. Understandably, sharing information about the further development and application of these detectors is one of the main purposes of this conference. The advantage of these dosimeters over most, if not all, of the *semi-3D* and *pseudo-3D* dosimetry systems is that they are tissue equivalent and can be molded into an anthropomorphic shape. They have the potential to be used as benchmarking tool for the commissioning of treatment plans of specific treatment techniques. Particularly when special IMRT or VMAT techniques or new treatments of a specific treatment site are clinically introduced, it would be helpful to have a *full-3D* dosimetry system in addition to the *semi-3D* and/or *pseudo-3D* systems presently used for that purpose. However, preparing these dosimeters is a time consuming process and the readout of dose after delivery is complicated. It is therefore obvious that at present *full-3D* dosimeters are at this moment not suitable for patient specific pre-treatment verification. The power of *full-3D* methods lies in their use for dosimetric end-to-end tests rather than for everyday use. In addition to end-to-end tests of clinically implemented irradiation techniques, *full-3D* dosimetry is also applied in pre-clinical studies, for instance for characterizing irradiations performed with small animal precision irradiators or synchrotron x-ray microbeams.

Regardless of its successful use for many applications, currently the technology of *full-3D* dosimetry is mainly restricted to research institutions. In other words, as mentioned by Schreiner in his IC3DDose 2014 presentation [20]: "However, despite its promise, true (BM: full) 3D dosimetry is still not widely practiced in the community. Its use has been confined primarily to select centres of expertise and to specialised quality assurance or commissioning roles where other dosimetry techniques are difficult to implement." Hopefully during this congress newly developed dosimeter preparation and scanning techniques will be presented that can also be employed in a relatively simple way in a non-academic environment.

4. Final comments
Patient specific pre-treatment verification ("QA") of IMRT/VMAT is often restricted to a check of the transfer of a plan from a TPS to the treatment unit and on its delivery, without performing a verification of the 3D dose distribution. Several studies have shown that there is a fundamental problem with such an approach, which is often unable to correctly differentiate an acceptable from an unacceptable plan [e.g., 18]. By comparing the various methods, EPID-based *in vivo* transit dosimetry might be a better approach to discover incorrect delivery of a specific treatment plan to a patient under clinical conditions.

Furthermore, in a number of institutions pre-treatment verification measurements are stopped after a certain number of checks, and replaced by an independent dose or MU calculation, in combination with a comprehensive QA program. For instance, the Netherlands Commission on Radiation Dosimetry recommends moving to calculation techniques after performing 100 or less experimental verifications of IMRT/VMAT for a specific treatment technique depending on the experience of the center, the complexity of the treatment, and if *in vivo* dosimetry is performed [21, 22]. In those situations the results of the dose/MU verification and *in vivo* dosimetry need to be evaluated before >20% of the total number of fractions is given.

EPIDs have a number of dosimetric characteristics that require a dedicated commissioning process to obtain highly accurate measurement results. It would be interesting if other types of transit dosimeters could be developed having, for instance, a less energy dependent response.

In vivo transit dosimetry should ideally be performed in real time so that errors can be caught immediately and the treatment interrupted if the delivered dose is not within pre-defined limits of acceptance. EPIDs can in principle do that but more research is needed to define criteria for halting the linac that should not miss serious errors but have at the same time a small number of false positives.

At this moment EPIDs are the only tools available for 3D *in vivo* dosimetry. Although there are a number of challenges related to their clinical use, 3D *in vivo* dose verification can be of great value in a department. The experience in some centers demonstrates that EPID-based 3D *in vivo* dosimetry requires a change in attitude to patient specific dose verification. It is no longer purely a physics or technical matter, but requires in addition clarifying dose differences observed in the clinic, i.e., incorporating many issues that may happen during the daily treatment of patients. As a result medical physicists will be much more involved in assuring the quality of the actual patient treatment than when only performing a pre-treatment dose verification measurement.

5. Acknowledgements

The author gratefully acknowledges the numerous discussions in the EPID research group at the Netherlands Cancer Institute on the many aspects of EPID-based 3D dosimetry and its relation to other dose verification methods.

6. References

[1] Baldock C *et al* 2010 *Phys. Med. Biol.* **55** R1-63
[2] ICRU 2010 Prescribing, recording, and reporting photon-beam intensity-modulated radiation therapy (IMRT). Report 83. International Commission on Radiation Units and Measurements, Oxford University Press, Oxford, UK.
[3] IAEA 2016 Accuracy requirements and uncertainties in radiation therapy. Human Health Series No. 31. International Atomic Energy Agency, Vienna, Austria.
[4] Hill R *et al* 2009 *Med. Phys.* **36** 3971-81
[5] IAEA 2016 Dosimetry of small static fields used in external beam radiotherapy: An IAEA-AAPM international code of practice for reference and relative dose determination. Technical Reports Series No. 483. International Atomic Energy Agency, IAEA, Vienna, Austria.
[6] Korreman SS 2013 *J. Phys.: Conf. Ser.* **444** 012007
[7] Greer P B *et al* 2007 *Med. Phys.* **34** 4389-98
[8] McCurdy BMC 2013 *J. Phys.: Conf. Ser.* **444** 012002
[9] van Elmpt W *et al* 2008 *Radiother. Oncol.* **88** 289-309
[10] Greer PB 2013 *J. Phys.: Conf. Ser.* **444** 012010
[11] Chytyk-Praznik K *et al* 2013 *Med. Phys.* **40** 031713
[12] Bedford JL *et al* 2014 *Med. Phys.* **41**: 021725
[13] Berry SL *et al* 2014 *Int. J. Radiat. Oncol. Biol. Phys.* **88** 289-309
[14] van Elmpt W *et al* 2008 *Radiother. Oncol.* **86**: 86-92
[15] Narayanasamy G *et al* 2015 *J. Appl. Clin. Med. Phys.* **16** 329–38
[16] van Elmpt WJ *et al* 2007 *Med. Phys.* **34** 2816-26
[17] Van Uytven E *et al* 2015 *Med. Phys.* **42** 6945-54
[15] Wendling M *et al* 2009 *Med. Phys.* **36** 3310-21
[16] Wendling M *et al* 2012 *Med. Phys.* **39** 367-77
[17] Schreiner LJ 2015 *J. Phys.: Conf. Ser.* **573** 012003
[18] Kry SF *et al* 2014 *Int. J. Radiat. Oncol.* **90** 1195-201
[19] NCS 2013 Code of practice for the quality assurance and control for intensity modulated radiotherapy. Netherlands Commission on Radiation Dosimetry Report 22. (http://radiationdosimetry.org)
[20] NCS 2015 Code of practice for the quality assurance and control for volumetric modulated arc therapy. Netherlands Commission on Radiation Dosimetry Report 24. (http://radiationdosimetry.org)

IOP Conf. Series: Journal of Physics: Conf. Series **847** (2017) 012025 doi:10.1088/1742-6596/847/1/012025

Convergence of SART + OS + TV iterative reconstruction algorithm for optical CT imaging of gel dosimeters

Yi Du[1, 2], Gongyi Yu[1, 3], Xincheng Xiang[1], Xiangang Wang[1], Yves De Deene[2]
[1]Institute of Nuclear and New Energy Technology, Tsinghua University, Beijing, 100084, China
[2]The Biomedical Engineering Laboratory, Department of Engineering, Macquarie University, Sydney, 2109, Australia
[3]Department of Radiation Oncology, Fudan University Shanghai Cancer Center, Shanghai, 200032, China

E-mail: wangxiangang@tsinghua.edu.cn

Abstract. Computational simulations are used to investigate the convergence of a hybrid iterative algorithm for optical CT reconstruction, i.e. the simultaneous algebraic reconstruction technique (SART) integrated with ordered subsets (OS) iteration and total variation (TV) minimization regularization, or SART+OS+TV for short. The influence of parameter selection to reach convergence, spatial dose gradient integrity, MTF and convergent speed are discussed. It's shown that the results of SART+OS+TV algorithm converge to the true values without significant bias, and MTF and convergent speed are affected by different parameter sets used for iterative calculation. In conclusion, the performance of the SART+OS+TV depends on parameter selection, which also implies that careful parameter tuning work is required and necessary for proper spatial performance and fast convergence.

1. Introduction

Gel dosimetry has received growing research interest for its outstanding three-dimensional dose measurement performance [1]. Upon irradiation, the radio-sensitive chemicals in gel dosimeters will react to the localized absorbed dose, which then induces changes in physical properties, such as relaxation time and optical density (OD). These changes can be read out by either MRI or optical CT, and, after calibration, three-dimensional dose maps can be obtained. Compared with MRI, optical CT has been proposed as an easily accessible and cost-efficient alternative for gel dosimeter readout [2-4].

In a typical optical CT scanning, optical transmission attenuated projections are captured by a detector, such as a photo-diode or CCD camera. When projections over 360 degrees are acquired, tomographic optical density images representing ODs can be reconstructed by either filtered backprojection (FBP for 2D, and FDK for 3D) or iterative algorithms. Subsequently, dose maps can be restored via dose-OD calibration. Optical CT traverse projections are often corrupted by noise and imperfection [5]. Compared with the filtered backprojection algorithm, iterative reconstruction has shown advantageous performance in noise and artefact suppression [6]. Based on our previous work, we have proposed a novel hybrid iterative optical CT reconstruction, i.e., simultaneous algebraic reconstruction technique (SART) integrated with ordered subsets (OS) iteration and total variance (TV) minimization regularization, i.e., *SART+OS+TV*. The method is detailed and fully evaluated in [7], which can be briefed as

IOP Conf. Series: Journal of Physics: Conf. Series **847** (2017) 012025 doi:10.1088/1742-6596/847/1/012025

$$f_j^{(k+1)} = f_j^{(k)} + \lambda \boldsymbol{R}^{\boldsymbol{T}} \left(g_j^{(k)} - \frac{\boldsymbol{R}_{ij} f_j^{(k)}}{||\boldsymbol{R}||_2} \right) \quad (1)$$

$$min[TV(f_j^{(k)})] = min \left[\int_x \int_y \sqrt{\left(\frac{\partial f_j^{(k)}}{\partial x}\right)^2 + \left(\frac{\partial f_j^{(k)}}{\partial y}\right)^2} \, dxdy \right] \quad (2)$$

where f is the tomographic image to reconstruct, g is the logarithmic projections, λ is a relaxation factor, and $\boldsymbol{R}$ is the system matrix (or projection operator) for optical transmission imaging geometry and physics; k stands for the kth ordered subset of projections, and j for the jth beamlet. In (1), we calculate the deviations between acquired projections and *reprojected* projections from the OD image f by forward projection, and use the projection deviations as error feedback to update the image by backprojection. In (2), the TV of OD image f is decreased iteratively to suppress noise and artefacts. In numerical implementation of this algorithm, three key parameters need to be chosen, which are the relaxation factor λ in (1) for SART iteration, the number of subsets N for OS iteration, and the relaxation factor α in primal-dual algorithm for TV minimization [8]. The details and evaluation of this hybrid algorithm for optical CT imaging is under draft for comprehensive discussion. In conclusion, this algorithm outperforms classical FDK in denoising, edge-preserving and artefact reduction.

As a significant but relatively independent section of the work on SART+OS+TV algorithm, the convergence of this hybrid algorithm is discussed using computational simulations.

2. Methods and Materials

2.1. Simulation Setup

2.1.1. Synthetic Phantom

A synthetic phantom is used for optical CT imaging simulation, which facilitates us to observe the iterative reconstruction process. The synthetic phantom is derived from [9] to mimic dose distribution in tissue equivalent gel irradiated by a 6-MV linac with 100cm SSD. The cylindrical phantom is 10 cm in diameter, and 10 cm in height, with a 2-cm wide square OD (irradiation) field offset 2-cm from the circular centre, as illustrated in Figure 1(a).

2.1.2. Scenarios

The geometric parameters used in optical CT imaging correspond to the specifications of our optical CT scanner [10]. In order to observe the influence of parameter selection, we launched a series of scenarios with as many as 36 different parameter sets of the relaxation factor λ, the number of subsets N in OS, the relaxation factor α in primal-dual algorithm for TV minimization. Compared with FDK, iterative reconstruction is time consuming. In this study, the calculation will be terminated once the iteration times reach 80.

2.2. Evaluation Metrics

The profiles of reconstructed images are given to see any bias from true OD values and dose gradient integrity performance. In the meantime, in order to describe the iterative convergence quantitatively, we choose the root-mean-square error (RMSE) for evaluation, which is defined as

$$RMSE^{(k)} = \sqrt{\frac{\sum_{i=1}^{M} \left(f_i^{(k)} - \hat{f}_i\right)^2}{M}} \quad (3)$$

where i is the ith pixel index in the image, and M is the total number of image pixels; $f^{(k)}$ is the reconstructed image at the kth iteration, and $\hat{f}$ is the true OD value of the phantom.

Meanwhile, since the synthetic phantom is cylindrical, we adopt the method in [11] to calculate the modulation transfer function (MTF) to describe the spatial resolution performance.

Table 1. Parameter sets for iterative reconstruction (IR).

Parameter Set No.	Parameter Item		
	λ	N	α
# 1	0.001	12	0.0001
# 2	0.005	12	0.0001
# 3	0.001	4	0.0001
# 4	0.001	360	0.0005

Table 2. RMSEs of different methods.

Method	RMSE
FDK(Hann)	0.0054
FDK(RL)	0.0049
IR # 1	0.0050
IR # 2	0.0021
IR # 3	0.0059
IR # 4	0.0044

3. Results and Discussion

Four different sets of reconstruction parameter sets are given in Table 1 for this study. Comparison of the results reconstructed with different parameter sets show that the convergence is influenced by parameter selection. The RMSE is calculated as a metric for a central place inside the phantom region. As reference, the results of FDK with Hann and RL filters respectively are shown as well.

3.1. Final Convergent Value

The profiles across the OD field of the synthetic phantom for different methods are plotted in Figure 1(b), and the RMSE results are given in Table 2. The small difference from the true ODs in both profiles and RMSEs implies that the results of the iterative method converge to true values without any significant bias.

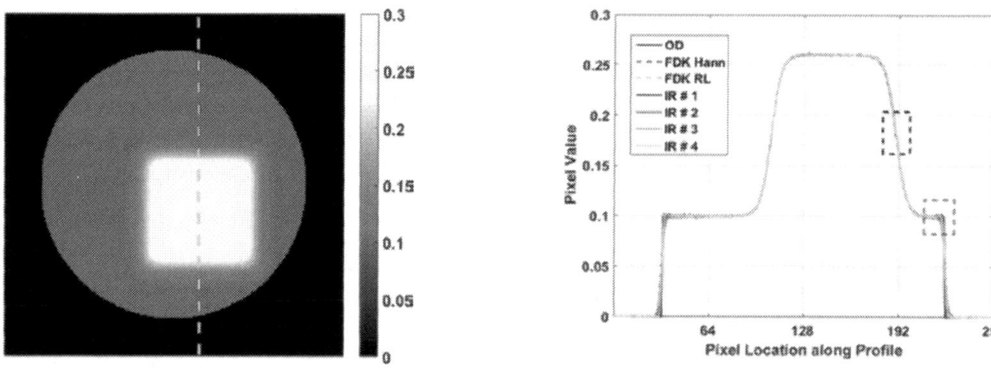

(a) (b)

Figure 1. Synthetic Phantom profiles(a) and profiles of convergent results of different parameter sets with FDK as reference (b).

3.2. Spatial Dose Gradient and Sharp Edge Integrity

Spatial dose gradients are important information for radiotherapy dosimetry, and the dose gradient integrity should be well preserved. In order to show the different performance in spatial gradient integrity, we selected a ROI inside the beam penumbra, and used the local profiles inside the ROI window for illustration. The ROI window we used is blue marked in Figure 1(b), and the profiles for different method are plotted in Figure 2(a).

Besides, in order to show the difference in spatial resolution, we used the circular edge of the synthetic phantom (in Figure 2(b)) for MTF calculation, the results of which are given in Figure 2(c). Comparing the dose gradients, edge profiles and MTF curves in Figure 2, we can see that in some case the results are smoother than that of FDK with a Hann filter, and in other case are sharper than FDK with a RL filter. The variable ability in edge response and MTF implies that the SART+OS+TV algorithm is quite versatile in spatial gradient integrity and spatial resolution, which depends on parameter selection.

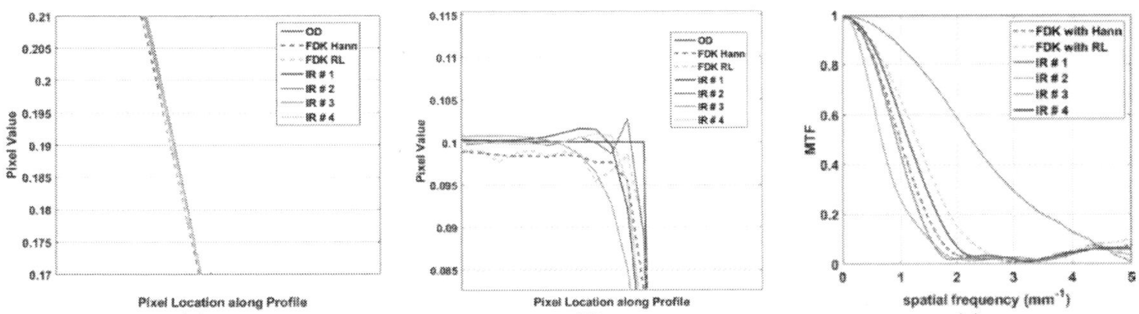

(a) (b) (c)

Figure 2. (a) Dose gradients of the blue ROI in Figure 1(b); (b) edge profiles corresponding to the pink ROI in Figure 1(b); (c) MTFs of different parameters with FDK as reference.

3.3. Rate of Convergence

The rate of convergence is a crucial characteristic in iterative reconstruction, which affects the computation cost. The MTF was calculated for different steps at intervals of 10 iterations to evaluate convergence. The MTF curves for different iterations and different sets of parameters are plotted in Figure 3. A clear difference in rate of convergence can be clearly observed for parameters.

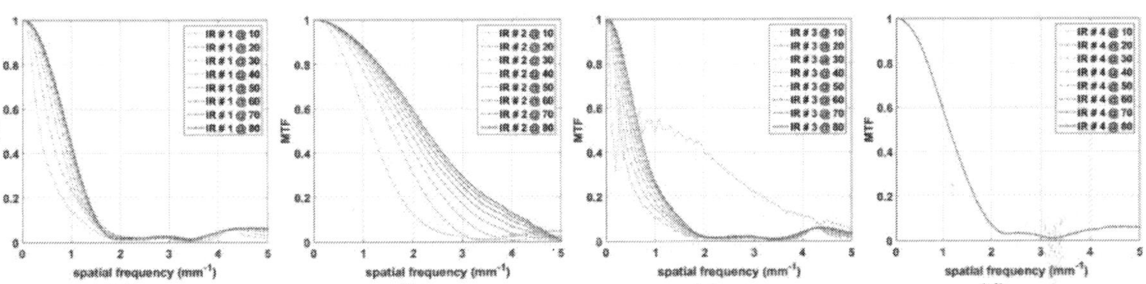

(a) (b) (c) (d)

Figure 3. MTF curve clusters of parameter set #1(a), #2(b), #3(c), and #4(d). Note that in (d) the MTFs for iterations 10-70 superpose each other for fast convergence.

4. Conclusions

For the SART+OS+TV algorithm, a clear dependence of the rate of convergence on the parameter set is observed. The comparative analysis shows that the results converge to the true values without significant bias, and MTF and rate of convergence are affected by different combinations of parameters used for iterative calculation. In conclusion, careful parameter tuning is necessary for proper spatial performance and fast convergence with SART+OS+TV iterative reconstruction.

5. Acknowledgement

Jointly Supported by the National Science Foundation of China (No. 61571262, No.11575095, No. 61171115), and the National Key Research and Development Program (No. 2016YFC0105406).

6. References

[1] Baldock C *et al* 2010 *Phys. Med. Biol.* **55** R1-63
[2] De Deene Y *et al* 2002 *Phys. Med. Biol.* **47** 2459
[3] De Deene Y and Baldock C 2002 *Phys. Med. Biol.* **47** 3117-41
[4] De Deene Y *et al* 2002 *Phys. Med. Biol.* **47** 3441-63

[5] Doran S J 2013 *J. Phys.: Conf. Ser.* **444** 012004
[6] Yi D *et al* 2015 *J. Phys.: Conf. Ser.* **573** 012063
[7] Du Y *et al* 2016 *Phys. Med. Biol.* **61** 8425-8439
[8] Chambolle A and Pock T 2011 *J. Math. Imaging Vis.* **40** 120-45
[9] De Deene Y 2015 *J. Phys.: Conf. Ser.* **573** 012076
[10] De Deene Y 2015 *J. Phys.: Conf. Ser.* **573** 012058
[11] Friedman S N *et al* W 2013 *Med. Phys.* **40** 051907

A prototype optical-CT system for PRESAGE 3D dosimeter readout

Devin Miles[1], Paul Yoon[1], Jacob Kodra[1], John Adamovics[2], Mark Oldham[1]
[1]Dept. of Radiation Oncology, Duke University Medical Center, Durham, NC, USA
[2]Dept. of Chemistry and Biology, Rider University, Lawrenceville, NJ, USA

Email: devin.miles@duke.edu, mark.oldham@duke.edu

Abstract. This work introduces the Duke Integrated-lens Optical Scanner (DIOS), a prototype optical-CT system designed for convenient and low-cost readout of PRESAGE 3D dosimeters. A key novelty of the DIOS is the incorporation of a multi-purpose light-collimating tank (the LC-tank). The LC-tank collimates light from a point source, maintains parallel ray geometry through a dosimeter mounted inside the tank, and refocuses emergent light onto a CCD detector. A second purpose is to dramatically reduce the amount of refractive matched fluid required in prior optical-CT scanners. This is achieved by substituting large quantities of refractive-matched fluid with solid RI-matched polyurethane. The advantages of DIOS include eliminating the need for expensive telecentric lenses, and eliminating the impracticality of large volumes of RI matched fluid. The DIOS is potentially more susceptible to stray-light artifacts. Preliminary phantom testing shows promising agreement between PRESAGE/DIOS readout and prior commissioned optical-CT scanners, as well as with Eclipse dose calculations.

1. Introduction

Prior works have demonstrated state-of-the-art 3D dosimetry [1] can be achieved with PRESAGE radiochromic plastic dosimeters and optical-CT readout [2-6]. The high transparency of Presage (low stray and scattered light contamination) enables compatibility with very fast broad-beam scanning approaches [7-9]. Given the high accuracy requirements for quantitative dosimetry, current broad beam scanners in our lab utilize expensive stray-light reducing telecentric lenses and substantial volumes of refractive matching fluid in order to attain best accuracy and results [10]. In an effort to develop a simpler, more practical, and less expensive optical-CT readout system, compatible with non-specialised clinics, we developed the Duke Integrated-lens Optical Scanner (DIOS). The DIOS is an economical alternative to our current systems (Duke Large-FOV Optical Scanner, DLOS). Here we present an evaluation of a prototype DIOS system.

2. Methods

The DIOS design consists of a filtered 3W red LED (633nm ± 10) incident on a solid light-collimating (LC) tank of refractive-index (RI) matched polyurethane, and a 12-bit monochromatic CCD camera (Figure 1). At a specific radius of curvature, the scanning tank is able to collimate the diverging LED light field into parallel-CT geometry as it transmits through a dosimeter sample. Any light that has not been absorbed or scattered within the system is re-converged onto the CCD for data acquisition. This system allows instantaneous acquisition of a large field-of-view optical projections through a sample. A dosimeter can be mounted onto a computer-controlled rotation stage to allow projections at varying angles, enabling tomographic imaging of measured optical density information. The addition of a diffusing film increases the uniformity of the incident light field, and a focusing lens with an aperture on the CCD is included to reject scattered light and reduce edge blurring.

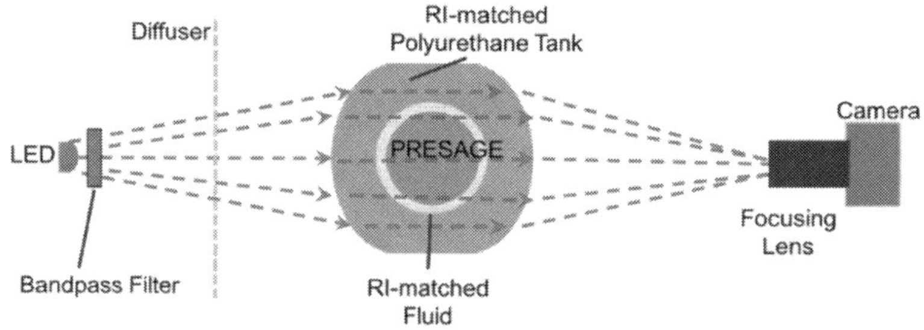

Figure 1. Basic DIOS schematic.

DIOS operation is functionally equivalent to the Duke systems (e.g. DLOS). However, unlike prior scanners, the DIOS design is able to omit costly telecentric lenses and large volumes of RI fluid (>10 L) by relying on the new RI-matched light-collimating polyurethane tank. A small amount of RI fluid is still required to limit heavy light scatter at interfaces, but this is typically less than 100 mL.

2.1. Light-collimating tank prototyping

To determine the optimal system parameters to attain parallel-ray CT geometry within the LC-tank, a Monte Carlo study was performed in ScanSim, an in-house optical-CT ray tracing software [11]. A diverging light source was modeled by a series of light rays originating from a single point, and their paths were tracked as each ray travels through different media at varying RI. Simulations show that a 25cm radius of curvature achieves sufficiently parallel-ray geometry for a typical 1kg PRESAGE dosimeter (10cm diameter, 10cm height cylinder), with focal distances at approximately 50cm.

Using ScanSim parameters, a positive mould of the LC-tank was modeled in AutoCAD (Autodesk, Inc., San Rafael, CA, USA) and 3D printed on a nominal fused-deposition modeling 3D printer. The light-incident surfaces were capped with a thin layer of stainless steel to minimize the impact of striation artifacts from 3D printing on scan quality. The finished positive mould was used to create a rubber negative mould and cast in transparent polyurethane (Figure 2).

Figure 2. Finished positive mould (left) and cast polyurethane LC-tank (right).

2.2. Benchmarking

A series of custom optical phantoms were produced to assess the imaging capabilities of the DIOS system. The modulation transfer function and resolution were measured by acquiring optical projection images featuring an attenuating edge phantom. A 3D 'spiral' phantom (1kg polyurethane cylinder, with 5mm channels drilled at varying radial distances) was used to verify geometric accuracy in tomographic reconstruction via comparison between DIOS optical-CT and conventional x-ray CT.

A 3D dosimetry study was performed to assess the DIOS's ability to accurately measure optical density. A 4-field box plan was chosen as a benchmarking test, to deliver regions of highly-uniform dose as well as sharp dose gradients. The treatment plan was created in Eclipse (Varian Medical

Systems, Palo Alto, CA) to deliver a maximum dose of 1 Gy onto a 1kg PRESAGE dosimeter. The dosimeter was irradiated using a Varian 600CD linac at 6MV, and OD was immediately read out using the DIOS. Accuracy was assessed using comparison between DIOS and the Eclipse treatment simulation, as well as readout from a separate commissioned optical-CT system (DLOS).

3. Results

MTF measurement shows the approximate cut-off frequency in the DIOS to be 4.5 mm^{-1}, implying maximum image resolutions of 0.22 mm can be achieved. Tomographic reconstruction of the spiral phantom shows great spatial agreement between optical-CT and x-ray CT (both reconstructed at 1mm isotropic voxel resolution).

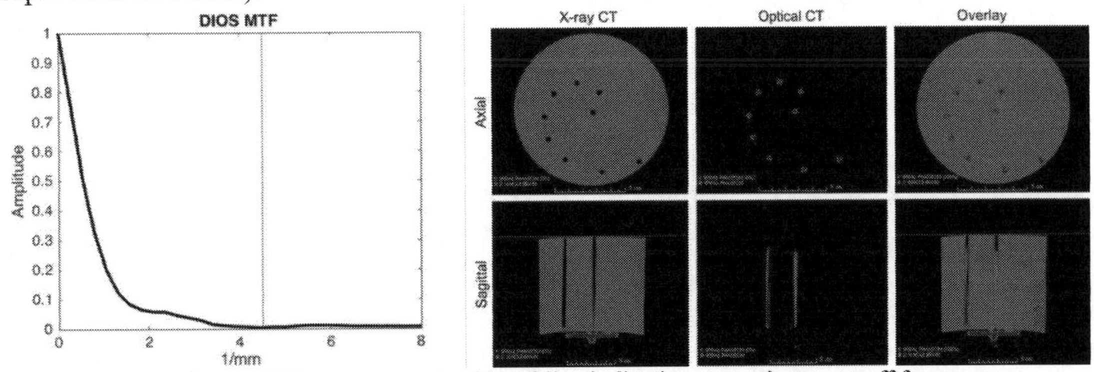

Figure 3. Left: 2D MTF measurement, with red line indicating approximate cut-off frequency. Right: Comparison of tomographic images of the 3D spiral phantom.

Dose comparison between Eclipse, DIOS, and the standard DLOS optical-CT system can be seen in figure 4. DIOS dose measurement shows good agreement between Eclipse and DLOS doses, only suffering from minor concentric ring artifacts and some speckling from slight movement of the diffuser. 3D gamma comparison between DIOS readout and Eclipse calculation shows 95.6% of voxels passing 3%/3mm criteria (Figure 5).

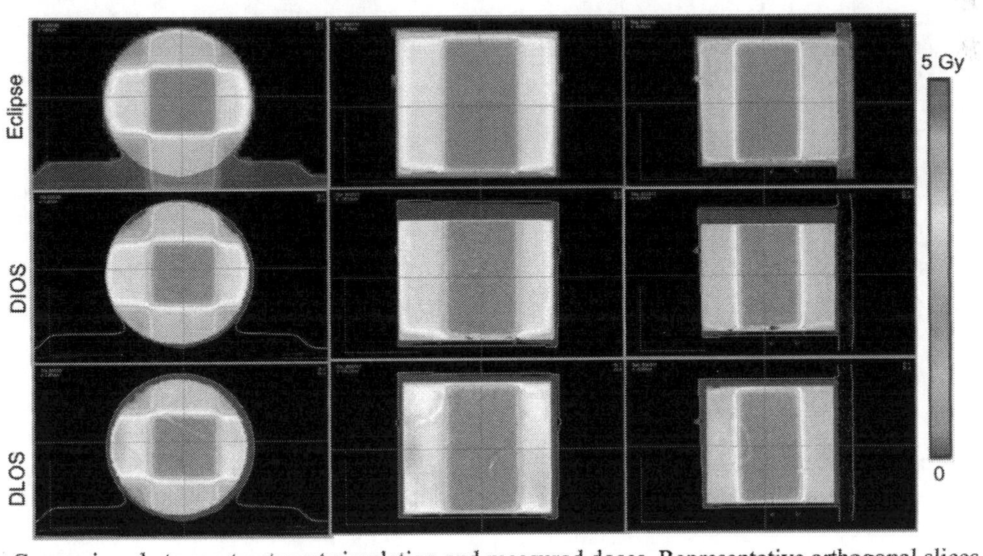

Figure 4. Comparison between treatment simulation and measured doses. Representative orthogonal slices of the Eclipse treatment simulation, DIOS readout, and DLOS readout are shown.

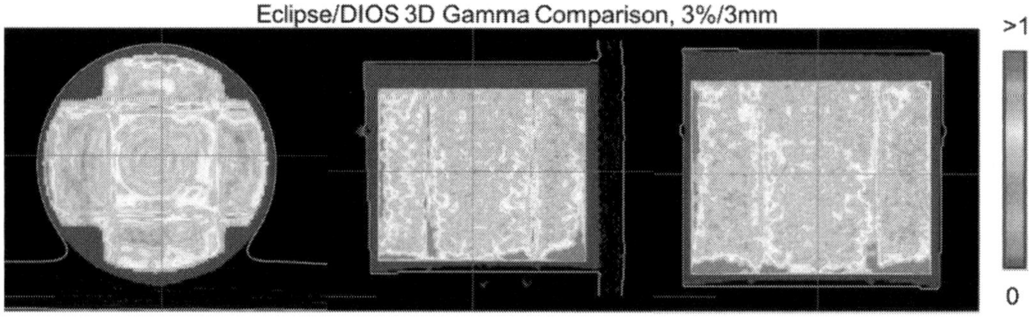

Figure 5. 3D gamma comparison between Eclipse treatment simulation and DIOS, where red voxels \ denote failure to meet criteria. 95.6% of voxels pass at 3%/3mm criteria.

4. Conclusions

The first prototype of the DIOS has demonstrated promising performance as an economical and practical optical-CT system for 3D dosimetry. By incorporating a solid LC-tank of RI-matched polyurethane, the DIOS is able to remove the need for costly and difficult components from prior systems, such as telecentric lenses and large volumes of refractive matching fluids. This system aims to enable comprehensive 3D dosimetry techniques more accessible to general clinical use. However, omission of telecentric lenses has the potential to introduce unwanted scattered light in data acquisition, resulting in possible feature blurring and overall reduction in spatial resolution. Future work will aim to commission the DIOS through more stringent dosimetry tests, and further prototypes will incorporate larger fields of view and more robust mechanisms of scatter rejection.

References

[1] Baldock C *et al* 2010 *Phys. Med. Biol.* **55** R1-63
[2] Bache S *et al* 2015 *Med. Phys.* 2015; **42** 846-55
[3] Juang T *et al* 2014 *Med. Phys.* 2014; **41** 071706
[4] Juang T *et al* 2015 *J.Phys.: Conf. Ser.* 2015; **573** 012039
[5] Iqbal K *et al* 2013 *J. Appl. Clin. Med. Phys.* **15** 363-74
[6] Gorjiara T *et al* 2011 *Med. Phys.* **38** 2265-74
[7] Oldham M 2006 *J. Phys.: Conf. Ser.* **56** 58-71
[8] Thomas A *et al* 2011 *Med. Phys.* **38** 4846-57
[9] Oldham M 2014 Methods and techniques for comprehensive 3D dosimetry. *Advances in Medical Physics* (Medical Physics Publishing) pp 69-86
[10] Hill R *et al* 2014 *Phys. Med. Biol.* **59** R183-231
[11] Oldham M 2010 *J. Phys.: Conf. Ser.* **250** 311-5

The quenching effect in PRESAGE® by a proton beam: Investigation of formulation dependence

M Carroll[1,2], M Alqathami[1] and G Ibbott[1]
[1]Department of Radiation Physics, University of Texas MD Anderson Cancer Center, 1515 Holcombe Blvd., Houston, TX 77030 USA
[2]University of Texas at Houston Graduate School of Biomedical Sciences, Houston, TX 77030 USA

E-mail: mcarroll1@mdanderson.org

Abstract: The radiochromic dosimeter PRESAGE® has shown potential in conventional radiotherapies, but it suffers from dose under-responding, or signal quenching, in proton therapy as a result of dependence on variations in the linear energy transfer (LET). Early investigations have shown the under-response is affected by the chemical concentration of the active components, but as yet this relationship has not been comprehensively measured. This study investigated the impact of PRESAGE® formulation changes on signal quenching to determine the magnitude by which the quenching can be minimized.

1. Background

Radiotherapy techniques have advanced significantly over the last two decades and dose planning has also become significantly more complex [1-3]. This is explicitly seen in proton therapy, which includes extremely steep dose gradients as a result of the Bragg peak. Conventional QA for complex field arrangements is performed using point dose measurements (usually ionization chamber) or planar systems (film and diode arrays) that are limited in their ability to record such gradients. Additionally, using conventional dosimetry systems to fully characterize complex dose distributions can be labor intensive. In 2015, an IROC-H phantom study showed that over 20% of participating proton institutions failed credentialing criteria, indicating that beam commissioning might have been inadequate [4]. A 3D dosimetry system could provide comprehensive volume measurements that are captured in a single irradiation and can potentially catch planning and delivery errors missed by conventional dosimetry systems.

PRESAGE® (Heuris Pharma, LLC, Skillman, NJ), a radiochromic polyurethane, is a three-dimensional (3D) dosimeter that has distinct advantages over many other 3D alternatives such as improved thermal and temporal stability, environmental robustness, and improved measurement accuracy through optical-CT imaging. The active ingredients in PRESAGE® that result in the dose signal are a halocarbon radical initiating activator and a leuco dye reporter which, when irradiated, undergoes a measureable darkening of color. Previous studies have demonstrated PRESAGE® can be successfully used to measure complex photon dose distributions [5-8].

Unfortunately, PRESAGE® suffers from a dose under-response, or signal quenching, in the Bragg peak region, making accurate proton dosimetry so far impossible [9-11]. This signal quenching has also been observed in polymer gels as well as most other 3D dosimetry alternatives and the source has been primarily attributed to LET-dependence [12]. Early proton irradiations in polymer gels found that high-LET dose deposition resulted in recombination of the radical initiators rather than signal activation resulting in dose under-response [13], local depletion of the signaling components [14], or some combination of the two. While conventional PRESAGE® irradiated with protons demonstrated maximum quenching greater than 30%, a newer formulation developed specifically for proton therapy was developed by Heuris Pharma, LLC which adjusted the concentrations of the active components. Our studies with this formulation measured a 20% under-response [15].

No known studies have comprehensively investigated relative concentrations of the radical initiator and leuco dye responder to chemically correct for PRESAGE® signal quenching in a proton beam. In this study, we investigated the relationship of the radical initiator component on LET-dependence to assess the potential of a formulaic correction of this quenching effect.

2. Methods and Materials

2.1 PRESAGE® Manufacturing

The PRESAGE® dosimeters used in this study were manufactured in-house using the chemical components described previously: Leuco Malachite Green (LMG) dye, chloroform ($CHCl_3$) radical initiator, dibutyltin dilaurate as a catalyst, and polyurethane resin (Parts A and B, Crystal Clear 204, Smooth-On). Ten formulation batches were made with radical initiator concentrations varying between 3.0-30.0% (w/w) using a method described by Alqathami *et al* [16]. A static 2.0% (w/w) LMG concentration was used in all formulations. The total polyurethane concentration between formulations was adjusted for changes to radical initiator concentration. Parts A and B maintained a 10:9 ratio as recommended by the supplier. The catalyst concentration was <<1% (w/w). After combining all ingredients, the formulations were poured into spectrophotometer cuvettes ($1x1x4.5\ cm^3$) then placed in a pressure chamber and allowed to cure for 24 hours at 70 psi to eliminate the formation of air bubbles.

2.2 Irradiation

The cuvettes were irradiated using a 225-MeV (26.9 cm range) passively scattered proton beam at the MD Anderson Proton Therapy Center (MDACC PTC, Houston, TX). The beam was modulated to deliver a 10 cm SOBP with a $10x10\ cm^2$ field size. A solid water phantom was used and cuvettes were positioned at four depths along the beam profile. The first depth was 10 cm and situated in the low LET dose plateau allowing dose normalization. Cuvettes were next irradiated at three points along the SOBP at the following depths to measure a uniform dose at varying LETs: 19 cm (the most proximal SOBP region), 22 cm (SOBP center), and 25 cm (the distal most SOBP region). Two cuvettes were independently irradiated at each position for each formulation. The cuvettes were irradiated to 193.6 MU (approximately 200 cGy to the SOBP). The beam profile was separately measured using a Zebra multilayer ion chamber (MLIC). Additionally, the dose-averaged LET (LET_d) was calculated analytically using a model developed by Wilkens *et al* [17].

2.3 Readout

The dose signal was measured by the change in optical density (ΔOD) of the cuvettes using a Thermo Scientific GENESYS™ 10S UV-VIS spectrophotometer. Optical measurements were made at the absorption peak wavelength of 632 nm approximately 24 hours after irradiation. The ΔOD was determined by subtracting the pre-irradiation measured OD from the post-irradiation measured OD. Cuvettes were protected from light exposure at all times and stored in a cold environment (<3°C) except when allowed to equilibrate to room temperature two hours prior to measurements and irradiation.

The ΔOD for each formulation was normalized to dose using the average of the measurements from cuvettes in the dose plateau region to the dose measured by the MLIC. The quenching magnitude for the measurements along the SOBP region was calculated as the difference of the normalized PRESAGE® signal to the ion chamber measurements made at each depth.

3. Results

Figure 1a shows the normalized dose responses of each PRESAGE® formulation relative to ion chamber measurements. All formulations demonstrated quenching along the SOBP. Figure 1b shows the magnitude of the quenching for better illustration. Formulations with radical initiator concentrations between 12-21% under-responded by less than 3% at the proximal SOBP dose point. The quenching magnitude for these concentrations increased slightly at the other measurement points. Formulations outside of this range had greater quenching magnitudes across all points. The 12% $CHCl_3$ formulation showed the lowest quenching across the whole dose profile with a maximum under-response of 14.6%. In contrast, the 30% $CHCl_3$ formulation showed a maximum under-response of 73.8%.

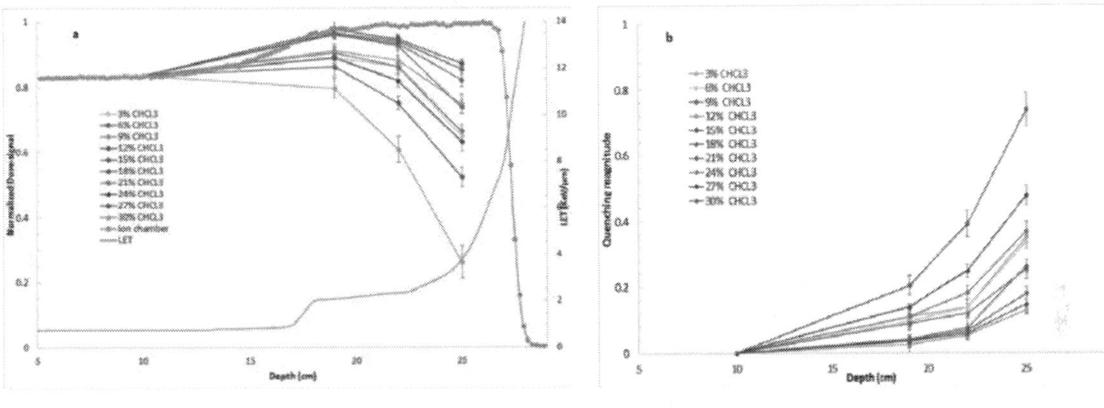

Figure 1. (a) The normalized dose-response curves for each PRESAGE® formulation irradiated by a proton beam compared with ion chamber measurements. The pointes are connected by straight lines to aid the eye. Additionally, the calculated LETd is illustrated. (b) Only the magnitude of the quenching is shown as a function of depth.

The calculated LET_d is shown in Figure 1a. Figure 2a correlates the quenching magnitude with the LET_d at each point. PRESAGE® formulations with radical initiator concentrations below 12% or above 18% demonstrated sharper quenching increase than intermediate concentrations. This resulted in a kind of minimized quenching plateau of similarly responding formulations between 12-18% in the lower LET measurement points. The relationship between dose quenching and LET_d is plotted in Figure 2b. This unexpectedly shows a linear relationship in formulations between 3-21% which becomes sublinear at higher radical initiator concentrations.

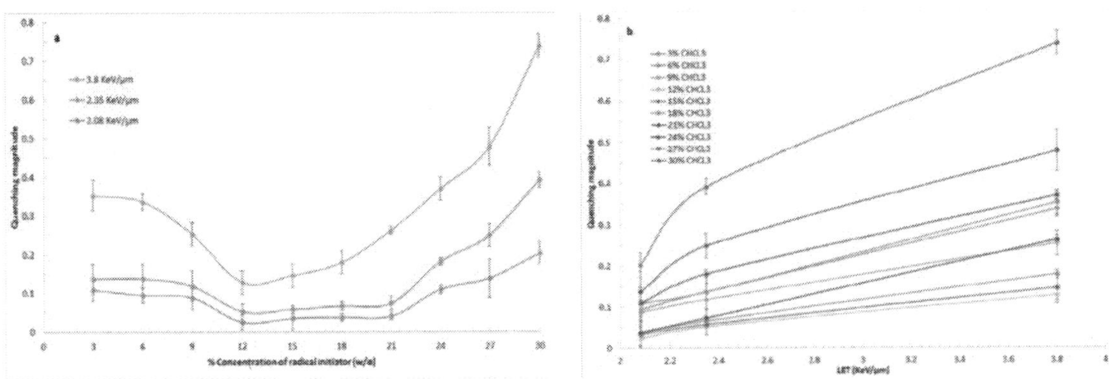

Figure 2. (a) The quenching magnitude at each measured LETd as a function of radical initiator concentrations. (b) The quenching as a function of LETd for each PRESAGE® formulation.

IOP Conf. Series: Journal of Physics: Conf. Series **847** (2017) 012027 doi:10.1088/1742-6596/847/1/012027

4. Discussion and Conclusions

The concentration of the radical initiator in PRESAGE® dosimeters is shown to affect the magnitude of the dose quenching in a proton beam. Unfortunately, a limit to this quenching reduction was reached before it could be fully eliminated, but this is sufficient to demonstrate that further formulation optimization can improve the standard proton formulation available today.

This study also offers some insight into the physical processes resulting in quenching. As radical initiator concentrations increase beyond a point, the quenching magnitude begins to sharply increase which demonstrates that radical recombination likely plays a role in the quenching effect, although it may not be the only contributor. Further studies are needed to determine the impact on quenching from leuco dye saturation as well as to determine if the quenching can be further reduced with chemical substitutions.

5. References

[1] Klein E E *et al* 2009 *Med. Phys.* **36** 4197-212
[2] Hill R *et al* 2009 *Med. Phys.* **36** 3971-81
[3] Hill R *et al* 2014 *Phys. Med. Biol.* **59** R183-231
[4] Summers P *et al* 2013 *Med. Phys.* **39** 3866
[5] Klawikowski S J *et al* 2014 *Phys. Med. Biol.* **59** N211-20
[6] Juang T *et al* 2014 *Med. Phys.* **41** 071706
[7] Grant R *et al* 2013 *J. Phys.: Conf. Ser.* **444** 012020
[8] Wuu C-S *et al* 2013 *J. Phys. Conf. Ser.* **444** 012055
[9] Zhao L et al 2010 *J. Phys.: Conf. Ser.* **250** 12035
[10] Gorjiara T *et al* 2011 *Med. Phys.* **38** 2265-74
[11] Zhao L *et al* 2012 *Phys. Med. Biol.* **57** N431-43
[12] Karger C P *et al* 2010 *Phys. Med. Biol.* **55** R193-234
[13] Gustavsson H *et al* 2004 *Phys. Med. Biol.* **49** 3847-55
[14] Schmid A I *et al* 2013 *J. Phys.: Conf. Ser.* **444** 012096
[15] Carroll M *et al* 2013 *J. Phys.: Conf. Ser.* **444** 012059
[16] Alqathami M A *et al* 2016 *Phys. Med. Biol.* **61** 813-24
[17] Wilkens J J and Oelfke U 2003 *Med. Phys.* **30** 806-15

IOP Conf. Series: Journal of Physics: Conf. Series **847** (2017) 012028 doi:10.1088/1742-6596/847/1/012028

Enhanced method for determining the low-LET saturation dose of PRESAGE®

Simon J Doran[1], Paolo Pellicioli[2], John Adamovics[3] and Elke Brauer-Krisch[2]
[1]CRUK Cancer Imaging Centre, Institute of Cancer Research, London, UK
[2]European Synchrotron Radiation Facility, Grenoble, France
[3]Department of Chemistry and Biology, Rider University, Lawrenceville, NJ, USA

E-mail: Simon.Doran@icr.ac.uk

Abstract. An accurate value for the saturation dose of PRESAGE® is important in elucidating the mechanism of the so-called *quenching* phenomenon in the 3-D dosimetry of proton beams. This quantity is difficult to obtain, because it involves measuring the optical density of a set of very highly attenuating samples. We present a novel method that makes use of the ability to irradiate microscopically-thin "slices" of a cuvette using microplanar beams from a synchrotron. Using the new method, we were able to measure optical densities of up to 50 cm^{-1} using a standard spectrophotometer, and we derived a value 14000 Gy for the saturation parameter D_{37}.

1. Introduction

Quenching of dosimeter response is a major issue in the measurement of the radiation dose from proton and other ion beams. "Quenching" is said to occur when, for a given physical dose, the dosimeter gives a lower output reading for radiation with high linear energy transfer (LET) than for the same dose of low-LET radiation. In the case of 3-D chemical dosimeters, that output reading might be R_2 value (polymer gel, read out with MRI [1]) or optical density (Fricke gel[2] or PRESAGE® [3] read out with optical-CT). Although it has been reported that some formulations of polymer gel do not exhibit quenching [4], most chemical dosimeters (and many "physical" dosimeters, too) are affected to some degree or other. The effects are particularly noticeable for measurements made at the Bragg peak in proton treatments and the implications of this for accurate dosimetry have been explored in [5].

The precise origin of the under-response in 3-D dosimeters still remains to be elucidated. Gustavsson *et al* [1] hypothesised that the effect was attributable to increased recombination of the free-radicals produced by the irradiation, before they could effect a chemical change in the dosimeter. A second possibility is local saturation of the dosimeter along the tracks of the protons or ions, where a very high dose is localised within a microscopic region. The latter situation may be described using the formalism of *track structure theory* [6]. A key parameter in this analysis is the dose required to saturate the dosimeter response at low LET. It can be shown that the expected dose-response relation is a single-exponential recovery curve of form $OD \propto 1 - \exp(-D/D_{37})$, where D_{37} is the dose required to convert a fraction $1/e$ (37%) of the active elements in the dosimeter. We have previously made several attempts to determine the value of D_{37} using a spectrophotometer and optical CT scanner [7, 8].

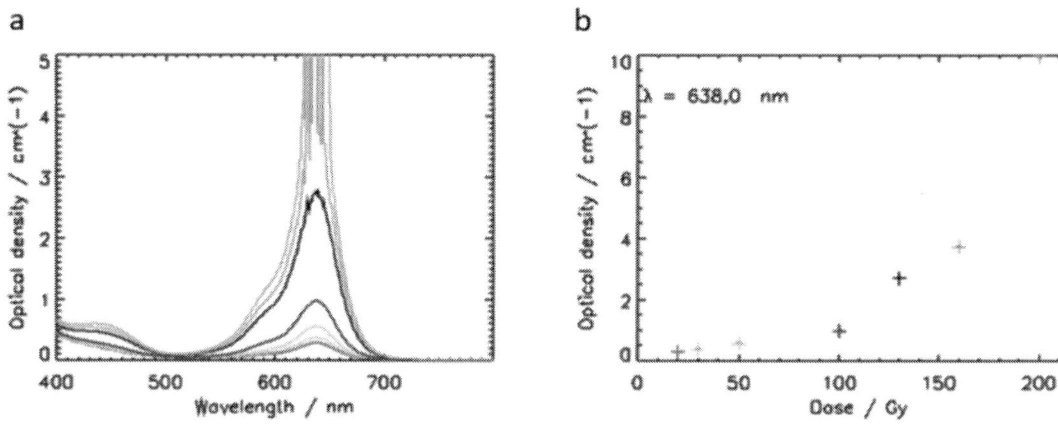

Figure 1. (a) Typical set of spectra obtained from spectrophotometer, illustrating the problems of a limited dynamic range in output optical density; (b) optical density values at peak of PRESAGE® spectrum (638 nm).

The significant experimental difficulty in measuring this parameter is illustrated in figure 1. Irradiated PRESAGE® cuvettes are typically analysed using a spectrophotometer, but all such instruments have a finite dynamic range: the minimum signal detectable above the noise threshold determines the maximum optical density that can be measured. On our current spectrophotometer (Cary 50, Agilent Technologies, Santa Clara, CA), this value is a little less than 3 cm^{-1}. As can be seen from the spectra in figure 1a, readings for the samples with the lowest four doses are continuous across the whole spectral range, but as the dose is increased further, severe distortions of the spectra occur in the region corresponding to the dose-dependent part of the PRESAGE® absorption and, consequently, the values in the graph of figure 1(b) are incorrect. Notice that the actual signal reaching the detector decreases exponentially with increasing optical density and so 1 cm cuvette samples that are highly irradiated pass virtually no light.

Our previous work [8] overcame this problem partially by employing a special type of cuvette with 2 mm path length. However, this proved unsuitable for formulation of PRESAGE® studied here, which had an extremely high saturation dose, resulting in very large optical densities. Hence, a new method was devised in which only a portion of each cuvette was irradiated, as described below. Because the ultra-high absorption can be confined to just a thin "slice" of the sample, the *overall* attenuation remains a level that can be measured. A knowledge of the geometry of the situation allows us to back-calculate what the attenuation would have been had the whole sample been irradiated.

2. Material and Methods

All measurements were performed on cuvettes with a 1 cm optical path length, filled with PRESAGE® (Heuris Pharma, Skillman, NJ). Samples were irradiated on beamline ID-17 at the European Synchrotron Radiation Facility (ESRF) and a full description of the general irradiation methodology and beamline characteristics has been given in previous publications, e.g., [9]. Cuvettes were placed immediately distal to a number of Perspex sheets to ensure appropriate electron equilibrium and such that the centre of the cuvette was located at the standard reference depth used in rest of the local dosimetry programme.

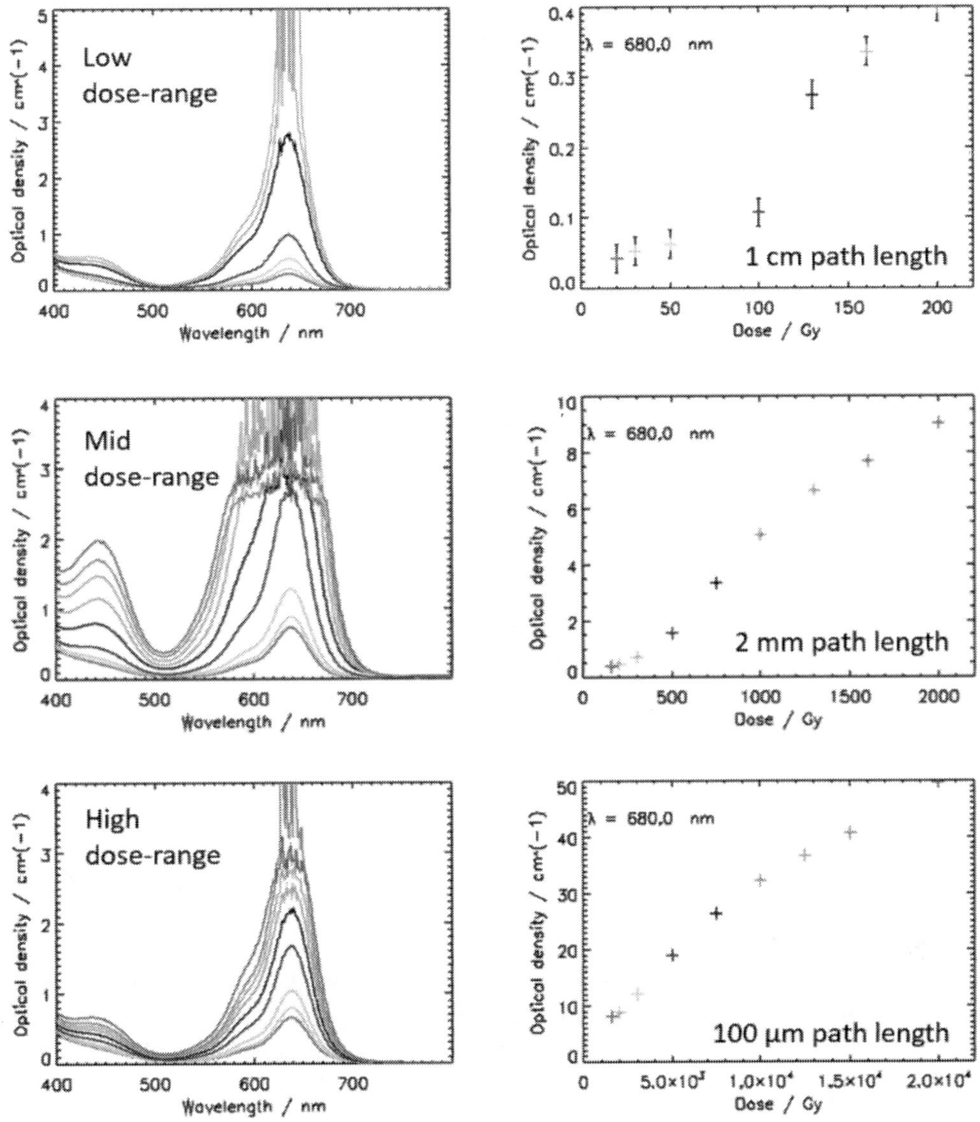

Figure 2. Spectra and graph of spectral value at a single wavelength (680 nm) for the complete set of samples. Note the difference between the vertical scales on the left and right. On the left are the values of optical density measured by the spectrophotometer, assuming a uniform 1 cm cuvette. On the right are the values of optical density corrected for the path length through the high-dose region of the sample. We are unclear of the reason why the linearity of the top-right graph is so poor (the point at 100 Gy is an unexpected outlier) or for the behavior of the first three points on the middle right graph).

Three separate sets of irradiations were performed: (i) Low-dose range: 7 samples were irradiated with a field size of 15 mm × 30 mm, thus covering the complete width of the cuvettes. The calculated doses at the reference point were 20, 30, 50, 100, 130, 160 and 200 Gy. (ii) Mid-dose range: 9 samples were irradiated with a "slit pattern" of 2 mm width (see [10]) to doses of 160, 200, 300, 500, 750, 1000, 1300, 1600 and 2000 Gy. (iii) High-dose range: 9 samples were irradiated with a microplanar beam of width 100 µm to doses of 1600, 2000, 3000, 5000, 7500, 10000, 12500, 15000 and 20000 Gy.

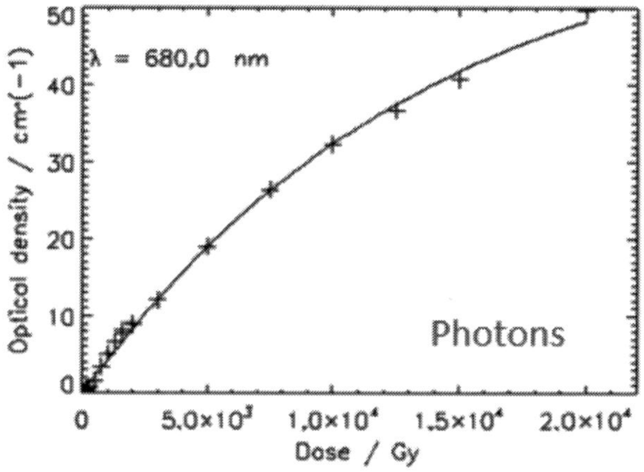

Figure 3. Final results for x-ray dose deposition at ESRF.

3. Results and discussion

The new method uses two techniques to read samples of high optical density. Different ranges of absorption coefficient α can be accommodated by irradiating different "slice thicknesses", d out of a total cuvette thickness of d_0. Since the spectrophotometer measures the total attenuation $A \propto \exp(-\alpha d)\exp(-\alpha_0[d_0 - d])$, it suffices to keep the product αd within the measureable range, with α_0, the absorption coefficient of unirradiated PRESAGE®, rapidly becoming trivial as the dose escalates. With prior knowledge of the dose-response relation, one could arrange for all spectra to be measured without the distortions illustrated in figure 2. However, since we did not "know the result in advance", our choice of thicknesses was not optimal. The second trick is thus to measure at a wavelength slightly away from the absorption maximum and, after experimentation, we chose 680 nm. Figure 2 shows the complete set of data obtained in the three ranges, whilst figure 3 displays all the data on a single graph, with a fit to the model. The fit parameter obtained was D_{37} = 14000 Gy. Further analysis at a number of different wavelengths is needed to determine the most appropriate method for calculating the error estimate in this value. However, a preliminary point of interest to note is that this is value is significantly in excess of the corresponding value obtained in our previous study [8], indicating that the PRESAGE® saturation dose is formulation-dependent.

4. Conclusion

With the new, enhanced method, it is possible to measure dose-response curves for 3-D dosimeters out to extremely high doses, thus simulating the "micro-dose" along a proton beam. In principle, if track structure theory is applicable and does indeed explain the quenching phenomenon, PRESAGE® formulations with higher values of D_{37} should exhibit reduced quenching.

5. References

[1] Gustavsson H *et al* 2004 *Phys. Med. Biol.* **49** 3847-55
[2] Bäck S Å J *et al* 1999 *Phys. Med. Biol.* **44** 1983
[3] Al-Nowais S *et al* 2009 *Appl. Radiat. Isotop.* **67** 415-8
[4] Zeidan O *et al* 2010 *Med. Phys.* **37** 2145-52
[5] Doran S *et al* 2015 *Phys. Med. Biol.* **60** 709-26
[6] Waligorski M *et al* 1986 *Int. J. Radiat. Appl. Instru. D.* **11** 309-19
[7] Al-Nowais S *et al* 2009 *J. Phys.: Conf. Ser.* **164** 012043
[8] Nowais S A *et al* 2010 *J. Phys.: Conf. Ser.* **250** 012034
[9] Doran S J *et al* 2010 *Phys. Med. Biol.* **55** 1531-47
[10] Rahman A A *et al* 2011 *Phys. Med. Biol.* **56** 4177

IOP Conf. Series: Journal of Physics: Conf. Series **847** (2017) 012029 doi:10.1088/1742-6596/847/1/012029

The IROC Houston Quality Assurance Program: Potential benefits of 3D dosimetry

D S Followill[1], H A Molineu[1], R Lafratta[2] and G S Ibbott[3]
[1]IROC Houston QA Center, MD Anderson Cancer Center, Houston, TX 77030, USA
[2]St Lukes Hospital, Houston, TX 77030, USA
[3]Department of Radiation Physics, MD Anderson Cancer Center, Houston, TX 77030, USA

Email: dfollowi@mdanderson.org

Abstract. The IROC Houston QA Center has provided QA core support for NCI clinical trials by ensuring that radiation doses delivered to trial patients are accurate and comparable between participating institutions. Within its QA program, IROC Houston uses anthropomorphic QA phantoms to credential sites. It is these phantoms that have the highest potential to benefit from the use of 3D dosimeters. Credentialing is performed to verify that institutions that are using advanced technologies to deliver complex treatment plans that conform to targets. This makes it increasingly difficult to assure the intended calculated dose is being delivered correctly using current techniques that are 2D-based. A 3D dosimeter such as PRESAGE® is able to provide a complete 3D measured dosimetry dataset with one treatment plan delivery. In our preliminary studies, the 3D dosimeters in our H&N and spine phantoms were found to be appropriate for remote dosimetry for relative dose measurements. To implement 3D dosimetry in IROC Houston's phantoms, the benefit of this significant change to its current infrastructure would have to be assessed and further work would be needed before bringing 3D dosimeters into the phantom dosimetry program.

1. Introduction

Even with current technologies and safety culture radiation delivery incidents continue to occur and to be reported on. During the past decade, a number of significant radiation therapy errors have been reported [1-3]. There have been several publications indicating that the introduction of advanced technology equipment has permitted errors to occur that might otherwise have been detected [4]. There are also indications that the demands of advanced technologies on department resources have drawn resources from simpler or basic functions [5]. If the basic dosimetry is not correct then it does not matter whether a clinic uses advanced technology or not, the dose delivery will still be inaccurate [6].

Clearly, larger errors have a significant impact on the success of radiation therapy clinical trials. But smaller errors, while having no discernable effect on the treatment of an individual patient, may influence the overall success of treatments of large numbers of patients [7-9] as found in phase III clinical trials. For multi-institutional clinical trials, the radiation treatments need to be comparable among all participating institutions in order to compare treatment arms with minimal uncertainty involved that might blur otherwise distinct differences. Data from Peters *et al* [10] demonstrated the benefit of good radiotherapy data indicating that even if the trial is not a radiation question, the radiotherapy component is still vital to the trial outcome. Thus the need for QA programs to monitor institutions participating in clinical trials where small deviations in the delivered dose to many patients from individual institutions may impact on the outcome of the trial is needed.

2. The Role of the IROC Houston QA Center (IROC Houston)

The IROC Houston QA Center was established in 1968 as the Radiological Physics Center (RPC) to contribute to the development, conduct, and QA of multi-institutional cooperative group clinical trials as an independent QA center. IROC-Houston currently monitors 1816 institutions (3700 megavoltage therapy machines) that participate in the National Cancer Institute's (NCI's) National Clinical Trial Network (NCTN) multi-institutional clinical trials. IROC Houston promotes six core support components to an effective clinical trial QA program: 1) institution qualification/ QA, 2) review of clinical trial protocols, 3) credentialing of institutions for specific protocols, 4) transfer of radiotherapy data, 5) clinical/technical review of the patient radiotherapy data, and 6) management of radiotherapy data for analysis. These methods within the six components help to ensure that radiotherapy data used for trial outcomes analysis has minimal uncertainty.

IROC Houston primarily focuses on institution qualification, clinical/technical review of patient radiotherapy data and credentialing of institutions. IROC Houston's QA program focuses on ensuring that radiation doses delivered are accurate and comparable between institutions participating in NCTN trials by auditing components of dose delivery from reference calibration to tumor dose calculation.

2.1. Institution qualification and ongoing QA

The qualifications of an institution to participate in clinical trials are initially discerned by the completion of a facility questionnaire (FQ) to understand the institution's current state of practice. As accelerator technology has become more complex and radiation delivery modalities change, demonstration of the expertise of use, appropriate QA program, equipment/resources necessary to deliver the radiotherapy treatment accurately is required. Institution qualification also includes remote verification of machine calibration with TLD/OSLD dosimeters. IROC Houston monitors approximately 14,000 megavoltage photon, electron and proton beams annually.

2.2 Clinical and technical review of the trial patient radiotherapy data

IROC Houston reviews the treatment plans (external beam and brachytherapy) prepared by participating institutions for patients registered on trials. These patient reviews may be 1) pre-treatment reviews, 2) on-treatment reviews and, 3) post-treatment reviews. Up to a nearly 1000 patient reviews are performed annually by IROC Houston. It is during the post-treatment reviews where the patient case is scored using clinical and dosimetric criteria, as per protocol or as a deviation.

2.3 Credentialing of institutions for specific protocols and treatment modalities

Clinical trials that require the use of advanced technologies such as IMRT, SBRT, IGRT, and proton therapy are considered challenging to implement in a comparable manner for clinical trials, so institutions are required to demonstrate their ability, through a credentialing process, before being permitted to register patients. The mission of IROC Houston is not to restrict participation but to assist institutions in any required remedial actions so that they meet the credentialing criteria and can accrue radiotherapy patients in a comparable manner to other participating institutions. Credentialing requirements include many components, but IROC Houston's primary mechanism is the irradiation of an anthropomorphic end-to-end QA phantom. It is with these QA phantoms that the potential benefits of 3D dosimetry can be realized.

2.3.1. IROC Houston's anthropomorphic end-to-end QA Phantoms

IROC Houston has developed multiple anthropomorphic end-to-end QA phantoms, beginning with the SRS and IMRT heads, to currently having seven different phantom types for photon and proton therapy. The phantoms, while not anatomically exact, are close approximations to the true anatomy of the various disease sites to be used in protocols. The phantoms currently use TLD and 2D planar radiochromic film to evaluate the dose delivered. These particular dosimeters are limited in their ability to fully evaluate the complex 3D dose distributions delivered using IMRT, IMPT and IGRT. However, in their present form, the phantoms provide a consistent test and criteria to evaluate each institution's ability to deliver a specific radiotherapy treatment ensuring comparability between participating institutions [11-12]. Current phantoms have been used to assess SRS, IMRT, SBRT, VMAT, moving targets, and proton therapy. Since 2000, a total of 5,671 phantoms have been irradiated by participating institutions in North America and elsewhere.

3. 3D dosimetry - Potential Benefits

Today's complex treatment plans are volumetric in design that may contain varying dose gradients depending on the prescription and normal tissue constraints. Independent peer review/verification of these complex treatments is becoming increasingly important in the clinic since these treatments comprise multiple steps, each of which could represent a failure mode resulting in a dose delivery error. Since these complex shaped dose distributions normally are designed to conform to irregularly shaped targets, traditional dosimetry methods, i.e. TLD and planar film measurements may not be sufficiently comprehensive to verify correct delivery [13]. It has been shown that IMRT QA of single IMRT fields may miss catching errors that appear when all fields are combined for the total tumor dose delivery [14]. Studies have shown this method of single planar dosimetry can produce gamma analysis results that are insensitive to the changes in measured and calculated IMRT dose distributions [15]. Volumetric dosimeters have that potential to measure a complete 3D measured dosimetry dataset against which to compare the calculated planned 3D dose distribution.

The IROC-Houston's phantoms are designed to yield only two or three planes of dose information, to compare to the treatment plan calculated data. The failure rate for IROC Houston's IMRT H&N phantom still remains at 10-15% even after 15 years of use [16]. A measured 3D dataset might even capture more errors and possibly provide clues as to the source of the error.

3.1 Preliminary studies

IROC Houston's H&N IMRT and spine phantoms were modified to accept 3D dosimeters to compare against currently used planar film measurements. Studies by Grant *et al* [17] and Lafratta *et al* [18] have shown that PRESSAGE® 3D dosimeters yield equivalent dose profile (figure 1) and gamma analysis (table 1) results as the radiochromic film in the spine and H&N phantoms, respectively.

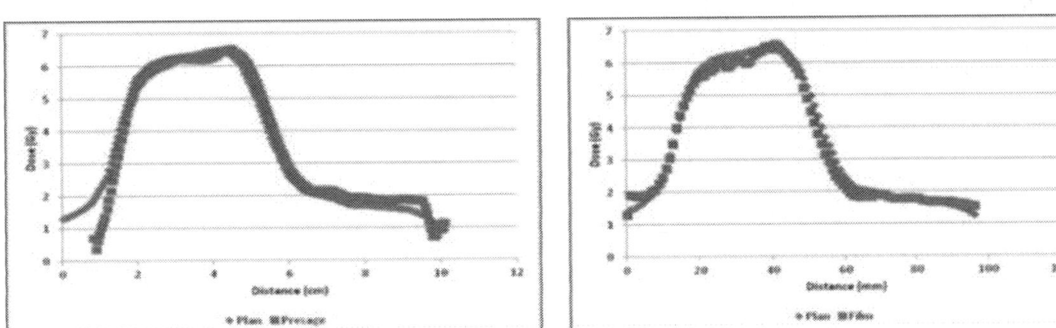

Figure 1. Comparison of dose profiles from PRESAGE® 3D dosimeter and radiochromic film to treatment plan calculated profile.

Table 1. Percentage of passing pixels between PRESAGE® and film in the IMRT H&N phantom

	Constraints	Film	PRESAGE®
Axial Plane	3%/3mm	85%	66%
	5%/3mm	93%	86%
	7%/4mm	99%	93%
Sagittal Plane	3%/3mm	85%	61%
	5%/3mm	90%	82%
	7%/4mm	90%	91%

4. Summary

There exists a definite potential for 3D dosimeters in the IROC Houston QA Center anthropomorphic phantom credentialing program. Results indicate near equivalency in measured data with the 3D dosimeters providing a more comprehensive volumetric assessment. To implement 3D dosimeters in the credentialing phantoms would require a significant change in the current IROC Houston infrastructure. This change needs to be fully evaluated.

5. Acknowledgement

This work was supported by a Public Health Service grant CA10953 and CA180803 awarded by the Department of Health and Human Services National Cancer Institute.

6. References

[1] *International Atomic Energy Agency. Safety Reports Series No. 17. Lessons Learned from Accidental Exposures in Radiotherapy* 2000 Vienna: IAEA
[2] Rogers L "Over 200 hurt or killed by botched radiation" *The Times (London) Online*: http://www.timesonline.co.uk/tol/news/uk/article711360.ece?token=null&offset=12&page=1
[3] WAToday, "Cancer patients 'given excess radiation'" http://www.watoday.com.au/breaking-news-national/cancer-patients-given-excess-radiation-20090131-7u9o.html
[4] Patton G A *et al* 2003 *Int. J. Radiat. Oncol. Biol. Phys.* **56** 50-57
[5] Marks L B *et al* 2007 *Int. J. Radiat. Oncol. Biol. Phys.* **69** 1579-1586
[6] Hill R *et al* 2014 *Phys. Med. Biol.* **59** R183-231
[7] Bentzen *et al* 2000 *Eur. J. Cancer* **36** 615-620
[8] Boyer A and Schultheiss T 1988 *Radiother. Oncol.* **11** 65-71
[9] Pettersen *et al* 2008 *Radiother. Oncol.* **86** 195-199
[10] Peters L J *et al* 2010 *J. Clin. Oncol.* **28** 2996-3001
[11] Ibbott G S *et al* 2006 *Technol. Cancer Res. Treat.* **5** 481-7
[12] Kry S *et al* 2012 *Int. J. Radiat. Oncol. Biol. Phys.* **84** S128-9
[13] Schreiner L J 2009 *J. Phys: Conf Ser.* **164** 012001
[14] McKenzie E M *et al* 2014 *Med. Phys.* **41** 121702
[15] Kruse J J 2010 *Med. Phys.* **37** 2516-24
[16] Molineu A *et al* 2013 *Med. Phys.* **40** 022101
[17] Grant R *et al* 2012 *Med. Phys.* **39** 3724
[18] Lafratta R *et al* 2014 *Med. Phys.* **41** 350

IOP Conf. Series: Journal of Physics: Conf. Series **847** (2017) 012030 doi:10.1088/1742-6596/847/1/012030

Automated four-dimensional Monte Carlo workflow using log files and real-time motion monitoring

P Sibolt[1,2], R O Cronholm[3], E Heath[4], C E Andersen[1] and C F Behrens[2]
[1]Center for Nuclear Technologies, Technical University of Denmark, Roskilde, Denmark
[2]Radiotherapy Research Unit, Herlev Hospital, Herlev, Denmark
[3]Department of Radiation Physics, Skåne University Hospital, Lund, Sweden
[4]Carleton Laboratory for Radiotherapy Physics, Carleton University, Ottawa, Canada.

Email: pasi@dtu.dk

Abstract. With emerging techniques for tracking and gating methods in radiotherapy of lung cancer patients, there is an increasing need for efficient four-dimensional Monte Carlo (4DMC) based quality assurance (QA). An automated and flexible workflow for 4DMC QA, based on the 4DdefDOSXYZnrc user code, has been developed in python. The workflow has been tested and verified using an in-house developed dosimetry system comprised of a dynamic thorax phantom constructed for plastic scintillator dosimetry. The workflow is directly compatible with any treatment planning system and can also be triggered by the appearance of linac log files. It has minimum user interaction and, with the use of linac log files, it provides a method for verification of the actually delivered dose in the patient geometry.

1. Introduction

Motion management in external beam radiotherapy is becoming increasingly sophisticated and the demands on quality assurance (QA) of advanced radiotherapy are therefore also increasing. One interesting example is the need for QA of emerging techniques for tracking and gating methods in radiotherapy of lung cancer patients. Many commercial treatment planning systems (TPS) have recognized difficulties to accurately calculate dose for dynamic treatments due to challenges related to breathing motion and heterogeneities. This has for example been shown for deep-inspiration breath-hold (DIBH) intensity-modulated radiotherapy (IMRT) of lung cancer patients, where Monte Carlo (MC) calculations revealed large inaccuracies in the dose calculated by the TPS [1]. Monte Carlo is considered to be the gold standard among dose calculation algorithms and the use of MC-based QA (MCQA) for verification of external beam radiotherapy is widely recommended, especially in the case of four-dimensional treatments [2]. However, implementation of MCQA often involves procedures including several steps of manual interaction or its integration into one specific TPS using a designated interface. An automated MCQA workflow with minimum user interaction is much more desirable. Preferably this MCQA workflow would enable four-dimensional Monte Carlo (4DMC) which models synchronously the dynamic beam configurations and the motion and deformation of the patient anatomy.

The purpose of this study was to incorporate a solution for 4DMC into an automated MCQA workflow with the possibility to use linear accelerator (linac) log files and motion monitoring signals for both pre-treatment and per-fraction dose verification.

IOP Conf. Series: Journal of Physics: Conf. Series **847** (2017) 012030 doi:10.1088/1742-6596/847/1/012030

2. Material and Methods

2.1. Workflow for four-dimensional Monte Carlo

In order to incorporate the synchronization between the dynamic beam configuration and the motion of the patient anatomy, MC simulations were carried out using 4DdefDOSXYZnrc. The 4DdefDOSXYZnrc code is an altered version of the EGSnrc [3] user code defDOSXYZnrc, where dose deposition is tracked in a deformed anatomy without altering the voxel grid [4,5]. It makes use of the source 20 of DOSXYZnrc for simulation of continuously varying beam configurations [6]. The 4DdefDOSXYZnrc user code samples a new geometry for each incident particle, which enables simulation of a continuously moving anatomy. The geometries are sampled by linearly interpolating a deformation vector field, determined from image registration between the reference phase and an extreme phase of the 4DCT, using the motion signal measured during treatment.

The use of linac log files, deformation vector fields, and motion monitoring signal as an input for the 4DMC simulations is incorporated within a workflow solution for automated MCQA. The workflow is built up of a number of different modules, all written in python, which are executed sequentially without user interaction (Table 1). The automation of the workflow depends on each module, at the end of execution, leaving data for the next module to process. The workflow is connected to the TPS by means of exports and imports done in the TPS. This implicitly means that the workflow is portable between TPSs. The first module reads and processes TPS exported DICOM files and the last module writes the resulting dose distributions as DICOM files. In addition to the original TPS DICOM files, the 4D workflow requires access to the deformation vector field and the motion monitoring signal. This access can be configured differently depending upon the deformable image registration software and motion monitoring system in use. Furthermore, the workflow can be initiated by the appearance of linac log files, which are used to write the DICOM input files needed to trigger the start of the workflow.

Table 1. A brief description of the different modules that constitutes the modified workflow enabling four-dimensional Monte Carlo based quality assurance.

Module	Brief description
I	Generates BEAMnrc/4DdefDOSXYZnrc input files replicating all plan and motion specific parameters. Initiates module II if CT data and RT Structure Set are exported from the TPS.
II	Builds a voxalized phantom based on the CT data and information form the RT Structure Set. The module is based on CTC-ask [7]. The distinct differences being that it is written in python, automated and includes patient support structures. Media selection rules are predefined by the user and can be differentiated for various structure types. Also handles the motion input and writes the vector field in the correct format for further use in the simulations.
III	Initiates treatment specific BEAMnrc simulation starting from a previously generated phase space scored above collimating devices.
IV	Concatenates phase spaces files (if parallel simulation). Computes number of histories required in order to achieve a fixed level of uncertainty. Initiates 4DdefDOSXYZnrc simulations.
V	Deletes auxiliary files (e.g. phase spaces)
VI	Converts to absolute dose. Writes DICOM RT objects using the exported files as templates.

One of the aims of the proposed workflow is that no simulation parameters should be hardcoded. Instead, the workflow uses initial phase space files together with templates, where only treatment specific parameters are overwritten. The workflow is controlled by extracting information from the provided DICOM RP files together with data from a global and a machine specific configuration file written in plain text. This makes the workflow flexible and independent of vendor, energy and fluence mode.

Sending a treatment plan through this workflow results in a set of DICOM RT objects (plan and dose), which are written using the TPS exported files as templates. This enables direct import to any TPS with automatic connection to the correct study.

2.2. Example with an in-house developed moving thorax phantom

A time-resolved plastic scintillator detector (PSD) dosimetry system was combined with a dynamic thorax phantom (both in-house developed), containing a PMMA sphere (tumor, ø = 5 cm) embedded in a balsa wood insert (lung) and laterally position in a hollow cylinder [8]. During irradiations, the cylinder containing the tumor was set in a controlled respiratory-like sinusoidal motion with a frequency of 0.25 Hz and peak-to-peak amplitude of 20 mm (corresponding to a clinically relevant motion with 15 breaths per minute). PSD measurements were performed in the center of the tumor for two half-arc 6 MV RapidArc plans (plan 1 and 2 optimized to give mean tumor doses of 1 Gy and 2 Gy, respectively) on a Varian TrueBeam linac. Trajectory log files and phantom motion profiles were obtained during the measurements and thereafter used for generating 4D MC input files. Deformation vectors corresponding to the phantom cylinder motion during treatments were manually generated and applied to the voxel grid at a reference phase. Monte Carlo simulations in the deformed anatomy were carried out, according to the workflow described above, with a calculation time of less than 24 hours on a standard CPU based cluster for a statistical uncertainty below 0.2 %. 4D MC input files were based on both the treatment plan beam configurations as planned in the TPS as well as the actually delivered dynamic beam configuration as extracted from the linac log files. A comparison between planned and delivered dose was conducted.

3. Results

3.1. Example with an in-house developed thorax phantom

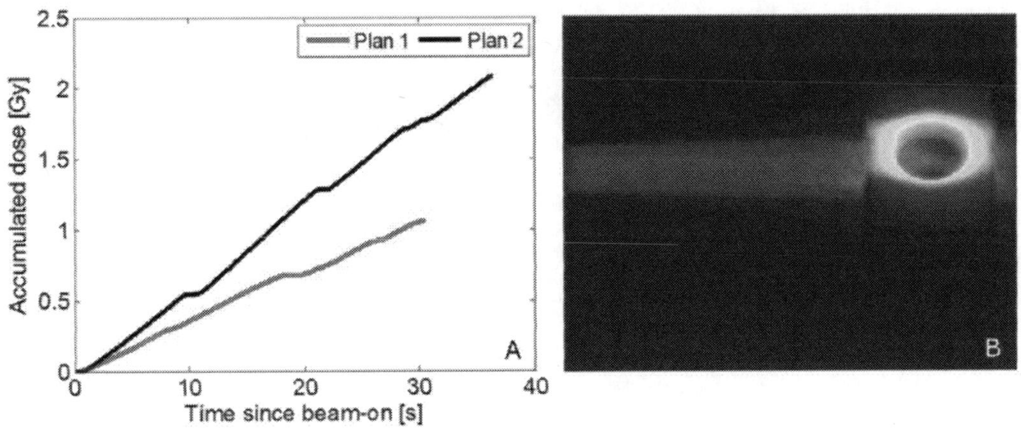

Figure 1. (A) The accumulated dose over time measured by the PSD in a dynamic thorax phantom for plan 1 and 2 and (B) the corresponding 4Ddefdosxyznrc calculated accumulated dose based on the actually delivered beam configuration obtained from the linac log file, presented for a slice centrally located in the tumor.

The time-resolved PSD measurements resulted in average accumulated doses (relative standard deviation of ~1 %; n = 6 per plan) deviating from corresponding TPS isocenter doses with -1.6 % and -2.8 % for plan 1 and 2, respectively (figure 1A). The hypothesis was that the deviations between measured and TPS calculated doses were true deviations as a result of the tumor motion and the difficulties for the TPS to accurately account for the lack of charged particle equilibrium (CPE). This

IOP Conf. Series: Journal of Physics: Conf. Series **847** (2017) 012030 doi:10.1088/1742-6596/847/1/012030

was verified by the 4D MC simulations (figure 1B). Simulations based on expected and actual beam configuration information resulted in differences in extracted accumulated tumor center doses of approximately 0.1 %, confirming that PSD measured and TPS calculated doses did not differ due to treatment delivery uncertainties.

4. Conclusions

An automated and flexible workflow for four-dimensional Monte Carlo QA, based on the EGSnrc user code 4DdefDOSXYZnrc, has been developed. The workflow is initiated from the TPS via export of files and thus directly compatible with any TPS. It can also be triggered by the appearance of linac log files. Enabling 4DMC requires an extra user interaction compared to 3D simulations due to the need for deformation and motion information. The end product is a set of DICOM RT objects that can be imported into, and analyzed in, the TPS. The major benefits of a solution for 4D dose verification like the one proposed here are the resource effectiveness, the fact that it requires no beam time and results in a dose in the patient geometry. Additionally, with the use of linac log files it provides a method for verification of the actually delivered dose.

5. References

[1] Ottosson W *et al* 2015 *Radiother. Oncol.* **117** 55-63
[2] Popescu I A *et al* 2015 *J. Phys.: Conf. Ser.* **573** 012004
[3] Kawrakow I *et al* 2016 The EGSnrc Code System: Monte Carlo Simulation of Electron and Photon Transport (http://nrc-cnrc.github.io/EGSnrc/doc/pirs701-egsnrc.pdfMore references)
[4] Heath E and Seuntjens J 2006 *Med. Phys.* **33** 434-5
[5] Heath E *et al* 2012 *Med. Phys.* **39** 3921
[6] Lobo J and Popescu I 2010 *Phys. Med. Biol.* **55** 4431-43
[7] Ottosson R and Behrens C F 2011 *Phys. Med. Biol.* **56** N263-74
[8] Ottosson W *et al* 2015 *J. Phys.: Conf. Ser.* **573** 012022

IOP Conf. Series: Journal of Physics: Conf. Series **847** (2017) 012031 doi:10.1088/1742-6596/847/1/012031

Evaluation of the respiratory motion influence in the 3D dose distribution of IMRT breast radiation therapy treatments

J C Lizar[1], L F Santos[1], F C Brandão[2], K C Volpato[2], F S Guimarães[2] and J F Pavoni[1]

[1]Departamento de Física, FFCLRP, Universidade de São Paulo, Av. Bandeirantes 3900, 14040-901, Monte Alegre, Ribeirão Preto, SP, Brazil

[2]Centro de Radioterapia de São Carlos, Rua Maestro João Seppe, s/nº, Jd Paraíso, São Carlos, SP, Brazil

Email: jessica_lizar@yahoo.com

Abstract. This study aims to evaluate the motion influence in the tridimensional dose distribution due to respiratory for IMRT breast planning technique. To simulate the breathing movement an oscillating platform was used. To simulate the breast, MAGIC-f phantoms were used. CT images of a static phantom were obtained and the IMRT treatment was planned based on them. One phantom was irradiated static in the platform and two other phantoms were irradiated while oscillating in the platform with amplitudes of 0.34 cm and 1.22 cm, the fourth phantom was used as reference in the MRI acquisition. The percentage of points approved in the 3D global gamma analyses (3%/3mm) when comparing the dose distribution of the static phantom with the oscillating ones was 91% for the 0.34cm amplitude and 62% for the 1.22 cm amplitude. Considering this result, the differences found in the dosimetric analyses for the oscillating amplitude of 0.34cm could be considered acceptable in a real treatment. The isodose distribution analyses showed a decrease of dose in the anterior breast region and an increase of dose on the posterior breast region, being these differences most pronounced for large amplitude motion.

1. Introduction

Breast cancer is one of the major malignant diseases that affect women in the world [1]. It is usually treated with a combination of conservative surgery and postoperative radiotherapy [2]. Traditionally, breast cancer radiotherapy is performed using conformal radiotherapy techniques (3D-RT) with two tangential opposed beams with the possibility of using physical wedges to homogenize the delivered doses.

Considering the modern techniques available nowadays in radiotherapy, breast cancer can be treated using Intensity Modulated Radiation Therapy (IMRT) by using several beams with non-uniform intensities that varies according to the irradiated volume, this technique allows a better dose coverage at the treatment target compared to 3D-RT [3]. In addition, breast IMRT significantly reduces doses to the surrounding organs at risk, such as contralateral breast, heart and lungs [4-6]. However, during these treatments, there is significant uncertainty in the dose received by the planned target volume (PTV) because of respiratory motion [7]. The average respiratory rate for an adult at rest is between 12 to 18 breaths per minute and the moving parts induced by breathing can introduce significant errors in several

IOP Conf. Series: Journal of Physics: Conf. Series **847** (2017) 012031 doi:10.1088/1742-6596/847/1/012031

steps of the radiation treatment planning and treatment itself. These errors are difficult to predict and quantified due to non-synchronization between the radiation beams and movements of the targets.

In this study, our goal is to evaluate the influence of the motion induced by breathing on the tridimensional dose delivered of IMRT breast treatments using MAGIC-*f* gel dosimeter [8] and an oscillating platform.

2. Materials and Methods

2.1. Gel preparation

MAGIC-*f* gel was prepared following the process described elsewhere [9].

Four phantoms with 0.5 L of volume and simulating the breast format were prepared (figure 1-a). Phantom 1 received the static IMRT treatment, phantom 2 received the IMRT treatments oscillating with amplitudes of 0.34 cm in the platform, phantom 3 received the same treatment oscillating with amplitudes of 1.22 cm in the platform and phantom 4 was used as a reference in the Magnetic Resonance Image (MRI) scanner. Seven calibration vials were also prepared from the same batch and were irradiated with doses up to 4 Gy.

(a) (b) (c)

Figure 1. (a) Phantom used to simulate the breast anatomy. (b) Oscillatory platform used to simulate breathing movement. (c) Phantom positioned on the platform to receive the IMRT radiation treatment.

2.2. Oscillatory Platform

An oscillatory platform made of acrylic material, supported by metallic bars that guide its movement and with a damping system for the maximum stability was used (figure 1-b). The platform allowed oscillations of 0.34 cm to 1.79 cm and 15 oscillations/minute.

2.3. Treatment planning and irradiations

CT images of the phantom positioned on the platform were acquired on a Phillips Brilliance Big CT Bore Oncology scanner with a voxel size of 1.0 x 1.0 x 1.0 mm^3, the platform was motionless and a point to be selected as the isocenter was marked with fiducials. It is important to mention that all the treatment planning was performed with the static CT images and, the evaluation of the motion influence on the 3D dose distribution will be considered as the variations of the measurements of the gel irradiated in the oscillating platform in relation to the gel irradiated in the static platform.

The CT data was imported on the Eclipse treatment planning system (TPS), version 10.0 (Varian Medical Systems), the isocenter was selected at the marked point, the phantom contour was delineated and a PTV was created considering a margin of 0.5cm inside the external contour. The IMRT treatment plan was created using 5 coplanar fields and a dose of 2 Gy was prescribed, simulating one fraction of the treatment.

This treatment plan was delivered to the phantom as described before: phantom 1 received the IMRT treatment positioned on the motionless platform and phantom 2 and 3 received the IMRT treatments oscillating in the platform.

2.4. Magnetic Resonance Imaging

MRI of the three phantoms were acquired 1 day after irradiation using a Phillips Achieva 3T scanner and the head coil. A multi-spin echo sequence was used consisting of 7 echo times multiples of 22.352 ms, repetition time of 14000 ms and voxel size of 2 x 2 x 2 mm³ with a space between slices of 2 mm. The MRI intensity voxels were converted to R2 values by using a software developed in MatLab®.

3. Results and Discussions

The calibration curve of this gel batch show a linear response (graph 1).

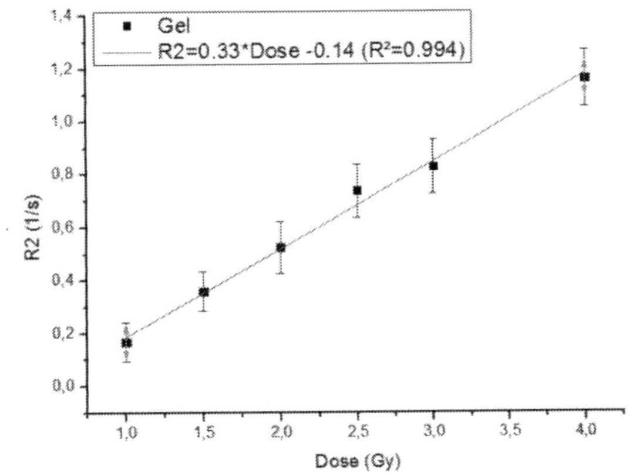

Graph 1. Magic-f calibration curve.

To avoid errors introduced by calibrating the gel dosimeter, but considering the linear behaviour of the calibration curve, each phantom 3D matrix data consisting of R2 values were normalized by its global maximum value (figure 2).

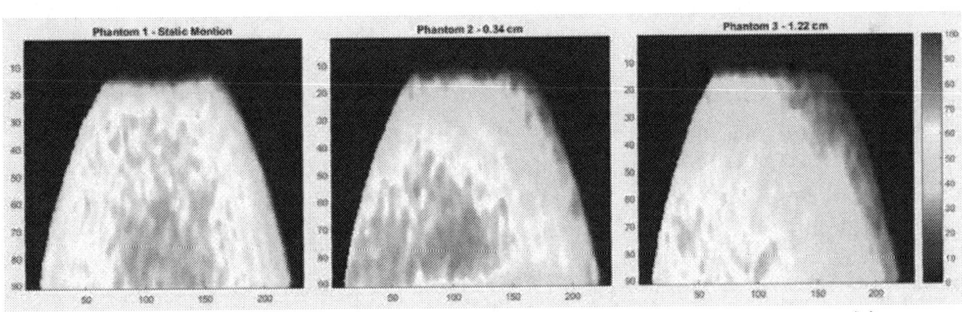

(a) (b) (c)

Figure 2. Normalized R2 distribution of isocenter images of the three phantoms treated with IMRT. (a) Phantom 1 was irradiated motionless. (b) Phantom 2 was irradiated oscillating with an amplitude of 0.34 cm. (c) Phantom 3 was irradiated oscillating with an amplitude of 1.22 cm.

To evaluate the motion influence on the dose distributions, a 3D global gamma analyses was performed using criteria of 3% of dose difference and 3mm in distance to agreement and a threshold of 15% was also used. The percentage of points approved in the 3D gamma analyses when comparing the dose distribution of phantoms 1 and 2 was 91% and for the dose distribution comparison of phantoms 1 and 3 was 62% (figure 3). These results show that the oscillation amplitude of 0.34mm, or a small respiratory breathing movement, could be considered acceptable in the clinical setting [10]. By analysing the isodose distribution, it is noticed a decrease of dose in the anterior breast region and an increase of dose on the posterior breast region. As expected, for both analyses the differences are most pronounced for large amplitude motions (figure 4) [11].

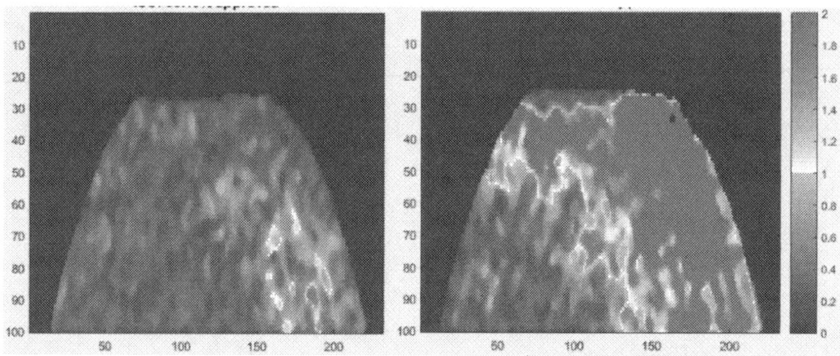

(a) (b)

Figure 3. 3D gamma analyses of isocenter images, colored points represent reproved values. (a) Gamma comparison of phantoms 1 and 2. (b) Gamma comparison of phantoms 1 and 3.

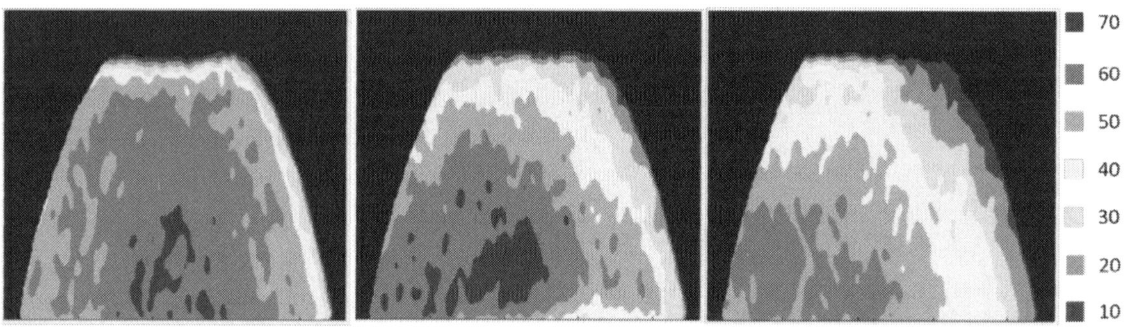

(a) (b) (c)

Figure 4. Isocenter isodose distribution for phantom 1 irradiated motionless (a), phantom 2 oscillating with amplitude of 0.34 cm (b) and phantom 3 oscillating with amplitude of 1.22 cm (c).

4. Conclusions

The study demonstrated that dose inhomogeneity generated by a breathing motion using IMRT technique, could not be a concern when simulating a moderate breathing movement (amplitude of 0.34 cm), however it was observed that the increase of respiratory motion amplitude increased the dose inhomogeneity (amplitude of 1.22cm) and this impacts on the dosimetric parameters of the target volume coverage.

5. References

[1] Stewart B and Wild P. World Cancer Report 2014. Lyon, France: International Agency for Research on Cancer. ISBN 978-92-832-0429-9
[2] National Comprehensive Cancer Network 2003 *J. Natl. Compr. Canc Netw. 1* 148
[3] Hill R *et al* 2009 *Med. Phys.* **36** 3971-81
[4] Barnett G C *et al* 2009 *Radiother. Oncol.* **92** 34-41
[5] Burmeister J *et al* 2008 *Med. Dosim.* **33** 6-3
[6] Landau D *et al* 2001 *Radiother. Oncol.* **60** 247-55
[7] Remouchamp V M *et al* 2003 *Int. J. Radiat. Oncol. Biol. Phys.* **55** 392-406
[8] Baldock C *et al* 2010 *Phys. Med. Biol.* **55** R1-63
[9] Fernandes J P *et al* 2008 *Phys. Med. Biol.* **53** N53-58
[10] Chi F *et al* 2015 *Cancer Radiother* **19** 180-6
[11] Menon G *et al* 2011 *Phys. Med. Biol.* **56** 7405

IOP Conf. Series: Journal of Physics: Conf. Series **847** (2017) 012032 doi:10.1088/1742-6596/847/1/012032

Dosimetric impact of a change in breathing period on VMAT stereotactic ablative body radiotherapy

T Olding[1,2], KM Alexander[1]
[1]Department of Physics, Queen's University, Kingston, Ontario, Canada, K7L3N6
[2]Cancer Centre of South Eastern Ontario at Kingston General Hospital, 25 King Street West, Kingston, ON, Canada, K7L5P9

E-Mail: tim.olding@krcc.on.ca

Abstract. The dosimetric impact of a change in breathing period during treatment was assessed for a volumetric modulated arc therapy (VMAT) stereotactic ablative radiotherapy (SABR) lung plan optimized according to our centre's planning protocol. Plan delivery was evaluated at three breathing rates ranging from 7 to 23 breaths-per-minute (BPM) against the planning anatomy (15 BPM) calculated dose. Dynamic ion chamber, EBT3 film and Fricke-xylenol orange-gelatin (FXG) gel measurements were acquired using a motion phantom with appropriate inserts for each dosimeter. The results show good agreement between measured and calculated plan dose within the internal gross tumour volume (IGTV) target.

1. Introduction

There has been a steady increase in the number of patients treated at our centre using the lung VMAT SABR technique since its adoption. Poor pulmonary function is a major reason for choosing radiation treatment over surgery in treating early stage (Stage I, T1/2, N0, <5 cm), non-small cell lung cancer. One of the associated challenges is that these patients often have a hard time maintaining regular breathing both at the time of 4DCT imaging and on treatment. The change in breathing amplitude can be monitored to some degree by assessment of the internal gross tumour volume (IGTV) via CBCT imaging at the time of treatment as compared to the planning CT IGTV [1]. However, breathing rate is not actively monitored during treatment for the motion-encompassing treatment technique used at our centre. Variations ranging from 7 up to 40 breaths-per-minute (BPM) have been observed at the time of planning CT, though most patients are reasonably consistent in breathing in the 10-20 BPM range.

Previous studies have looked at the dosimetric effects that accompany a change in breathing pattern, including ion chamber [2] and 2D film measurements for 3D conformal deliveries and ion chamber measurements for VMAT deliveries [3, 4]. Our centre's VMAT SABR planning approach aims to minimize the motion of the multi-leaf collimator (MLC), keeping an open beam aperture around the planning target volume (PTV) during delivery in order to produce the desired dose distribution and reduce the effects of interplay [5] in the delivery. However, limited MLC motion is allowed within the PTV volume (beam's eye view) for plan optimization, which raises a greater possibility of dosimetric variations with change in breathing rate compared to open aperture 3D conformal deliveries. The purpose of this work is to evaluate the impact of a change in breathing rate (period) on our centre's VMAT SABR technique using different dosimetric modalities in a similar manner as before [6], including 0D ion chamber, 2D EBT3 Gafchromic film, and 3D FXG gel dosimetry.

2. Materials and Methods

2.1. Delivery Validation

The dosimetry tools used for delivery validation in this investigation were: a) point single ion chamber (PR-05P 0.07 cc, Capintec, Ramsey, NJ), b) 2D Gafchromic EBT3 film (ISP, Wayne, NJ), and c) 3D FXG gel dosimeter with optical CT readout [6]. Dynamic 4D computed tomography (4DCT) scans were acquired of a Quasar motion phantom (Modus Medical Devices Inc, London, ON, Fig. 1a) with a cedar insert containing a 3 cm diameter acrylic sphere 'tumour' and a hole for insertion of the ion chamber. A Philips Brilliance Big Bore CT scanner (Philips Medical Systems, Cleveland, OH) was used for CT data acquisition, with the addition of a bellows motion tracking system for 4DCT phase binning. A similar cedar insert was used for the EBT3 film (Fig. 1b). The scans were then repeated with a custom gel insert (Fig. 1c) in the Quasar phantom (Fig. 1d). A 10 mm peak-to-peak amplitude 15 BPM sinusoidal breathing cycle was used for all 4DCT scans. This 10 mm amplitude was chosen as it is the limit for inclusion in the associated SABR clinical trial that our centre participates in.

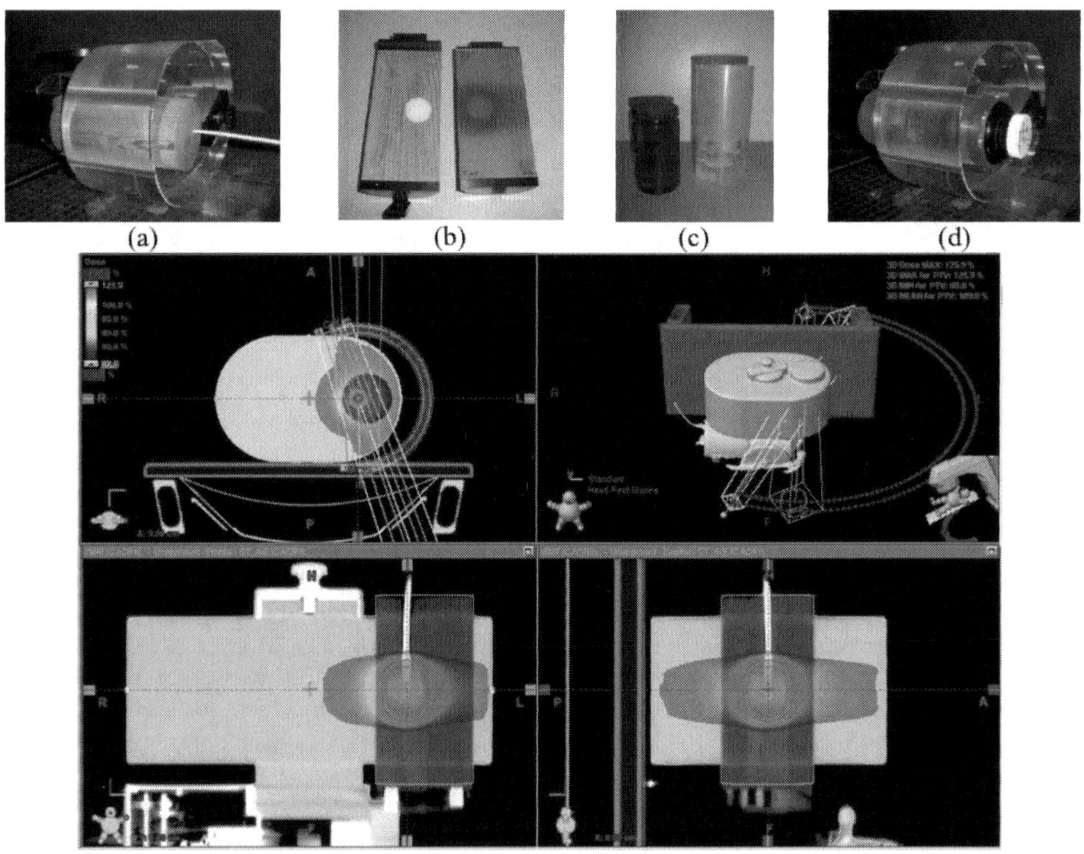

(a) (b) (c) (d) (e)

Figure 1. (a) Quasar motion phantom with ion chamber insert, (b) photo of cedar film insert with overlaid EBT3 film from dynamic delivery, (c-d) gel insert and FXG gel in Quasar phantom, and (e) Eclipse screen capture of the high monitor unit dynamic ion chamber plan used in this study.

For VMAT SABR planning, dual coplanar clockwise/counterclockwise 6MV partial arcs were employed, extending from just outside the approximated edge of the contralateral lung anteriorly to 180 degrees posteriorly (Fig. 1e). The collimator angle on the two arcs was set to ±30 degrees to minimize discrepancies arising from the interplay effect [5]. A RapidArc™ VMAT plan was optimized and calculated in Eclipse™ v.10 (Varian Medical Systems, Palo Alto, CA) on the 2 mm slice thickness

4DCT-derived average scan of the film-Quasar phantom. The plan had a target objective placed on the total number of monitor units (MU) in the optimization stage of treatment planning to limit MLC motion and realize open MLC apertures throughout the arc. This plan was then re-calculated (but not re-optimized) on the ion chamber and gel 4DCT-derived average phantom datasets. The film and ion chamber plans were normalized to a prescription dose of 2 Gy (100% prescription dose covers 95% of PTV target volume), while the gel plans were re-normalized (i.e. monitor units scaled down) to a prescription of 1.8 Gy to accommodate the optimal range for FXG gel dose readout accuracy. A screen capture of the ion chamber plan is shown in Figure 1e.

All treatment plans were delivered on a Varian Trilogy 2100iX linear accelerator (Varian Medical Systems, Palo Alto, CA), with Advanced OBI™ imaging cone beam CT (CBCT) setup verification. Each dose measurement was compared against calculated Eclipse treatment planning system (TPS) dose using software appropriate for that QA tool. EBT3 film measurements were acquired using an in-house built CCD-lightbox film scanner and analyzed in MATLAB (Mathworks, Newark, NJ). The reconstructed, calibrated gel dose data from the Vista™ optical CT scanner (Modus Medical Devices Inc., London, ON) was compared to Eclipse dose in 3D Slicer [7] (www.slicer.org).

3. Results

3.1. Dosimeter-Plan Dose Comparison

Results for the dosimeters used in this work are summarized in Table 1. The reported ion chamber values are the average of 5 readings. Point measurements are in similar agreement with Eclipse as previous work [6]. 2D Film and 3D gel results in Table 1 are reported for gamma comparison within the IGTV volume. These results are consistent with the point ion chamber readings, although the gel measurements were lower than expected along the motion direction near the edges of the IGTV.

Table 1. Summary: measured vs. Eclipse doses.

Dosimeter	Breathing Rate (BPM)	Prescription (Gy)	Point Dose Difference (%)			% Gamma ≤1 (3%/3mm)
			CW Arc	CCW Arc	Total	
Ion Chamber	7	2	99.4 ± 0.9	98.7 ± 0.1	99.2 ± 0.6	-
Ion Chamber	15	2	99.2 ± 0.3	98.9 ± 0.1	99.0 ± 0.2	-
Ion Chamber	23	2	99.3 ± 0.4	98.8 ± 0.1	99.1 ± 0.2	-
EBT3 Film	7	2	-			94.7
EBT3 Film	15	2	-			99.5
EBT3 Film	23	2	-			94.5
FXG Gel	7	1.8	-			95.9
FXG Gel	15	1.8	-			95.1
FXG Gel	23	1.8	-			85.3

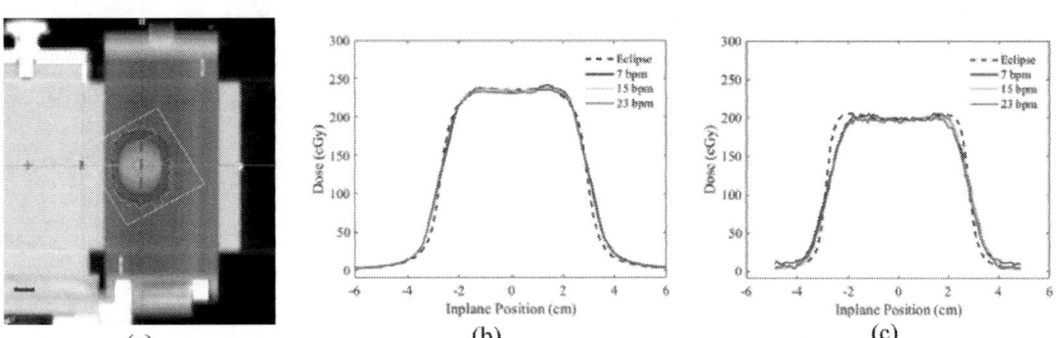

Figure 2. (a) Eclipse screen capture of the film plan with red dashed line along axis of motion showing location of profile measurements for the (b) EBT3 film and (c) FXG gel deliveries.

4. Discussion & Conclusions

With reference to the results summarized in Table 1, a negligible variation was observed in mean value of the ion chamber measurements from the different breathing rate deliveries. However, the standard deviation in ion chamber readings was noticeably higher for the 7 BPM delivery, and in particular, the clockwise arc. This is related to the fact that there is more MLC motion within the PTV for this arc (orange outline in Fig. 2a shows the MLC aperture with the great incursion into the PTV outline on beams-eye view). For the slower BPM tumour motions, there is greater potential for the ion chamber to spend an increased fraction of the delivery time either directly behind the MLC leaves or not blocked by the MLC leaves, depending on the starting point of the motion. This leads to greater variability in ion chamber readings. The standard deviation on the point dose was observed to be on the order of 1% at the lower BPM, so the result is not clinically significant. However, it does show that limiting the MLC motion during plan optimization reduces day-to-day variability in the dose delivery. The 2D film results show good agreement with Eclipse. The 7 bpm film dose profile in Figure 2b shows some evidence of the variability introduced by tumour motion at slower breathing rates, with the dose being greater than the Eclipse dose on the right side of the central axis profile and less than the Eclipse dose on the left side of the profile. The gel results (Table 1, Fig. 2c) show evidence of the effects of diffusion, with the gradients at the edges of the dose distribution being less than expected. This leads to a reported larger dose difference between Eclipse and gel in these regions and lower IGTV volume gamma agreement. Even with this effect being present, the relative comparison of 3D gel results to each other at the different breathing rates still supports the conclusion that the variability in delivery at different breathing rates is clinically insignificant. Again, the 7 bpm results show the greatest difference between the left and right side of the profile along the central axis.

Our Eclipse VMAT SABR planning strategy limits MLC incursion into the PTV volume by a combination of a high priority on reducing the dose in the ring region outside the target volume and an objective setting an appropriate upper limit on the number of monitor units (MU) delivered by the plan, which correspondingly opens up the MLC aperture and heats up the plan. From the results reported in this work, it can be concluded that the dose delivery from this planning approach is not greatly affected by variations in patient breathing rate at the time of treatment, with the caveat that the amplitude of the motion is properly monitored at CBCT setup verification and is subsequently stable throughout the radiation treatment.

5. Acknowledgements

Research funding for this work has been provided by the Canadian Institutes of Health Research (CIHR) and the Cancer Centre of Southeastern Ontario.

6. References

[1] Clements N *et al* 2013 *Med. Phys.* **40** 021904
[2] Hill R *et al* 2009 *Med. Phys.* **36** 3971-81
[3] Pham D *et al* 2013 *Med. Dosim.* **38** 304-8
[4] Caloz M *et al* 2015 *J. Phys.: Conf. Ser.* **573** 012023
[5] Ong C *et al* 2011 *Int. J. Radiat. Oncol.* **79** 305-11
[6] Olding T *et al* 2013 *J. Phys.: Conf. Ser.* **444** 012073
[7] Alexander K M *et al* 2015 *J. Phys.: Conf. Ser.* **573** 012042

IOP Conf. Series: Journal of Physics: Conf. Series **847** (2017) 012033 doi:10.1088/1742-6596/847/1/012033

New developments in EPID-based 3D dosimetry in The Netherlands Cancer Institute

B Mijnheer, R Rozendaal, I Olaciregui-Ruiz, P González, R van Oers, A Mans
Department of Radiation Oncology, The Netherlands Cancer Institute, Plesmanlaan 121, 1066 CX Amsterdam, The Netherlands

E-mail: b.mijnheer@nki.nl

Abstract. EPID-based offline 3D *in vivo* dosimetry is performed routinely in The Netherlands Cancer Institute for almost all RT treatments. The 3D dose distribution is reconstructed using the EPID primary dose in combination with a back-projection algorithm and compared with the planned dose distribution. Recently the method was adapted for real-time dose verification, performing 3D dose verification in less than 300 ms, which is faster than the current portal frame acquisition rate. In this way a possibility is created for halting the linac in case of large delivery errors. Furthermore, a new method for pre-treatment QA was developed in which the EPID primary dose behind a phantom or patient is predicted using the CT data of that phantom or patient in combination with in-air EPID measurements. This *virtual* EPID primary transit dose is then used to reconstruct the 3D dose distribution within the phantom or patient geometry using the same dose engine as applied offline. In order to assess the relevance of our clinically applied alert criteria, we investigated the sensitivity of our EPID-based 3D dose verification system to detect delivery errors in VMAT treatments. This was done through simulation by modifying patient treatment plans, as well as experimentally by performing EPID measurements during the irradiation of an Alderson phantom, both after deliberately introducing errors during VMAT delivery. In this presentation these new developments will be elucidated.

1. Introduction

The use of EPID-based 3D dosimetry in external beam radiotherapy has been addressed in several studies [e.g., 1-3]. The first two papers describe the use of the incident fluence, determined from a measured EPID transit image, followed by Monte Carlo techniques to assess the delivered 3D patient dose distribution. Van Uytven *et al* [3] have developed an algorithm that converts also a measured EPID transit image to incident fluence, and employs thereafter a collapsed-cone convolution calculation to determine the delivered 3D patient dose distribution.

In the Netherlands Cancer Institute (NKI) a back-projection algorithm has been implemented for the 3D dose verification of almost all IMRT/VMAT and 3DCRT treatments using *in vivo* EPID dosimetry [4]. Details of our method and its clinical implementation have been described elsewhere [5]. Briefly, our back-projection algorithm uses EPID measurements behind a patient or a phantom to determine the EPID primary dose, i.e. the dose component at the level of the EPID resulting from the primary radiation coming from the accelerator head. Using this EPID primary dose, a fast pencil beam-type of dose calculation algorithm reconstructs the 3D dose distribution in a patient or phantom in combination with the CT data of that patient or phantom. Reconstructed 3D dose distributions are then compared with planned dose distributions using 3D γ-evaluation within the 50% isodose surface using global 3%/3mm indicators. This rather high threshold was chosen to avoid artificial improvement of the gamma evaluation results by including low dose volumes, and to avoid the build-up region. In this presentation we will elucidate the new developments in the field of EPID-based 3D dosimetry in our institution since our presentation at IC3DDose 2014 [5].

2. Real-time 3D dosimetry

The offline EPID-based 3D dosimetry system clinically employed at NKI is in principle suited for online treatment verification, provided the system is able to complete 3D dose reconstruction and verification within 420 ms, the present acquisition time of a single EPID frame. For this purpose the current 3D dose reconstruction algorithm was sped up via a new multithreaded implementation, and by performing all computations that are not dependent on portal image acquisition separately, thus removing the need for doing these calculations in real time. The real-time dosimetry system was tested by irradiating an anthropomorphic phantom with various VMAT plans. Two types of serious delivery errors were introduced: the number of monitor units per control point was doubled, and all moving MLC leaves and jaws were retracted 10 cm on both sides. Figure 1 shows the result of these real-time verifications of a VMAT prostate cancer treatment. The complete processing of a single portal frame, including dose verification, took 266 ±11 ms on a standard CPU. The introduced delivery errors were detected after 5–10 s irradiation time. The system is now tested clinically, particularly with respect to criteria to halt the linac.

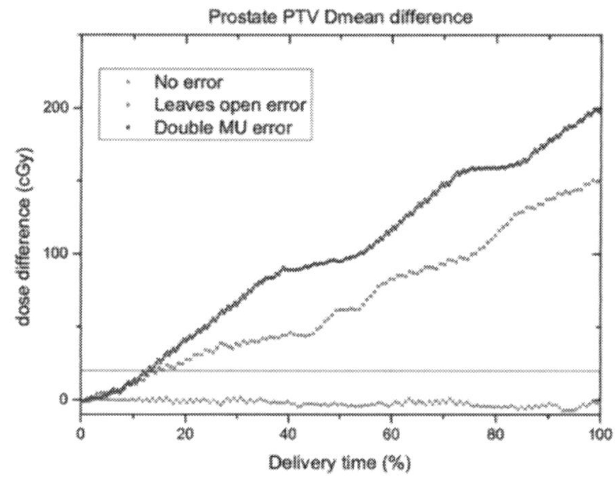

Figure 1. Difference between reconstructed and planned mean dose in the PTV for three scenarios (no error, all leaves open error, double MU error), as a function of delivery time. The horizontal line indicates the detection threshold.

3. Pre-treatment 3D dose reconstruction

EPID-based pre-treatment 3D dose verification is only performed at NKI for high dose single fraction treatments and for treatments where *in vivo* verification is not possible. In these situations, the clinical plans are verified with an EPID measurement behind a slab polystyrene phantom. These phantom measurements are also performed if *in vivo* dosimetry verification shows an alert for which no clear explanation can be given, for instance after inspecting the CBCT scan made prior to the *in vivo* dose measurement. In order to eliminate the need for phantom re-planning and the accurate positioning of a phantom for such a type of measurement, a new method for pre-treatment EPID-based 3D dose verification was developed. In this method our current algorithm was adapted to calculate the EPID primary dose behind the phantom using in-air EPID measurements of the treatment and CT data. Using this *virtual* EPID primary dose, the algorithm then reconstructs virtual phantom 3D dose distributions in combination with the CT data of the phantom.

Table 1. Comparison of EPID-reconstructed virtual patient and real phantom dose distributions with the corresponding planned dose distributions using γ-evaluation data. A dose difference of 3% of the maximum planned dose and a distance-to-agreement of 3mm were used as criteria in the γ-evaluation. Results are presented as average value ± 1 standard deviation.

	#fields		γ_{mean}	$\gamma_{99\%}$	%γ<1
Virtual patient	91	IMRT	0.38 ± 0.13	1.13 ± 0.43	95.6 ± 6.0
	105	VMAT	0.48 ± 0.09	1.34 ± 0.33	94.1 ± 4.7
Real phantom	70	IMRT	0.39 ± 0.13	1.08 ± 0.32	95.2 ± 7.6
	53	VMAT	0.45 ± 0.09	1.23 ± 0.33	95.2 ± 4.8

An even more interesting application of this novel method is the possibility to reconstruct virtual patient 3D dose distributions using the in-air EPID measurements in combination with patient CT data. Virtual patient dose distributions were determined and compared to the corresponding planned dose distributions using γ-evaluation (3% global dose/3mm), yielding γ-values that are very similar to the phantom verification data (see Table 1). In this way delivered dose distributions can also be analyzed in terms of dose-volume histogram (DVH) parameters, providing a useful indication to radiation therapy staff of the impact of detected deviations. By combining virtual patient dose distributions with *in vivo* data it is also possible to separate patient related errors from other types of uncertainties.

4. Sensitivity of EPID-based 3D dosimetry

The usefulness of different γ-evaluation indicators and DVH parameters in detecting errors, as well as the values for the alert criteria used in the clinic for these indicators, are continuously evaluated. We therefore investigated the sensitivity of our EPID-based 3D dose verification system to detect delivery errors in VMAT treatments. This was done in two ways; first the effect of delivery errors was simulated by modifying different VMAT plans (H&N, lung, prostate and rectum). These treatment plans were modified by adding a 3% and 6% offset to the given number of monitor units, and by systematic shifting leaf bank positions by various distances. The 3D *in vivo* dose distributions were then compared to the unmodified and modified treatment plans. To determine the detectability of the various delivery errors, we made use of a receiver operator characteristic (ROC) methodology. The results of the determination of the area under the ROC curve showed that the parameter γ_{mean} is the best discriminator for the detection of leaf position and monitor unit errors. Compared to γ_{mean} the DVH parameter ΔD_{50} for the PTV performed worse as a discriminator in all cases.

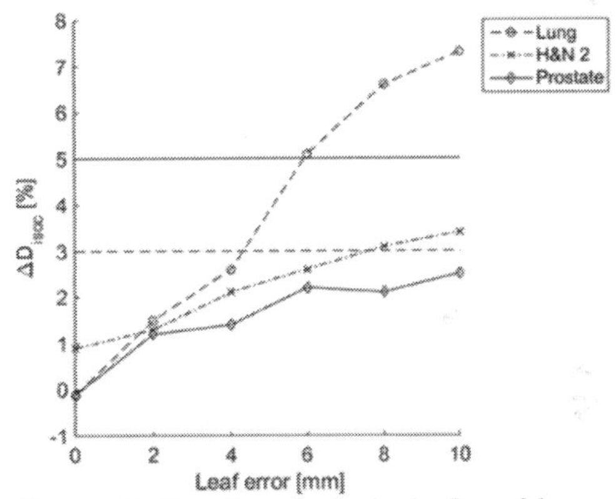

Figure 2. Results of single leaf position error measurements during various VMAT irradiations of an Alderson phantom shown as dose difference at isocenter. The dashed and solid lines indicate the tolerance and action level, respectively, as currently applied in the clinic.

In a second study the sensitivity of our clinically used EPID-based 3D transit dosimetry approach in detecting deliberately introduced errors during VMAT delivery, was investigated experimentally. For this purpose EPID measurements were performed during the irradiation of an Alderson phantom in order to mimic the *in vivo* situation as close as possible. 3D dose distributions were then reconstructed using our back-projection model, and compared with planned dose distributions by means of 3D γ-evaluation (3% global dose/3mm). Quantification of individual and multiple leaf position errors (not shown) indicated that the lung plan was more sensitive to these errors than the prostate and head-and-neck plans (see Figure 2). A horizontal shift of the phantom of 2 cm showed that no alerts were raised for the prostate and lung case, but a very large difference was observed in the reconstructed 3D dose distribution

IOP Conf. Series: Journal of Physics: Conf. Series **847** (2017) 012033 doi:10.1088/1742-6596/847/1/012033

for the head-and-neck cases. This can be expected because a shift of 2 cm moved the isocenter much closer to the outer contour of the head-and-neck cases compared to the lung and prostate plans, thus causing a large change in the transit dose.

Out of a total of 26 serious delivery errors in these plans, the number of errors detected using the Alderson phantom and our current alert criteria was 23 (88%) using γ_{mean} and 15 (58%) with $\gamma_{1\%}$, thus confirming the results of the ROC study.

5. Conclusions

Our method of EPID-based 3D dosimetry was adapted for real-time dose verification, resulting in 3D dose verification in less than 300 ms. In this way a possibility is created for halting the linac in case of large delivery errors.

The results of the tests with the novel pre-treatment dose reconstruction algorithm showed that virtual patient dose reconstruction performs as accurate as real phantom dose reconstruction thus making pre-treatment QA faster and less cumbersome. This method, which uses planning CT data, allows also the use of DVH metrics for comparison with planned 3D dose distributions.

Determinations of the sensitivity of our EPID-based 3D dose verification method showed that the parameter γ_{mean} might be the best discriminator for the detection of errors during VMAT delivery when using 3D γ-evaluation. The clinical use of DVH parameters still needs further investigation.

6. References

[1] Van Elmpt W *et al* 2009 *Int. J. Radiat. Oncol. Biol. Phys.* **73** 1580-7
[2] Persoon LC *et al* 2013 *Acta Oncol.* **52** 1484-9
[3] Van Uytven E *et al* 2015 *Med. Phys.* **42** 6945-54
[4] Mijnheer BJ *et al* 2015 *Pract. Radiat. Oncol.* **5** e679-87
[5] Mijnheer B *et al* 2015 *J. Phys.: Conf. Ser.* **573** 012014
[6] Spreeuw H *et al* 2016 *Med. Phys.* **43** 3969-75

IOP Conf. Series: Journal of Physics: Conf. Series **847** (2017) 012034 doi:10.1088/1742-6596/847/1/012034

Real-time 3D dose imaging in water phantoms: reconstruction from simultaneous EPID-Cherenkov 3D imaging (EC3D)

P Bruza[1], J M Andreozzi[1], D J Gladstone[1,2], L A Jarvis[2], J Rottmann[3] and B W Pogue[1,2]
[1]Thayer School of Engineering, Dartmouth College, Hanover NH 03755 USA.
[2]Geisel School of Medicine, Dartmouth College, Hanover NH 03755 USA.
[3]Dana Farber Cancer Institute, Brigham and Women's Hospital, Boston MA 02115 USA.

E-mail: petr.bruza@dartmouth.edu

Abstract. Combination of electronic portal imaging device (EPID) transmission imaging with frontal Cherenkov imaging enabled real-time 3D dosimetry of clinical X-ray beams in water phantoms. The EPID provides a 2D transverse distribution of attenuation which can be back-projected to estimate accumulated dose, while the Cherenkov image provides an accurate lateral view of the dose versus depth. Assuming homogeneous density and composition of the phantom, both images can be linearly combined into a true 3D distribution of the deposited dose. We describe the algorithm for volumetric dose reconstruction, and demonstrate the results of a volumetric modulated arc therapy (VMAT) 3D dosimetry.

1. Introduction

Using visible Cherenkov radiation as a surrogate of the dose distribution in water phantoms is a promising method to perform fast quality assurance (QA) and relative dosimetry of megavoltage photon and electron beams [1,2]. Since the first demonstration of direct visualization of dose deposition in a clinical setting [3], Cherenkov imaging has been tested in many applications of clinical radiation dosimetry [4,5]. In a previous study [6] we demonstrated a tomographic 3D reconstruction of dose in a water phantom via tomographic acquisition of the Cherenkov light distribution in a 180° scan. This required several minutes of acquisition time which is prohibitive for imaging dynamic beam modalities such as volumetric modulated arc therapy (VMAT).

We have addressed the challenges of fast 3D dosimetry [7] by developing a technique that allows for real-time dose imaging in water phantoms. As shown in Figure 1, the method – EPID-Cherenkov 3D dosimetry (EC3D) – relies on simultaneous acquisition of EPID images and lateral Cherenkov images. While the EPID provides a 2D transverse distribution of transmitted beam, the Cherenkov imaging provides an accurate lateral view of the dose. Assuming a homogeneous density and composition of the phantom, as true in water tanks, both images can be linearly combined into a consistent 3D distribution of the deposited dose. Most importantly, both EPID and Cherenkov techniques provide images with high frame rates (~10 fps) which permits real-time 3D beam reconstruction. In this contribution we present the means of dynamic volumetric dose reconstruction by combining the EPID and Cherenkov image information in a straightforward manner. The reconstruction is demonstrated using a custom VMAT plan for the AAPM Task Group 119 (TG-119) C-Shape geometry [8].

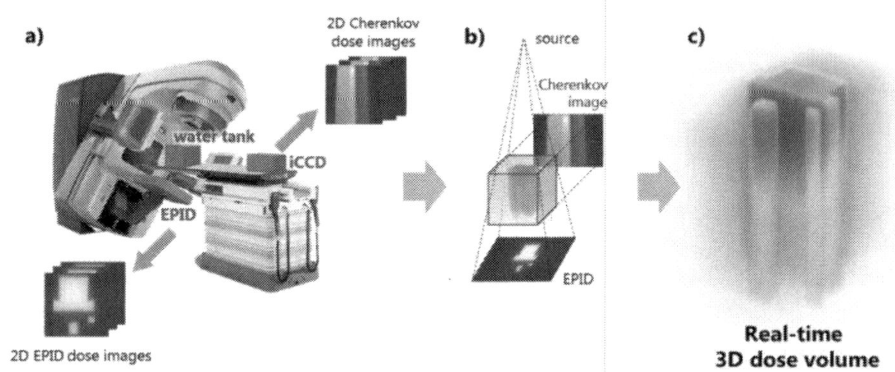

Figure 1. a) Schematic experimental setup for the EPID-Cherenkov 3D dosimetry (EC3D); b) illustration of the EPID and Cherenkov imaging geometry; c) measured 3D relative dose map produced by a static 6MV X-ray beam in a water phantom.

2. Methods

2.1. Experimental setup

The experiment was conducted in a clinical radiotherapy setting using a Varian TrueBeam™ LINAC (Varian Medical Systems, Palo Alto CA, USA), equipped with an HD120™ high definition multi-leaf collimator (MLC) and an AS1000 EPID detector. As a target we used a 30x40x30 cm water phantom, consisting of 1.0 g/L solution of Quinine Sulfate (99% Quinine hemisulfate monohydrate, Alfa Aesar, Ward Hill, MA) solution in tap water. The quinine fluorophore was used to increase the isotropic emission of Cherenkov radiation through absorption of UV light and fluorescence re-emission in the blue wavelengths [6]. The EPID was positioned at a source-to-EPID distance of *SED* = 150 cm during both static beam and VMAT delivery. Cine-mode EPID images were acquired with 10fps and saved in DICOM format. The EPID featured a symmetric pixel pitch of 0.384 mm in the detector plane. The Cherenkov images were captured by a time-gated intensified Charge-Coupled Device (ICCD) camera PI-MAX4 1024i (Princeton Instruments, Acton, MA) equipped by 135 mm f/2 lens, yielding a 0.257 mm/pixel theoretical resolution. The camera was located on the rotational axis of the LINAC gantry at a distance of 3 m from isocenter. The intensifier gate of 3.8 μs duration was triggered by the target current ("TargI") signal of the LINAC. The CCD was cooled to -20°C and operated at 50 accumulations-on-chip, resulting in a 5 fps video stream. A VMAT plan for TG-119 C-Shape geometry was created in the Eclipse treatment planning system (TPS) (Varian Medical Systems, Palo Alto CA, USA) and was designed to deliver 200 cGy at D_{max}.

2.2. 3D image reconstruction

As illustrated in Figure 2, the 3D reconstruction consisted of three consecutive steps. First, a 3D perspective projection of the EPID image was generated by mapping the EPID intensity onto a perspective-transformed 3D coordinate system (shown in Fig 2a). Second, a 3D correction factor was created from the ratio of projected EPID and Cherenkov images (shown in Fig 2b). Third, the EPID projection was multiplied by the correction factor yielding a relative dose distribution (shown in Figure 2c). The reconstruction was performed for each frame of the EPID and Cherenkov image pair.

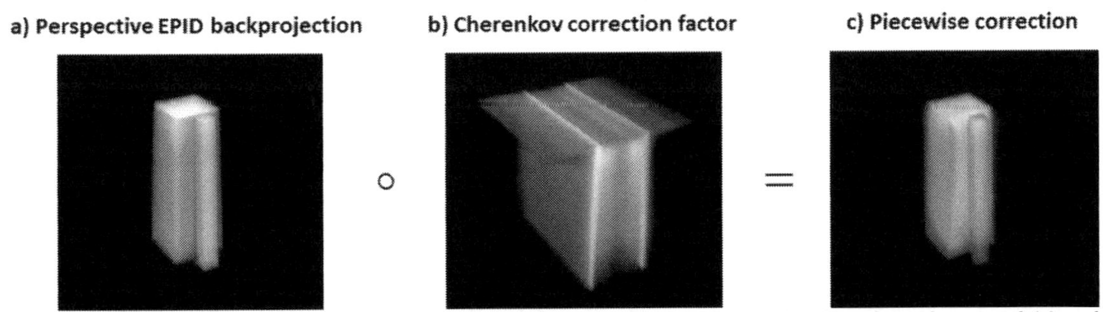

Figure 2. Steps involved in the 3D dose reconstruction process. 1) A 3D EPID volume is created (a) and piecewise multiplied by Cherenkov correction factor (b), yielding a 3D dose volume (c).

2.2.1. *EPID perspective projection*

The 2D EPID image, represented by values $E(x,y)$, as well as the true spatial coordinates (x,y) and gantry angle α were read from the individual EPID images. To prepare the EPID image for 3D transformation, a 3D EPID volume $E(\mathbf{X}) = E(x,y,z)$ was created by replicating the EPID image normal to its plane at incremental z distances. Then, a perspective transform $T:(x,y,z) \rightarrow (x',y',z')$ was applied to $E(\mathbf{X})$ by forward-mapping it's coordinates **X** via transformation matrix **T**

$$\mathbf{X}' = \mathbf{TX} \tag{1}$$

yielding a perspective-transformed EPID volume:

$$E(\mathbf{X}') = E(\mathbf{TX}) \tag{2}$$

The transformation matrix **T** must include the setup geometry (source-to-EPID distance *SED*, source-to-isocenter distance *SSD*, and size of the EPID volume) as well as the gantry rotation Θ relative to the horizontal (x',y') plane. The matrix was thus composed of sub-matrices $\mathbf{T}_1$ to $\mathbf{T}_5$:

$$\mathbf{T} = \mathbf{T}_5\mathbf{T}_4\mathbf{T}_3\mathbf{T}_2\mathbf{T}_1 \tag{3}$$

whose order follows the logical steps during reconstruction. These steps include $\mathbf{T}_1$ translation of the geometrical center (x_c, y_c) of EPID image to the geometric origin; $\mathbf{T}_2$ perspective projection towards a point at the source-to-EPID distance *SED*; $\mathbf{T}_3$ translation to the center of gantry rotation (*SED-SSD*), $\mathbf{T}_4$ rotation around gantry rotation axis y by angle Θ, and $\mathbf{T}_5$ translation to the isocenter (x_c, y_c, z_c):

$$\mathbf{T}_1 = \begin{bmatrix} 1 & 0 & 0 & -x_c \\ 0 & 1 & 0 & -y_c \\ 0 & 0 & 1 & 0 \\ 0 & 0 & 0 & 1 \end{bmatrix} \quad \mathbf{T}_3 = \begin{bmatrix} 1 & 0 & 0 & 0 \\ 0 & 1 & 0 & 0 \\ 0 & 0 & 1 & SED-SSD \\ 0 & 0 & 0 & 1 \end{bmatrix} \quad \mathbf{T}_5 = \begin{bmatrix} 1 & 0 & 0 & x_c \\ 0 & 1 & 0 & y_c \\ 0 & 0 & 1 & z_c \\ 0 & 0 & 0 & 1 \end{bmatrix}$$
$$\mathbf{T}_2 = \begin{bmatrix} 1 & 0 & 0 & 0 \\ 0 & 1 & 0 & 0 \\ 0 & 0 & 1 & 0 \\ 0 & 0 & -1/SED & 1 \end{bmatrix} \quad \mathbf{T}_4 = \begin{bmatrix} \cos\Theta & 0 & -\sin\Theta & 0 \\ 0 & 1 & 0 & 0 \\ \sin\Theta & 0 & \cos\Theta & 0 \\ 0 & 0 & 0 & 1 \end{bmatrix} \tag{4}$$

A cubic interpolation between neighboring points was used to calculate the final voxel intensity values.

2.2.2. Cherenkov correction

The Cherenkov correction factor acts on the projected EPID volume. It was defined for each voxel row in superior-inferior (y) direction, and its magnitude was defined as an intensity ratio of the Cherenkov image and summed EPID image:

$$C(x',z') = I_{CH}(x',z')\left(\sum_{y} E(\mathbf{X}')\right)^{-1} \tag{5}$$

The ratio results from the fact that the Cherenkov image represents a measured dose, summed in y-direction. The resulting 2D correction factor $C(x',z')$ was then replicated along the y-axis over the range of the EPID volume. A piecewise multiplication of EPID volume and 3D correction factor finally yielded the 3D dose distribution within the reconstructed volume:

$$D(\mathbf{X}') = C(\mathbf{X}') \circ E(\mathbf{X}') \tag{6}$$

Here the $\circ$ operator symbolizes a Hadamard product.

3. Results

To demonstrate the feasibility of our technique, we acquired and reconstructed a dynamic 3D dose map from the delivery of a VMAT plan in TG-119 C-Shape geometry. In Figure 3 we present the intermediate results of EPID back projection, raw Cherenkov images and Cherenkov correction factors, as well as the final 3D accumulated dose volume for 10 out of 320 frames from the VMAT plan delivery.

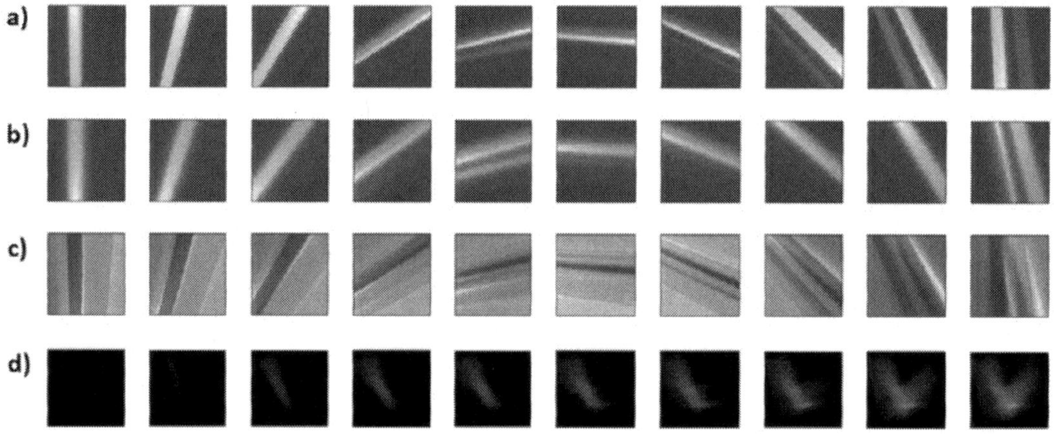

Figure 3. Sample of 10 intermediate and final results from the full course of VMAT delivery. Rows top to bottom: a) lateral view of back projected EPID; b) Cherenkov image; c) lateral view of Cherenkov calibration factor; d) final 3D view of accumulated dose.

4. Conclusions and discussion

We presented a dynamic volumetric dose reconstruction by combining information from EPID and Cherenkov images. Namely, this process described the means of 3D post-processing, and demonstrated the feasibility of our technique. Known issues of EPID image fidelity due to internal scattering and readout artifacts [9] are being addressed. In future work we will validate the EC3D technique further by quantitative comparison with ionization chamber measurements, as well as simulated dose distribution from the treatment planning system and Monte-Carlo calculations. It is hypothesized that the proposed technique may have a large impact in clinical pre-treatment plan verification and quality assurance due to the high spatial and temporal resolution of the measured 3D dose distributions it provides.

5. Acknowledgement

Financial support by National Institutes of Health (NIH) grants R01 CA109558 and R43 CA199681, and F31 CA192473.

6. References

[1] Glaser A K *et al* 2013 *Opt. Lett.* **38** 634
[2] Glaser A K *et al* 2014 *Med. Phys.* **41** 062102
[3] Axelsson J *et al* 2011 *Med. Phys.* **38** 4127-32
[4] Zhang R *et al* 2013 *J Biomed Opt.* **18** 110504
[5] Andreozzi J M *et al* 2016 *Med. Phys.* **43** 993
[6] Glaser A K *et al* 2015 *Med. Phys.* **42** 4127
[7] Low D 2015 *J.Phys.: Conf. Ser.* **573** 012009
[8] Ezzell G A 2009 *Med. Phys.* **36** 5359
[9] Berry S L *et al* 2010 *Med. Phys.* **37** 2425

IOP Conf. Series: Journal of Physics: Conf. Series **847** (2017) 012035 doi:10.1088/1742-6596/847/1/012035

Measurements of superficial dose distributions in radiation therapy using translucent cryogel dosimeters

M M Eyadeh[1], M Wierzbicki[2], K R Diamond[2]
[1]Physics Department, Faculty of Science, Yarmouk University, Irbid, 21163 Jordan
[2]Department of Medical Physics and Applied Radiation Sciences, McMaster University and Juravinski Cancer Centre, 699 Concession St, Hamilton, ON, L8V 5C2, Canada

Email: molhem.e@yu.edu.jo

Abstract. Superficial dose distributions were measured using radiochromic translucent poly(vinyl alcohol) cryogels. The relationship between dose to the skin surface and dose measured with the bolus (cryogel) was established using a series of oblique irradiations. Gafchromic film was placed under the bolus, and the ratio of bolus-film dose was determined for angle ranging from 0° to 90°. The average ratio over 0-67.5 degrees (0.800 ± 0.064) was used as the single correction factor to convert dose in bolus to dose to the skin surface, and applied to bolus measurements of skin dose from head and neck intensity modulated radiation therapy (IMRT) treatments delivered to a RANDO phantom. The resulting dose distributions were compared to film measurements using gamma analysis with a 3%/3mm tolerance and a 10% threshold. The minimum gamma pass rate was 95.2%. This study is the first report on the use of a poly(vinyl alcohol) cryogels based dosimeter to provide an accurate estimation of superficial dose distributions in megavoltage photon beams.

1. Introduction

In radiation therapy, the dose deposited at the skin surface by megavoltage photon beams may be substantially lower than the underline tissue [1]. However, treatment of superficial disease requires that the prescribed dose be delivered up to the skin surface. A layer of bolus may be placed on the skin to increase the dose deposited in the superficial tissues [2, 3]. Accurate knowledge of dose to superficial tissues is important, but this confounded due to the inaccuracy of most treatment planning systems (TPSs) in the buildup region [4]. TPSs are able to calculate skin dose within ± 25% [5-10]. Therefore, *in vivo* measurements are desirable to verify the skin surface dose.

Dosimetric systems such as thermoluminescent detectors (TLDs) [11] metal-oxide semiconductor field-effect transistors (MOSFETs) [12] and radiochromic film [13] are used during radiotherapy for dose verification purposes [14]. Film is difficult to form to surfaces that contain both convex and concave regions, which complicates dosimetry [15]. Another approach would be to use gel dosimeters [16, 17]. Dosimetric cryogels are flexible materials that conform easily to curved regions of the body. It may provide an accurate estimation of superficial dose distributions [18]. Recently, these cryogels were used to monitor chest wall radiation therapy treatment [19].

The dose estimated using TPS at depths of 0.5 to 1.0 cm lacks the accuracy desired during radiotherapy. For megavoltage photon beams, the dose increases up to 60% within the first 0.5 cm depth, making the measured surface dose sensitive to the buildup thickness [20-22]. Due to the rapid dose build up, the dose will not be uniform throughout the 0.5 cm thick bolus material, and not equal to the actual

IOP Conf. Series: Journal of Physics: Conf. Series **847** (2017) 012035 doi:10.1088/1742-6596/847/1/012035

surface dose. It may be possible to derive a correction factor from the dose measured in the gel dosimeter to the surface dose.

The purpose of the presented here work is to evaluate the ability of a translucent poly(vinyl alcohol) cryogel (PVA-C) containing ferrous benzoic xylenol orange (FBX) radiosensitive bolus material [23] to measure skin dose during radiotherapy. The concept is demonstrated using clinical IMRT fields delivered to the head and neck region of a RANDO phantom (Phantom Laboratory, Salem, NY, United States).

2. Materials and Methods

In this study, translucent FBX-PVA-C was used as radiochromic bolus using PVA concentration of 15% by weight. A detailed description of its production was described elsewhere [19, 23]. The hydrogel was decanted into custom plastic moulds with interior dimensions of 15 x 15 cm² and 0.5 cm thickness; the hydrogels were subjected to 3 cycles of 18 hour freezing at -80 °C and 6 hours thawing at room temperature. The cryogel is flexible and can conform to most parts of a patient's body. The bolus samples were imaged pre- and post-irradiation using 2D camera system using charge coupled device (CCD) camera, a Lumen-Essence BK-600 uniform red light emitting diode (LED) array, and a light tight box. 2D images were collected by the 16-bit CCD and stored as 'TIFF gray image' files. 2D Absorption coefficient maps were generated using an in-house MATLAB program.

The relationship between bolus absorption coefficient and delivered dose was established using a Varian iX linear accelerator (Varian Inc., Palo Alto, United States) under full scatter condition. The expected doses in cGy were computed using the Pinnacle v9.2 TPS (Philips, Amsterdam, Netherlands). The same procedure was employed to relate optical density and dose in pieces of EBT-2 Gafchromic film (lot #A052810-01) [International Specialty Products, Wayne, New Jersey, United States]. The EBT-2 film was scanned using an Epson 11000XL Scanner (Proscan, Avision, Australia) and analyzed using Film QA™ Pro (Ashland, Wayne, New Jersey, United States).

Open field irradiations were used to examine the relationship between the dose distribution recorded by the radiochromic bolus and the true surface dose, which was estimated using EBT-2 film by using a stack of radiochromic bolus and film on the surface of a polystyrene phantom. The configuration of these measurements is shown in figure 1.

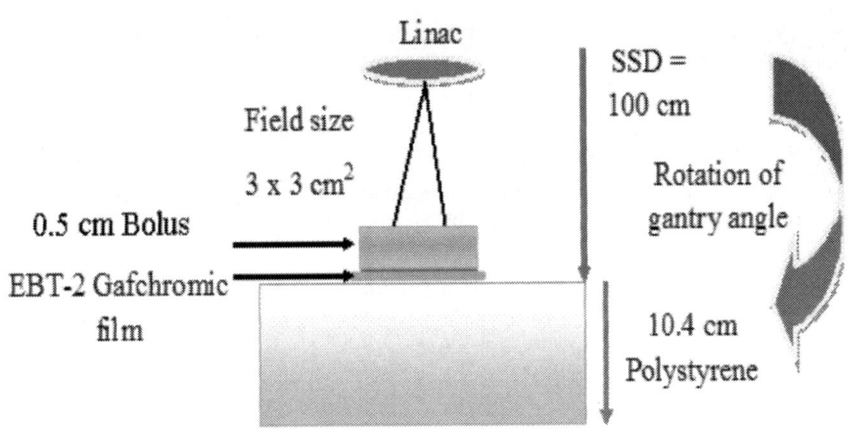

Figure 1. Schematic of the radiochromic bolus and EBT-2 film irradiation. The irradiation was repeated with gantry angles of 0°, 22.5°, 45°, 67.5°, and 90°. At each angle, a 3 x 3 cm² field was formed using the jaw collimator and 1000 MU were delivered with a rate of 600 MU/min.

Proof-of-principle *in vivo* dosimetry measurements were performed on clinical head and neck IMRT treatment plans that were delivered to a RANDO phantom. A Philips Brilliance Big Bore scanner was used to acquire 1 mm thick CT slices of the head and neck portion of a RANDO phantom. Two simple static parallel-opposed-pair (POP) beam arrangements were planned for the neck region; two previously treated clinical cases with bolus including a 3-field larynx and a 9-field head and neck IMRT were selected and positioned on the RANDO data to approximate the arrangement on the patient. The bolus was manually cut to size on the phantom surface using orthogonal light field projections as a guide. These plans were then delivered to the phantom with EBT-2 film placed on the phantom surface below a layer of 0.5 cm radiochromic bolus to evaluate the skin surface dose directly.

3. Results and Discussion

The dose measured in the radiochromic bolus and the dose at the underlying surface both increased with increasing gantry angle. This is consistent with previous studies that reported increasing surface doses with incident beam angles [11, 24, 25].

Film QA Pro software was used to visually align dose maps. The point-by-point ratio was computed. The mean ratios for the irradiated areas are summarized in Table1. If we restrict ourselves to a range of angles, 0° to 67.5° for example, we find that the average ratio between radiochromic bolus and film is 0.800 ± 0.064. Using this correction factor, the agreement between the film and the corrected bolus was improved at all gantry angles with average differences ranging from 1.4 to 1.9%. This suggests that a 0.5 cm radiochromic cryogel should be able to predict dose deposited at the bolus-skin interface. This factor was used to correct all subsequent bolus images.

Table 1. The mean ratio between surface dose and the dose measured in the radiochromic bolus at different gantry angles ranging from 0° to 90°.

Gantry angle	Mean ratio ± Standard deviation
0°	0.749 ± 0.005
22.5°	0.760 ± 0.005
45°	0.802 ± 0.009
67.5°	0.890 ± 0.016
90°	0.930 ± 0.002

For all gantry angles, the TPS overestimated the surface dose measured with film by 14.3 to 25.6%. This is consistent with related studies. Previously, it was reported that two TPSs overestimated surface dose by up to 18.5% when compared to Gafchromic film measurements [7].

In this study, gamma analysis was used to evaluate the agreement between the corrected radiochromic bolus and EBT-2 film measurements. A 2D gamma analysis was performed using the Film QA Pro software using 3%/3mm criteria and a 10% dose threshold. The percentage of pixels passing the gamma evaluation ranged from 95.2% to 96.4% as shown in Table 2.

Table 2. Gamma pass rates for comparisons of corrected radiochromic bolus and Gafchromic film measurements (3%/3mm, 10% threshold) for different field arrangements.

Irradiation Geometry	Gamma passing rate (%)
0° / 180° POP, static beams	96.1
90° / 270° POP, static beams	96.4
3-field beams larynx, IMRT	95.2
9-field beams neck, IMRT	95.5

The results suggest that the radiochromic bolus measures skin dose with sufficient accuracy for clinical use. Furthermore, the cryogel material is more flexible than film, providing improved skin contact. It can be wrapped easily around curved surfaces and provide *in vivo* dosimetry in areas where skin dose verification is desired. Different types of treatments with different cryogel thicknesses may yield different results and would have to be tested. However, for head and neck treatments with a typical bolus thickness the results of this paper should be valid.

4. Conclusion
A comparison of EBT-2 Gafchromic film and FBX-PVA-C radiochromic bolus suggests that the cryogel may provide an accurate estimation of skin surface dose distribution using a simple correction factor. Radiochromic bolus may then be used in place of more traditional forms of bolus to perform *in vivo* dosimetry in regions where the skin dose is important. The main advantage of this system over film is that it is more flexible, allowing it to be wrapped around complex curved surfaces.

5. Acknowledgement
This work has been supported by the NSERC Discovery Grant program and the Yarmouk University Physics Department.

6. References
[1] Hill R *et al* 2014 *Phys. Med. Biol.* **59** R183-231
[2] Khan F M 2010 *The Physics of Radiation Therapy*: 4th ed, Lippincott Williams & Wilkins, Baltimore, MD, USA.
[3] Vyas V *et al* 2013 *Med. Dosim.* **38** 268-73
[4] Fraass B *et al* 1998 *Med. Phys.* **25** 1173-822
[5] Mutic S and Low D A 2000 *Med. Phys.* **27** 163-5
[6] Dogan N and Glasgow G P 2003 *Med. Phys.* **30** 3091-6
[7] Chung H *et al* 2005 *Med. Phys.* **32** 2682-9
[8] Court L E *et al* 2008 *J. Appl. Clin. Med. Phys.* **9** 29-35
[9] Penettierei V *et al* 2009 *Radiother. Oncol.* **93** 94-101
[10] Kry S F *et al* 2011 *J. Appl. Clin. Med. Phys.* **13** 20-34
[11] Hsu S H *et al Phys. Med. Biol.* **53** 2593-606
[12] Falco M D *et al* 2015 *J. Appl. Clin. Med. Phys.* **16** 298-10
[13] Morales J E *et al* 2014 *Austral. Phys. Eng. Sci. Med.* **37** 303-9
[14] Hill R *et al* 2009 *Med. Phys.* **36** 3971-81
[15] Nakano M *et al J. Appl. Clin. Med. Phys.* **13** 83-97
[16] Johansson S Å *et al* 1997 *Acta. Oncol.* **36** 283-90
[17] Baldock C *et al* 2010 *Phys. Med. Biol.* **55** R1-63
[18] Chu K C *et al* 2000 *Phys. Med. Biol.* **45** 955-69
[19] Eyadeh M M *et al* 2016 *J. Appl. Clinic. Med. Phys.* **17** 308-19
[20] Devic S *et al* 2006 *Med. Phys.* **33** 1116-24
[21] Xiang H F *et al* 2007 *Med. Phys.* **34** 1266-73
[22] Qi Z Y *et al* 2009 *Med. Phys.* **36** 59-70
[23] Eyadeh M M *et al* 2014 *Phys. Med. Biol.* **59** 1773-87
[24] Lin J P *et al* 2001 *Appl. Radiat. Isot.* **55** 383-91
[25] Rinker G and Grussel E 1987 *Med. Phys.* **14** 870-73

IOP Conf. Series: Journal of Physics: Conf. Series **847** (2017) 012036 doi:10.1088/1742-6596/847/1/012036

Inorganic scintillator detectors for real-time verification during brachytherapy

G Kertzscher[1] and S Beddar[1,2]
[1]Department of Radiation Physics, University of Texas MD Anderson Cancer Center, Houston, TX 77030, USA
[2]The University of Texas Graduate School of Biomedical Sciences at Houston, Houston, TX 77030, USA

E-mail: gakertzscher@mdanderson.org

Abstract Widespread use of real-time dose measurement technology to verify brachytherapy (BT) treatments is currently limited because only few detectors exhibit the large dynamic range and signal intensities that is required to accurately report the data. Inorganic scintillator detectors (ISDs) are promising for real-time BT verification because they can exhibit large signal intensities. Luminescence properties of ISDs based on ruby, Y_2O_3:Eu and CsI:Tl were compared with BCF-60 plastic scintillators to determine their potential for BT verification. Measurements revealed that ISDs can exhibit signal intensities 1800 times larger than BCF-60 and that the Čerenkov and fluorescence light contamination is negligible. The favourable luminescence properties of ISDs opens the possibility to manufacture simplified detector systems that can lead to more widespread real-time verification during BT treatment deliveries.

1. Introduction

The dosimetric and clinical conditions in BT prevent the widespread use of scintillator detectors for *in vivo* dosimetry (IVD) [1,2]. The absorbed dose rate can shift by 2 orders of magnitude, e.g. if the BT source moves away from the detector from 0.5 to 5 cm, which requires the detector to be sensitive in a wide dynamic range. Furthermore, IVD in the tumor region requires insertion of the detector into narrow BT catheters which is possible only if the cross-section of the scintillator volume is small; preferably $<$ 1 mm. However, small scintillator volumes limit the signal intensity and can lead to large statistical uncertainties which reduces the error detection capacity [1,3].

Inorganic scintillator detectors (ISDs) are promising for real-time IVD during BT because several materials exhibit light yields (photons/keV) ~1 order of magnitude greater than organic scintillators [4]. ISDs with large signal intensities may lead to small scintillator volumes that are sensitive in a wide dynamic range and that measure dose rates with small statistical uncertainties. Large signal intensities also reduce the impact of the stem signal, which is the contaminating Čerenkov and fluorescence light induced in the fiber-optic cable [5]. The stem signal is prominent for BT source positions near the fiber-optic cable and far from the scintillator volume, and requires complexity in the read-out system to suppress it for plastic scintillator detectors (PSDs) [6] because their signal intensity is limited [7,8].

The objective of this study was to assess the potential of ISDs for real-time verification during BT by comparing the signal intensity of inorganic scintillator materials with an organic scintillator that is commonly used in radiotherapy, and by evaluating the relevance of the stem signal contribution.

2. Material and methods

2.1. Detector systems

The ISDs and PSD consisted of a 1 mm-size scintillator that was optically coupled to a 1 mm-diameter and 15 m-long fiber-optic cable (ESKA GH-4001, Mitsubishi Rayon Co. Ltd., Japan) made of poly(methyl methacrylate) (PMMA). The fiber-optic cable transmitted the scintillation light to the photodetector system which consisted of a charge-coupled device (CCD) camera or a spectrometer spectrograph (Luca S 658M and Shamrock SR-163i, respectively, Andor Technology Ltd., UK).

The ISDs (see table 1) were based on a ruby crystal (Al_2O_3 doped with Cr^{3+}; 49558, Edmund Optics Inc., USA), Y_2O_3:Eu in powder form (QK63/N-C1, Phosphor Technology Ltd, England) or CsI:Tl (SICCAS, Shanghai, China). The PSD was based on the organic scintillator BCF-60 (Saint-Gobain Crystals, France) which is made of polystyrene. The BCF-60, ruby and CsI:Tl volumes were coupled to the fiber-optic cable with transparent glue (NOA68, Norland Products Inc., USA) that was cured with ultraviolet light. The Y_2O_3:Eu volume was prepared by mixing 2 mass units Y_2O_3:Eu powder with 1 mass unit of the transparent glue. The mix was cured with ultraviolet light inside a transparent 1 mm-inner diameter plastic tubing attached to the fiber-optic cable.

Table 1. Physical characteristics of the scintillator materials.

Scintillator material[a]	Emission peak [nm]	Density [g/cm^3]	Electron density [electrons/cm^3]	Effective atomic number [Z_{eff}]	Shape	Volume [mm^3]
BCF-60 (O)	530	1.05	3.4×10^{23}	5.7	Cylinder	0.79
Ruby (I)	694	3.98	1.2×10^{24}	11.3	Half-sphere	0.26
Y_2O_3:Eu (I)	611	5.0	1.4×10^{24}	36.1	Cylinder	0.79[b]
CsI:Tl (I)	550	4.51	1.1×10^{24}	54.0	Cube	1.0

[a] One organic (O) and 3 inorganic (I) scintillator materials were included in this study.
[b] The volume of the Y_2O_3:Eu crystals inside the 0.79 mm^3 cylinder-shaped scintillator was ~0.26 mm^3.

The scintillator volume of the ISDs were covered with a reflecting material to enhance the signal intensity. The ruby and Y_2O_3:Eu volumes were covered with reflective paint (EJ-510, Eljen Technology, USA) and the CsI:Tl crystals enclosed by manufacturers with polytetrafluoroethylene. The influence of the photoluminescence (PL) of the ISDs was suppressed by placing a thin plastic optic filter (Rosco Laboratories Inc., USA) between the scintillator volume and the fiber-optic cable (adapted from [9]).

2.2. Measurements

A Nucletron MicroSelectron v2 HDR afterloader was used throughout all experiments, and the air-kerma strength of the ^{192}Ir source was between 17.6 and 29.4 mGy m^2 h^{-1}.

2.2.1. Scintillation intensities. The relative signal intensities of the ISDs and PSD were determined based on measurements of 3 different detector samples of each scintillator material. The signal intensity of the bare fiber without a scintillator volume was subtracted from the ISD and PSD intensities to remove the influence of the stem signal. The signal intensities were measured individually for each detector placed 2 cm from the BT source in a rigid polymer-based phantom (Solid Water, Gammex, USA). The scintillation intensities and electronic noise were recorded per unit time with the CCD camera setup, which incorporated a 25-mm (F1.8) lens (Computar, USA) and a 25-mm spacer. The total measurement variability due to the positional uncertainty of the detectors and BT source, repeated connections between the detectors and CCD camera, and statistical uncertainty was $< 5\%$. The measurements were corrected for differences in scintillator volumes, the quantum efficiency of the CCD camera, and the attenuation in the PMMA fiber-optic cable.

2.2.2. Stem signal. The stem signal contamination for the ISDs and PSD was determined based on scintillation intensity and emission spectra measurements, and on absorbed dose calculations [10]. The signal intensities of the detectors and the bare fiber were measured under identical BT source irradiation conditions, for which the absorbed dose in the scintillator volume and fiber-optic cable were calculated. The signal intensity per absorbed dose was obtained for each scintillator material and the fiber-optic cable by dividing the measured intensities with the calculated absorbed doses.

The emission spectra of the scintillator materials and the stem signal were measured with the spectrometer. The emission spectra were normalized to the corresponding signal intensity per absorbed dose value, such that each wavelength bin value was proportional to the absorbed dose in the scintillator volume or the fiber-optic cable. The emission spectra and the signal intensities were corrected for the attenuation in the PMMA fiber-optic cable and the quantum efficiency (QE) of the photodetectors.

The normalized emission spectra were multiplied by calculated absorbed doses in the scintillator volume and fiber-optic cable corresponding to various BT source positions relative to the detectors. The calculations allowed determining the stem signal contamination for the ISDs and PSD, e.g. for BT source positions near the fiber-optic cable and far from the scintillator volume.

3. Results and Discussion

Emission spectra of the scintillator materials and the PMMA fiber-optic cable that were induced by an ^{192}Ir BT source are shown in figure 1 (left-hand plot). The spectra have been corrected for the quantum efficiency of the spectrometer and the attenuation in 15 m PMMA fiber-optic cable.

The measurements revealed higher signal intensities for all ISDs compared with the BCF-60-based PSD (figure 1, right-hand log-scale plot). The signal intensity of the ruby-based ISD was 16 times larger than that of the PSD for 15 m PMMA attenuation and 39 times larger for 0 m attenuation. The corresponding values for the Y_2O_3:Eu-based ISD were 63 and 92, and 1700 and 1800 for the CsI:Tl-based ISD. The wavelength-dependence of the attenuation in PMMA governs the signal increase when the fiber-optic cable is shortened. If shortened from 15 to 0 m, the signal intensities would increase by a factor 1.5, 3.7, 2.2, and 1.6 for the BCF-60-, ruby-, Y_2O_3:Eu-, and CsI:Tl-based detectors, respectively.

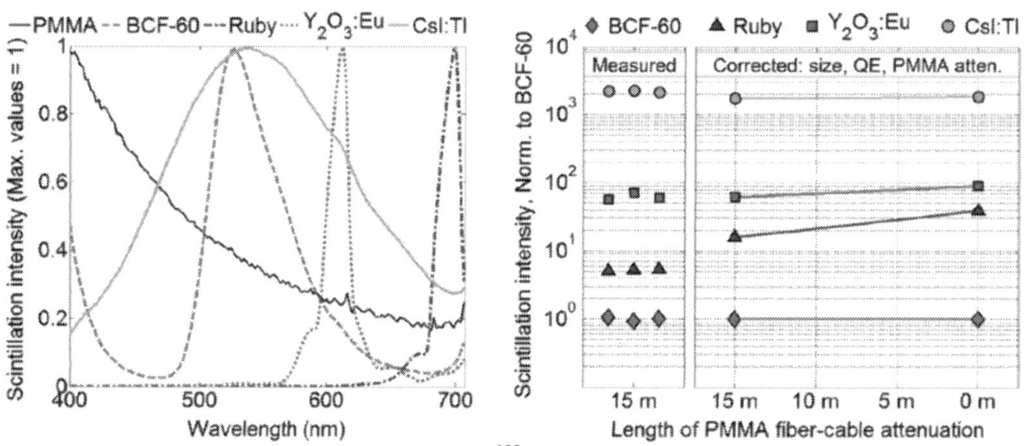

Figure 1. (Left) Emission spectra induced by an ^{192}Ir BT source of the scintillators used in this study and of the PMMA fiber-optic cable. **(Right)** Measured and corrected scintillation intensities of the detectors. The ISD intensities were normalized to the average value of the 3 BCF-60-based PSDs. The intensities were corrected for the size of the detector volume, the quantum efficiency (QE) of the CCD camera, and the attenuation in the fiber-optic cable.

The signal components for each detector for BT source positions 1 cm from the axis defined by the fiber-optic cable are shown in figure 2. The coordinate-convention defined for the TG-43 protocol is assumed [10]. The ISDs incorporated a long-pass filter between the scintillator and the fiber to suppress the PL background. The left-hand plot shows the superposition of the scintillation and stem signal for each detector when the source is placed 1 cm from the fiber-optic cable and 5 cm from the scintillator. The right-hand plots show that the stem signal is prominent for BCF-60-based PSDs and less prominent for the ISDs. The stem signal is ~3% of the total signal for ruby-based ISDs when the

BT source is 8 cm away from the scintillator, and 0.7% and 0.2% for the Y_2O_3:Eu- and CsI:Tl-based ISDs, respectively.

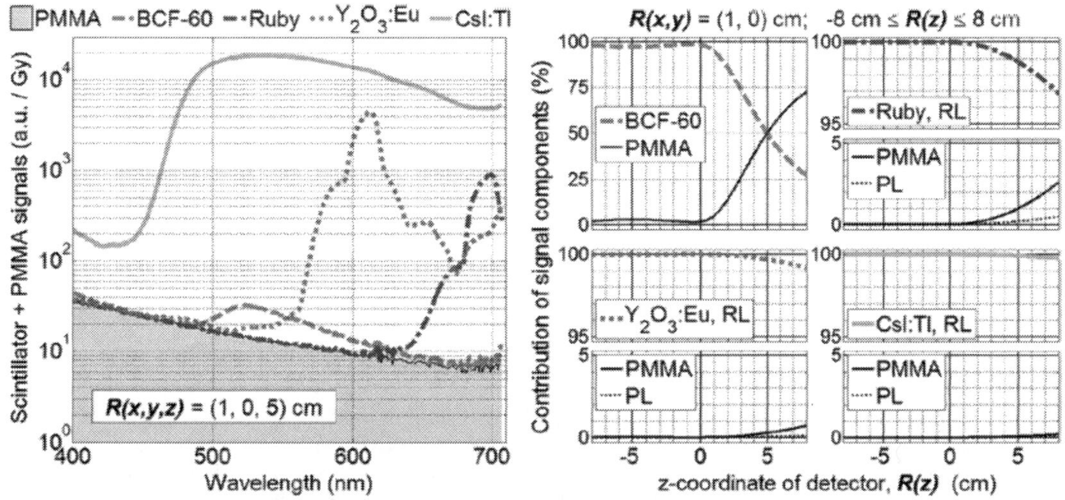

Figure 2. (Left) The sum of the scintillator and stem signal (PMMA) for each scintillator material used in this study. The emission spectra are proportional to the signal intensities when an ^{192}Ir BT source is placed 1 cm from the fiber-optic cable and 5 cm from the scintillator volume. **(Right)** The contribution of the different signal components in the PSD and the 3 ISDs for BT source positions 1 cm from the axis defined by the fiber-optic cable.

4. Conclusion

Inorganic scintillator detectors (ISDs) based on ruby, Y_2O_3:Eu and CsI:Tl are strong candidates for real-time verification for brachytherapy (BT), primarily because of their large signal intensities that were up to 39, 92 and 1800 times larger, respectively, than that of BCF-60-based plastic scintillator detectors. The stem signal is negligible for CsI:Tl-based ISDs, which establishes that complexity in the read-out system to suppress it can be avoided. Correction methods for the temperature- and energy-dependence of the ISD response are necessary. ISDs can open new frontiers for simplified, cost effective detector systems that can lead to more widespread real-time verification during BT.

5. Acknowledgment

The research reported in this publication was supported by the National Cancer Institute of the National Institutes of Health under Award Number R01 CA153824 from the National Cancer Institute. The content is solely the responsibility of the authors and does not necessarily represent the official views of the National Cancer Institute or the National Institutes of Health. The authors are thankful to Phosphor Technology Ltd, England, and SICCAS, China, for providing free crystal samples.

6. References

[1] Tanderup K *et al* 2013 *Med. Phys.* **40** 1-15
[2] Kertzscher G *et al* 2014 *Br. J. Radiol.* **87** 1-16
[3] Andersen C E *et al* 2009 *Med. Phys.* **36** 5033-43
[4] Derenzo S *et al Lawrence Berkeley Natl. Lab.*
[5] Beddar A S *et al* 1992 *Med. Phys.* **37** 925–35
[6] Beddar A S *et al* 1992 *Phys. Med. Biol.* **37** 1901-13
[7] Fontbonne J M *et al* 2002 *IEEE Trans. Nucl. Sci.* **49** 2223-7
[8] Therriault-Proulx F *et al* 2011 *Med. Phys.* **38** 2176-9
[9] Jordan K J 1996 *Proc. SPIE* **2705** 170-8
[10] Rivard M J *et al* 2004 *Med. Phys.* **31** 633-74

IOP Conf. Series: Journal of Physics: Conf. Series **847** (2017) 012037 doi:10.1088/1742-6596/847/1/012037

3D dose computation algorithms

T Knöös[1,2]
[1]Department of Haematology, Oncology and Radiation Physics, Skåne University Hospital, SE-205 02 Malmö, Sweden
[2]Department of Medical Radiation Physics, Clinical Sciences, Lund University, SE-221 85 Lund, Sweden

Email: tommy.knoos@mod.lu.se

Abstract. The calculation of absorbed dose within patients during external photon beam radiotherapy is reviewed. This includes the modelling of the radiation source i.e. in most cases a linear accelerator (beam modelling) and examples of dose calculation algorithms applied within the patient i.e. the dose engine. For the first part - the beam modelling, the different sources in the treatment head as target, filters and collimators etc are discussed as well as their importance for the photon and electron fluence reaching the patient. The consequences of removing the flattening filter, which several vendors now have made commercially available, is also shown. The pros and cons regarding different dose engines ability to consider density changes within the patient will is covered (*type a* and *b* models). Engines covered are, for example, pencil-beam models, collapsed cone superposition/-convolution models and combinations of these, as well as a glimpse on Monte Carlo methods for radiotherapy. The different models' ability to calculate dose to medium (tissue) and or water is. Finally, the role of commissioning data especially measurements in today's model based dose calculation is presented.

1. Introduction
Clinically used dose planning systems have for many years used calculation algorithms for X-rays and γ-beams, which make use of empirically, determined inhomogeneity corrections. These corrections have been applied to either measured dose distributions using e.g. film or simple analytical models describing the distribution in homogeneous water. Modelling of the dose distribution has not relied on basic photon and electron interaction (basic principles). The methods in use today are often model based algorithms where one separates the modelling of the radiation source from the in-patient dose calculation, the so-called dose engine [1].

2. Historical algorithms
A short description of older algorithms seems to be in place and we divide then depending on their ability to model scattered radiation. These models can be summarised as correction models applied to the dose distribution in water.

2.1. Algorithms without Scatter Modelling
In this group of algorithms, e.g. effective attenuation, isodose shift, tissue air ratios (TAR), effective source skin distance (SSD) [2] the only influence of the patient geometry is thru a scaling of the radiological depth for the calculation of absorbed dose. More or less sophisticated versions exist - approximations of homogeneous media, regional corrections or pixel by pixel corrections. These

methods only account for inhomogeneities with respect to their density along the fan line. A better method was introduced with the algorithm by Batho where the position of the inhomogeneity also is considered [3]

These methods were all, in principle, developed prior to the use of CT. The resulting dose distributions from these models can in some clinical cases differ up to 20 % from measurements, especially in low density media irradiated with narrow (5 x 5 cm^2) high energy photons beams [4]. Deviations of the same magnitude between measurements and calculations were found for a tangential beam geometry using the effective attenuation method. Of these methods, the Batho method estimates the absorbed dose with the highest accuracy. Lulu and Bjärngard have extended the Batho algorithm to account also for the lateral extension of inhomogeneities [5, 6].

2.2. *Algorithms with Scatter Modelling*

The next group of methods all use the radiological depth to determine the primary dose component but also accounts for scatter produced in the irradiated volume.

The Equivalent Tissue Air Ratio, ETAR algorithm relies on a calculation of an effective homogeneous density which is assigned to each point using weighting factors applied to the surrounding points (voxels) [7]. Geometric dimensions (e g depth, field size) are scaled from unit density to the effective density using O'Connor's theorem [8, 9]. The weighting factors used are calculated using inelastic scattering cross sections for water, i.e. basic principles are applied. The agreement with measurements are better for this method than for the simple algorithms. An accuracy of ± 3 % can be expected in most cases [7], however, in some situations, the deviation for the ETAR method can increase, for example, tangential beams, where deviations up to 5 % can be found [10].

An extension of the ETAR method using TAR's calculated by Monte Carlo have been proposed [10]. The new TAR's are more accurate especially for small fields where they fall to zero for zero field size. Therefore, the absence of lateral electronic equilibrium for narrow beams in low density media is modelled better. The final equation used is a superposition integral similar to the methods discussed by Mackie *et al* [11] and Mohan *et al* [12].

In general, none of the methods discussed deals with the lack of electronic equilibrium for narrow beams of high energy X-rays in low density media. All these models assume that the energy transferred to electrons is absorbed locally i.e. the collision kerma is equal to the absorbed dose.

3. Modern dose calculation algorithms

3.1. *Multi source models*

The limited accuracy of the methods described above has initiated much work on methods based on physical principles. The approach used mainly today is fluence or energy fluence multi source models. The latter is describing the radiation source, usually the treatment head of the accelerator down to different detailed levels. For example, the target generating x-rays from the impinging electron which is considered the primary radiation beam, scattered radiation from primary collimator, beam flattening filter (if present), monitor chamber, multi leaf collimator and conventional jaw system. A common simplification is to reduce all scatter sources to one single usually placed at the position of the flattening filter. This position is chosen since the flattening filter accounts for the majority of the scattered radiation from the treatment head.

Once the model of the radiation source is set up, one calculates the output from it, as an energy fluence distribution. In the case of using Monte Carlo this can be a phase space of the energy fluence i.e. a description of type of particles, their position, direction, and energy. The energy fluence distribution is modulated by field shaping parts of the treatment head. The sources lateral distribution and relative magnitude is usually adjusted based on measurements of conventional measurements e.g. lateral profiles, depth doses, output factors in air and water.

IOP Conf. Series: Journal of Physics: Conf. Series **847** (2017) 012037 doi:10.1088/1742-6596/847/1/012037

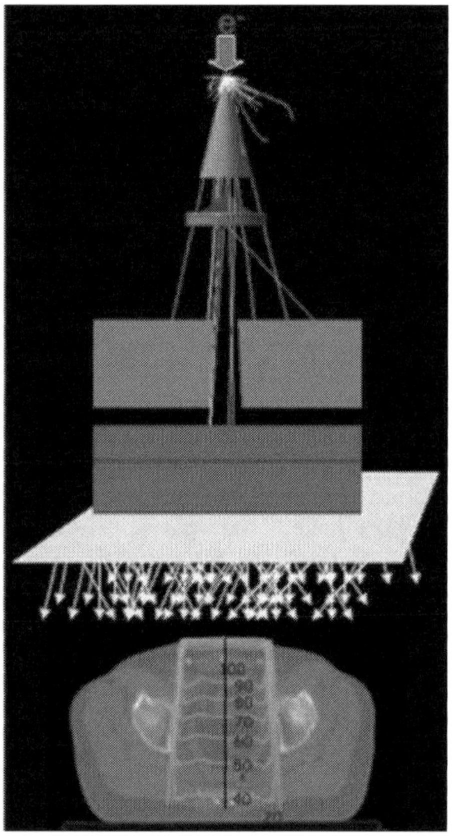

Figure 1. A schematic describing the separation of the modelling into a multi-source model for the treatment head and an in-patient dose calcualtion. The plane below the jaw system represents the phase space of out-coming particles. The patient is represented with a single transversal cross-section and isodose lines are added for illustration.

3.2. In-patient or dose engine

Dose calculation is based on the previous energy fluence impinging on and into the patient. The patient is represented by a 3D voxel matrix where each voxel is representing the density (relative physical or electron density depending on system). For some models, also a medium may be given for each voxel.

3.2.1. Point kernels

The energy fluence is ray traced thru the patient matrix and attenuated considering the density of each voxel and when appropriate also the medium of the voxel. For each voxel, the TERMA (Total Energy Released in Mass is calculated as the product of the energy fluence Ψ and the mass energy absorption coefficient μ/ρ. The following equation gives the TERMA T differential in energy E at an arbitrary point in the 3D room given by the vector $\boldsymbol{r}$.

$$T_E(\boldsymbol{r}) = \frac{\mu_E}{\rho} \cdot \Psi_E(\boldsymbol{r})$$

From the TERMA distribution one gets the absorbed dose by a convolution with a point kernel describing the transport of energy by primary electrons release (collision kerma) and scattered photons. (The differential in energy has been excluded for clarity, the integral has to be performed over all energies).

$$D(\boldsymbol{r}) = \int T(\boldsymbol{r}') \cdot P(\boldsymbol{r} - \boldsymbol{r}')d\boldsymbol{r}$$

The differential in energy has been excluded for clarity, the integral has to be performed over all energies. It is, however, quite common to perform the integral over energy before the convolution, thus only one TERMA matrix and one single point kernel is required.

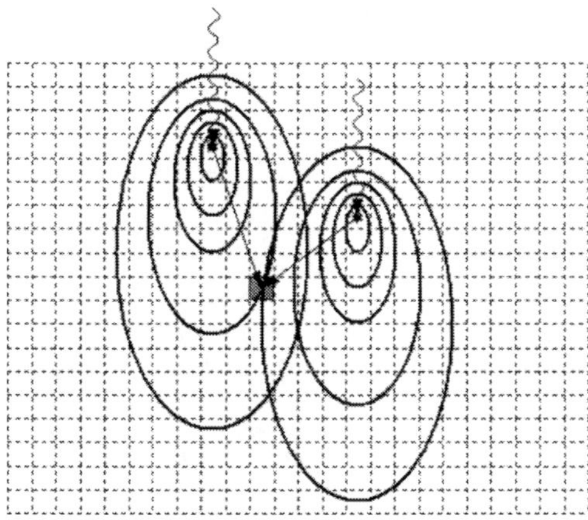

Figure 2. A matrix of TERMA (dotted lines) is convolved with a point kernel, shown as isolevels (solid lines and broken lines respectively) representing the pattern of absorbed dose around an interaction point. The absorbed dose to the shaded voxel (dose point) is determined as the TERMA at the interaction point times the level of the kernel at the dose point. This is repeated for all interaction sites (only two is shown for clarity) to get the total absorbed dose to a single voxel.

This time-consuming integral can be solved using Fourier transforms where the convolution integral is replaced by the inverse Fourier transform of the product between the transforms of T and P. This approach is fundamental in the work by Boyer who used Fast Fourier Transform (FFT) to solve the convolution of dose distribution kernels with photon fluence distributions to give the final dose in 3D [13, 14]. However, if the kernel function P varies with position due to changes in density, an analytical superposition integral has to be solved instead.

The point dose kernels used, which includes electrons released in the first interaction as well as single and multiple scattered photons, are generally calculated using Monte Carlo simulations but analytical methods are also applicable.

3.2.2. Convolution in inhomogeneous medium
Inhomogeneous media can be included in the convolution integral either using a large number of dose distribution kernels, one per density, or a single kernel combined with the following *scaling* theorems:

"In a medium of a given composition exposed to a uniform flux of primary radiation (such as X-rays or neutrons) the flux of secondary radiation is also uniform and independent on the density of the medium as well as of the density variations from point to point." [15]

"In a given irradiation system the ratio of scattered photons to primary photons remains unchanged when the density of the irradiated material is changed if all the linear dimensions of the system are altered in inverse proportions to the change in density." [8, 9]

Density in this context should be interpreted as interaction sites per unit volume, e g electron density for incoherent scattering of photons. It is also assumed that the amount of secondary radiation produced at a point is proportional to the local number of interaction sites.

Mackie *et al* used superposition kernels calculated for several densities with interpolation and scaling between them [11]. A similar approach using kernels for only a *single* density combined with the scaling theorems has been applied [12, 16-18].They all use ray tracing of scatter between interaction and absorption point and employ the scaling theorem for inhomogeneities. Transportation of energy by electrons is included in the kernels, therefore, electronic disequilibrium will be modelled.

3.2.3. Pencil beam kernels
A method to increase the calculation speed to make it clinical feasible was the introduction of pencil beam kernels where the dose calculation problem has decreased by going from 3D to 2D convolution. By integrating the point kernel along the depth direction, we get a 2D description of the dose distribution.

Figure 3. Illustration of the principle of pencil beams showing creation of pencil beams by convolving/integrating TERMA along a ray with the point kernel describing the energy transport and absorption. Left panel - TERMA along the central ray of a very narrow photon beam, middle panel - point kernel and right panel - the resulting pencil beam.

Pencil beams have been created in various ways for example from measurements by radial differentiation of relative dose on central axis from broad beam dosimetric quantities [19]. Another method was the differentiation of radial beam data [20]. Monte Carlo methods have also been used to calculate pencil beam kernels in water [16, 21, 22]. Usually the pencil beams are parametrized and the actual values of these are found by a fitting process to measured data [22].

Applying the pencil beam convolution is described in principle by the following equation:

$$D(x,y,z) = \iint \Psi(x',y') \cdot P(x-x',y-y',z)dxdy$$

Here the energy fluence distribution Ψ at a certain specified plane is convolved with the pencil beam kernel P. In heterogeneous media, the pencil beam kernel is scaled along the propagation direction by replacing the z with the radiological depth z_{radiol} considering the density of the voxels along the ray. No scaling is performed, in the initial implementations, perpendicular to the propagation direction. Thus, the pencil beam convolution is a so-called *type a* algorithm [23]. There is, however, today one pencil beam model where a lateral scaling of the kernel has been added [24]. The commercial implementation is the analytical anisotropic algorithm (AAA) [25, 26]. The performance of pencil beams algorithms have been reported in several reports [27-32].

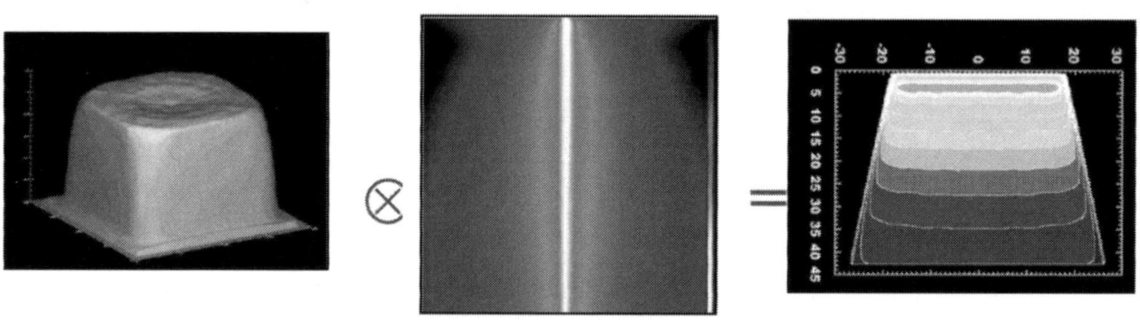

Figure 4. Illustration of the 2D energy fluence distribution (left) convolved with the pencil beam (middle) to get the dose distribution represented by a tranvese plane (right).

3.2.4. Collapsed cone convolution

Another approximation to increase speed (which maybe is not a problem today) as well as managing inhomogeneities is to discretize the point kernel. The kernel is discretized in about 100 directions and by letting each direction representing a cone where all energy is collapsed to the center of the cone [33, 34].

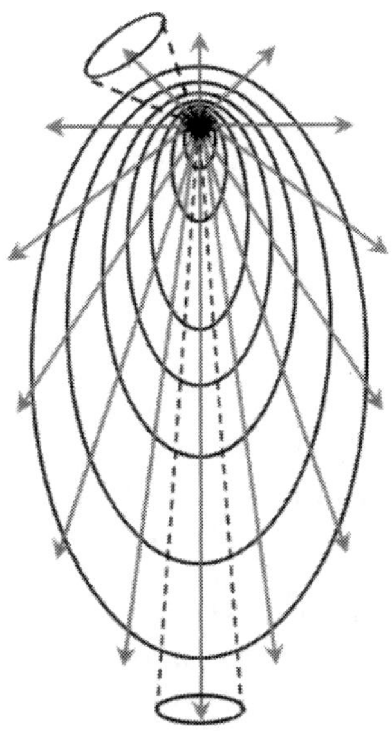

Figure 5. The point kernel (illustrated by the isolevels) is discretised along the arrows which each represent a cone where energy is transported. All energy in the cone is collapsed to the central ray.

When collapsing the energy in the cone into the central ray the inverse square dependence is removed. The distribution of cones is not isotropic, instead a higher density is used in the forward direction because the majority of energy in high energy beams is transported in this direction. A few cones take care of the backscatter. Density of cones should be high enough where significant amount of energy is deposited such that each voxel is passed by one or more collapsed cones.

The convolution integral is commutative thus changing the order of the functions does not change the result

$$D(\boldsymbol{r}) = \int T(\boldsymbol{r}') \cdot P(\boldsymbol{r} - \boldsymbol{r}')d\boldsymbol{r} \text{ is equal to } D(\boldsymbol{r}) = \int P(\boldsymbol{r}') \cdot T(\boldsymbol{r} - \boldsymbol{r}')d\boldsymbol{r}$$

This property can be interpreted as shown in the following figure:

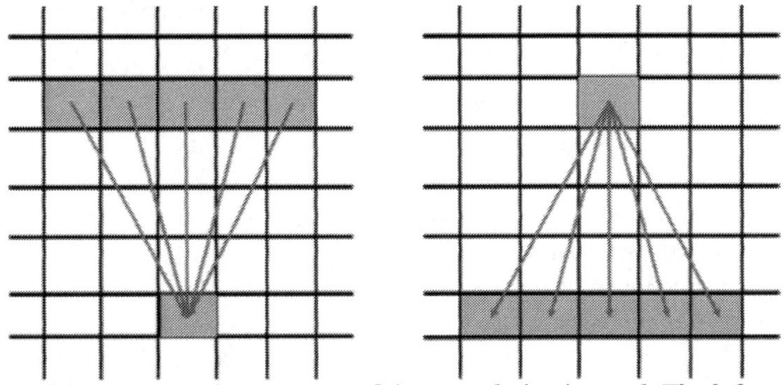

Figure 6. Illustration of the commutative property of the convolution integral. The left panel shows the "dose deposition view" and the right panel the "interaction view" [4].

For the dose deposition view, energy or TERMA from the surrounding voxels is summed up to the deposition voxel. When all contributions are summed up, the total dose in the voxel is determined. In the interaction view, the TERMA in a voxel is spread out to the surrounding voxels and all contributions from interaction points have to be covered. In principle, the first approach can be used to calculate the absorbed dose in a single point/voxel only. The implementations based on this approach utilize the possibility to have a dose grid with different spacing to speed up calculations. For example, in areas with small gradients in TERMA or patient density a coarse grid can be used and in high gradients a finer one.

Another approach used to efficiently perform the convolution is to align the point kernels such that the cones in the same direction overlaps. This makes it possible to ray trace through the TERMA matrix along each cone direction and recursively pick up energy from each traversed voxel, transport, attenuate and then deposit energy along the axis of the cone [33].

3.3. Linear Boltzmann transport equation

A more direct way of performing in-patient dose calculations is to start with the linear transport equation by Boltzmann (LBTE). For many years one have used Monte Carlo (MC) techniques to statistically solve the problem [35-38]. MC methods use stochastic distributions describing the physics to simulate the transport of individual particles (photons and charge particles) through the density and medium matrix describing the patient geometry. The limitation, at least up today, is the large number of particles is required to get a sufficient high uncertainty of the deposited dose in each voxel of interest. MC can be seen as a microscopic solution of the LBTE.

Recently analytical solutions have been applied to LBTE for radiotherapy planning problems in a macroscopic approach i.e. instead of individual particles we create, follow, and attenuate energy fluencies in a matrix describing the patient. An exact analytical solution is not possible, mainly due to its complexity, thus a numerical have to be used, i.e. discrete ordinate method or grid based solvers [39-42]. Evaluation of a commercial implementation have been presented by e.g. Fogliata *et al* [43, 44] and Hoffman *et al* [45].

In short, the grid based method can be described as; the properties of the radiation source and the absorbing medium (patient) are described, the multi-source model (e.g. by a phase space) is used to get the energy fluence to the patient and then ray traced through the irradiated volume (observe this is very similar to the point kernels methods discussed above). In the next step the discrete ordinates is applied to solve the LBT equation giving photon energy fluence and the electron sources in each voxel. Finally, the electrons are transported to obtain electron fluence distribution in each voxel and by applying mass stopping power the absorbed dose is determined. The energy of photons and electrons are discretized, 25 and 49 levels, retrospectively. Transport directions is discretized based on the energy involved between 32 and 512 directions (These data are from the Acuros implementation [46]).

Interesting to comment is that both solutions to the LBTE show inherent errors, for MC we have stochastic uncertainties. For discrete ordinate solvers, the concept of discretization may produce uncertainties. As always this can be improved by prolonging the computation by increasing the number of particles in the MC case and for the solver using a finer discretization.

4. Dose to medium or dose to water

For many years, there have been a discussion on how to report absorbed dose during radiotherapy procedures [47]. No consensus really exists if the dose should be reported as absorbed in water or in the actual medium. With new algorithms as MC and grid based LBTE solvers getting more common this problem is growing. Both these approaches can report either.

The issue can be divided into two parts, a) if the energy/TERMA during the ray trace considers the physical properties of the transversed voxels or all media is considered to be water, and b) the other step is how the deposition of energy is accomplished. Introducing the formalism $D_{\text{transport, deposit}}$ where transport and deposit are the medium used respectively. They can either take the value water (w) or medium (m).

Considering the kernel approaches, pencil beam and point kernels, are all determined in water, analytical, MC or from experiments. Looking in detail to pencil beam algorithms, scaling according to the transversed density (radiological depth scaling) assumes all media is water and no corrections are performed during dose deposition. Thus pencil beam models transports and deposit dose in water, $D_{w,w}$.

For the point kernel models available, no consistent handling exists. In principle, the TERMA or equivalent can be ray-traced considering the medias physical properties or not. The kernels are determined in water but when dose deposition occurs a correction using mass stopping power ratio between medium and water may be applied. No single conclusion on what is reported can be given, however, we probably have one of the following situations: $D_{w,w}$, $D_{m,m}$ *or* $D_{m,w}$.

MC models have the possibility to transport in water or medium and the same is valid for dose deposition. Most common is to transport in medium and deposit in medium. If the user want dose to water, a mass stopping power ratio between water and medium is applied to each voxel [48]. Thus we have $D_{m,m}$ *or* $D_{m,w}$. When applying the mass stopping power ratio one usually have to use a macroscopic value for the energy deposit in the voxel since the electron energy fluence differentiated in energy in the voxel is not known. Alternatively, this could be done microscopically if done for each single energy deposition.

The discrete ordinate solvers for the LBTE determines the electron energy fluence differentiated in energy for each voxel and either the mass stopping power for the medium or for water is used to determine the absorbed dose. All transport is accomplished in the medium thus we have $D_{m,m}$ *or* $D_{m,w}$. The choice is user selectable in the only existing commercial implementation [49, 50].

5. Summary

Today we have very accurate tools available for estimating the absorbed dose within the patient during radiotherapy [51-57]. We still have, however, several problems to solve. The models discussed here are in principle for static patients without movements i.e. intrafractional and/or interfractional movements. The latter can probably be solved by daily imaging and adapting the today's plan. This can be accomplished by e.g. a library of plans or on-line re-optimisation and calculation. For intrafractional movements, we have used PTV, ITV etc. and margin recipes to assure that the target get the correct dose. This will in many cases lead to an over-irradiation of healthy tissues surrounding the tumour. Techniques with synchronisation of the patient with the treatment delivery e.g. gating and tracking is close to be clinical routine at many departments.

Another problem that has been noticed when introducing more accurate algorithms is a tendency of optimizers to put in too much dose in the vicinity of the tumour to assure an adequate dose in the whole tumour. This is especially noted when working with the PTV concept and tumours in low density regions such as the lung. In these cases, one should probably use a combination of the *type a* and *b* algorithms where optimisation is performed using *type a* and then followed by a recalculation with an appropriate *type b* model. More physically one can describe this as an optimisation based on energy fluence (pencil beam model) followed by dose calculation with full lateral electron modelling.

6. References

[1] Ahnesjö A and Aspradakis M A 1999 *Phys. Med. Biol.* **44** R99-155
[2] ICRU 1976 **24** Bethesda, Maryland, USA
[3] Batho H F 1964 *J. Canadian Ass. Radiologists* **15** 79-83
[4] Mackie T R *et al* 1985 *Med. Phys.* **12** 188-96
[5] Lulu B A and Bjärngard B E 1982 *Med. Phys.* **9** 372-7
[6] Lulu B A and Bjärngard B E 1982 *Med. Phys.* **9** 907-9
[7] Sontag M R and Cunningham J R 1978 Radiology 129 787-94
[8] O'Connor J E 1957 *Phys. Med. Biol.* **1** 352-69
[9] O'Connor J E 1984 *Med. Phys.* **11** 678-80
[10] Wong J W and Henkelman R M 1983 *Med. Phys.* **10** 199-208
[11] Mackie T R *et al* 1985 *Med. Phys.* **12** 327-32
[12] Mohan R *et al* 1986 *Med. Phys.* **13** 64-73

[13] Boyer A 1984 *Med. Phys.* **11** 552-4
[14] Boyer A and Mok E 1985 *Med. Phys.* **12** 169-77
[15] Fano U 1954 *Radiat. Res.* **1** 237-40
[16] Ahnesjö A *et al* 1987 *Acta. Oncol.* **26** 49-56
[17] Knöös T and Nilsson M 1987 *Proc. 9th Int. Conf. on the Use of Comp. in Rad. Ther.*
[18] Nilsson M and Knöös T 1992 *Phys. Med. Biol.* **37** 69-83
[19] Ceberg C P *et al* 1996 *Med. Phys.* **23** 505-11
[20] Storchi P and Woudstra E 1996 *Phys. Med. Biol.* **41** 637-656
[21] Mohan R and Chui C S 1987 *Med. Phys.* **14** 70-7
[22] Ahnesjö A *et al* 1992 *Med. Phys.* **19** 263-73
[23] Knöös T *et al* 2006 *Phys. Med. Biol.* **51** 5785-807
[24] Ulmer W *et al* 2005 *Med. Biol.* **50** 1767-1790
[25] Van Esch A *et al* 2006 *Med. Phys.* **33** 4130-48
[26] Tillikainen L *et al* 2008 *Phys. Med. Biol.* **53** 3821-39
[27] Knöös T *et al* 1994 *Phys. Med. Biol.* **39** *1609-28*
[28] Knöös T *et al* 1995 *Phys. Med. Biol.* **40** 1411-20
[29] Knöös T *et al* 1995 *Radiother. Oncol.* **37** 74-80
[30] Weber L *et al* 1996 *Med. Phys.* **23** 307-16
[31] Hurkmans C *et al* 1996 *Phys. Med. Biol.* **41** 1277-90
[32] Hansson H *et al* 1999 *Phys. Med. Biol.***44** N201-7
[33] Ahnesjö A 1989 *Med. Phys.* **16** 577-92
[34] Miften M *et al* 2000 *Phys. Med. Biol.* **45** 817-33
[35] Nelson W R *et al* 1985 *Stanford Linear Accelerator Report* SLAC-265
[36] Fippel M 2004 *Med. Phys.* **31** 1235-42
[37] Rogers D W 2006 *Phys. Med. Biol.* **51** R287-301
[38] Chetty I J *et al* 2007 *Med. Phys.* **34** 4818-53
[39] Wareing T A *et al* 2001 *Nucl. Sci. Eng.* **138** 256-68
[40] Gifford K A *et al* 2006 *Phys. Med. Biol.* **51** 2253-65
[41] Vassiliev O N *et al* 2008 *Int. J. Radiat. Oncol.* **72** 220-7
[42] Vassiliev O N *et al* 2010 *Phys. Med. Biol.* **55** 581-98
[43] Fogliata A *et al* 2011 *Radiat. Oncol.* **6** 82
[44] Fogliata A *et al* 2011 *Med. Phys.* **38** 6228-37
[45] Hoffmann L *et al* 2012 *Acta Oncologica* **51** 376-85
[46] Failla G A *et al* 2015 *Varian white paper RAD* 10156 10/10 350
[47] Ma C M and Li J 2011 *Phys. Med. Biol.* **56** 3073-89
[48] Siebers J V *et al* 2000 *Phys. Med. Biol.* **45** 983-95
[49] Mampuya W A *et al* 2016 *J. Appl. Clin. Med. Phys.* **8** 6338
[50] Rana S *et al* 2014 *Int. J. Cancer Ther. Oncol.* **2** 020421
[51] Mather M L *et al* 2003 *Ultrasonics* **41** 551-9
[52] Baldock C *et al* 2010 *Phys. Med. Biol.* **55** R1-63
[53] Vial P *et al* 2006 *Phys. Med. Biol.* **51** 5517-38
[54] Hurley C *et al* 2006 *Nucl. Instrum. Meth. A* **565** 801-11
[55] Baldock C 2009 *J. Phys.: Conf. Ser.* **164** 01200
[56] Hill B *et al* 2005 *Med. Phys.* **32** 1589-97
[57] Hill R *et al* 2009 *Med. Phys.* **36** 3971-81

IOP Conf. Series: Journal of Physics: Conf. Series **847** (2017) 012038 doi:10.1088/1742-6596/847/1/012038

Clinical management of tumour volume changes in VMAT head & neck radiation treatment

K M Alexander[1], J Gooding[2], L J Schreiner[1,2] and T Olding[1,2]
[1]Department of Physics, Queen's University, Kingston, Ontario, K7L3N6, Canada
[2]Cancer Centre of South Eastern Ontario at Kingston General Hospital, 25 King Street West, Kingston, ON, K7L5P9, Canada

E-Mail: kevin.alexander@krcc.on.ca

Abstract. The impact of changing anatomy due to tumour shrinkage was assessed for a VMAT head & neck plan optimized according to our centre's planning protocol. A custom-built wax phantom accommodating ion chamber, MOSFET, EBT3 film and Fricke-xylenol orange-gelatin (FXG) gel dosimeters and a variable size bolus 'tumour' was used in the investigation. Results indicate that the practice of initiating a patient re-scan and verification plan is appropriate when a change in external body contour greater than 1 cm compared to the original anatomy is observed.

1. Introduction

Radiation treatment of head and neck cancers often involves significant tumour volume shrinkage over the course of treatment, particularly in the parotid, oral cavity and neck regions. At our centre, when daily cone beam computed tomography (CBCT) imaging of a head and neck patient on treatment indicates a change in external body contour greater than 1 cm compared to the planning CT anatomy, the oncologist may decide to initiate a new planning CT and have the treat approved plan calculated and verified for target coverage and organ-at-risk (OAR) sparing on the new CT anatomy. If the dose distribution is acceptable, then treatment continues as before; if not, a re-plan is initiated.

The purpose of this work is to develop a head-and neck phantom that can accommodate different types of dosimeters and perform an end-to-end validation of this clinical process step by evaluating the 3D dosimetry [1] of a variable volume tumour in the head-and neck region. In this investigation, the dosimetric impact of a shrinking tumour is assessed for a VMAT head-and neck plan using ion chamber, MOSFET, EBT3 Gafchromic film, and Fricke-xylenol orange-gelatin (FXG) gel dosimeters incorporated into a custom-built wax head and neck phantom.

2. Materials and Methods

2.1. Phantom Design

A plaster cast of a Rando phantom (figure 1a) was used to create a wax head and neck phantom (figure 1b) that could accommodate different dosimeters (figure 1c). The phantom included a cavity for insertion of either i) a gel dosimeter or ii) a wax-filled jar with a hole for ion chamber insertion. A slot was also cut for placement of EBT3 film on an axial plane within the region of the target volume (figure 1b). A representative 'tumour' was made in two parts that fit together to form a single tumour conforming to the neck region below the ear of the wax phantom (figures 1a,b). Each part of the wax tumour measured a maximum thickness of approximately 1.3 cm near the location of the slot for film insertion for a combined total maximum tumour thickness of approximately 2.6 cm.

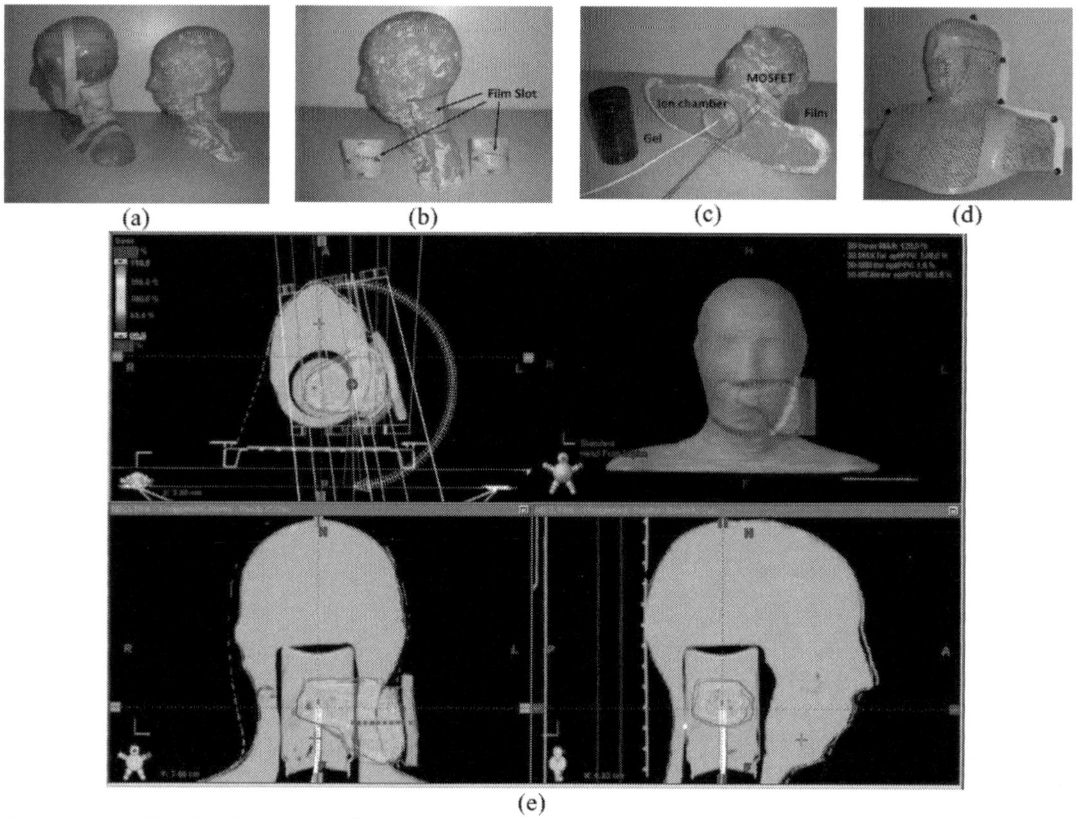

Figure 1. (a) Rando phantom used to create wax phantom, (b) two part wax bolus 'tumour', (c) Photo showing all four dosimeters used in the study: FXG gel jar fitting in an interior cavity, ion chamber in wax-filled jar in the same interior cavity, EBT3 film inserted in slot through 'tumour into wax Rando neck, and MOSFET dosimeter affixed to phantom surface. (d) Head and neck mask used for wax Rando immobilization with custom wax bolus on mask to bring full dose to target volume surface. (e) VMAT plan delivered to wax Rando with ion chamber insert, with red dotted line showing location of film on the coronal plane (lower left window).

2.2. Delivery Validation

The dosimetry tools used for delivery validation in this investigation were: a) a single point ion chamber (PR-05P 0.07 cc, Capintec, Ramsey, NJ), b) a MOSFET (TN-502RDM-H, Thomson Nielson, Division of Best Medical Canada Ltd, Ottawa, ON) c) 2D Gafchromic EBT3 film (ISP, Wayne, NJ), and c) 3D FXG gel dosimeter with optical CT readout [2]. CT scans of the two phantom geometries (gel insert and ion chamber insert) were acquired according to our centre's standard head and neck imaging protocol (120kV, 2mm slice thickness, 2 mm spacing) using a Philips Brilliance Big Bore CT scanner (Philips Medical Systems, Cleveland, OH). The phantom was immobilized in a thermoplastic mask (figure 1d) on a Type-S Overlay head and neck board (Civco Medical Solutions, Coralville, IA) during CT planning and treatment delivery.

For VMAT planning, dual coplanar clockwise/counterclockwise 6MV partial arcs were employed, extending from just beyond midline anteriorly to 180 degrees posteriorly (figure 1e). The collimator angle on the two arcs was set to ±30 degrees to minimize overlap of multi-leaf collimator (MLC) leakage. A RapidArc™ VMAT plan was optimized and calculated in Eclipse™ v.10 (Varian Medical Systems, Palo Alto, CA) on the 2 mm slice thickness scan of the wax phantom containing the ion chamber wax jar insert. This plan was then re-calculated (but not re-optimized) on the CT dataset containing a gel dosimeter jar. The optimized plan was normalized to a prescription of 1.8 Gy to accommodate the optimal range for FXG gel dose readout accuracy. A screen capture of the ion chamber plan is shown in figure 1e.

All treatment plans were delivered on a Varian Trilogy 2100iX linear accelerator (Varian Medical Systems, Palo Alto, CA) with Advanced OBI™ imaging cone beam CT (CBCT) setup verification. Each dose measurement was compared against calculated Eclipse treatment planning system dose using software appropriate for that QA tool. EBT3 film measurements were acquired using an in-house built CCD-lightbox film scanner and analyzed using the Film Dosimetry Analysis tool in 3D Slicer [3]. The reconstructed, calibrated gel dose data from the Vista™ optical CT scanner (Modus Medical Devices Inc., London, ON) was compared to Eclipse dose in 3D Slicer [4] (www.slicer.org).

3. Results

3.1. Dosimeter-Plan Dose Comparison

Results for the dosimeters used in this work are summarized in Table 1. The first value in the point dose difference column compares the measurement to the Eclipse calculated value on the associated full/half/no tumour anatomy. The second (bracketed) value is the comparison of the Eclipse calculated dose on the full tumour anatomy to the measurement delivered to the half/no tumour anatomy.

Table 1: Summary: measured vs. Eclipse doses.

Dosimeter	Tumour	Prescription (Gy)	Point Dose Difference (%)	% Gamma ≤1 (3%/3mm)
Ion Chamber	Full	1.8	-1.2	-
Ion Chamber	Half	1.8	-1.5 (-0.5)	-
Ion Chamber	No	1.8	-1.4 (+1.1)	-
MOSFET	Full	1.8	-1.0	-
MOSFET	Half	1.8	-2.6 (+6.6)	-
MOSFET	No	1.8	-3.5 (+6.4)	-
EBT3 Film	Full	1.8	-	78.9
EBT3 Film	Half	1.8	-	83.2
EBT3 Film	No	1.8	-	60.7
FXG Gel	Full	1.8	-	98.5
FXG Gel	Half	1.8	-	87.8
FXG Gel	No	1.8	-	56.1

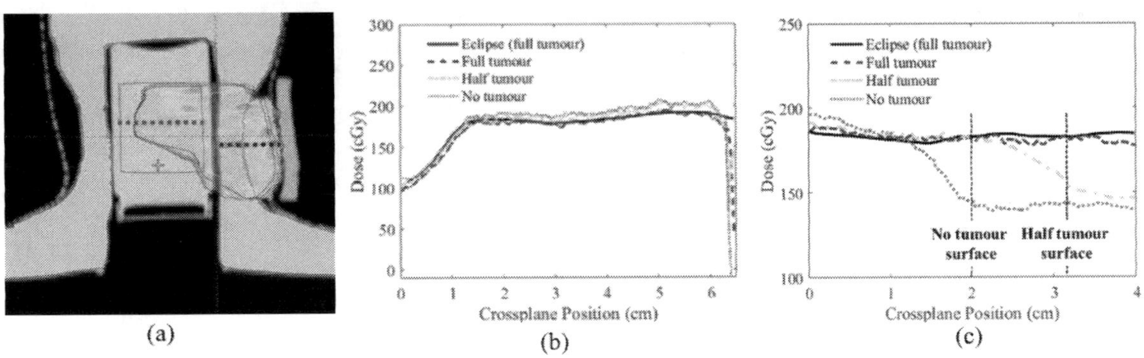

Figure 2. (a) Eclipse screen capture of the gel plan with red and blue dashed lines showing the location of profile measurements for the (b) FXG gel and (c) EBT3 film deliveries.

4. Discussion & Conclusions

The ion chamber results in Table 1 indicate good agreement between Eclipse and delivery measurement at depth for all anatomies studied. And, as MOSFETs have a greater uncertainty, the MOSFET measurements (located 1 cm superior to isocentre at tumour surface) suggest good agreement between calculation and measurement at the tumour surface, albeit at only one point. The ion chamber readings at depth increased as expected when the half/no tumour measurements were compared to the calculated Eclipse values on the full tumour anatomy. The MOSFET results for the half/no tumour anatomies are surprisingly high given the gap between the tumour surface and the bolus placed in the exterior of the mask (figure 1d). If these single point results were the only measurements available, they would suggest that the surface and depth dose coverage is adequate and that treatment could continue as per normal even with an external body contour reduction on the order of 2 cm.

To provide an additional surface dose measurement, a 3.5 cm x 5 cm piece of film was placed in a slot on an axial plane within the tumour (figure 1b). The analysis for the full tumour film measurement shows reasonable agreement with Eclipse at depth, but lower agreement than expected near the surface (Table 1, figure 2c), affecting the gamma comparison results. Note that the gamma comparison was limited on all three anatomies to the region within the phantom/tumour. For the half /no tumour anatomy, the film dose at depth again agrees with Eclipse, but severe underdosing on the order of 20% is observed in the measurement near the surface. The underdose is likely related to the accuracy of Eclipse calculations within and a short distance downstream of air cavities [5]. There is a small gap between the tumour surface and mask bolus, creating a cavity in the full tumour anatomy upstream of the tumour volume. As the bolus tumour layers are removed for the half/no tumour anatomy, the size of the upstream air cavity increases, which could be the source of the discrepancy between Eclipse calculation and surface film measurement underdose. This does not explain the good agreement between Eclipse and surface MOSFET measurements, so further investigation is warranted.

Looking at the gel results, the 3%,3mm gamma agreement between Eclipse (full tumour anatomy) and measurement is 87.8% for the half tumour anatomy (max. tumour shrinkage ~1.3 cm) and 56.1% for the no tumour anatomy (max. tumour shrinkage ~2.6 cm), neither of which would be clinically acceptable at our centre. The location of the ion chamber measurement is closer to the left side of the profile in figure 2b, at a deeper depth in phantom where better agreement is observed between Eclipse and the gel dosimeter delivery data. This indicates that the practice at our centre of having a 1 cm change in external body contour being the point of initiating a patient re-scan, verification of the original plan on the new planning CT anatomy and possible re-plan is therefore quite reasonable.

While not discussed, the design/manufacture of a suitable phantom incorporating different dosimeter types at appropriate locations was a significant challenge of critical importance. Also, the development of proper tools for registering/analyzing multiple types of dosimetry data [3, 4] was a key element in enabling a clinically meaningful end-to-end validation of the treatment process step in question. Both of these challenges are perhaps understated in the 3D dosimetry literature.

5. Acknowledgements

Research funding for this work has been provided by the Canadian Institutes of Health Research (CIHR) and the Cancer Centre of Southeastern Ontario.

6. References

[1] Baldock C *et al* 2010 *Phys. Med. Biol.* **55** R1-63
[2] Olding T *et al* 2010 *Phys. Med. Biol.* **56** 1259-79
[3] Alexander K M *et al* 2016 (presented at IC3DDose 2016)
[4] Alexander K M *et al* 2015 *J. Phys.: Conf. Ser.* **573** 012042
[5] Joshi C *et al* 2010 *J. Med. Phys.* **35** 73-80

IOP Conf. Series: Journal of Physics: Conf. Series **847** (2017) 012039 doi:10.1088/1742-6596/847/1/012039

Dosimetric validation study of a flattening filter free SABR treatment by use of a normoxic PAG gel dosimeter with MRI readout

A Venning[1], M Chandroth[1], B Chick[1] and G Livingston[1]
[1]Mid-North Coast Cancer Institute, Port Macquarie Base Hospital, Port Macquarie, NSW, Australia

E-mail: tony.venning@ncahs.health.nsw.gov.au

Abstract. This paper describes clinical studies of the normoxic PAG gel dosimeter with MRI readout for validation of complex flattening filter free (FFF) treatment delivery techniques. Recently our department implemented the Elekta FFF technology for stereotactic ablative radiotherapy (SABR) treatments using dynamic conformal arc (DCAT) and volumetric arc therapy (VMAT). In this study 6MV FFF DCAT and VMAT plans were analysed using 3D gel dosimetry techniques and compared against standard clinical dosimetry QA tools. Some challenges with this validation study included high dose per fraction for SABR, MRI access time and R_2-dose calibration errors. A study of the dependency of the dose response on dose rate revealed insignificant dose rate dependence of the normoxic PAG dosimeter up to 20 Gy/min for both photon energies (6 MV and 10 MV).

1. Introduction

Three-dimensional dosimeters are of great interest in a progressive radiotherapy facility utilising complex and advanced treatments techniques such as high dose rate dynamic SABR. In our Centre, we recently introduced three-dimensional polymer gel dosimeters for encapsulating and validating 6MV FFF VMAT and DCAT treatments. Treatments of this nature, however, are difficult and complex to validate and mainstream quality assurance tools are often limited in their ability to satisfy clinical demands [1-7]. Three-dimensional polymer gel dosimeters offer significant advantages for validating steep 3D dose distributions of this complexity. This paper discusses some of the challenges encountered while using the normoxic PAG gel dosimeter [8, 9] in the clinical setting such as high dose short fractionation regimes, dose rate response and geometric and dosimetric accuracy. Clinical examples are presented for 6MV FFF VMAT and DCAT treatments planned using Monaco™ v 5.10.02 and delivered using the Elekta Agility™ head.

2. Methods and Materials

2.1. Phantom simulation and treatment planning

The gel container was initially filled with water and imaged using a Siemens Somatom™ Helical CT scanner. Our department uses the Monaco™ Monte Carlo treatment planning system. The 6MV FFF DCAT and VMAT patient plans were copied onto the phantom data set retaining segment shapes and monitor units using a 2 mm calculation grid with 1% statistical uncertainty in dose *per plan*. The resulting transverse dose distributions were exported with 2 mm centre slice spacing separation to compare with the measured gel dose distributions.

2.2. Dosimeter manufacture

The normoxic PAG dosimeters were manufactured on the bench top in a fume cupboard using a method described elsewhere [5]. The formulations consisted of 3.5% *N,N*–methylene-bis-acrylamide (bis), 3.5% acrylamide (AA), 5% gelatine, 5 mM tetrakis (hydroxymethyl) phosphonium chloride (THPC), 1 mM hydroquinone (HQ) and 88% deionised H_2O.

2.3. Phantom & calibrations vials

In each study the liquid dosimeter was poured into eleven 15 ml Pyrex^{Tm} glass calibration vials and a 3 litre container of similar type that has been used elsewhere [10] (see Figure 2). The dosimeter was left on the bench top to cool for an hour and then subsequently placed in a refrigerator to set overnight.

2.4. Phantom and calibration vial irradiation

Temperature control measures were implemented before irradiation to ensure the dosimeters were in equilibrium with the ambient air temperature. The gel phantom was placed on the treatment couch at the linac iso-centre and irradiated according to the patient's treatment plan. The calibration vials were irradiated to different known doses in water according to the TRS-398 protocol [11].

2.5. MRI readout

After equilibrating with the ambient air temperature, the dosimeters were imaged using a Siemens Avanto 1.5T clinical MRI scanner with a head coil. An interleaving multi-echo-pulse sequence consisting of 16 echoes with TE of 40 ms, TR of 3240 ms, FOV 128 x 128 mm^2, 2 mm slice thickness, pixel bandwidth of 130, 1 acquisition with 21 slices separated by 2 mm centres was used resulting in an imaging time of 42 minutes [12]. Due to MRI access constraints only the minimal imaging parameters were used. It was determined that a TE of 40ms with 16 echoes was sufficient to cover the range of R_2-dose response within the dosimeter.

2.6. Processing and analysis

MRI images were processed using in-house software created in Matlab^{TM}. R_2 maps of each axial slice in the phantom were calculated and converted to dose maps using the R_2-dose response calibration curve generated from the calibration vials. The dose maps were imported into the SunNuclear Patient Software^{TM} v6.2 for gamma map analysis with the TPS's dose distributions. Only the normalised/geometric dose distribution was studied in these examples. Gamma maps were subsequently assessed using relative mode in the SNC Patient software with a pass/fail criterion of 3%/3mm and a 10% lower dose threshold.

2.7. Clinical QA measurement using SNC ArcCheck and CC04 chamber

Both DCAT and VMAT plans were measured using our standard clinical QA dosimetry tools for comparison purposes which included the SNC ArcCheck^{TM} and a point dose measurement using a CC04 ionisation chamber at the isocentre.

2.8. FFF Dose rate study

Using the calibration vials a dose rate dependence study was performed to assess the response of the normoxic PAG dosimeter for FFF beams up to a maximum dose rate of 2000 cGy/min for a dose of 5Gy at each dose rate. R_2-dose response was plotted as a function of dose rate.

3. Results and Discussion

3.1. R_2-dose response

Figure 1 shows the R_2-dose response curves for the two batches of gels made in this clinical study. The R_2-dose sensitivities were 0.31029 $s^{-1}Gy^{-1}$ for the DCAT study and 0.29175 $s^{-1}Gy^{-1}$ for the VMAT study both with linear regions that extended up to 5 Gy.

3.2. Gel phantom measurement compared to TPS with gamma analysis

Figure 2 shows the irradiated DCAT gel phantom along with the MRI dose map. Only central axis transverse dose maps are shown in figures 3 and 4 for each delivery technique compared against the TPS with the associated gamma values. Adjacent to each dose map comparison is a profile taken through the central axis of the dose distribution and the associatred gamma values for the above criteria. In the transverse plane 21 slices at 2 mm separation can be analysed and by interpolation, 128 sagital and coronal dose planes are available to analyse at 1 mm centre separations, which allows the dose distribution in each plane to be assessed.

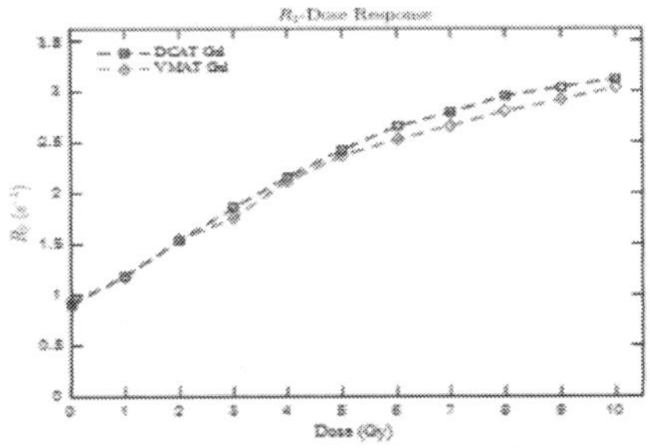

Figure 1. R_2-dose response of the two batches of normoxic PAG gel.

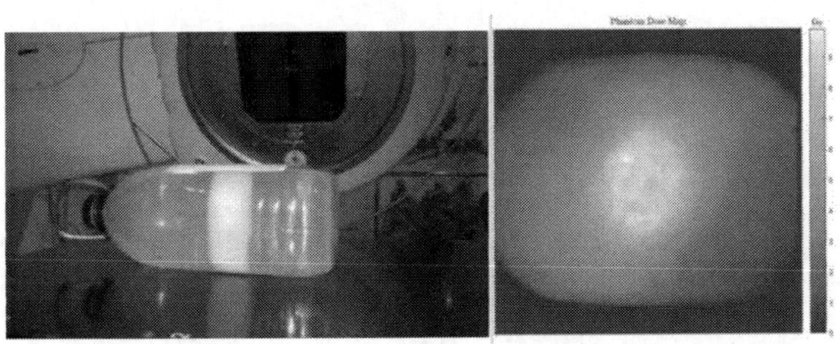

Figure 2. Irradiation of the normoxic PAG gel dosimeter with the associated colour rendered MRI dosemap.

The identified errors in R_2 in the phantom are potentially related to reasons discussed elsewhere, such as, over response of R_2 in a large phantom volume compared with relatively small calibration vials [13]. Normalisation was performed on the central axis with a small shift applied to locate a better overall fit of the dose distribution. Off–axis dose distributions were also assessed, however due to the limited scope within this abstract these will not be covered in this article.

3.2.1. DCAT plan. The overall pass rate for the central axis slice was 93.6% with 776 points failing the criteria and 11388 passing. No gamma score was above 2 for this dose plane.

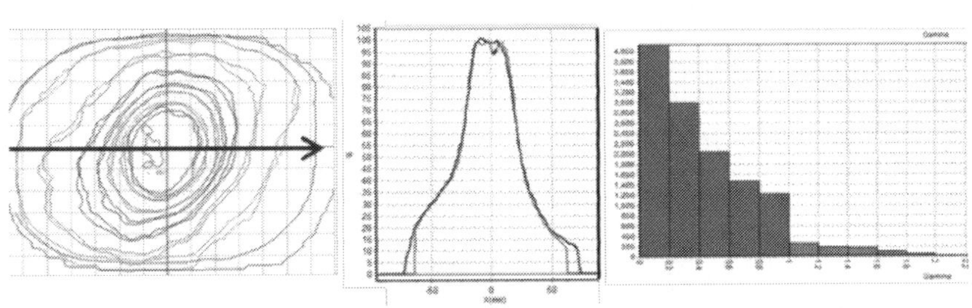

Figure 3. Gamma map analysis for DCAT plan with central axis profile.

3.2.2. VMAT plan. The overall pass rate for the central axis transverse slice was 87.7% with 1383 points failing the criteria and 9865 passing. The gamma scores were all below 2 for this dose plane.

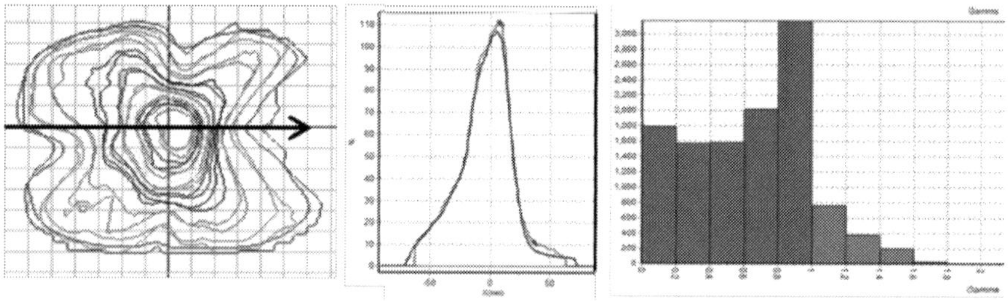

Figure 4. Gamma map analysis for VMAT plan with central axis profile.

3.3. Clinical QA measurement using SNC ArcCheck and CC04 ionisation chamber

The VMAT plan passed the specified gamma criteria at 97.6% with 9 points failing and 360 passing. The highest gamma score was 2.45. The dose at the isocentre measured -2.2% below the TPS value. The DCAT plan passed the specified gamma criteria at 95.5% with 21 points failing and 441 passing. The highest gamma score was 2.7. The point dose measured 2.4% above the TPS value.

3.4. FFF dose rate study

Figure 5 shows the R_2- response as a function of dose rate for a dose of 5Gy. The plot shows a weak dependence with increasing dose rate with a linear fit having a gradient of $-6\times10^{-5}\ s^{-1}cGy^{-1}min$ for 6MV and $-9\times10^{-5}\ s^{-1}cGy^{-1}min$ for 10MV.

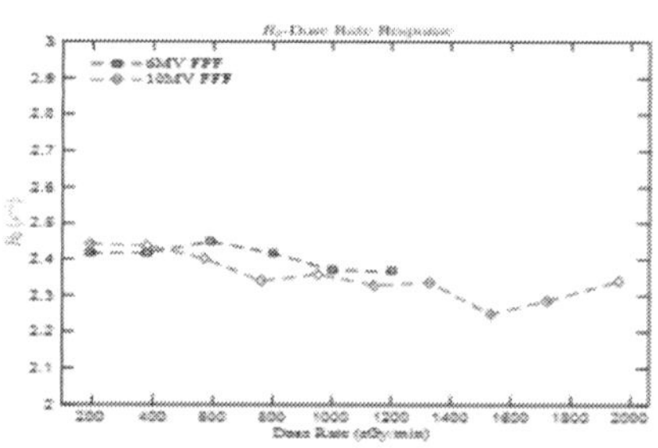

Figure 5. R_2-dose rate response.

4. Conclusion

There are many challenges involved in setting up a gel dosimetry program [14-16] using MRI [17-23]. Expertise, time and access constraints are limiting factors in imaging optimization. Clinical requirements dictate that for short fractionation high dose regimes that the geometric and dosimetric parameters of a plan are well known and within tight tolerances. In this study, due to an R_2 over response in the dosimeter only the normalised dose distribution was studied. The errors in R_2 are potentially related to reasons discussed elsewhere [13]. While the normoxic PAG dosimeter has highly favourable characteristics for radiotherapy applications, it has a limited useful dose range [24] which is problematic for SABR doses. A typical fractionation regime would be 10Gy/fx, however when calculated on the gel phantom doses can be in excess of 20Gy. Ideally a gel will receive doses within the range of its optimal dose resolution [25]. The normoxic PAG dosimeter manufactured in this study exhibited a weak dose rate dependence at FFF dose rates and further work should be conducted to fully characterise this observation. Further development work will continue on our dosimetric and calibration techniques for high dose high dose rate SABR fractionation regimes.

5. References

[1] Baldock C *et al* 2010 *Phys. Med. Biol.* **55** R1-R63
[2] Hill R F *et al* 2008 *Radiat. Meas.* **43** 1258-64
[3] Hill R *et al* 2014 *Phys. Med. Biol.* **59** R183-231
[4] Hill R *et al* 2009 *Med. Phys.* **36** 3971-81
[5] Hill B *et al* 2008 *Phys. Med.* **24** 149-58
[6] Healy B J *et al* 2003 *Med. Phys.* **30** 2282-91
[7] Hill B *et al* 2002 *Phys. Med. Biol.* **47** 4233-46
[8] Venning A J *et al* 2005 *Phys. Med. Biol.* **50** 3875-88
[9] De Deene Y *et al* 2006 *Phys. Med. Biol.* **51** 653-73
[10] Hill B *et al* 2005 *Med. Phys.* **32** 1589-97

[11] Gábor K *et al* 2015 *Mag. Onk.* **59** 119-23
[12] De Deene Y and Baldock C 2002 *Phys. Med. Biol.* **47** 3117-41
[13] De Deene Y 2006 *J. Phys.: Conf. Ser.* **56** 72-85
[14] Mather M L *et al* 2003 *Phys. Med. Biol.* **48** N269-75
[15] Trapp J V *et al* 2004 *Phys. Med. Biol.* **49** N139-46
[16] Hurley C *et al* 2006 *Nucl. Instrum. Meth. A* **565** 801-11
[17] Baldock C 2006 *J. Phys.: Conf. Ser.* **56** 14-22
[18] Baldock C 2009 *J. Phys.: Conf. Ser.* **164** 012002
[19] Murry P *et al* 2000 *Austral. Phys. Eng. Sci. Med.* **23** 44-51
[20] Mather M L *et al* 2003 *Ultrasonics* **41** 551-9
[21] Davies J B *et al* 2008 *Radiat. Phys. Chem.* **77** 690-96
[22] Venning A J *et al* 2004 *J. Phys.: Conf. Ser.* **3** 155-8
[23] Baldock C *et al* 1998 *Phys. Med. Biol.* **43** 695-702
[24] De Denne Y *et al* 2006 *Phys. Med. Biol.* **51** 653-73
[25] Baldock *et al* 2001 *Phys. Med. Biol.* **46** 449-60

IOP Conf. Series: Journal of Physics: Conf. Series **847** (2017) 012040 doi:10.1088/1742-6596/847/1/012040

Validation of 4D dose calculation using an independent motion monitoring by the calypso tracking system and 3D polymer gel dosimetry

P Mann[1,2], N Saito[1,2], C Lang[1,2], A Runz[1,2], W Johnen[1,2], M Witte[1,2], D Schmitt[2,3] and C P Karger[1,2]

[1]Department of Medical Physics in Radiation Oncology, German Cancer Research Center, INF 280, 69120 Heidelberg, Germany

[2]National Center for Radiation Research in Oncology (NCRO), Heidelberg Institute for Radiation Oncology (HIRO), INF 400, 69120 Heidelberg, Germany

[3]Department of Radiation Oncology, University Hospital Heidelberg, INF 400, 69120 Heidelberg, Germany

E-mail: p.mann@dkfz.de

Abstract. This study aims to evaluate an in-house developed 4D dose calculation algorithm that uses Calypso motion tracking data and to compare the results against 3D polymer gel dosimetry measurements. For this, a cylindrical water phantom was constructed that allows to insert (i) the polymer gel, (ii) a PinPoint ® ionization chamber and (iii) Calypso *beacons*™ for motion tracking. A treatment plan covering a *gel flask* in the center of the static phantom plus a 1 mm margin homogeneously with dose was generated. During irradiation, however, the phantom was moved periodically by means of a robot with a peak-to-peak amplitude of 2.5 cm. The results of the 4D dose calculations show good agreement with the gel-dosimetric measurements in most of the volume. Remaining small deviations have to be evaluated in further experiments. The developed experimental setup allows for 3D-dosimetric validation of 4D dose calculations algorithms prior to application in patients.

1. Introduction

Modern adaptive radiotherapy techniques (e.g. gating or tracking) have the potential of significant normal tissue sparing. Clinical implementation of these techniques, however, requires validation of the intended workflow using both, phantom experiments and 4D dose calculations. Several experimental settings with dynamic phantoms have been proposed showing the feasibility of 3D dose measurements and 4D dose calculation algorithms [1, 2]. To further push 4D dose calculation towards clinical application, real-time target data from clinically accepted tracking devices need to be integrated into the 4D dose calculation. The aim of this study was to validate an in-house developed 4D dose accumulation algorithm, which uses motion tracking data from the Calypso system, by a comparison between the calculated dose distribution and a 3D polymer gel dosimetry [3] measurement in a moving phantom.

2. Material and Methods

2.1. *Cylinder phantom*

The phantom (figure 1) consists of a double-walled cylindrical polymethyl-methacrylate (PMMA) shell (diameter 16 and 18 cm, respectively, height 10 cm). In this experiment, both cavities were filled with water. To perform 3D dose measurements, a BAREX™ container filled with dosimetry gel (subsequently referred to as *gel flask*) was placed in the center of the phantom. An additional bore allows inserting a PinPoint® Ionization Chamber (31014, PTW, Freiburg, Germany) into the *gel flask* to perform absolute dose measurement (see figure 1). In this case, the dosimetry gel serves only as an absorber and is not evaluated.

2.2. *Motion robot and tracking system*

During irradiations, the phantom was moved periodically with a dedicated motion robot perpendicular to the incident beam in cranio-caudal direction with 2.5 cm peak-to-peak amplitude and a period of 7.5 seconds. A $\cos^4$ motion trajectory was applied as it has been demonstrated to provide a good fit to the population data for respiratory motion [4]. Both motion-tracking and beam-on/off monitoring during the irradiation was realized with the Calypso system (Varian Medical Systems Inc., Palo Alto, California, USA). The system uses three electro-magnetic markers, referred to as *beacons*™, emitting a stimulated response signal, which is then recorded by a sensor array and used for localization of the *beacons* within an accuracy < 1 mm at a repetition rate of 10 Hz [5]. The *beacons* were inserted into the *gel flask* adapter with a distance of approx. 4 cm to the center of the gel sample (see figure 1a).

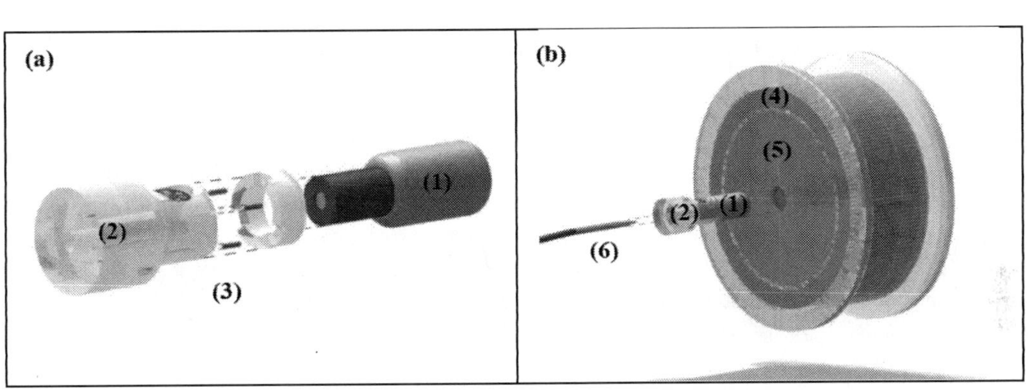

Figure 1. (a) The *gel flask* (1) is inserted into a dedicated adapter (2), which is also loaded with three Calypso *beacons* (3). (b) The cylinder phantom consists of a water-filled outer- (4) and inner- (5) PMMA ring. For dose measurements, the *gel flask* and Calypso *beacons* can be inserted into the phantom. To acquire a reference point-dose measurement, a PinPoint® chamber (6) can be placed at the center of the phantom by means of an additional bore.

2.3. *3D-polymer gel dosimetry*

3D dose measurements were performed with the PAGAT dosimetry gel (PolyAcrylamide Gelatin gel fabricated at ATmospheric conditions). After irradiation, the gel polymerizes locally and alters the relaxation rate R_2 of the transverse magnetization depending on the absorbed radiation dose. This effect can be measured by magnetic resonance imaging (MRI). The PAGAT dosimetry gel consists of monomers (3% acrylamide and 3% N, N-methylene-bis-acrylamide) as active components embedded within a gelatin matrix (6% Gelatin Sigma Aldrich) and 10 mM THPC (bis[tetrakis(hydroxymethyl)-phosphonium] chloride) acting as an antioxidant. The production of the gel followed the protocol described elsewhere [6]. An absolute 3D-dose distribution can be determined after establishing a mono-exponential calibration curve [7]. Due to its chemically inert properties [8] a BAREX™ container (distributed by VELOX GmbH, Hamburg, Germany) was used as gel dosimetry storage container (length l = 45 mm, outer diameter d = 28 mm, wall thickness t = 0.6 mm and volume V = 25 ml). Prior to the MRI evaluation, both calibration- and *gel flask* were stored for 2 h at room temperature inside a water-flow phantom keeping the temperature constant within ± 0.1 °C during the MR measurement. Calibration- and *gel flask* were then scanned simultaneously on a 3T Biograph mMR (Siemens Healthineers, Erlangen, Germany) using the following imaging parameters: TR = 3000 ms, 32 equidistant echoes, TE = 22.5 – 720 ms, resolution = 1 × 1 × 1 mm³, BW = 130 Hz/pixel with a scan time of t = 6 min 24 s for each of the 27 slices. To acquire sufficient information on the 3D dose distribution of the *gel flask*, 27 slices were measured and co-registered with the calculated treatment plan using dedicated MR-markers (Beekly Medical, Bristol, USA).

2.4. Phantom irradiation experiments

Treatment planning CTs for the static phantom were acquired on a SOMATOM Definition Flash (Siemens Healthineers, Erlangen, Germany). 3D conformal radiation therapy (3DCRT) treatment plan optimization was realized with the Raystation TPS (RaySearch Laboratories, Stockholm, Sweden) using a collapsed cone dose calculation algorithm with a dose grid resolution of 1.0 × 1.0 × 1.0 mm³. The *gel flask* plus a 1 mm margin was considered as planning target volume (PTV) and was covered homogeneously with a median dose of D_{50} = 4 Gy delivered from 7 beam directions (0, 40, 80, 200, 240, 280, 320) using only one beam segment per gantry angle. Outside the PTV, the dose gradient was 2 Gy / cm. The treatment plan was delivered to the moving phantom using a 6 MV linear accelerator (Artiste, Siemens Healthineers, Erlangen, Germany) with photons and a dose rate of 300 MU/min. The calibration flasks were irradiated under reference conditions with doses between 0 and 7 Gy.

2.5. 4D Dose calculation

A 4D dose distribution has been calculated retrospectively using an in-house developed software by shifting the target point within the planning CT based on the motion information and beam status recorded with the Calypso system. For each target point position, a new sub-beam was assigned together with a weighting factor corresponding to the duration of the irradiation at the respective target point position. In total, 1227 sub-beams were used for the 4D dose calculation. As the 4D dose calcu-lation uses the pencil beam algorithm of the in-house developed VIRTUOS treatment planning system (TPS) [9], the dose distribution for the static case was additionally recalculated with VIRTUOS using the same treatment parameters. To assess the accuracy of the dose calculation algorithms of both TPS, a PinPoint® chamber measurement was performed in the homogeneous region.

3. Results

3.1. Accuracy of dose calculation algorithms

Both dose calculation algorithms revealed very similar dose distributions (figure 2). With respect to the PinPoint® chamber measurement, only minor deviations of +0.7 % and -1.0 % were obtained for the Raysearch system and VIRTUOS, respectively.

3.2. 4D Dose calculation

A representative sagittal slice of the MR-based dose evaluation is displayed in figure 2a. Homo-geneous target coverage was measured for approx. 80 % of the evaluated volume while an expected decrease of up to 62.5 % was seen towards the neck of the flask. Figure 2b compares measured and calculated dose profiles located at the central region of the *gel flask*. In all cases, measured and calculated dose agree well within the homogeneous area. The dose gradient region (position <15 mm), however, was better described by the retrospectively performed 4D than by the static dose calculation. Evaluation of the profile at the bottom-region (figure 2c) could not completely reproduce the measured dose distribution. The dose gradients were shifted by 3 mm against each other. Nevertheless, the measurement of this profile agreed better with the 4D than with the static dose calculation.

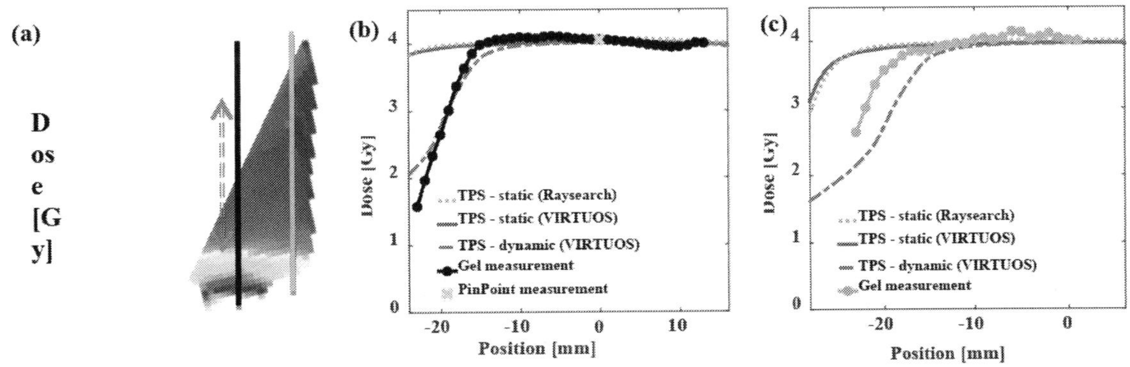

Figure 2. (a) Sagittal slice of the gel-based dose evaluation. The direction of the phantom motion is indicated (green arrow). (b) and (c) Comparison of measured and planned dose. Position and orientation of the profiles within the *gel flask* are indicated in (a). The profile for the static phantom is shown for both planning systems.

4. Discussion

The developed phantom setting allows (i) performing irradiations under dynamic conditions with arbitrary pre-defined motion patterns, (ii) measuring dose in 1D and 3D and (iii) recording motion trajectories with the Calypso system to retrospectively calculate a 4D dose distribution. In addition, absolute 3D gel dosimetry was performed. Due to the non-symmetrical $\cos^4$ motion-pattern and an amplitude of only 2.5 cm, a dose reduction was only visible at the neck of the *gel flask*. This decrease was also calculated with the in-house developed 4D calculation algorithm and the result showed a good agreement with measurements. At the bottom of the *gel flask*, however, the deviations were larger than at the center. Due to the applied workflow, positioning errors as well as dose differences resulting from temperature changes in the gel can be excluded. To further evaluate the source of this dose discrepancy, future measurements should also include a *gel flask* irradiation under static conditions to distinguish gel-dosimetric from motion-related error sources. In addition, a larger motion amplitude will be applied, which leads to a dose decrease on both sides of the flask and which would allow identifying potential artefacts at the neck of the flask. Finally, a dosimetric impact of the robot table in terms of beam attenuation and scattering has to be investigated.

Future experiments will also include patient-based motion-patterns in 3D as well as more complex irradiation techniques such as intensity modulated radiotherapy (IMRT) and gated beam delivery.

5. Conclusion

The combination of the moving phantom, the Calypso motion tracking system, 3D polymer gel dosimetry and the in-house developed 4D dose calculation algorithm are valuable tools for validating intrafractional adaptive radiotherapy by both, calculations as well as measurements. Especially, the Calypso system could also be used to reconstruct motion-related changes of the dose distribution in patients.

6. Acknowledgements

This work was partly supported by the Federal Ministry of Education and Research (BMBF) Germany, grant number 01IB13001B (SPARTA), the European Metrology Programme for Innovation and Research (EMPIR), grant number 15HLT08MRgRT and the Transregional Collaborative Research Center 125 C01 of the German Research Foundation (DFG). The authors would like to thank Yves DeDeene and Bianca Berndt for the valuable advice on fabrication of the polymer dosimetry gel.

7. References

[1] Ceberg S *et al* 2008 *Phys. Med. Biol.* **53** N387-96
[2] Niu C J *et al* 2012 *Med. Phys.* **39** 765-76

[3] Baldock C *et al* 2010 *Phys. Med. Biol.* **55** R1-63
[4] George R *et al* 2005 *Med. Phys.* **32** 2850
[5] Franz A M *et al* 2014 *Phys. Med. Biol.* **59** 6797-810
[6] De Deene Y *et al* 2006 *Phys. Med. Biol.* **51** 653-73
[7] Venning A J *et al* 2005 *Phys. Med. Biol.* **50** 3875-88
[8] Vergote K 2005 University of Gent (PhD thesis)
[9] Schlegel W *et al* 1992 *Int. J. Radiat. Oncol. Biol. Phys.* **24** 781-7

IOP Conf. Series: Journal of Physics: Conf. Series **847** (2017) 012041 doi:10.1088/1742-6596/847/1/012041

Leuco-crystal-violet micelle gel dosimeters: Component effects on dose-rate dependence

J C Xie[1], E A B Katz[1], K M Alexander[2], L J Schreiner[2,3] and K B McAuley[1]
[1]Dept. Chemical Engineering, Queen's University, Kingston, Canada K7L 3N6
[2]Dept. of Physics, Queen's University, Kingston, Ontario, Canada, K7L 3N6
[3]Cancer Centre of Southeastern Ontario, Kingston, Canada, K7L5P9

Email: kim.mcauley@queensu.ca

Abstract. Designed experiments were performed to produce empirical models for the dose sensitivity, initial absorbance, and dose-rate dependence respectively for leucocrystal violet (LCV) micelle gel dosimeters containing cetyltrimethylammonium bromide (CTAB) and 2,2,2-trichloroethanol (TCE). Previous gels of this type showed dose-rate dependent behaviour, producing an ~18% increase in dose sensitivity between dose rates of 100 and 600 cGy min^{-1}. Our models predict that the dose rate dependence can be reduced by increasing the concentration of TCE, CTAB and LCV. Increasing concentrations of LCV and CTAB produces a significant increase in dose sensitivity with a corresponding increase in initial absorbance. An optimization procedure was used to determine a nearly dose-rate independent gel which maintained high sensitivity and low initial absorbance. This gel which contains 33 mM CTAB, 1.25 mM LCV, and 96 mM TCE in 25 mM trichloroacetic acid and 4 wt% gelatin showed an increase in dose sensitivity of only 4% between dose rates of 100 and 600 cGy min^{-1}, and provides an 80% greater dose sensitivity compared to Jordan's standard gels with similar initial absorbance.

1. Introduction

Leucocrystal violet (LCV) micelle gel dosimeters developed by Babic *et al* [1] can be read optically and produce accurate 3-D dosimetry with excellent spatial stability, but have low dose sensitivity compared to alternatives such as Fricke gel dosimeters which use ferric ions as their reporter molecule [2]. LCV gels are preferable to leucomalachite green (LMG) gels, also developed by Jordan's research group, because LCV is less soluble in water causing it to preferentially dissolve in micelles, thereby reducing diffusion [1,3,4]. Nasr et al. attempted to optimize Jordan's LCV gel recipe by adjusting concentrations of LCV and Triton X-100 (Tx100) surfactant and adding 2,2,2-trichloroethanol (TCE) as a sensitizing agent [5, 6]. They found that increasing LCV concentration beyond ~1.25 mM leads to precipitation and formation of turbid gels [5, 6]. Increasing surfactant concentration can raise this limit at the cost of increased initial absorbance [5, 6] and viscosity during manufacture. More sensitive gels could not be produced using these ingredients without an undesirable increase in absorbance of the gel prior to irradiation. Because high initial absorbance was correlated with high Tx100 concentration, alternative surfactants were considered [5, 6]. Optimized gels using the surfactant cetyltrimethylammonium bromide (CTAB) showed higher dose sensitivity and excellent spatial stability while still maintaining low initial absorbance. But, unlike Jordan's LCV gel, the proposed gel exhibited significant dose-rate dependence [6]. Dose-rate dependence has been observed in LMG micelle gels studied by Vandecasteele et al. who used sodium dodecyl sulfate as the surfactant. They reduced the undesirable dose-rate

IOP Conf. Series: Journal of Physics: Conf. Series **847** (2017) 012041 doi:10.1088/1742-6596/847/1/012041

dependence from 75% to 33% between dose rates of 50 and 400 cGy min^{-1} by manipulating concentrations of LMG, surfactant, and gelatin [7]. Høye et al. performed similar manipulations on silicone-based LMG dosimeter recipes, producing a dose-rate independent phantom by increasing the dye concentration at the cost of spatial stability [8].

2. Materials and Methods

2.1. Gel preparation

1- Add gelatin (Type A, 300 Bloom porcine skin) to 75 wt% of the total deionized water preheated to 60 °C (Mix A). Stir until gelatin dissolves and a final temperature of 45 °C is reached.
2- Add surfactant (CTAB or Tx100), trichloroacetic acid (TCAA), and TCE (sensitizer) to the remaining deionized water (Mix B) and stir at room temperature until dissolved.
3- Add LCV into Mix B. Cover with a box and stir at room temperature for 5 minutes.
4- Pour Mix B into Mix A and stir for 5 minutes.
5- Pour this gel into 14 polystyrene cuvettes (4.5 mL, 10 mm path length) and cap. Record pH.
6- Refrigerate at 4 °C for 24 h, then place in a 22 °C water bath for 20 min. before irradiation.

2.2. Irradiation and scanning procedure

A Varian Clinac 6EX linear accelerator was used to irradiate the cuvettes to doses of 1, 2, 3, 5, 10, 15, and 20 Gy; half at 100 cGy min^{-1} and the rest at 600 cGy min^{-1}, with 6 MV photons. Irradiation was conducted using an isocentric setup with a 100 cm source-to-axis distance, 1.5 cm plastic water build-up, 10 cm backscatter, and a 15x15 cm^2 field size. Absorbance measurements were taken at 590 nm before (A_0) and 10 mins. post-irradiation (A_1) using a SpectoVis Plus spectrophotometer over the 380–900 nm wavelength range. Spectra were calibrated by subtracting the spectrum of a reference cuvette containing deionized water. The change in attenuation coefficient ($\Delta\mu$) between the absorbances was calculated using:

$$\Delta\mu = \ln(10) \cdot x^{-1}(A_1 - A_0)$$

where $x = 1$ cm is the cuvette path length.

2.3. Designed experiments to optimize LCV micelle gels made with CTAB and TCE

A two-level three-factor full factorial design (runs 1.1 to 1.8 in table 1) was conducted in random order along with 9 centre-point runs (averaged and displayed as run 1.9) and 3 replicates of the Jordan gel (displayed as run 1.0). Three replicates of Nasr's gel optimized for high sensitivity (displayed as run 1.10) were also produced. Gelatin and TCAA concentrations were held at 4 wt% and 25 mM, respectively, in all gels. Linear regression (forced through the origin) was used to fit dose sensitivities (DS) for all gels at both dose rates (100 and 600 cGy min^{-1}). The sensitivity ratio (SR) was computed using:

$$\mathrm{SR} = \mathrm{DS}(600) \cdot \mathrm{DS}(100)^{-1}$$

Models of the form:

$$\mathrm{Y} = \beta_0 + \beta_1[\mathrm{CTAB}] + \beta_2[\mathrm{LCV}] + \beta_3[\mathrm{TCE}] + \beta_{12}[\mathrm{CTAB}][\mathrm{LCV}] + \beta_{13}[\mathrm{CTAB}][\mathrm{TCE}] + \beta_{23}[\mathrm{LCV}][\mathrm{TCE}] \quad (1)$$

were fitted with JMP® statistical software using runs 1.1 to 1.10 in table 1 (where Y = A_0, DS(100), DS(600), SR) and the resulting parameter estimates are reported in table 2. Replicate runs were used individually (not averaged) during the model fitting. Recipe optimization was performed in Matlab® using sequential quadratic programming.

3. Results and Discussion

3.1. Designed experiments to optimize LCV micelle gels made with CTAB and TCE

Dose sensitivities in table 1 are consistent with results of Nasr *et al* [6], with a similar variance observed at the centre points for dose sensitivity and initial absorbance. The initial absorbances in the current study are higher by ~0.01, potentially due to batch differences in chemicals and cuvettes used.

As apparent from the parameter estimates in table 2, increasing the concentration of CTAB and LCV has a significant influence on the dose sensitivity of the gels. This result was expected because increasing micelle and reporter molecule concentrations increases the probability that free radicals encounter LCV in micelles to produce the response. The concentration of the sensitizer (TCE) shows a significant influence on sensitivity at lower dose rates but may have less influence at higher dose rates.

In agreement with previous results from Nasr *et al* [6], increasing LCV concentration tends to increase the initial absorbance of the gel, as indicated by the positive value of 27.9 mM^{-1} in table 2 (significant at the 85% confidence level, but not at 95%). Note that, unlike gels manufactured with Tx100, increasing CTAB surfactant does not cause a significant increase in initial absorbance [6].

Estimated values of parameters β_1, β_2 and β_3 in the SR model are all negative (except the CTAB-TCE interaction, and only β_3 significant at the 95% confidence level), suggesting that recipes with more micelles, LCV and sensitizer tend to be less dose-rate dependent. This effect could be the result of a higher probability of free radicals encountering micelles and reacting with LCV to induce colour change.

The model equations of the form shown in equation 1 were used to search for an improved gel recipe, using the parameter values in table 2. The optimization process minimized the objective function:

$$J = -\mathrm{DS}(100) + \omega_1 A_0 + \omega_2(\mathrm{SR} - 1)^2 \tag{2}$$

where ω_1 and ω_2 are positive weighting factors (Gy^{-1} cm^{-1}) to set the relative importance of dose sensitivity, initial absorbance, and sensitivity ratio. Bounds on the species concentrations were set at $9\text{ mM} \leq [\text{CTAB}] \leq 33\text{ mM}$, $0.75\text{ mM} \leq [\text{LCV}] \leq 1.25\text{ mM}$, and $40\text{ mM} \leq [\text{TCE}] \leq 120\text{ mM}$ to ensure reasonable gel recipes would be selected. Base values for ω_1 and ω_2 were set at 0.2 and 0.066, respectively, so that all three terms on the right-hand side of equation 2 are similar in magnitude. The resulting optimal gel is displayed as run 1.11 in table 1. Optimization was performed again with $\omega_1 = 0$ and $\omega_2 = 1$ to weigh heavily towards a dose-rate independent gel, resulting in the recipe for run 1.12. Run 1.11 produced a gel which maintained a reasonable sensitivity and initial colour compared to the other gels, however would still not be attractive for use in dosimetry with its 14% dose-rate dependence. Run 1.12 did produce a gel which had a more acceptable dose-rate dependence of 4% and a sensitivity similar to that of Nasr's most sensitive gel (run 1.10) [6].

4. Conclusions

Radio-chromic LCV micelle gels manufactured with CTAB surfactant exhibit improved dose sensitivity and comparable initial absorbance when compared to the LCV gel proposed by Babic *et al* [1]. Unfortunately, these new gels can exhibit undesirable dose-rate dependence (as high as 32% between dose rates of 100 and 600 cGy min^{-1}). The degree of dose-rate dependence varies with the gel recipe. Linear regression results suggest that high concentrations of CTAB, LCV and TCE may alleviate this effect. An optimized gel recipe was developed containing 33.0 mM CTAB, 1.25 mM LCV, 96.0 mM TCE, and 25.0 mM TCAA in 4 wt% gelatin. This gel showed only a small dose-rate dependence of 4% between the dose rates of 100 and 600 cGy min^{-1}, which may be acceptable for practical 3-D dosimetry applications [9]. However, batches do show a large amount of variance and further work reproducing these results is necessary. The proposed gel has a dose sensitivity 80% higher than the original LCV micelle gel developed by Babic *et al* [1] while still at a similar initial absorbance. In future, we will explore the possibility to develop gels with lower dose-rate dependence and even higher dose sensitivity by expanding the bounds used for recipe optimization, and attempt to modify the preparation methods to produce more consistent results.

Table 1. Influences of CTAB, LCV, and TCE on the initial absorbance, dose sensitivities, and sensitivity ratio of LCV micelle gels. Gels shown use CTAB as their surfactant, except gel 1.0 which used Tx100. Errors shown are one standard deviation from the mean in runs where replicates were performed.

Run	Surfactant [mM]	LCV [mM]	TCE [mM]	Initial Absorbance	Dose Sensitivity [x10^{-3} Gy^{-1} cm^{-1}] (100 cGy min^{-1})	Dose Sensitivity [x10^{-3} Gy^{-1} cm^{-1}] (600 cGy min^{-1})	Sensitivity Ratio	pH
1.0	4.0	1	0.0	0.041 ± 0.003	6.8 ± 0.6	6.8 ± 0.5	0.99 ± 0.02	3.37
1.1	9.0	0.75	40.0	0.034	6.31	8.05	1.27	3.33
1.2	9.0	0.75	80.0	0.033	7.83	8.73	1.12	3.33
1.3	9.0	1.25	40.0	0.056	7.21	9.21	1.28	3.27
1.4	9.0	1.25	80.0	0.046	7.51	9.17	1.16	3.33
1.5	25.0	0.75	40.0	0.035	7.44	9.85	1.32	3.31
1.6	25.0	0.75	80.0	0.044	9.83	10.92	1.11	3.32
1.7	25.0	1.25	40.0	0.047	9.30	10.42	1.12	3.29
1.8	25.0	1.25	80.0	0.054	12.37	13.85	1.12	3.31
1.9	17.0	1.0	60.0	0.042 ± 0.005	8.6 ± 0.5	10.0 ± 0.5	1.16 ± 0.06	3.33
1.10	33.0	1.25	120.0	0.055 ± 0.005	12.6 ± 0.5	12.8 ± 0.9	1.05 ± 0.08	3.32
1.11	9.0	0.90	120	0.040	9.80	11.2	1.14	3.24
1.12	33.0	1.25	96.0	0.047	12.4	12.9	1.04	3.22

Table 2. Parameter estimates from regression analysis of the data in table 1 to model initial absorbance, dose sensitivity and sensitivity ratio; with units consistent with those provided in table 1. **Bolded values** are significant at a 95% confidence level.

Parameter Coefficient	Parameter Estimates [x10^{-3}] A {Initial Absorbance}	DS(100) {Sensitivity}	DS(600) {Sensitivity}	SR {Sensitivity Ratio}
β_0	**40.3**	**0.246**	**3.08**	**1592**
β_1	-2.45	**0.0940**	**0.201**	-10.3
β_2	27.9	**3.41**	**2.79**	-238
β_3	-0.113	**0.0727**	0.0277	**-7.74**
β_{12}	1.09	0.0695	-0.0286	-11.5
β_{13}	0.0247	-0.000516	-0.000516	-0.0320
β_{23}	-0.336	-0.029	0.00478	5.64

5. References

[1] Babic S *et al* 2009 *Phys. Med. Biol.* **58** 6791-808
[2] Baldock C *et al* 2001 *Austral. Phys. Eng. Sci. Med.* **24** 19-30
[3] De S *et al* 2002 *Spectrochimica Acta* A **58** 2547-55
[4] Jordan K and Avaakumov N 2009 *Phys. Med. Biol.* **54** 6773-89
[5] Nasr A T *et al* 2015 *Phys. Med. Biol.* **60** 4665-83
[6] Nasr A T *et al* 2015 *Phys. Med. Biol.* **60** 4685-704
[7] Vandecasteele J *et al* 2011 *Phys. Med. Biol.* **56** 627-51
[8] Høye E M *et al* 2015 *Phys. Med. Biol.* **60** 5557-70
[9] Baldock C *et al* 2010 *Phys. Med. Biol.* **55** R1-63

IOP Conf. Series: Journal of Physics: Conf. Series **847** (2017) 012042 doi:10.1088/1742-6596/847/1/012042

Real time radiotherapy verification with Cherenkov imaging: development of a system for beamlet verification

B W Pogue[1,2,3], V Krishnaswamy[1,3], M Jermyn[3], P Bruza[1], T Miao[3], William Ware[3], S L Saunders[1], J M Andreozzi[1], D J Gladstone[1,2] and L A Jarvis[2]
[1]Thayer School of Engineering, Dartmouth College, Hanover NH 03755 USA
[2]Geisel School of Medicine, Dartmouth College, Hanover NH 03755 USA
[3]DoseOptics LLC, Lebanon NH 03766 USA

E-mail: brian.w.pogue@dartmouth.edu

Abstract. Cherenkov imaging has been shown to allow near real time imaging of the beam entrance and exit on patient tissue, with the appropriate intensified camera and associated image processing. A dedicated system has been developed for research into full torso imaging of whole breast irradiation, where the dual camera system captures the beam shape for all beamlets used in this treatment protocol. Particularly challenging verification measurement exists in dynamic wedge, field in field, and boost delivery, and the system was designed to capture these as they are delivered. Two intensified CMOS (ICMOS) cameras were developed and mounted in a breast treatment room, and pilot studies for intensity and stability were completed. Software tools to contour the treatment area have been developed and are being tested prior to initiation of the full trial. At present, it is possible to record delivery of individual beamlets as small as a single MLC thickness, and readout at 20 frames per second is achieved. Statistical analysis of system repeatibilty and stability is presented, as well as pilot human studies.

1. Introduction

Cherenkov imaging has been in embryonic testing in the past 5-6 years, with the first clinical images demonstrated in 2014 [1]. This trial demonstrated for the first time that it was possible to record the delivery of radiation to human tissue with a camera in the treatment room. This imaging was achieved by gating the Intensified CCD camera (ICCD) to the short linac pulses, 3-4 microsecond bursts with a repetition rate of 200-360 Hz, such that the signal could be recorded with the room lights on and not interfering too much with the image. Background room light subtraction and median filtering were added into the image processing to produce images of Cherenkov emission which are isolated to the beam entrance and exit surfaces. The signal appears to come from 4-6mm average depth, based upon Monte Carlo studies and tissue phantom tests. The nest phase of this work has been to develop a clinical viable approach to testing this in whole breast radiotherapy, where the cameras were permanently mounted in the treatment room, and the image capture was automated and analysis simplified. This report discusses these latest developments.

2. Methods

2.1 Hardware setup

The new Cherenkov imaging system was developed by DoseOptics LLC and customized for installation at the Norris Cotton Cancer Center in the 10MV LINAC Varian treatment room used for 6MV and 10MV irradiation in whole breast and chest wall cases. Previous work, published in 2014[1], is illustrated in the top two images of Figure 1, where the ICCD camera (Princeton Instruments, PI-MAX4) [2] was used on a tripod for temporary deployment in the room, synchronized to the LINAC output pulses and capturing Cherenkov emission images of either entrance or exit beam. The processed image of one patient is shown at top right, overlaid on the 3D profile from the CT from the treatment plan. This matched the planned dose quite well, as recently published[3]. The new system utilizes an intensified CMOS camera design, with two such cameras installed in the ceiling of the treatment room (bottom left). Initial pilot testing of one of the cameras demonstrated the ability to resolve a single MLC stripe (5mm thickness) on a patient's tissue (bottom right in Figure 1).

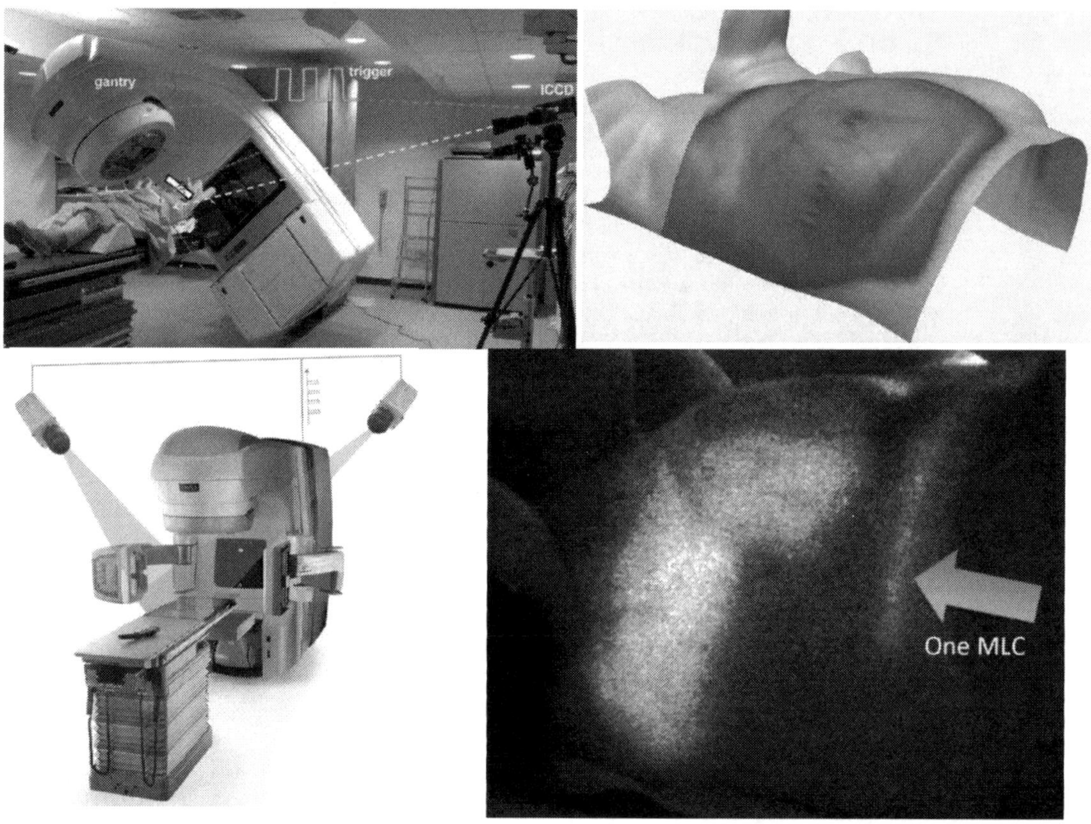

Figure 1. Illustration of the original trial (top left) with an ICCD camera on the tripod, synced to the LINAC and capturing a single view of the patient getting whole breast irradiation. This image is overlaid on the CT scan surface (top right) for visual comparison to the dose. The new system (bottom left) using two ICMOS cameras (DoseOptics LLC) is illustrated mounted to the ceiling and the image from an entrance beam showing a single MLC beamlet is shown (bottom right) for the field in field part of the treatment.

2.2 Image Processing

Image processing is required to ensure clean image data results, as are shown from a sequence of images in Figure 2. The raw intensified images have significant numbers of saturated pixels across the image (see Fig 2a), This noise is thought to come from stray gamma rays in the room, coming from both the linac head and the tissue being treated. The noise increases significantly with decreasing distance to the target, and at the distance of the cameras positioned, encompasses less than 10% of the pixels on the camera. After median filtering the image is significantly cleaned up, Fig 2b. Then when a background image is captured and subtracted from the original, the image is predominantly only containing the Cherenkov image data, Fig 2c.

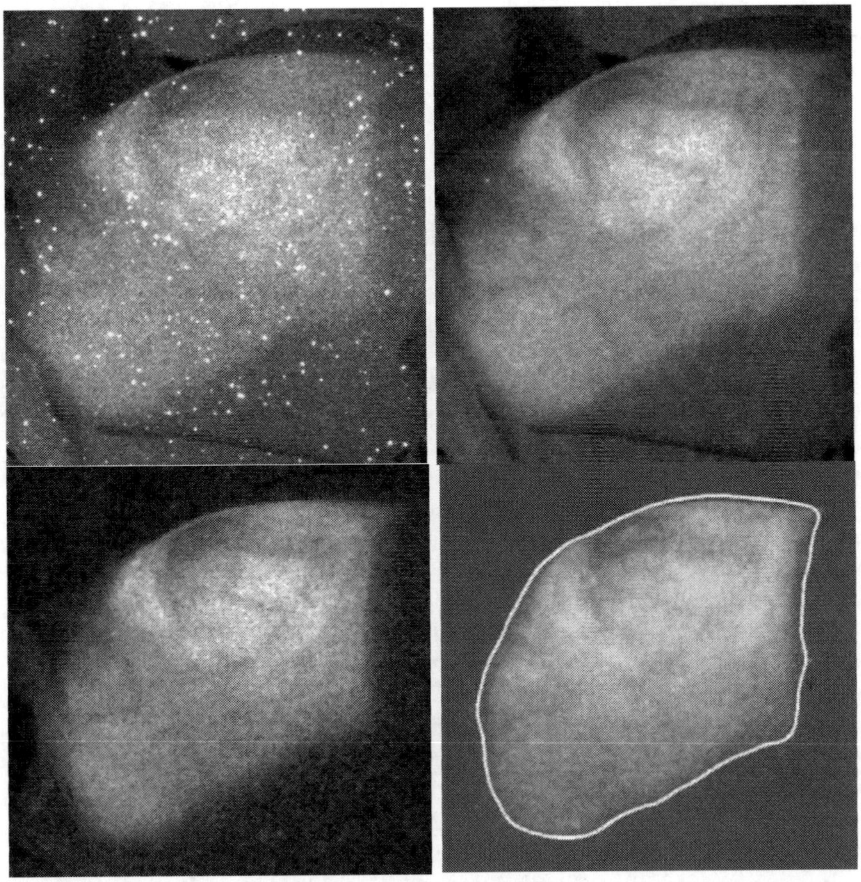

Figure 2. Images of raw Cherenkov imaged with full noise and background signal (a) and then with median filtering to remove the saturated pixels of noise (b). The background subtracted image is then shown in (c), and a threshold contoured version of the beam shape in (d).

2.3 Software Analysis of beam position

A custom contouring algorithm was developed for extraction of the beam shape from the observed Cherenkov beam, as illustrated in Figure 2d, for the processed image in Figure 2c. This contour has a range of features required, but the most important being the ability to robustly outline the viewed Cherenkov in a way which is a faithful representation of the image, even in the presence of moderate signal to noise, and rapidly changing beam shapes such as field in field, or dynamic wedge treatment. The algorithm to compare to dose reads in the treatment plan and extracts out the predicted dose contour at a depth comparable to the origin of the Cherenkov signal, calculated to be approximately 4mm below the surface, as illustrated in Figure 3. A range of depths were examined for superior match based upon patient studies, and the ideal depth was chosen.

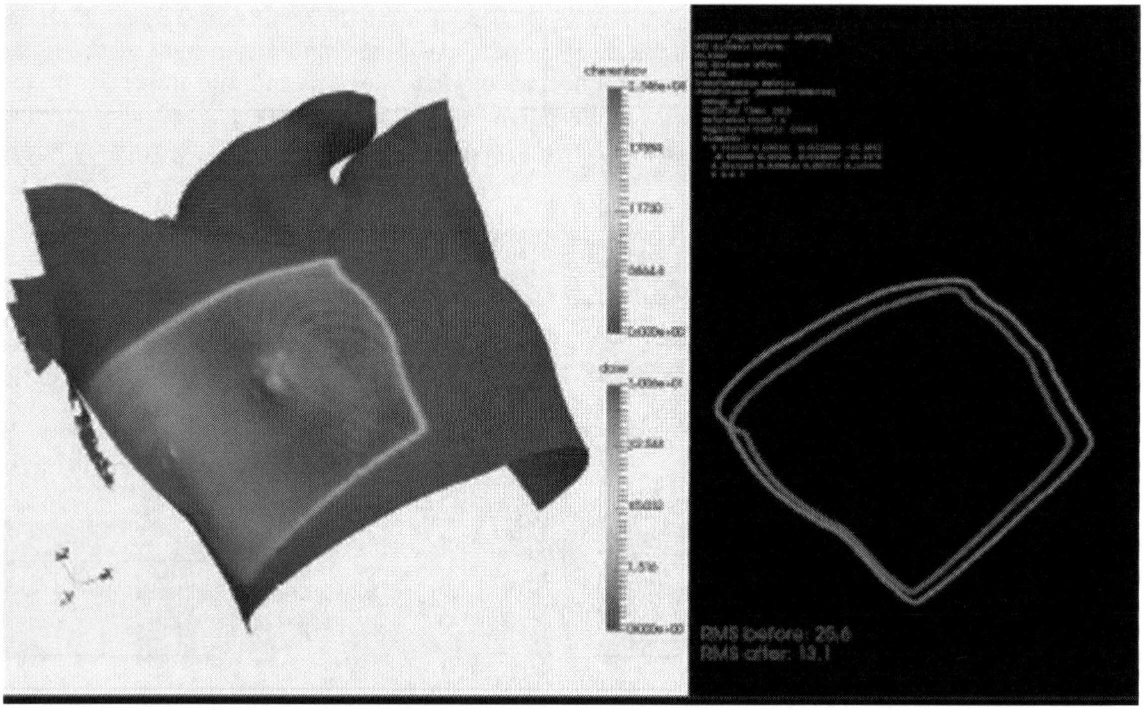

Figure 3. A single breast treatment case is examined with the contouring feature, to directly compare the Cherenkov image outline to the planned dose outline on the chest surface, estimated several millimetres below the surface where the Cherenkov signal originates from.

3. Results

The results of the preliminary set up and testing have demonstrated that the new cameras are capable of achieving high signal to noise level, with onboard processing for first pass noise removal. The gain levels and signal to background trade off was optimized for the expected imaging setting for the clinical trial. The ability to image a single 5mm beamlet from one leave of the MLC was taken as our metric of success, to ensure that we could contour and image the smallest feature expected in a dynamic treatment. The signal level from a breast treatment imaged at 20 frames per second is consistent with signal levels from single photon imaging in each pulse. Since the LINAC pulse rate is 360 Hz, a readout of 20 fps means that each frame has 360/20 = 18 pulses integrated together in it. The signal is integrated on the CMOS camera and read out, and successive readout frames are median filtered to suppress the highest salt and pepper noise features from the images.

4. Discussion & Conclusions

This work is a preliminary report of the set-up of an ongoing trial, with the goal of providing real time 3D surface verification of beam position, and eventually surface dosimetry. The choice of camera design was developed after extensive testing in previous studies, and through optimization of the cost-performance trade off of what can be achieved. The constraints in imaging Cherenkov during external beam radiotherapy are:

1) Low signal level: near single photon counting in each LINAC pulse
2) High noise environment with substantial salt and pepper type noise from stray gamma radiation
3) Background levels which are comparable to the Cherenkov signal requiring subtraction
4) Limited perspective view of the treated tissue, based upon line of sight

The strengths of the approach proposed here are:

1) High clean intensifier gain, and fast CMOS based readout
2) Multiple camera imaging allows simultaneous readout of entrance and exit surfaces
3) Mounting of the cameras lateral to the bed on the ceiling provides a reasonably good line of sight, with close proximity to the patient to maximize signal detected.
4) Contouring of the treatment dynamically allows extraction of features to compare to the treatment plan and to compare between daily fractions.

The full results of the clinical trial will be completed by the time of the conference and preliminary analysis of performance will be presented.

5. Acknowledgments
This work has been financially supported by NIH grant R44CA199836, as well as the Norris Cotton Cancer Center Pilot Funds support.

6. References
[1] Jarvis L A *et al* 2014 *Int. J. Radiat. Oncol. Biol. Phys.* **89** 615-22
[2] Andreozzi J M *et al* 2015 *Med. Phys.* **42** 994-1004
[3] Zhang R *et al* 2015 *Phys. Med. Biol.* **60** 1-14

Preliminary investigation of a luminescent colloidal quantum dots-based liquid scintillator

Marie-Ève Delage[1,2], Marie-Ève Lecavalier[3], Dominic Larivière[3], Claudine Nì Allen[4] and Luc Beaulieu[1,2]

[1]Département de physique, de génie physique et d'optique et Centre de recherche sur le cancer, Université Laval, Québec, Canada

[2]Centre de recherche du CHU de Québec – Université Laval, Québec, Canada

[3]Laboratoire de radioécologie, Département de chimie, Université Laval, Québec, Canada

[4]Département de physique, de génie physique et d'optique et Centre d'optique, photonique et laser (COPL), Université Laval, Québec, Canada

E-mail: marie-eve.delage.1@ulaval.ca, beaulieu@phy.ulaval.ca

Abstract. Nanoparticles are appealing materials because of their versatility in addition to the uniqueness of their properties enabling one to use them in different applications. Radiation detection with colloidal quantum dots (cQDs) is one of the domains where the nanoscience entered lately. The luminescent nanocrystals that are cQDs are of particular interest in scintillation dosimetry, where they could play the role of the fluorophore in a liquid scintillator. The study presented in this paper investigates the response of a cQD-based liquid scintillator to X-ray radiation in order to characterize the dose and cQDs concentration dependence of the radioluminescence (RL) signal. For a beam energy of 180 kVp, the latter was found to be linear as a function of dose, with a majority of the signal (~80%) coming from the cQDs and not the solvent, in this case hexane. Even with an ultra low concentration (μM), cQDs emit sufficient light to be detected. The RL intensity followed also a linear trend as a function of cQDs concentration, independently of the exposure time.

1. Introduction

The interest for nanotechnology has been constantly growing over the last decades because of the opportunities offered by nanoscale physics. This young science has first reached the medical sciences when nanoparticles (NPs) were used for drug delivery and optical imaging. It gained popularity in medical physics too where, nowadays, gold NPs are used as dose enhancers during radiotherapy treatments. Different declinations of NPs are available and allow one to choose the right type for the intended application. One of those types is particularly interesting for applications in scintillation dosimetry: colloidal quantum dots (cQDs), luminescent nanocrystals (NCs) of semiconductors. cQDs benefit from the quantum confinement of their charge carriers to have discrete energy levels. Their properties are then proper to the NC size and shape due to a modulation of the bulk medium properties of the same composition. One example of these properties is that they are brighter light emitters than their bulk counterpart. They also have a size-dependent broad absorption and narrow emission spectra [1]. The latter is of interest when considering the match of the scintillation wavelength to the photodetector sensitivity range. Finally, the surface chemistry of the NCs, and their small size, allows for the dispersion of cQDs in many physical supports (matrices), including water, in an ultra low concentration, of the order of the ppb.

Consequently, the motivation of using a cQD-based liquid scintillator for 3D dosimetry is to get a liquid scintillator with easy control on the scintillation peak wavelength, on the type of solvent and to get a scintillator with low energy dependence. Even if the scintillating cQDs are composed of high-Z elements, the really low concentration of NCs in the liquid could allow one to get rid of this dependence. Lecavalier *et al* [2] have already investigated preliminarily the response of cQDs to cobalt irradiation and obtained promising results concerning the use of cQDs dispersed in hexane and in water for liquid scintillation applications. The study presented in this paper describes further investigation of the response to ionizing irradiation of a cQD-based liquid scintillator. The objective of this particular work is to characterize the scintillation of cQDs dispersed in hexane as a function of dose and concentration.

2. Materials and methods

2.1. cQDs synthesis and composition

The NCs are synthesized in a three-neck flask: the cores are first grown with precursors heated at 250°C to get CdSe, than the shells are synthesized using the successive ion layer adsorption and reaction (SILAR) method [2]. Multiple shells surround the cQDs' core in order to passivate the dangling bonds responsible for the reactivity of the cQDs with their environment. They are successively composed of CdS, $Cd_{0.5}Zn_{0.5}S$ and ZnS as depicted in figure 1. The cQD diameter and elementary composition dictates the peak emission wavelength. Figure 2 presents different colors available for CdSe cQDs. The cQDs were dispersed in hexane and four dilutions were prepared at 1/5, 1/10, 1/20 and 1/30 of the initial concentration, which was 22.5 micro molar (µM). These dilutions fractions correspond respectively to 588, 294, 147 et 98 parts per billion (ppb).

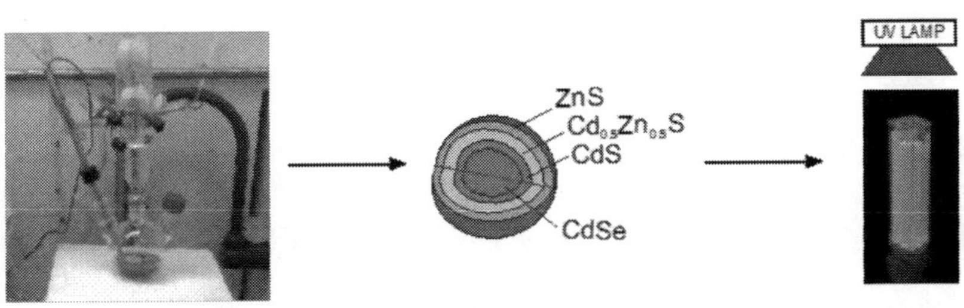

Figure 1. At the end of the synthesis, the cQDs are dispersed in hexane and already have a bright color. When exposed to a UV lamp, the an intense visible photoluminescence is seen. The scheme in the middle of the figure depicts the different layers of the cQDs totalizing a radius of 3 to 4 nm.

Figure 2. CdSe cQDs peak emission wavelengths available sweep widely the visible spectrum as a function of the NCs size (picture from our laboratory LabOMC, Laboratoire d'Optique de la Matière Condensée,).

2.2. Irradiation conditions

The cQDs liquid preparations were irradiated with an orthovoltage device Xstrahl 200 at 180 kVp with a dose rate of 369 MU/min. The field had a diameter of 5 cm and a SSD to D_{max} of 20 cm. The cQDs were also irradiated at 6 MV (Varian Clinac iX) with a 10 x 10 cm^2 field. The 6 MV measurements were also achieved with a commercial liquid scintillator (Ultima Gold, Perkin Elmer) to get a basis of comparison of the cQDs' intensity.

2.3. Detection set-up
A CCD camera was used to image the vial/cuvette with the cQDs in hexane placed 1 cm away from the field applicator. The set-up was covered with black blankets to cut the residual ambient light contamination. A vial containing only hexane was also irradiated in the same conditions to account for the Cherenkov light production in the solvent.

3. Results and Discussion

3.1. Dose and energy dependence
The radioluminescence (RL) signal collected at 180 kVp was found to be linear with exposure time, hence with dose deposited in the cQDs. Figure 3 presents this linear dependence of the RL intensity over a dose range up to 2.5 Gy. It also shows that the majority (~80%) of the RL intensity is due to cQD scintillation with the remaining 20% from Cherenkov (or fluorescence) from hexane. This implies that the energy transfer between the solvent and the cQDs is good, which is promising since the solvent could be eventually changed to optimize this energy transfer. There is no need for a third component acting in the energy transfer because the cQDs have a broad absorption spectrum, which can be tuned by their size. Even if the NCs have a concentration in the μM range, the signal is sufficient for scintillation measurement. This ultra low concentration lets us believe that cQDs will not perturb the beam for dose measurement purposes.

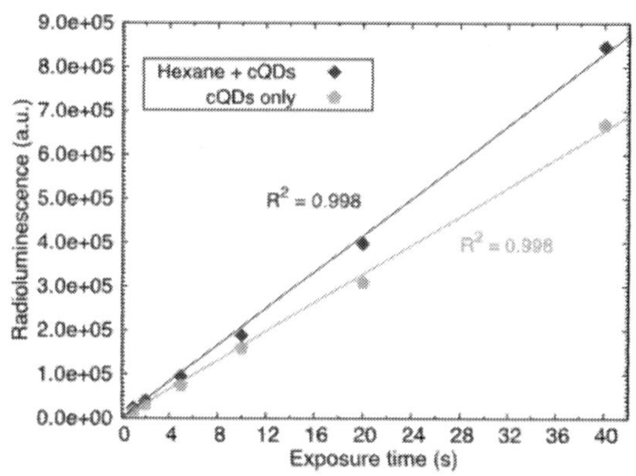

Figure 3. Radioluminescence intensity of the cQD-based liquid scintillator follows a linear trend on a dose range up to 2.5 Gy, with the majority of the signal resulting from the cQDs scintillation. The "cQDs only" curve represents the total signal of cQDs in hexane from which was subtracted the signal from hexane only irradiated in the same conditions.

The collected signal at 6 MV for the cQDs in hexane represents about 3% of the total signal of the Ultima Gold scintillator. At first, this proportion may seem low, but one has to keep in mind that the concentration of fluorophore in the Ultima Gold is far more important than that of the cQDs in hexane. When comparing the proportion of Cherenkov produced in hexane only to the total scintillation signal of cQDs in hexane, we get that the scintillation signal is 120% that of the Cherenkov's. Since the cQDs emission wavelength is at 615 nm, we looked up only at the red channel of the CCD to see the change in that proportion: the percentage reaches up to 500%.

3.2. RL intensity as a function of cQDs concentration
Each dilution was irradiated in the same conditions to characterize the dependence of the RL intensity as a function of the cQD concentration. At the same time, we also collected the RL signal as a function of dose, which gave similar results to figure 3 for all concentrations. As for the concentration

dependence, we found a linear trend of the signal as shown in figure 4, which is valid for the 4 exposure times tested.

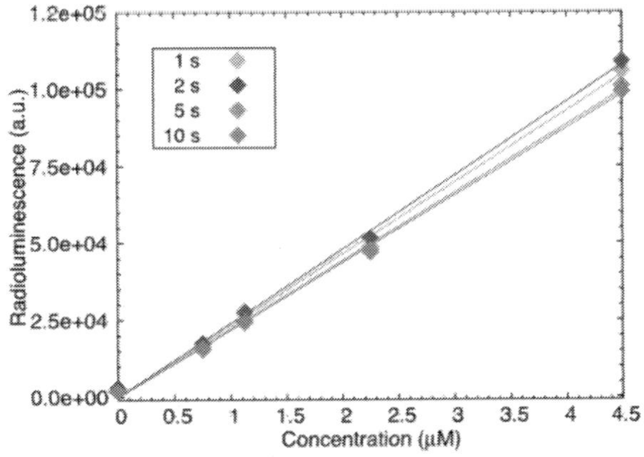

Figure 4. Changing the cQDs concentration of the liquid scintillator affect linearly the radioluminescence intensity. No saturation is seen in this concentration range, the reabsorption of the scintillation light is not significant enough to lower the overall emission quantum yield.

No saturation point, where the RL intensity starts to decrease, was reached so far as it could be observed in photoluminescence [3]. Beyond differences in sample concentration, the excitation sources, ionizing vs. visible, do not have the same energy fluence and energy values (eV vs keV). This could affect the energy transfer mechanism that stays optimal for ionizing radiation. Overall, the linearity observed offers a great way of normalizing the scintillation signal and a possibility to optimize the efficiency by increasing the concentration while remaining in the ppb range.

4. Conclusion

The first preliminary results presented here for cQD-based liquid scintillators are promising. The scintillator fulfills an important requirement for dosimeters to be suitable for applications, that is the linearity of its signal as a function of the dose. Also, it was shown that the concentration of the liquid samples can be taken as a normalization tool and could be chosen to be higher to get a better light production. Further investigation will look into the energy transfer dependence as a function of the solvent, in particular in toluene, alkyl benzene and water, the last two making it easier to manipulate as 3D dosimeters.

5. Acknowledgements

MED acknowledges partial support by the CREATE Medical Physics Research Training Network grant of the Natural Sciences and Engineering Research Council (Grant number: 432290).

6. References

[1] Smith A M and Nie S 2010 *Acc. Chem. Res.* **43** 190-200
[2] Lecavalier M-E *et al* D 2013 *Chem. Commun.* **49** 11629-31
[3] Rempel S V *et al* 2014 *Phys. Solid State* **56** 568-71

IOP Conf. Series: Journal of Physics: Conf. Series **847** (2017) 012044 doi:10.1088/1742-6596/847/1/012044

Three-dimensional radiation dosimetry based on optically-stimulated luminescence

M Sadel[1], E M Høye [2], P S Skyt[2], L P Muren[2], J B B Petersen[2] and P Balling[1]
[1]Dept of Physics and Astronomy, Aarhus University, Denmark
[2]Dept of Medical Physic, Aarhus University / Aarhus University Hospital, Denmark

E-mail: msadel@phys.au.dk

Abstract. A new approach to three-dimensional (3D) dosimetry based on optically-stimulated luminescence (OSL) is presented. By embedding OSL-active particles into a transparent silicone matrix (PDMS), the well-established dosimetric properties of an OSL material are exploited in a 3D-OSL dosimeter. By investigating prototype dosimeters in standard cuvettes in combination with small test samples for OSL readers, it is shown that a sufficient transparency of the 3D-OSL material can be combined with an OSL response giving an estimated >10.000 detected photons in 1 second per 1mm^3 voxel of the dosimeter at a dose of 1 Gy. The dose distribution in the 3D-OSL dosimeters can be directly read out optically without the need for subsequent reconstruction by computational inversion algorithms. The dosimeters carry the advantages known from personal-dosimetry use of OSL: the dose distribution following irradiation can be stored with minimal fading for extended periods of time, and dosimeters are reusable as they can be reset, e.g. by an intense (bleaching) light field.

1. Introduction

Modern radiotherapy (RT), such as e.g. intensity-modulated RT or volumetric modulated arc therapy, as well as the emerging techniques using heavy charged particles, e.g. proton therapy, employ complex spatially modulated 3D radiation fields to deliver therapeutic doses during treatment procedures. Thus, quality assurance (QA) before patient treatment is considered as a necessary prerequisite [1]. The basic dosimetric tools in RT are ionization chambers [2, 3]. On the other hand, methods based on luminescent passive detectors such as optically stimulated luminescence (OSL) detectors are also widely applied, especially for personal dosimetry and phantom measurements [4, 5]. However, it seems that only 3D dosimeters (consisting of a radiosensitive volume) meet the expectations of complex 3D dose measurements. The currently known materials used for 3D dosimetry are based mostly on polymerizing [6-11] and radiochromic gels or plastics [12-23]. However, the clinical use of these dosimeters has so far been limited, possibly due to different shortcomings of these materials, e.g. in terms of stability, environmental sensitivity or cost of use [24].

Dosimetry based on OSL relies on the ability of using light for interrogating trapped states in certain crystalline materials. Upon RT, the population in the trapped states carries information on the local dose and when subsequently subjecting the OSL material to light, the trapped electrons are released to the conduction band and recombine with holes to emit photons with an energy exceeding that of the stimulating light. Based on the intensity of this OSL light it is possible to evaluate the absorbed dose [4]. It has previously been demonstrated that two-dimensional (2D) dose distributions can be obtained by imaging the OSL light from a sample during stimulation, e.g. for use in RT [5].

IOP Conf. Series: Journal of Physics: Conf. Series **847** (2017) 012044 doi:10.1088/1742-6596/847/1/012044

In this paper, a new approach to 3D dosimetry will be presented. We propose a new reusable 3D radiation dosimetry system, based on OSL material embedded homogenously inside a transparent silicone elastomer matrix, a so-called 3D-OSL dosimeter. With an appropriate optical setup, 3D dose distributions can be obtained from the OSL emission.

2. Materials and methods

2.1. The matrix material
The matrix consists of a transparent polymeric material based on polydimethylsiloxane silicones (PDMS), which is a commercially available silicone consisting of pre-weighed monomer and curing agent. In the current investigation the SYLGARD® 184 Silicone Elastomer Kit from Dow Corning with a curing agent in 1:10 weight ratio was used.

2.2. The OSL material
The OSL material used is lithium fluoride (LiF) doped with magnesium, copper and phosphorus (LiF:Mg,Cu,P - MCP). This material is in widespread use as a very well-established and highly sensitive TL detector e.g. in personal dosimetry, but it has also very good OSL properties [5, 25-27]. The MCP powder is embedded homogenously inside the silicone matrix which acts as a host of OSL grains.

2.3. Preparation
Prototype 3D dosimeters were cast in 1 x 1 x 5 cm^3 cuvettes containing 0.3 g of MCP microcrystals mixed homogenously in 4 g of silicone elastomer. The dosimeters were obtained by mixing dry MCP particles into a silicone matrix. In order to check the anticipated OSL signal levels from each 1 mm^3 voxel of a 3D dosimeter, standard OSL-reader aluminum trays carrying amounts of OSL material corresponding to those found in 1 mm^3 of the 3D dosimeter were prepared. Two concentrations of OSL powder, which were embedded inside the silicone matrix and poured into the trays, were investigated. Sample no. 1 used approximately 0.06 mg of pure OSL powder, while sample no. 2 used 0.2 mg of OSL powder. A reference sample containing clear silicone without powder was also prepared.

2.4. The read-out system
The samples were read-out using the standard Risø TL/OSL DA-20 reader. The samples were irradiated with beta radiation to a dose at the level of 1 Gy in all cases. The samples were stimulated with blue light emitting diodes (LEDs), with emission centered at 470 nm and an intensity of ~80 mW/cm^2 at the sample. A Hoya UV-340 filter of 7.5 mm thickness and Ø = 45 mm were used as a detection filter. For the analyses of the OSL signal, the period of stimulation was 100 s.

3. Results and discussion
Although both the PDMS matrix and the individual OSL particles are transparent, the 3D-OSL dosimeter material appeared slightly misty, see Figure 1. The transparency depended on the amount of MCP powder used, so the concentration of OSL particles must be optimized as a compromise between signal level per volume and overall transparency. Figure 1 shows a prototype fulfilling both requirements, as discussed in more detail below. One of the advantages of using MCP material in combination with the Sylgard 184 silicone elastomer is that the refractive-index match between LiF and Sylgard 184 is quite good for visible wavelengths, which minimizes light scattering from the embedded particles.

During the first second of a standard OSL read-out procedure, ~10,000 and ~40,000 counts were observed from samples 1 and 2, respectively. This corresponded to the anticipated signal levels from each 1mm^3 voxels from the 3D dosimeter (at a concentration similar to that of figure 1). We also

IOP Conf. Series: Journal of Physics: Conf. Series **847** (2017) 012044 doi:10.1088/1742-6596/847/1/012044

noted that the silicone matrix in itself – as expected – did not add to the OSL signal (see curve labelled "Empty probe" in Figure 2).

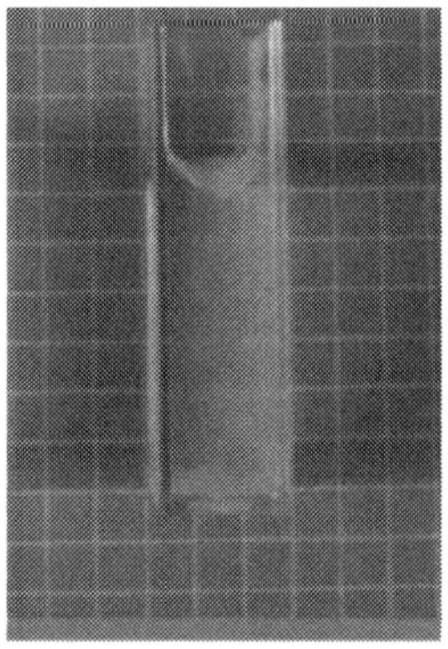

Figure 1. A 3D dosimeter in the form of a 1 x 1 x 5 cm^3 cuvette containing 4 g of silicone matrix combined with 0.4 g of MCP powder.

As mentioned, the data presented in the figure 2 were obtained using a commercial OSL reader equipped with UV filters and a PMT-tube. In order to obtain 3D distributions for dosimeters as shown in figure 1, an appropriate optical setup should be applied. For example, by stimulating the OSL dosimeter with a light sheet (e.g. from a laser source), and imaging the luminescence intensity across that sheet (e.g. by a combination of optical filters and a camera), the dose distribution in a plane can be measured. When this plane is shifted across the dosimeter, a 3D dose distribution is directly obtained without the need for inversion algorithms.

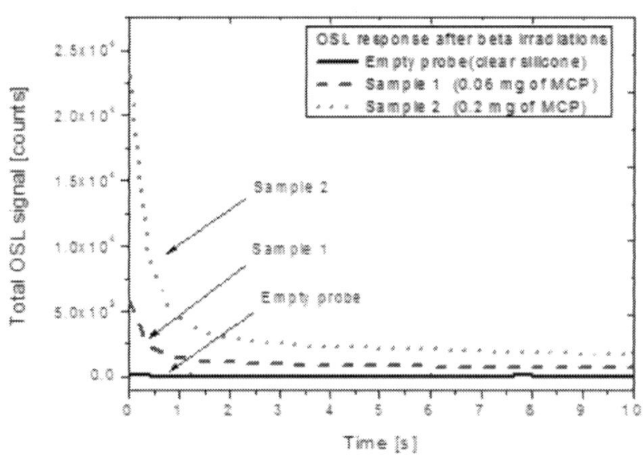

Figure 2. OSL decay-curves showing the number of counts in 0.4 s time bins from different 3D-OSL samples obtained by irradiating with a beta source to a dose of 1 Gy. Sample 1 (blue dashed line) contains 0.06 mg of MCP powder, sample 2 (red dotted line) contains 0.2 mg of MCP powder and the last sample contains silicone without MCP powder (black solid line). The OSL was stimulated with a commercial OSL Risø TL/OSL DA-20 reader as described in the text.

An advantage of using the OSL mechanism for dosimetry is that the dosimeters become reusable. It is well known that OSL dosimeters can be reset by temperature or light-bleaching stimulation. The same approach can be used for the OSL material embedded in the silicone host. The new 3D dosimeter also benefits from the well-established dosimetric properties of the MCP material which comprise a wide dynamic range and a linear dose response, together with the potential ability to measure the linear energy transfer (LET) in therapeutic beams. [4, 5, 25-28]. The use of PDMS as a host material makes it simple to cast dosimeters into anthropomorphic shapes and the flexibility of the material even allows for simulating organ deformations during RT in realistic geometries [23].

4. Conclusion

We proposed a new reusable 3D dosimeter system based on OSL material embedded homogenously in a transparent silicone matrix and have documented the initial good OSL signal levels for the amount of

MCP anticipated in a 1mm^3 voxel. The new 3D-OSL dosimeter has the potential to verify complex 3D RT doses with high spatial resolution while maintaining the attractive properties of OSL dosimetry.

5. Acknowledgements
This research project was supported by the Danish Cancer Society, award numbers R72-A4486-13-S2 and R56-A2927-12-S2.

6. References
[1] Followill D S *et al* 2007 *Med. Phys.* **34** 2070-6
[2] Hill R *et al* 2009 *Med. Phys.* **36** 3971-81
[3] Hill R *et al* 2014 *Phys. Med. Biol.* **59** R183-231
[4] Yukihara E G and McKeever S W S M 2011 *Optically Stimulated Luminescence: Fundamentals and Applications* (West Sussex: Wiley and Sons)
[5] Olko P 2008 *Radiat. Meas.* **43** 864-869
[6] Baldock C *et al* 1998 *Phys. Med. Biol.* **43** 695-702
[7] Novotny J *et al* 2001 *Med. Phys.* **28** 2379
[8] Heufelder J *et al* 2003 *Med. Phys.* **30** 1235
[9] Gustavsson *et al* 2004 *Phys. Med. Biol.* **49** 3847
[10] Zaidan O A *et al* 2010 *J. Phys.: Conf. Series* **250** 012032
[11] Daniel A *et al* 2011 *Med. Phys.* **38** 1313
[12] Adamovics J *et al* 2006 *Rad. Prot. Dosim.* **120** 107
[13] Al-Nowais S et al 2009 *Appl. Rad. and Isot.* **67** 415-8
[14] Zhao L *et al* 2010 *J. Phys.: Conf. Series* **250** 012035
[15] Skyt P *et al* 2011 *Med. Phys.* **38** 2806
[16] Skyt P *et al* 2012 *Med. Phys.* **39** 7232
[17] Skyt P *et al* 2013 *Acta Oncolgica* **52** 1445-50
[18] Zhao L *et al* 2012 *Phys. Med.Biol.* **57** N431
[19] Gorjiara *et al* 2011 *Med. Phys.* **38** 2265-74
[20] Juang T *et al* 2013 *J. Phys.: Conf. Series* **444** 012080
[21] Alqathami M *et al* 2012 *Rad. Phys and Chem.* **81** 867
[22] Doran S *et al* 2015 *J. Phys.: Conf. Series* **573** 012043
[23] Høye E M *et al* 2015. *Phys. Med. Biol.* **60** 5557
[24] Hill B *et al* 2005 *Med. Phys.* **32** 1589-97
[25] Bos A J J *et al* 1996 *Radiat. Prot. Dosim.* **65** 199-202
[26] Bilski P *et al* 2011 *Radiat. Meas.* **47** 1728-31
[27] Sadel M *et al* 2015 *Radiat. Meas.* **82** 8-13
[28] Granville D A *et al* 2016. *Phys. Med. Biol.* **61** 1765-79

IOP Conf. Series: Journal of Physics: Conf. Series **847** (2017) 012045 doi:10.1088/1742-6596/847/1/012045

In vivo verification of particle therapy: how Compton camera configurations affect 3D image quality

D Mackin[1], E Draeger[2], S Peterson[3], J Polf[2] and S Beddar[1]
[1]The Department of Radiation Physics, University of Texas MD Anderson Cancer Center, 1515 Holcombe Boulevard, Unit 94, Houston, Texas 77030, USA
[2]The Department of Radiation Oncology, University of Maryland School of Medicine, Baltimore MD 21201, USA
[3]The Department of Physics, University of Cape Town, Rondebosch 7700, South Africa

Email: dsmackin@mdanderson.org

Abstract. The steep dose gradients enabled by the Bragg peaks of particle therapy beams are a double edged sword. They enable highly conformal dose distributions, but even small deviations from the planned beam range can cause overdosing of healthy tissue or under-dosing of the tumour. To reduce this risk, particle therapy treatment plans include margins large enough to account for all the sources of range uncertainty, which include patient setup errors, patient anatomy changes, and CT number to stopping power ratios. Any system that could verify the beam range *in vivo*, would allow reduced margins and more conformal dose distributions. Toward our goal developing such a system based on Compton camera (CC) imaging, we studied how three configurations (single camera, parallel opposed, and orthogonal) affect the quality of the 3D images. We found that single CC and parallel opposed configurations produced superior images in 2D. The increase in parallax produced by an orthogonal CC configuration was shown to be beneficial in producing artefact free 3D images.

1. Introduction

Though many methods have been proposed for *in vivo* verification of particle therapy, most methods utilize the secondary radiation produced by nuclear scattering of the beam particles. Early research into treatment verification focused on post treatment PET imaging of the beta decay of ^{11}C and ^{15}O. Though this method can utilize commercial PET scanners, the positron emission is not closely correlated with the beam range, and it suffers from poor resolution due to biological washout [1, 2]. Prompt gamma (PG) radiation does not suffer from these limitations, and it has garnered more of the recent research attention. However, because of the high energy (2 – 8 MeV) of PG radiation, detection system designed for PET and SPECT are not suitable. Novel detection systems, designed for PG radiation emitted during particle therapy must be developed.

The purpose of our research is to develop a clinical imaging system capable of *in vivo* verification of the delivery of particle therapy treatments. To this end, we have been collaborating with H3D Inc. (An Arbor, Michigan, USA) to develop a custom, multi-stage cadmium, zinc, telluride (CZT) Compton camera (CC) gamma detection system and a novel image reconstruction algorithm capable of producing 3D images of gamma emission. In this article, we begin by describing the gamma emission produced during proton and the hardware components of our Compton camera, PolarisJ2. We next investigate how single camera, parallel opposed, and orthogonal CC configurations affect the quality of the reconstructed 3D images using both a back projection and an iterative reconstruction algorithm.

2. Materials and Methods

2.1. *The prompt gamma spectrum*

The spectrum of the secondary particles produced by nonelastic scattering during particle therapy is complex. However, for PG verification of particle therapy, two gamma energies are of particular importance, 4.44 and 6.13 MeV which are excited nuclear states of carbon and oxygen respectively [3, 4]. These gammas are the most prevalent above 3 MeV, and the 6.13 MeV gamma ray emission is closely correlated with the Bragg peak.

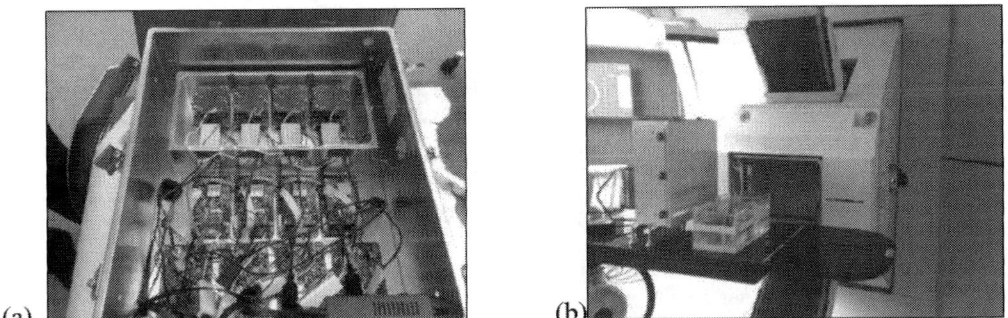

Figure 1. (a) The inside of a single stage of a PolarisJ2 Compton camera. (b) A single PolarisJ2 CC stage in a proton therapy treatment vault.

2.2. *The PolarisJ2 Compton Camera Imaging System*

The PolarisJ2 from H3D (An Arbor, Michigan), has been updated from the PolarisJ system [5] to provide a larger detector volume and much faster readout. The current iteration of the detector consists of 2 independent detector stages, each with 4, 4x4 arrays of CZT crystals. Each crystal is pixelated into an 11x11 grid, and each pixel is read out independently by an application specific integrated circuit (ASIC) controlled a CPU running a Linux operating system. The four crystal arrays and controllers are shown in Fig. 1 (a). A single PolarisJ2 CC stage in a proton therapy treatment vault is shown in Fig. 1 (b).

2.3. *Compton camera configurations and simulations*

CC imaging systems rely on parallax to achieve resolution in the direction orthogonal to detector faces, and the finite width of the detector limits the orthogonal resolution [6]. One way to reduce this problem would be to simply make the detector much bigger. However, multiple smaller detectors may provide more parallax, and, therefore, higher quality images. For this study we simulated a 200 MeV proton beam on a water phantom using a Geant4.9.4 Monte Carlo model described in detail in [7]. The two simulated CC detector stages were $8 \times 8 \times 1$ cm^3 and centered 9.5 cm and 17.3 cm from isocenter. The simulations were performed multiple times with the Compton camera positioned at 0, 90, 180, and 270 degrees. The independent simulations were then combined to produce parallel opposed and orthogonal configurations. Each simulation used 2×10^9 particle histories.

To reconstruct the images we used a back projection algorithm that counts the intercepts between the Compton cones and the voxels in the region of reconstruction. We also used an iterative reconstruction algorithm derived based on stochastic origins ensembles method [8, 9]. Only two scatters were used for each detected gamma, and the energies were assumed to be either 4.44 MeV or 6.13 MeV. The known proton spot position was used as prior information for both reconstruction methods. All images were produced using 20k detected gammas.

3. Results

The back projection images are instructive because they visualize the raw information that is the input to the iterative reconstruction algorithm. In the top rows of Fig. 2 and Fig. 3, artifact of the Compton scatter cones projected away from the CC are apparent in the direction orthogonal to the CC.

IOP Conf. Series: Journal of Physics: Conf. Series **847** (2017) 012045 doi:10.1088/1742-6596/847/1/012045

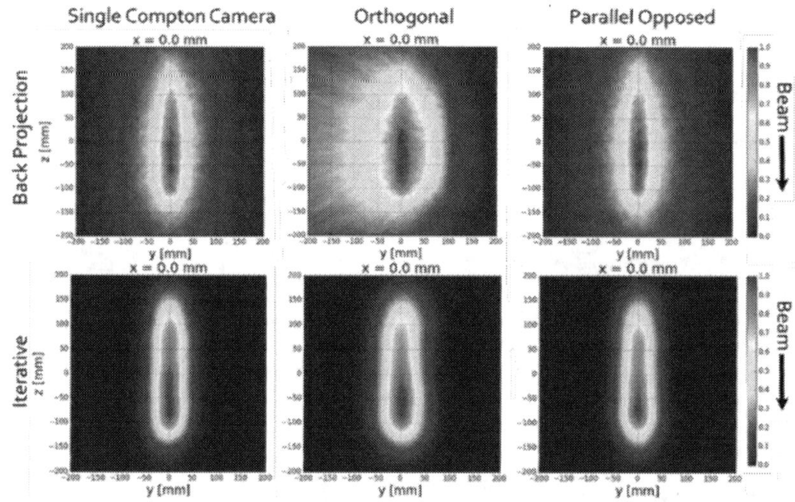

Figure 2. A comparison of 2D slice images of a simulated 200 MeV proton beam incident on a water phantom. The columns are three detector configurations: a single Compton camera (left), two Compton cameras aligned orthogonal to each other (middle), and parallel opposed (right). The rows show images created using back projection (top) and stochastic origin ensembles iterative (bottom) algorithms. The color scale represents relative gamma emission intensities.

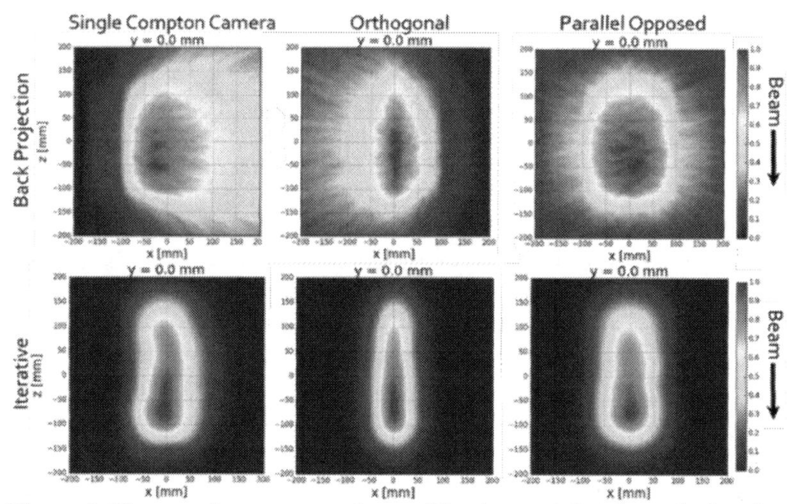

Figure 3. The same image comparison as Fig. 4 except that the x=0 plane has been replaced with the y=0 plane.

For the single and parallel opposed CC configurations, the artifacts appear only in the $y = 0$ plane, orthogonal to the detector faces (Fig. 3). For the orthogonal CC configuration, the artifacts appear in both planes (Fig. 2 and Fig 3, top rows). However, the parallax provided the orthogonal enables the iterative reconstruction to remove the artifacts in both planes (Fig. 2 and Fig 3, bottom rows). Due to the limited parallax in the $y = 0$ plane for the single and parallel opposed CC configurations, the iterative reconstruction produced an inaccurate curvature in the single CC images and a widening of the images in the parallel opposed configuration (Fig. 3, bottom row).

IOP Conf. Series: Journal of Physics: Conf. Series **847** (2017) 012045 doi:10.1088/1742-6596/847/1/012045

4. Discussion and Conclusions

We have demonstrated the effects of CC configuration on the 3D images of the PG emission for proton beams. Similar results can be expected for particle therapy beams except that the PG emission is much greater for heavier ions [10]. Practical considerations such as cost and physical space will limit the size and number of CCs that can be used in a particle therapy vault. However, we have demonstrated that the parallax information gained from multiple CCs can improve image quality. In this study, we focused on the increased parallax, and, therefore, we used equal numbers of detected gammas for each configuration. When used clinically, multiple detectors will also increase the number of gamma ray detections available for the image reconstruction. In this study, we looked only at beams along the central axis. Both the parallel opposed and orthogonal configurations are likely to show biases for beams that are closer to one CC than the other. These biases will have to be corrected as part of the image reconstruction.

There is not one ideal detector configuration for CC imaging systems. In this study, parallel opposed and single CC configurations produced superior back projection images in 2D. The increase in parallax provided by an orthogonal CC configuration was shown to be beneficial in producing artefact free 3D images.

5. References

[1] Knopf A *et al* 2008 *Phys. Med. Biol.* **53** 4137
[2] Parodi K *et al* 2007 *Int. J. Radiat. Oncol.* **68** 920-34
[3] Kozlovsky B *et al* 2002 *Astrophys. J. Suppl. S.* **141** 523
[4] Verburg J M *et al* 2012 *Phys. Med. Biol.* **57** 5459
[5] McCleskey M *et al* 2015 *Nucl. Instrum. Meth. A* **785** 163
[6] Maxim V 2014 *IEEE T. Image Process.* **23** 332-41
[7] Polf J C *et al* 2014 *Phys. Med. Biol.* **59** 2325
[8] Andreyev A *et al* 2011 *Med. Phys.* **38** 429
[9] Mackin D *et al* 2012 *Phys. Med. Biol.* **57** 3537 - 53
[10] Pinto M *et al* 2014 *Phys. Med. Biol.* **60** 565

IOP Conf. Series: Journal of Physics: Conf. Series **847** (2017) 012046 doi:10.1088/1742-6596/847/1/012046

Cherenkov imaging in the potential roles of radiotherapy QA and delivery

B W Pogue[1,2], R Zhang[1], A Glaser[1], J M Andreozzi[1], P Bruza[1], D J Gladstone[1,2] and L A Jarvis[2]
[1]Thayer School of Engineering, Dartmouth College, Hanover NH 03755 USA
[2]Geisel School of Medicine, Dartmouth College, Hanover NH 03755 USA

E-mail: brian.w.pogue@dartmouth.edu

Abstract. Cherenkov emission has a direct proportionality to the deposited dose at the local level, and capture of these emitted light signals allows visualization of real time maps of dose in vivo. Mapping the Cherenkov signals through water tanks illustrates how 3D Cherenkov can be achieved, either as 2D plus time, or 3D in static imaging. Imaging Cherenkov from patients shows how signals can be acquired which map out radiation dose in real time. The signals are affected by several factors, each of which will take some calibration to resolve, yet intrinsically the signal is shown to be a linear reporter of dose delivered. Development of calibration methodologies is ongoing in both research and development work.

1. Introduction

Relatively recently, the discovery of Cherenkov light emission from radionuclides has been exploited to allow optical imaging of PET agents and therapy monitoring, (Fig 1(a)) in small animal work) [1-3]. The emission is dominant in the blue range (Fig 1(b)) and falls off significantly with increasing wavelength, yet most of the light emitted from tissue is highly attenuated in the UV-blue-green, due to hemoglobin absorption, and so most of the detected light is above 600 nm when imaging tissue. The intensity of emission is a strong function of particle energy (Fig 1(c))[4], however, it is relatively constant in the megavoltage energy range used in radiation therapy. Attenuation within tissue is due to scattering as well as high absorption in the UV-blue-green wavelengths, and so the detected Cherenkov light from animals originates from superficial interactions.

Despite the limitations of Cherenkov light (scattering and absorption limit emission to immediately sub-surface regions) it is apparent that every patient undergoing radiation therapy emits detectable Cherenkov light, and this is a signal which directly reports on the radiation dose delivery. It reports on the combination of patient position, and surface dose delivered. Examples of human imaging of Cherenkov are shown below in Figure 2. Thus, the signal is freely available, and provides potential video coverage of treatment.

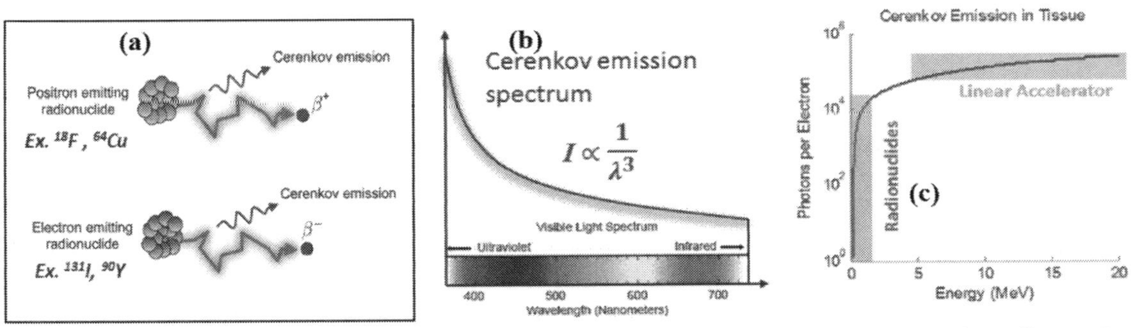

Figure 1. The use of Cherenkov light to image e$^+$ or e$^-$ emission is illustrated, as exploited recently to allow optical imaging of nuclear medicine agents. The emission spectrum (b) is dominated by UV-blue light, falling off with a λ^{-2} dependence. Emission intensity versus e$^-$ energy (c) increases dramatically in 0.2-2 MeV and then saturates above 20 MeV[5]. Emission yield in radiotherapy >10X higher than PET Cherenkov.

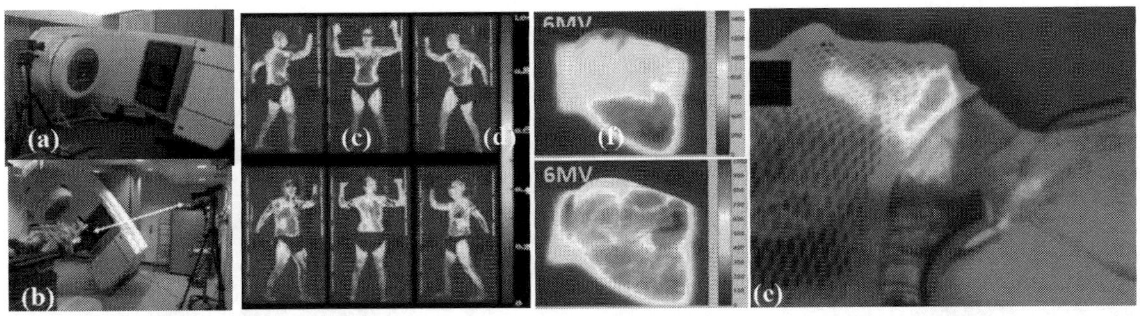

Figure 2. Imaging of Cherenkov emission is shown from a camera beside the LINAC for total skin irradiation (a) and off on the adjacent wall (b) for breast irradiation imaging. Cherenkov images (false color) are shown from total skin irradiation electron therapy (c). The total skin treatment was delivered and imaged using all 6 positions in the Stanford technique (Andreozzi et al, Med Phys 2016), and breast imaging, (d) was shown for both entrance (top) and exit dose (bottom) (Zhang et al, Med Phys 2015). One example of Head and neck treatment beam imaging is shown in (f).

2. Methods

Imaging of Cherenkov emission has been accomplished originally with an ICCD camera (Princeton Instruments, PI-MAX4)[6] mounted on a tripod in the treatment room. This camera is synchronized to the LINAC output current pulse from the target, I_{Targ}, to capture intensified images only during the 3.5 microsecond pulses of the device. The intensifier has a short gate with nanosecond time resolution, allowing the on/off with sufficient temporal resolution to capture the signal only during the pulse on. The pulses are integrated on the CCD camera for approximately 20-40 shots, and then read out as a single image. The LINAC pulses at 360 Hz, and the frame rate from the camera has been 6-20 frames/per second. Cherenkov intensity from both water and tissue is strong in this acquisition procedure, with average lateral irradiance approximately several mW/cm^2. This is also near the irradiance of background room lights as well, so an acquisition of background light without the Cherenkov must be captured, and the two subtracted to yield the contribution of just the Cherenkov. This process works well because they are similar intensities and one does not dominate over the other. Further reductions in background can be achieved through careful choice of the lights used in the room or subdued lighting.

Water-beam imaging has been accomplished with a standard ionization chamber calibration water tank, with the back and side walls painted black, or with black cloth taped to them. A mixture of tap water and 1 g/L of quinine sulphate is used in the tank for a fluorescent reporter of the Cherenkov signal [7-9]. This quinine is there to diminish the acute angular direction at which photons are emitted when charged particle radiation is traveling directly from the beam. Much of the Cherenkov intensity is

IOP Conf. Series: Journal of Physics: Conf. Series **847** (2017) 012046 doi:10.1088/1742-6596/847/1/012046

directly downwards at a 41^0 angular cone around the direction of travel of the incident photons and electrons. Secondary scatter blurs this cone a bit, but it is still quite present. This quinine absorbs the UV Cherenkov photons and re-emits them in an isotropic spatial pattern, increasing the laterally directed signal from the beam. Imaging of the beam depth dose 2D profile was examined as was the temporal imaging of the beam in a treatment plan from IMRT and VMAT.

Images on tissue or tissue phantoms are available from Cherenkov emission at either entrance or exit beam. For total skin electron therapy, the camera was positioned beside the linac output (Fig 2(a)), and for breast imaging, the camera was positioned at the side wall, lateral to the patient couch (Fig 2(b)), and just inferior to the beam plane, to avoid penumbra from a lateral beam being incident upon the camera. A 50 mm focal length lens was used with f/1.8 aperture. The processed image of one patient from each of total skin treatment (c) and whole breast irradiation (d) are shown in Fig 2. In the case of the total skin, there were 6 poses for each patient and images of each pose are shown. In the case of total breast irradiation with 6 MV photons, the entrance beam dose (top) and the exit beam dose (bottom) could be captured independently on one side of the patient.

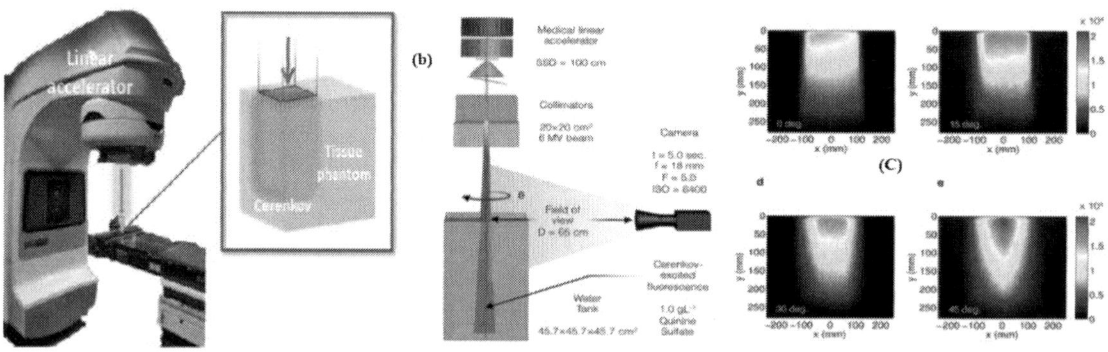

Figure 3. The LINAC beam geometry used experimentally is shown (a), with simulation geometry (b). Experimental images of beam at angles 0°, 15°, 30° and 45° (c) around the square beam. [8, 12]

3. Results

3.1 Water Tank Imaging

The collisional energy loss of secondary electrons leads to dose in tissue, and the soft-collision radiative energy loss due to Cherenkov emission are independent processes, but both are largely energy independent above 220 keV. This proportionality in energy loss per unit path length at the majority of the electron energy spectrum used in clinical radiotherapy is the fundamental basis for use of Cherenkov emission imaging as a dosimetry surrogate. Despite disproportionality at energies less than 220 keV, the range of these low energy electrons is less than 1 mm. The dose distribution is largely dictated by energetic electrons above the threshold value for Cherenkov emission. Indeed this is well known, as Monte Carlo simulations of dose often do not track secondary photons <200 keV because they contribute so little to the overall dose[4]. A Dose/Cherenkov comparison has been systematically examined in water, using the geometry of Figure 3, and showing measured data from depth dose and profile curves in Figure 4.

IOP Conf. Series: Journal of Physics: Conf. Series **847** (2017) 012046 doi:10.1088/1742-6596/847/1/012046

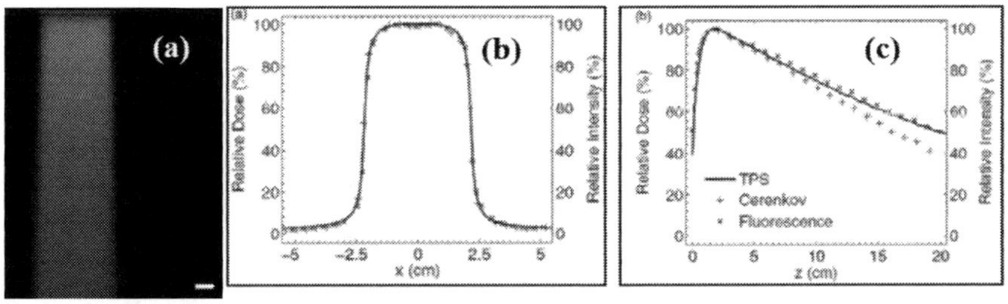

Figure 4. (a) Photograph of Cherenkov-induced fluorescent emission from quinine in water, the linearity of image intensity with applied dose is nearly perfect, and (b) a lateral beam profile is shown (crosses) with the known beam profile (ionization chamber), and (c) depth dose curve shown Cherenkov and Cherenkov-fluorescence relative to treatment planning system (TPS). [11]

While Cherenkov light is emitted at exactly 41^0 from the direction of travel of the electrons, we have found that by adding fluorescent dye quinine sulphate (used in tonic water) to water tank experiments, we could then image the fluorescent light which was generated by Cherenkov excitation of the dye, and this fluorescent emission is isotropic [7, 11]. Images of emission from the water tank are shown in Figure 6. In Figure 7, we demonstrate the ability to use these lateral profiles at multiple angles to reconstruct a 3D map of Cherenkov emission, which is a direct surrogate for 3D dose. We have recently demonstrated the ability to verify IMRT and VMAT treatment plans in water tanks [8], and also for imaging of large area complex beams in water tanks [8, 12].

The images and data in this water imaging clearly indicate there is an outstanding match between the observed Cherenkov fluorescence and the deposited dose in water. Monte Carlo simulations indicated Cherenkov-fluorescence to dose has linear proportionality and that the agreement was better than 5% accuracy in complete IMRT and VMAT treatment plans [8].

3.2 Patient imaging – delivery verification

Human piilot studies were completed comparing Cherenkov intensities on the patient to predicted doses from the treatment planning system and to thermoluminescent diode (TLD) measurements on the same patients. Some preliminary data taken from patients illustrate that, on a case study, there is good linearity between dose per field planned and Cherenkov intensity recorded experimentally. In this case, shown in Figure 5, the doses for a collection of subjects at the 6 MV entrance field match the planned fraction dose very well with a linear regression between the two. The match between these data is still under investigation, with comparisons being studied on exit doses and 10 MV doses, for those subjects given the split dose regimine. The data in Figure 5(c) clearly shows a linear dependence of Cherenkov intensity with dose. The figure shows correlation exists between planned skin dose and Cherenkov imaged dose for 10MV tangent beams. Other calibration factors are likely needed in some circumstances, such as pigmentation, incident angles, energy level and exit vs. entrance dose to correct for Cherenkov emissions compared to delivered dose. A full validation of these factors in phantom work has been tested and calibration factors determined [13],

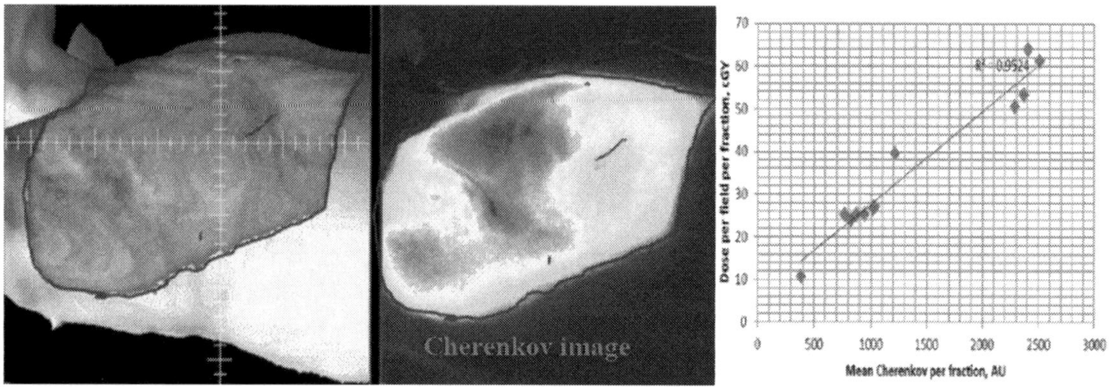

Figure 5. Visualization of surface dose (5mm depth) in a treatment plan is shown in (a) and the Cerenkov image captured during the patient's treatment is shown in (b). Maps of dose (cGy) per fraction versus imaged Cherenkov intensity were used from 6 subjects to create a linear calibration line (c). This type of imaging could allow real-time detection of delivery and verification of delivered dose, at video rate.

4. Discussion

Cherenkov video imaging is early in the development process, but it has the potential to pave the way for a number of innovations in radiation therapy imaging, such as in the areas of quality audit, and treatment verification. Applications of in vivo imaging to date have largely focused on whole breast irradiation because of the match to large soft tissue surfaces which need verification. However less used applications such as total skin irradiation are also good applications of this methodology, because of the need for large complex surface area dosimetry, in a setting where few tools exist to capture the signal.

The most immediate proposed development areas from this work include:

1) The demonstration that radiation dose delivery could be **visualized,** naturally in real time, implicitly verifying what is happening. This has both scientific and real life implications.
2) Direct ability to monitor **beam shape** in **real-time on the patient's skin** throughout the treatment by the radiation therapy team in comparison to that predicted by the planning system, tracking treatment fields with dynamic multi-leaf collimator movement. Cherenkov images can be superimposed on the normal video stream that the radiation therapists typically view. Our experience to date suggests that it increases the radiation team's comfort in knowing the plan is being delivered as intended, by directly viewing it throughout the duration of each treatment.
3) The treatment can be **recorded for permanent archive** to determine what happened in the cases of complex treatment QA audits.
4) The high sensitivity to **blood vessel location** provides internal patient soft tissue anatomy, which could be used for highly accurate treatment verification and positioning to an <u>accuracy of 1mm</u>. These internal biologic "fiducials" can indicate inaccurate alignment or when fiducials are distorted in combination with change in measured exit could indicate anatomy changes and need to re-plan the patient's treatment.
5) The Cherenkov intensity is **generally proportional to dose**, so the images could be interpreted as **surface dose maps**, once more fully calibrated as proposed here.

These are innovative roles for Cherenkov imaging, and will be tested in future research and development of the methods for image capture and calibration. Ongoing work in calibration is perhaps the most important part of determining the role for Cherenkov imaging in radiotherapy.

5. Acknowledgments
This work has been financially supported by NIH grant R44CA199836, as well as the Norris Cotton Cancer Center Pilot Funds support.

6. References
[1] Spinelli A E and Boschi F 2015 *Phys. Med.* **31** 120-9
[2] Kotagiri N *et al* 2015 *Nat. Nanotechnol.* **10** 370-9
[3] Czupryna J *et al* 2015 *J. Nucl. Med.* **56** 483-8
[4] Glaser A K *et al* 2014 *Phys. Med. Biol.* **59** 3789-811
[5] Ross H H 1969 *Analytical Chemistry* **41** 1260
[6] Andreozzi J M *et al* 2015 *Med. Phys.* **42** 994-1004
[7] Glaser A K *et al* 2013 *Med. Phys.* **40** 012101
[8] Glaser A K *et al* 2014 *Med. Phys.* **41** 062102
[9] Glaser A K *et al* 2015 *Med. Phys.* **42** 4127
[10] Zhang R *et al* 2015 *Phys. Med. Biol.* **60** L1-14
[11] Glaser A K *et al* 2013 *Phys. Med. Biol.* **58** 601-19
[12] Glaser A K *et al* 2013 *Opt. Lett.* **38** 634-6
[13] Zhang *et al* 2016 J. *Biophotonics* 1-12. doi: 10.1002/jbio.201500344

IOP Conf. Series: Journal of Physics: Conf. Series **847** (2017) 012047 doi:10.1088/1742-6596/847/1/012047

Accuracy and reproducibility in x-ray computed tomography polymer gel dosimetry

E Maynard[1], M Hilts[2,3], E Heath[4] and A Jirasek[5]
[1]Department of Physics and Astronomy, University of Victoria, Victoria, British Columbia, V8W 2Y2, Canada
[2]Department of Physics, University of British Columbia - Okanagan Campus, Kelowna, British Columbia V1V 1V7, Canada
[3]Medical Physics, BC Cancer Agency, Centre for the Southern Interior, Kelowna, BC, V1Y 5L3, Canada
[4]Carleton Laboratory for Radiotherapy Physics, Physics Department, Carleton University, Ottawa, Ontario K1S 5B6, Canada
[5]Department of Physics, I.K. Barber School of Arts and Sciences, University of British Columbia, Kelowna, British Columbia V1V 1V7, Canada

E-mail: andrew.jirasek@ubc.ca

Abstract. This work assesses the overall reproducibility and accuracy of an x-ray computed tomography (CT) polymer gel dosimetry (PGD) system using a N-isopropyl-acrylamide (NIPAM) based polymer gel and investigated what effects the use of generic, inter-batch, and intra-batch gel calibration have on dosimetric and spatial accuracy. Overall excellent spatial and dosimetric accuracy was found with this system for generic, inter-batch calibration methods.

1. Introduction

In polymer gel dosimetry (PGD), a carefully tuned recipe of radiosensitive chemicals is dissolved in gelatin to form a three-dimensional (3D) dosimeter. When irradiated, a gel dosimeter will undergo chemical changes that can be read out and related to the 3D dose absorbed within the gel. Polymer gels have natural advantages in dosimetry as they are inherently 3D, dose integrating, tissue equivalent and deformable. Traditional PGD has been used with MRI as the readout method [1-4], with other read out modalities such as optical CT [5, 6] and x-ray CT [7-11] imaging emerging more recently.

X-ray CT PGD is limited by the small size of the density changes (approximately 1 mg/cm^3 per 1 Gy of absorbed dose [12]) that occur due to the polymerization of the gel. X-ray CT PGD is, therefore, hampered by low signal-to-noise ratio and low contrast in its dose resolution. Research has focused on maximizing the signal by increasing the sensitivity of the gel recipe [13, 14], using filtering to decrease noise and remove artifacts [15, 16] and improving calibration techniques [17]. To enable clinical implementation, the reproducibility, dosimetric accuracy and spatial accuracy that can be expected of an x-ray CT PGD system needs to be established.

The aim of this work is to assess the overall reproducibility and accuracy of an x-ray CT polymer gel dosimeter system initially characterized by Johnston *et al* [9]. This study also investigates the effect that calibration across different gel batches has on the accuracy of the system. Specifically, we investigate the use of inter-batch gel calibration, average or generic calibration curves and intra-gel calibration across different locations within the gel dosimeter.

2. Materials and Methods

2.1. Gel Fabrication

All gel dosimeters consisted of, by weight, 75.5% deionized water, 5% gelatin (Sigma-Aldrich Canada, Oakville, ON), 15% N-isopropyl-acrylamide (NIPAM, TCI Chemicals, OR, USA), 4.5% N,N'-methylenebisacrylamide (BIS, Sigma-Aldrich) and 5 mM tetrakis hydroxylmethyl phosphonium chloride (THPC, Sigma-Aldrich) and were fabricated using a procedure initially described by Chain *et al* [13]. Blank gels were also fabricated for background subtraction of CT images. In total, four 1 L gels were fabricated for analysis in this work with all four gels manufactured from a single lot of NIPAM.

2.2. Treatment planning and irradiation

All gels were irradiated with three 3 x 7 cm^2 fields, at 45°, 270° and 315° in a calibration designed using the ECLIPSE treatment planning system (Varian Medical Systems, Palo Alto, CA) which covered a full range of doses up to a maximum of approximately 27 Gy. This treatment plan was delivered to two separate locations within the gel, at the top and bottom of the gel container. When irradiated each gel was setup using an anthropomorphic head and neck phantom filled with water to ensure complete immobilization of the gel and provide sub-millimeter position reproducibility [18]. The dose for this treatment plan was calculated using the Vancouver Island Monte Carlo (VIMC) system which has been thoroughly validated for the calculation of complex dose deliveries [19]. The VIMC dose calculation was used for calibration of the gel dosimeter and as a comparison to assess the accuracy of the gel dosimeter.

2.3. X-ray CT imaging

Irradiated gels and blank gels were imaged using an Optima CT580 multislice CT scanner (GE Medical Systems, Milwaukee, WI). All CT images were acquired using imaging settings based off previous work by Baxter et al [20] and Jonhston *et al* [9] to optimize gel dosimetry scanning protocols. Each gel was set up on the CT scanner bed for imaging using the head and neck phantom to ensure synchronized localization of the gel for imaging and irradiation [18].

2.4. Image processing and calibration

The gel images were processed using image averaging, background subtraction of blank gels, adaptive mean filtering and remnant artifact removal as described by Jirasek *et al* [16]. Calibration of the gels was performed using the techniques described by Johnston *et al* [9] and refined by Jirasek and Hilts [1517].

The gels were calibrated in 4 different ways to assess any changes in accuracy when using different calibration techniques. *Self-Calibration* is where gel image is calibrated using the calibration curve generated by its own image; it is essentially calibrated with itself. This eliminates any error introduced by differences between different gel batches and is a "best case scenario" for the dosimeter. *Intra-gel Calibration* is where the gel image at the bottom of the container is calibrated using the calibration pattern at the top of the gel container. *Average Calibration* is where the gel image is calibrated using the average calibration curve of the four gels in the sample. The results from this type of calibration will determine the plausibility of using a generic calibration curve for this dosimetry system. *Most Divergent Calibration* is where the gel image is calibrated using the single calibration curve from the sample that produces the least accurate dosimetric results. The gel dose generated by each calibration method was compared to the VIMC calculated dose across the sample of four gels using (i) dose difference in low dose gradient region (with a gradient threshold = 0.2), (ii) distance-to-agreement (DTA) in high dose gradient regions, and (iii) gamma distributions and gamma pass rates.

3. Results and Discussion

Figure 1 is an example of the different types of dose analysis used on an individual gel dosimeter. The dose difference map between Monte Carlo and a self-calibrated gel measured dose is shown in figure 1a where good agreement is seen in the low dose gradient regions. The gel dose is lower on average than calculated by Monte Carlo at the highest doses but is more accurate at other dose levels. The gel in figure 1a has an absolute mean dose difference of 0.46 Gy across all points within the low dose gradient region. Figure 1b shows the distance-to-agreement (DTA) between measured gel dose and Monte Carlo calculated dose in the high dose gradient region. In this gel, more than 90% of points in the high dose gradient region have a DTA of less than 1 mm and more than 99% of points have a DTA of less than 2 mm. Figure 1c combines both the high and low gradient regions together and a gamma analysis gives a sense of the dosimetric and spatial accuracy through the entire gel dosimeter. In this gel fewer than 2% of points fail the gamma metric when using a 3%/3mm criterion.

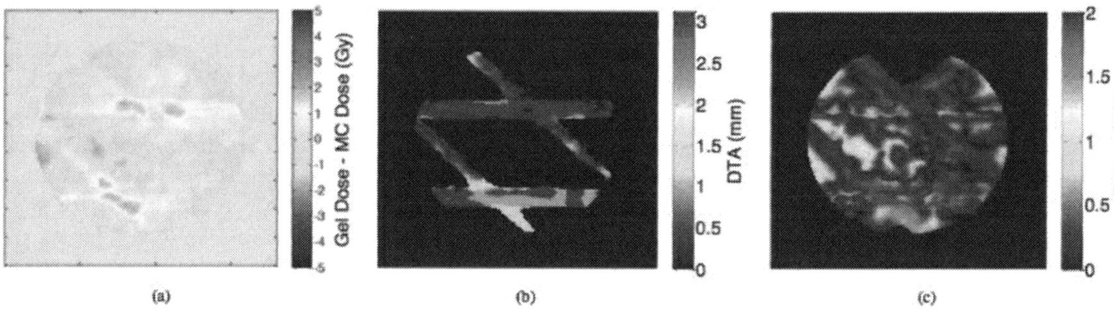

Figure 1. (a) Measured gel dose – Monte Carlo calculated dose for an individual self-calibrated gel dosimeter, (b) DTA map of the high dose gradient regions comparing gel dose to Monte Carlo dose and (c) gamma index map with a 3%/3mm criterion and 10% dose threshold.

Table 1 summarizes the results across all gel dosimeters used in the sample for each of the described calibration methods. The average calibration method produces a mean dose error similar to the "best case" self-calibration method which demonstrates the effectiveness of using an average calibration. Even when using the most divergent calibration the average dose error remains below 3%. The intra-gel calibration method produces an average dose error of approximately 5% due to differences in the dose response at the top and bottom of the gel container. A dose dependent correction was calculated using an average of this difference in response and with this correction, the accuracy of the intra-gel calibration method improved to within 2%.

Table 1. Summary of results in the reproducibility study. Dose error values are measured in the low dose gradient region and relative to the maximum dose. Distance-to-agreement (DTA) values are measured in the high dose gradient region. Gamma analysis is performed across the entire gel with a 10% dose threshold and a 3%/3mm dose criterion.

Calibration method	Mean dose error	Mean DTA (mm)	Mean pass rate DTA < 2 mm	Mean pass rate $\gamma < 1$ (3%/3mm)	Mean gamma value
Self-calibration	1.62% ± 0.05%	0.608 ± 0.03	99.8% ± 0.2%	98.2% ± 0.4%	0.320 ± 0.005
Average calibration	1.82% ± 0.25%	0.628 ± 0.07	99.6% ± 0.3%	96.8% ± 3.6%	0.348 ± 0.040
Most divergent calibration	2.57% ± 0.50%	0.718 ± 0.17	97.2% ± 2.6%	90.9% ± 7.0%	0.454 ± 0.074
Intra-gel calibration	4.83% ± 1.10%	1.340 ± 0.21	83.3% ± 5.9%	76.6% ± 11.7%	0.706 ± 0.145
Intra-gel calibration (corrected)	1.82% ± 0.26%	0.628 ± 0.07	99.6% ± 0.3%	96.8% ± 3.7%	0.348 ± 0.041

DTA analysis in the high dose gradient regions complements the dose difference values measured in the low gradient regions. Overall the spatial accuracy is excellent for the self, average, and corrected intra-gel calibration methods. The average calibration results again indicate good reproducibility within the sample set and even when using the most divergent calibration within the sample, more than 97% of points in the high gradient region have a DTA of less than 2 mm.

Average passing rates ($\gamma < 1$) at the 3%/3mm criterion across the sample gels are greater than 95% for the average calibration and intra-gel calibration when corrected. Again we see the effectiveness of using an average calibration curve when compared to self-calibration and the effectiveness of the dose dependent correction for intra-gel calibration.

4. Conclusion

Comparison of gel measurements with Monte Carlo dose calculations found excellent dosimetric accuracy when using an average (or generic) calibration and minimal accuracy was lost when compared to a "best case scenario" self-calibration method. An intra-gel calibration method was also investigated but required a dose dependent correction to avoid large dose discrepancies. Spatial accuracy was also found to be excellent for the average calibration method and little accuracy was lost when compared to the self-calibration method. Gamma analysis using a 3%/3mm criterion also found good agreement between the gel measurement and Monte Carlo dose calculation when using either the average calibration or self-calibration methods (96.8% and 98.2%, respectively).

Overall, this work has established the accuracy and reproducibility of an x-ray CT PGD dosimetry system and demonstrated the effectiveness of inter-batch calibration and the use of a generic calibration method.

5. Acknowledgements

The authors would like to acknowledge financial support from the Natural Sciences and Engineering Research Council of Canada. The authors would also like to acknowledge Dave Yardley and Steven Gray for help with phantom design and construction as well as Sergei Zavgorodni and Reid Townson for their help in using the VIMC system.

6. References

[1] Maryanski *et al* 1993 *Magn. Reson. Imaging* **11** 253-8
[2] Mather M L *et al* 2003 *Ultrasonics* **41** 551-9
[3] De Deene Y *et al* 2009 *J. Phys.: Conf. Ser.* **164** 012033
[4] Vandecasteele J *et al* 2013 *Phys. Med. Biol.* **58** 19-42
[5] Rankine L *et al* 2013 *Med. Phys.* **40** 051701
[6] Chang Y *et al* 2014 *Radiat. Phys. Chem.* **104** 180-7
[7] Hilts M *et al* 2000 *Phys. Med. Biol.* **45** 2559-71
[8] Kakakhel M *et al* 2011 *Med. Phys.* **38** 5130-5
[9] Johnston H *et al* 2012 *Phys. Med. Biol.* **57** 3155-75
[10] Chiu C *et al* 2014 *Appl. Radiat. Isot.* **90** 245-250
[11] Hill B *et al* 2005 *Med. Phys.* **32** 1589-97
[12] Baldock C *et al* 2010 *Phys. Med. Biol.* **55** R1-63
[13] Chain J *et al* 2011 *Phys. Med. Biol.* **56** 2091-102
[14] Jirasek A *et al* 2010 *Phys. Med. Biol.* **55** 5269-81
[15] Hilts M *et al* 2008 *Med. Phys.* **35** 344-55
[16] Jirasek A *et al* 2012 *Phys. Med. Biol.* **57** 3137-53
[17] Jirasek A *et al* 2014 *Phys. Med. Biol.* **59** 597-614
[18] Hilts M *et al* 2011 *Med. Phys.* **38** 3507
[19] Bush K *et al* 2008 *Phys. Med. Biol.* **53** N359-70
[20] Baxter P *et al* 2007 *Med. Phys.* **34** 1934-43

IOP Conf. Series: Journal of Physics: Conf. Series **847** (2017) 012048 doi:10.1088/1742-6596/847/1/012048

Tetrazolium salt monomers for gel dosimetry I: Principles

Kalin I Penev[1,2], Meng Wang[1] and Kibret Mequanint[1,3]
[1]Department of Chemical and Biochemical Engineering, University of Western Ontario, London, ON, N5A 5B9, Canada
[2]Modus Medical Devices Inc., 1570 N Routledge Park, London, ON, N6H 5L6, Canada
[3]Biomedical Engineering Program, The University of Western Ontario, London, Ontario, N6A 5B9, Canada

E-mail: kmequani@uwo.ca

Abstract. Tetrazolium salts (TS) have been previously used for radiochromic dosimetry in solutions, films and three dimensional (3D) gelatine-based gels. However, widespread application for 3D dosimetry has not been achieved due to the required high concentrations and associated high costs of the TS dimer used in prior research. Through careful selection of TS monomer, sensitivity-enhancing additives and inert gel forming material, we report the preparation of a non-diffusing, chemically stable, 3D dosimeter with linear sensitivity between 0 and 80 Gy with submillimolar requirements for the active TS.

1. Introduction

The use of tetrazolium salts (TS) in liquid and film dosimetry of ionizing radiation is well-established [1, 2]; whereas, three-dimensional (3D) dosimetry has been reported in a single PhD thesis [3] and the resulting patent [4]. The underlying chemistry in all TS dosimeters is the radiochromic reduction of a colourless TS into an intensely coloured formazan dye. To increase the radiation chemical yield of the reaction, addition of a hydroxyl radical quencher is necessary. Previously, various radical quenchers such as alcohols, sodium formate and Triton X-100 have been used [1, 4]; where Triton X-100 may also affect the sensitivity by micelle formation [4, 5]. TS film and gel dosimeters showed: (i) no apparent signal diffusion as the formazan dye is insoluble in water and remains trapped within the film or gel matrix, (ii) good chemical stability, and (iii) limited or no inhibitory effect of the dissolved oxygen [1-4]. However, widespread use TS for 3D dosimetry has not been established, due to the perceived low sensitivity and high relative cost of TS [6]. Particularly, the cost of the compound of choice in recent reports [2-4] – a tetrazolium dimer, known as nitrotetrazolium blue chloride (NBT) – is prohibitive at the required concentrations of up to 4 g/L (5 mM). By contrast, the original research on TS dosimeters in solution [1] used less expensive tetrazolium monomers. Here we report the first use of TS monomers for gel dosimetry. Figure 1 presents the chosen compounds: nitrotetrazolium chloride (NTC) (**1**) and bisnitrotetrazolium chloride (BNC) (**2**). NTC has been used in solution and film dosimetry [1]; it is not commercially available, but can be prepared relatively easily [7]. BNC has not been tested previously, and is more sensitive to reducing agents than NTC [8]. As gel-forming material, both gelatin and gellan gum were tested. Gellan gum is a microbially produced polysaccharide which forms clear hydrogels with high mechanical strength and thermal stability [9]. As hydroxyl radical quenchers, we used sodium formate and propylene glycol. Monovalent alcohols were excluded due to their high volatility and Triton X-100 may be investigated in future research. The effects of the concentrations (C) of the chosen TS and additives on the dose sensitivity (S) were fitted to a first-order saturation equation (1), under the assumption that the S reaches a maximum:

$$S = S_0 + S_{max} \frac{C}{K_{1/2} + C} \tag{1}$$

Where S_0 is the sensitivity of the in the absence of additives ($S_0 = 0$ at $C_{TS} = 0$), S_{max} is the maximum increase of sensitivity and $K_{1/2}$ is the half-saturation constant, at which $(S - S_0) = 0.5S_{max}$.

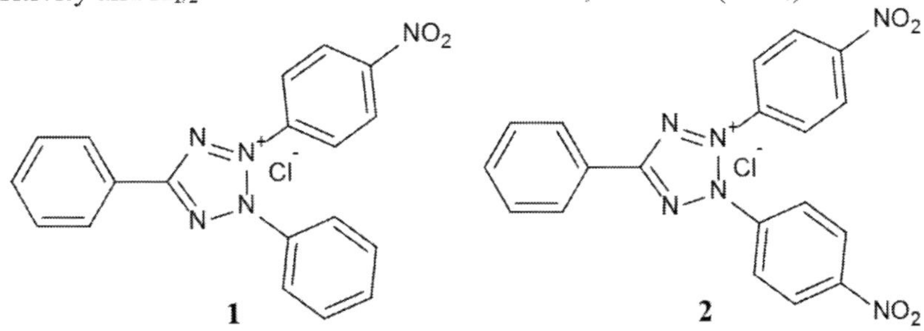

Figure 1. Evaluated tetrazolium salts: **1** = 2-(4-nitrophenyl)-3,5-diphenyl-2*H*-tetrazolium chloride = nitrotetrazolium chloride (NTC), **2** = 2,3-bis(4-nitrophenyl)-5-phenyl-2*H*-tetrazolium chloride = bisnitrotetrazolium chloride (BNC).

Here (part I), we present the initial characterization of NTC- and BNC-gel dosimeters in cuvettes, using gelatine and gellan gum, while part II of the paper discusses the characterization of a specific BNC-gellan gum composition, studied in clinically relevant volumes.

2. Materials and Methods

Gels were prepared using the following materials: NTC (synthesized in-house [7]), BNC (TCI America, USA, cat. no: B1047), propylene glycol (PG) (Caledon Laboratories, Canada, cat. no: 6840), sodium formate (SF) (Caledon, cat. no: 7840), gelatin (Gl) (Sigma-Aldrich, cat. no: G1890), and gellan gum (Gn) (Alfa-Aesar, cat. no: J63423, lot no: C01Z014, gel strength: 434 g/cm^2) to final concentrations of 0.05 – 2.50 mM NTC or BNC, 0 – 20 % (v/v) PG, 0 – 200 mM SF, and 5% (w/v) Gl or 1.25% (w/v) Gn. Samples were poured into 4 mL poly(methyl methacrylate) cuvettes, closed with polyethylene caps, and placed overnight in a refrigerator. Cuvettes were irradiated the following day on a Co-60 source (at 100 cGy/min) at the London Regional Cancer Program in London, Canada to 20 Gy, at 21 – 23 °C, in triplicates for the initial testing, and between 0 and 80 Gy in subsequent tests. Absorbance measurements were performed at 22 °C on a Cary 60 spectrophotometer (Agilent). To create dose-gradients, cuvettes were semi-shielded by a lead block, irradiated to 20 Gy, and scanned on a microplate reader (Infinite 200 Pro, Tecan). The dose gradient data was fitted to an analytical approximation of the complementary error function, erfc(x) [10]. All optical measurements are reported as the linear attenuation coefficient, μ (cm^{-1}). Individual compositions are designated as **#$_1$TS/#$_2$XX/#$_3$Gx** meaning: #$_1$ mM of the respective TS, #$_2$ mM or % (v/v) of the additive, and #$_3$ % (w/v) of the gel-forming agent.

3. Results and Discussion

Figures 2 and 3 show the effects on concentration of the TS and two additives on the dose sensitivity of NTC and BNC dosimeters in gelatin and gellan gum, respectively, at the wavelengths of maximum attenuation. BNC-containing gels showed higher sensitivity to dose, even at lower concentrations of the tetrazolium salt, as $K_{1/2}$ for BNC is significantly lower than $K_{1/2}$ for NTC (Figure 2A). Addition of radical quencher led to a two- to three-fold increase of sensitivity (Figure 2B). Sodium formate showed a stronger positive effect on the sensitivity than propylene glycol. However, sodium formate could not be tested in gellan gum at high concentrations due to strong cross-linking by the sodium ion (Na^+) resulting in formation of translucent gels.

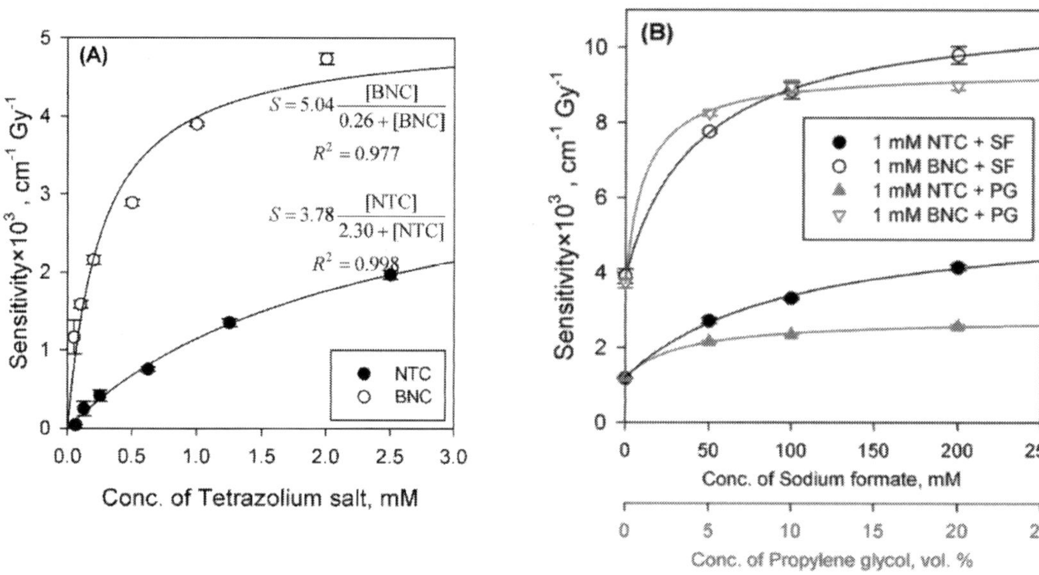

Figure 2. Dose sensitivity of NTC (494 nm) and BNC (535 nm) in 5% (w/v) of gelatine as a function of: **(A)** tetrazolium salt concentration ([TS]), and **(B)** radical scavenging additiveconcentrations at a constant [TS]. Data is fitted to Equation 1.

As shown in Figure 3, in the absence of additives the dose sensitivity in gellan gum was linear in relation to the TS concentration, indicating very high values for $K_{1/2}$. However, multi-fold increase of sensitivity was seen in the presence of 10% (v/v) propylene glycol. Additionally, the background attenuation of all NTC-gelatin gels was stable for at least 24 h (data not shown). BNC-gelatin gels, however, became darker at rates as high as 6.0×10^{-3} $cm^{-1}h^{-1}$ for 1.0BNC/20PG/5Gl. By contrast, the BNC-gellan gum gels showed excellent stability (Figure 4).

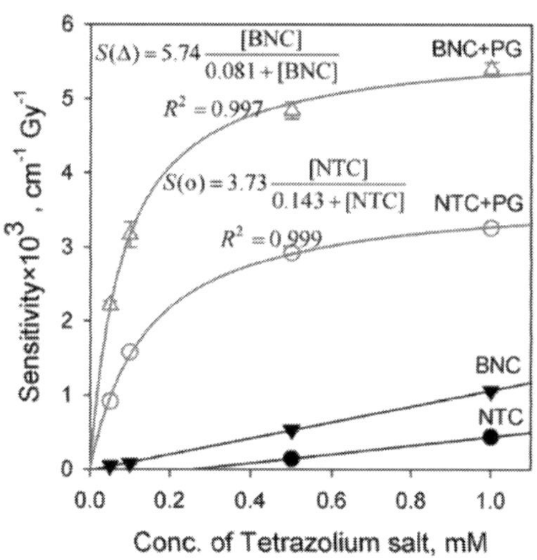

Figure 3. Dose sensitivity of NTC (at 494 nm) and BNC (at 535 nm) in 1.25% (w/v) gellan gum gel in: water (solid symbols) or 10% (v/v) PG (open symbols). Data is fitted to Equation 1.

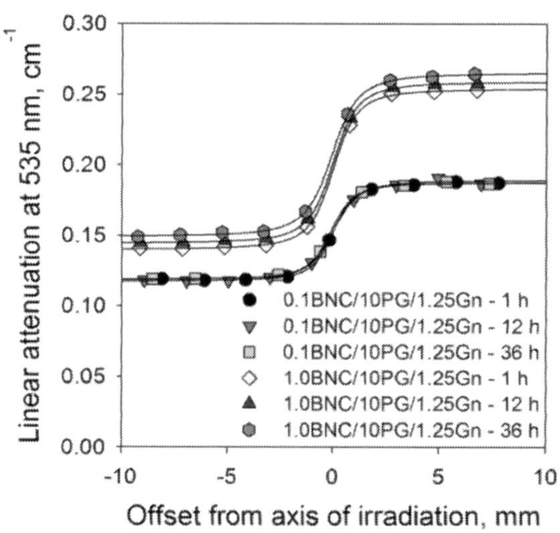

Figure 4. Profiles of dose signal in semi-shielded cuvettes of BNC-gellan gels, stored at room temperature for 36 h after irradiation to 20 Gy. Data is fitted to erfc(x).

BNC-gellan gum compositions were chosen for the remainder of the study. As seen in Figure 4, the 0.1BNC/10PG/1.25Gn composition showed stable background attenuation and dose response with no diffusion observed for at least 36 hours at room temperature. However, at 1.0 mM BNC some increase of attenuation was observed over the course of the 36 hours, but the diffusion remained undetectable. Excluding the spontaneous darkening at 1.0 mM BNC, the signal development for both compositions took less than 30 minutes for doses up to 80 Gy (data not shown), and the observed colour changes were analogous – from colourless to purple. Figure 5 gives the attenuation spectra at 0.1 mM BNC, and the dose response for both compositions between 0 and 80 Gy. Maximum dose response was observed at 535 nm. The achieved dose sensitivity (~4×10^{-3} $cm^{-1}Gy^{-1}$) is on par with previously reported minimally-diffusing gel dosimeters [5]; however, our gels exhibit no apparent diffusion and have very high chemical stability.

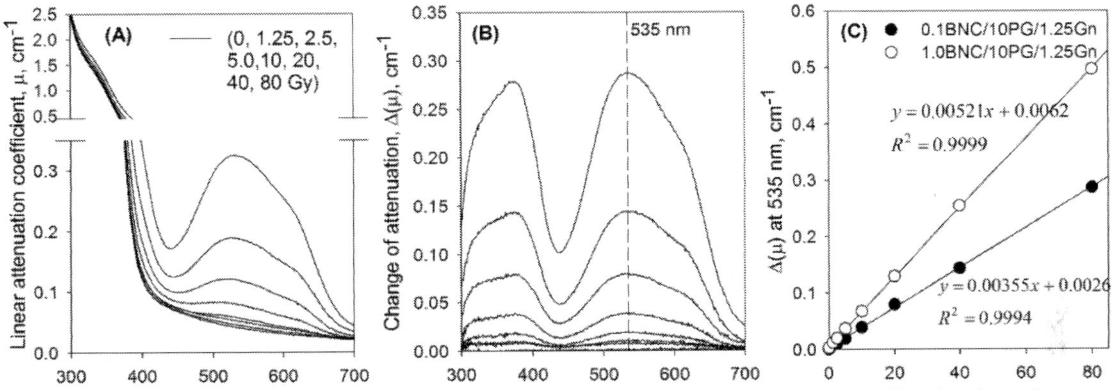

Figure 5. A, B: attenuation spectra and change of attenuation at doses from 0 to 80 Gy in 0.1BNC/10PG/1.25Gn; **C**: linearity of the dose response for two compositions at λ_{max} = 535 nm.

4. Conclusions

We have successfully demonstrated the use of tetrazolium salts for preparation of 3D, non-diffusing, radiochormic gel dosimeters, with linear dose response in the range of 0 to 80 Gy, sensitivity on par with other minimally diffusing gel dosimeters, and greatly enhanced chemical stability. In part II, we discuss the use of one such composition in bulk-volume samples and clinically-relevant settings. Our future research includes the synthesis and testing of novel tetrazolium salts for enhanced sensitivity of the non-diffusing gel dosimeters in either gelatin or gellan gum.

5. Acknowledgements

The authors would like to acknowledge the financial support of the Mitacs Elevate fellowship, the Ontario Centres of Excellence and Modus QA.

6. References

[1] Pikaev A K and Kriminskaya Z K 1998 *Rad. Phy. Chem.* **52** 555
[2] Basfar A A, Rabaeh K A, Moussa A A and Msalam R I 2011 *Rad. Phy. Chem.* **80** 763
[3] Marsden C M 2000 PhD Thesis, Rutgers University, Ann Arbor, NJ, USA
[4] Appleby A 2003 *U.S. Patent* 6621086
[5] Babic S, Battista J J and Jordan K J 2008 *J. Phys.: Conf. Ser.* **146** 012044
[6] Adamovics J A 2004 *U.S. Patent* 20040211917
[7] Penev K I and Mequanint K 2016 *J. Het. Chem.* **53** 1655
[8] Seidler E 1980 *Histochem. J.* **12** 619
[9] Shungu D, Valiant M *et al.* 1983 *Appl. Envir. Microbio.* **46** 840
[10] Kron T, Jonas D and Pope J M 1997 *Magn. Reson. Imaging* **15** 211

IOP Conf. Series: Journal of Physics: Conf. Series **847** (2017) 012049 doi:10.1088/1742-6596/847/1/012049

Tetrazolium salt monomers for gel dosimetry II: Dosimetric characterization of the ClearView™ 3D dosimeter

Yudy Ascención[1], Jennifer Dietrich[1], Kibret Mequanint[2,3] and Kalin I Penev[1,2]
[1]Modus Medical Devices Inc., 1570 N Routledge Park, London, ON, N6H 5L6, Canada
[2]Department of Chemical and Biochemical Engineering, The University of Western Ontario, London, ON, N5A 5B9, Canada
[3]Biomedical Engineering Program, The University of Western Ontario, London, Ontario, N6A 5B9, Canada

E-mail: kpenev2@uwo.ca, kpenev@modusqa.com

Abstract. ClearView is a colourless radiochromic 3D dosimeter that is chemically and mechanically stable before and after irradiation. Upon irradiation, colour is generated from the radiochemical reduction of a tetrazolium salt suspended within a gellan gum gel. Here we present the dosimetric characterization of small and large volume samples (4 mL and 1 L, respectively) from three different batches of ClearView. The dose sensitivity was linear up to 80 Gy and constant within a batch but varied among batches between 3.4×10^{-3} and 4.1×10^{-3} $Gy^{-1}cm^{-1}$. Once generated, the radiation signal within the large samples remained unchanged for at least 58 days. Both electrons and photons produced the same signal at different energy levels; however, at dose rates below 400 cGy/min, a drop in sensitivity was observed. Overall, ClearView is best suited for relative dosimetry up to 80 Gy at high dose rates.

1. Introduction

Modern radiotherapy relies on advanced techniques (beam modulation, arc delivery, *etc.*) to deliver high radiation doses to irregularly shaped, small volume targets [1]. At present, commissioning of equipment and quality assurance of treatments are most often performed by a combination of one- or two-dimensional detectors, *e.g.* ion chambers, diode arrays and films [2]. However, these methods can provide only sparse three-dimensional (3D) data, which complicates the quality assurance process and adds uncertainty to its results [3]. To overcome this issue, the use of "true 3D" dosimetry has been proposed and multiple 3D dosimeters have been developed. These include gel, plastic and elastic (*e.g.* silicone-based) dosimeters with signal detection *via* magnetic resonance imaging or optical computed tomography [4]. Notwithstanding the wide variety of materials, true 3D dosimetry still has a limited clinical application, because of multiple drawbacks: poor chemical and/or image stability due to spontaneous colour change and/or signal diffusion within the dosimeter, complex and complicated preparation and use, imaging and data analysis.

The radiochromic gel discussed here was designed to provide chemically stable 3D dosimetry with no apparent signal diffusion. This allows dosimeter manufacturing, pre-irradiation and post-irradiation imaging to be handled at an off-site facility, as a commercial product. The dosimeter is based on the radiochromic reduction of a colourless, water-soluble tetrazolium salt into an insoluble formazan dye suspended in gellan gum gel, as discussed in part I of this paper. The composition has been licensed to Modus QA (Modus Medical Devices Inc.) and developed as a product under the name ClearView.

Here, the dosimetric characterization of ClearView gel in small (4mL) and large (1 L) volumes is presented.

2. Materials and Methods

2.1. Gel manufacturing

A 20-litre jacketed process reactor, described previously [5], was used for the manufacturing of 12 to 18-litre batches of ClearView with the following composition per litre: 12.50 g gellan gum (Alfa Aesar, USA, cat. no.: J63423), 103.60 g propylene glycol (AMRESCO, USA, cat. no.: 0575), 42.5 mg (0.10 mmol) 2,3-bis(4-nitrophenyl)-5-phenyl-2*H*-tetrazolium chloride (BNC) (Expert Synthesis Solutions, London, Canada) and deionized water. The gellan gum was dissolved at temperatures over 85 °C but the active ingredients were added at 70 °C. The hot solution was filtered through a 1.0 or 0.5 μm filter and poured into one-litre clear polyethylene terephthalate jars. Controlled cooling allowed gel formation before storage at 4-6 °C. Simultaneously, poly(methyl methacrylate) (PMMA) cuvettes (1.0 cm square cross-section, 4 mL volume) were filled. The density of the gels at room temperature was determined with a 250-mL volumetric flask using three samples from two batches.

2.2. General set-up and irradiations

Optical measurements for the cuvettes were performed on a GENESYS 10S spectrophotometer (Thermo Fisher Scientific, Canada) at 535 nm, while the jars were imaged on a Vista 15 optical cone beam CT scanner with a green LED source (520-535 nm) (Modus QA, London, Canada). Irradiations took place within 2 to 5 days of preparation of the gels, which were pre-scanned on the day of irradiation, and post-scanned the following day with follow-up of up to 58 days. Irradiations were performed on either of two linear accelerators from Varian Medical Systems (USA): Clinac 21iX and TrueBeam at the London Regional Cancer Program (London, Canada). Cuvettes were placed three to six at a time on a PMMA block and submerged into water to a depth of 5 cm to the centre of the cuvette (for photons) or d_{max} (for electrons) and irradiated using 10x10 cm square beams. The 1-litre jars were irradiated using 6x6 cm square beams of photons or electrons, unless otherwise noted. The water surface over the cuvettes and the top surface of the gels were positioned at 100 cm source to surface distance. Percentage depth dose curves (PDD) were obtained as described elsewhere [6], and referenced to CC13 (IBA Dosimetry America, USA) ion chamber measurements.

To determine the linearity and stability of the dose response, cuvettes were irradiated up to 80 Gy at 600 cGy/min using 6 MV photons and 12 MeV electron beams. For the sensitivity in large samples, jars from three different batches were irradiated with 12MeV electrons to 40 Gy (at d_{max}) at 600 cGy/min [7]. The energy sensitivity to dose response was tested at 20 Gy, using 6 to 18 MV photons and 6 to 20 MeV electrons at 600 cGy/min in cuvettes from two different batches of gel. To test the sensitivity of the dose response to a wider range of energies and dose rates, cuvettes and jars of different batches were irradiated to 20 Gy at a range of electron and photon energies and dose rates, including flattening filter free (FFF) deliveries. Finally, a dose fractionation experiment was performed using arc treatments to 40 Gy at the isocenter using 6 MV photons at 600 cGy/min in two different settings: (i) in a single fraction, (ii) in five fractions of 8 Gy delivered sequentially.

3. Results and discussion

3.1. Physical characteristics

ClearView appears colourless or pink-tinted before irradiation and becomes purple upon irradiation. The gel is strong and inelastic at room temperature with a volumetric mass density of 1009±1 g/L at 20 °C. The atomic composition (in mole percent) is 65.94% H, 31.35% O and 2.71% C.

3.2. Dosimetric characteristics

According to Figure 1A both photon and electron beams gave identical response in the cuvettes, while dose calibration in a jar showed lower but linear sensitivity. The decreased sensitivity in jars is likely due to the wider bandwidth of the LED source relative to the spectrophotometer used for the cuvette samples. The attenuation profile within the jar remained constant for at least 58 days, showing neither diffusion nor changing magnitude (Figure 1B), while both non-irradiated and irradiated cuvette samples remained stable for at least 21 days with noticeable drying of the gel thereafter (Figure 1C).

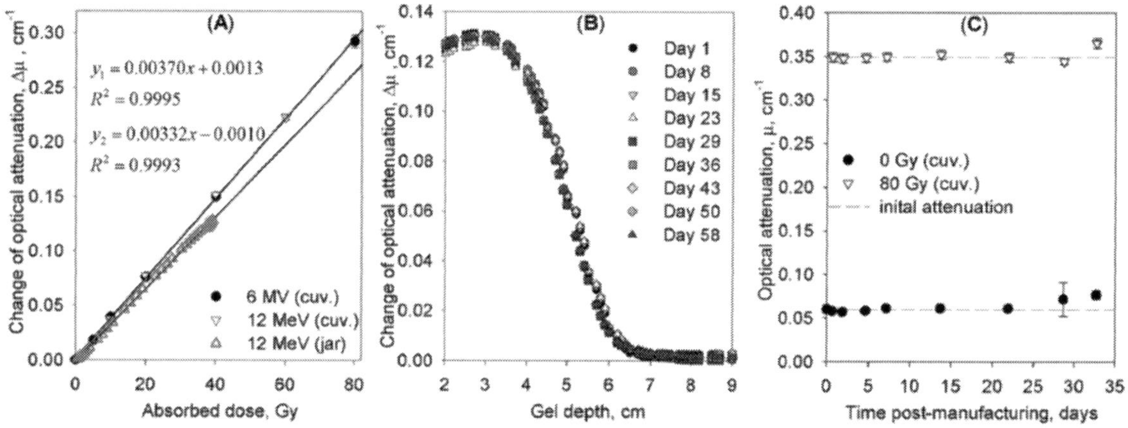

Figure 1. Calibration curves for cuvettes and a bulk sample at 600 cGy/min (**A**), stability of the background and dose response in a one-litre jar (**B**) and cuvette samples (**C**). Error bars show standard deviations for three cuvette samples.

Figure 2 shows the sensitivity and the PDD curves in jars from three different batches. Whereas the sensitivity remains linear in all cases, it varies between 3.4×10^{-3} and 4.1×10^{-3} $Gy^{-1}cm^{-1}$. The sensitivity variability can be attributed to two main factors: differences in the properties of the raw materials and changing process conditions. Here, the gellan gum used in the fabrication of the first batch (lot no. R13B050) had a different gel strength than the gellan gum used in batches 2 and 3 (lot no. Z22B034): 497 g/cm^2 and 606 g/cm^2, respectively, even though the supplier and catalog numbers were identical (Alfa Aesar, J63423). Further, the filtration process was updated between batches 2 and 3: from 1.0 to 0.5 μm filtration. Regardless of the different sensitivities, all tested batches showed consistent PDD curves, in agreement with the ion chamber measurements (Fig. 2B).

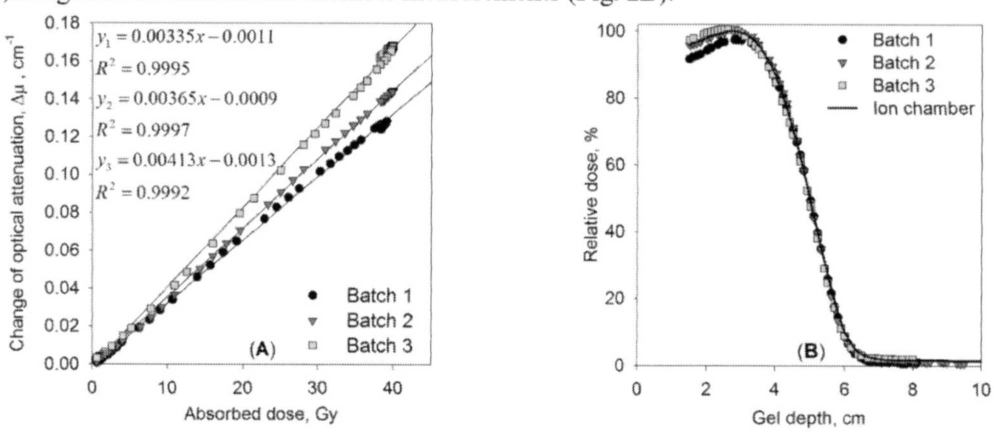

Figure 2. Batch-to-batch variability of sensitivity of three gel batches, irradiated to 40 Gy at d_{max} at 600 cGy/min by a 12 MeV electron beam: calibration curves (**A**) and PDD curves (**B**).

Figure 3 presents the dose response as a function of the beam energy and dose rate. No significant change of the sensitivity at different energies was seen in either electron or photon irradiations (Fig. 3A). Based on the jar data, which is less noisy and more clinically relevant than the cuvette data, the sensitivity is constant within samples of the same batch at dose rate above 400 cGy/min. Additionally, the fractionation experiment showed identical responses in jars irradiated in a single fraction of 40 Gy (6 MV photons, 600 cGy/min) and in five consecutive fractions of 8 Gy (data not shown). However, at dose rates below 400 cGy/min the sensitivity dropped (Figure 3B).

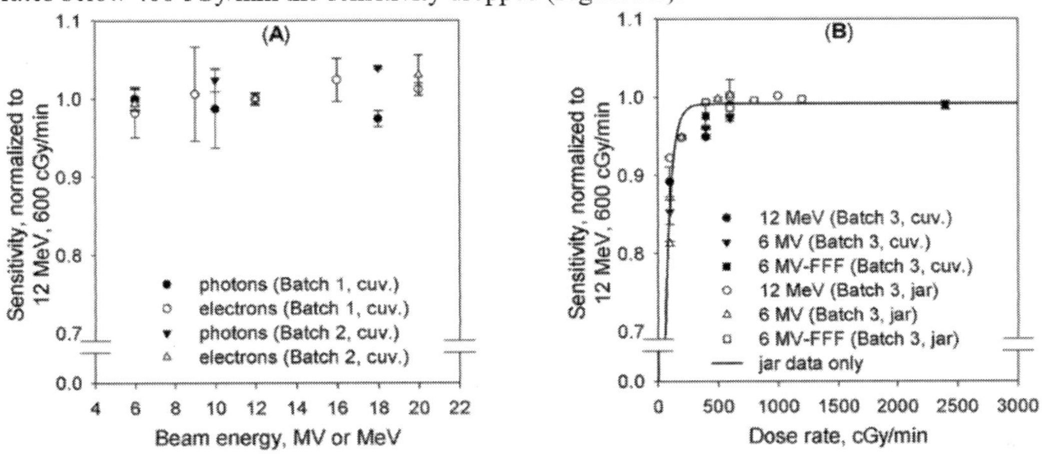

Figure 3. Energy (**A**) and dose rate (**B**) effect on the dose response in cuvettes and jars from different batches. For direct comparison the data is normalized to the response of a 12 MeV electron beam irradiation at 600 cGy/mim. Error bars for cuvette data show standard deviations for three samples. Error bars for jar data show the error of the slope of the calibration curves.

We hypothesise that the decreased sensitivity below 400 cGy/min is associated with the low concentration of the active compound (BNC) in ClearView. Future research will focus on finding the mechanism of the dose rate effect with the goal of minimizing or eliminating it.

4. Conclusions

The ClearView radiochromic gel showed excellent stability for both irradiated and non-irradiated samples, without apparent optical changes or diffusion within a period of at least two months from manufacturing in large-volume samples. Batch-to-batch variability was observed in regards to the sensitivity to the delivered dose (3.35×10^{-3} and 4.1×10^{-3} $Gy^{-1}cm^{-1}$); however, the dose response was linear in all tested batches, identical for photons and electrons and unaffected by the energy in the range of 6 to 18 MV, and 6 to 20 MeV, respectively. However, there was a noticeable drop of the sensitivity at dose rates below 400 cGy/min. Thus, ClearView is best suited for relative dosimetry up to 80 Gy at high dose rates. Future research and product development will focus on minimizing the dose-rate effect and increasing the sensitivity of the ClearView gel.

5. Acknowledgements

The authors would like to acknowledge the technical assistance of Matt Mulligan and Linada Kaci from the London Regional Cancer Program (London, Canada).

6. References

[1] Baldock C *et al* 2010 *Phys. Med. Biol.* **55** R1-63
[2] Hill R *et al* 2009 *Med. Phys.* **36** 3971-81
[3] Schreiner L J *et al* 2013 *J. Phys.: Conf. Ser.* **444** 012016
[4] Schreiner L J 2015 *J. Phys.: Conf. Ser.* **573** 012003
[5] Penev K I and Mequanint K 2015 *J. Phys.: Conf. Ser.* **573** 012078
[6] Alexander K M *et al* 2015 *IFMBE Proceedings* **51** 521
[7] Jordan K 2013 *J. Phys.: Conf. Ser.* **444** 012009

IOP Conf. Series: Journal of Physics: Conf. Series **847** (2017) 012050 doi:10.1088/1742-6596/847/1/012050

Active heating of large gel dosimeters

K J Jordan and K H Dekker
London Regional Cancer Program (London Health Sciences Centre)
Department of Medical Biophysics, Western University
London, ON, Canada

E-mail: kevin.jordan@lhsc.on.ca

Abstract. Temperature uniformity of 3D dosimeters is key factor for accurate measurements. For gel dosimeters larger than 1 litre the time to change from storage to room temperature can be several hours. Local thermal histories within certain dosimeter materials can also cause measurable differences in ionizing radiation response. Passive heating by immersion in a water bath may not be optimum. Active uniform heating of 15 cm diameter, hydrogel, dosimeters using a near-infrared, incandescent light source reduced the time required to heat a hydrogel from 4 to 20°C uniformly from greater than five hours to less than 3 hours.

1. Introduction

Temperature is an important variable in chemical radiation dosimeters. Thermal gradients inside 3D dosimeters can cause artifacts in measured dose distributions [1]. Temperature dependent dose response has been observed in virtually all gel dosimeters. Babic *et al* [2] found a 4% per degree increase in dose response during irradiation of leuco crystal violet micelle gel at 20°C. Olding *et al* [3] reported a 3% per degree increase in attenuation coefficient for ferrous-xylenol orange gel (FX) during readout at 590 nm. Certain versions of the radiochromic plastic dosimeter PRESAGE® have the lowest temperature sensitivity: 2.5% per degree at 20°C [4]. Exothermic polymerization gel dosimeters are likely even more sensitive to thermal gradients during and after irradiation. Also, the local thermal history of gel dosimeters may have a spatially dependent effect on dose sensitivity. Babic *et al* [5] demonstrated how thermal history and auto-oxidation can lead to apparent variations in low dose response (<0.5 Gy) for FX gels. These effects suggest cooling and heating gels up uniformly is prudent. It is common practice to let gels acclimate to room temperature passively, in order to achieve a uniform temperature distribution prior to irradiation experiments. Experiments to validate dose response uniformity of 3D dosimeters are crucial for validating a specific measurement protocol. If thermal gradients are present during irradiation but not accounted for, deviations from the actual dose distribution will occur. As gel dosimeters become larger, the warm-up time increases, lengthening measurement sessions and reducing practicality. In this study we present, proof of concept results from an active heating system to accelerate uniform gel warming. Incident near-infrared light is absorbed by the water within the hydrogel as the gel is rotated in a water bath.

2. Methods

2.1 Gel Fabrication

A 15cm diameter, 2.2 litre, cylindrical PETE vessel was filled with a 4% gelatin (300 Bloom, porcine, Type A, Sigma) hydrogel (without radiochromic dye) and stored in a refrigerator at ~4°C. The gel was cast with transparent plastic drinking straws (outer/inner diameter = 5.0/4.7 mm) held vertically, with the bottoms sealed, placed at the rotation axis and at 5cm radius, to form channels for insertion of a thermistor probe. The straws were filled with water to the height of the gel surface.

2.2 Passive Heating

The gel was removed from the refrigerator and placed in a water bath initially at a temperature of 20 °C. The water bath temperature and the gel temperatures at the 0 and 5cm radii were measured over approximately 5 hours. The water bath was continuously stirred.

2.3 Active Heating

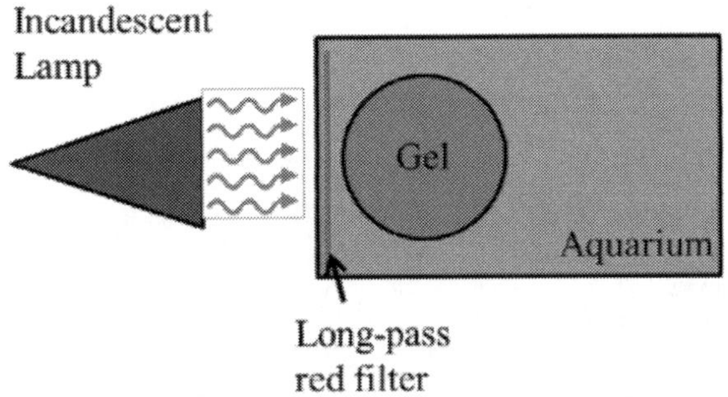

Figure 1. Top view schematic of active heating apparatus. Gel rotated manually.

Figure 1 shows the schematic for the active heating device. The gel cylinder was placed 1 cm inside the entrance window of a glass aquarium which was illuminated by a Sylvania 250 W infrared heat lamp from one side. The water bath provides additional sample surface heating and filtration of longer wavelengths by absorption. A piece of red transparent acetate film was placed in the water on the lamp side to absorb shorter visible wavelengths for warming of gels that will be already coloured. The aquarium was initially at a temperature of 21.4°C. The gel was manually rotated by 90° approximately every 15 minutes, and the water bath temperature was lowered back to 21.5°C roughly once per hour. The temperature at the center was measured over a 3 hour timespan. The temperature at 5 cm radius was measured periodically after approximately 1 hour 50 minutes of warming.

3. Results and Discussion

Figure 2 shows the water bath, 5cm radius, and central axis temperatures at mid height of gel as a function of time for the passive heating experiment. It required approximately 5 hours to reach equilibrium within ~1°C.

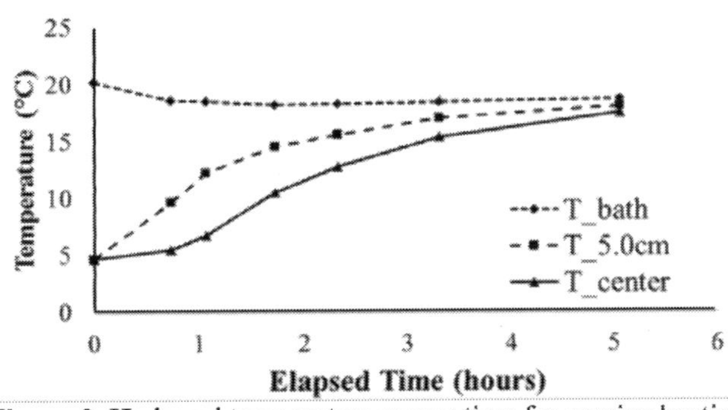

Figure 2. Hydrogel temperature versus time for passive heating.

Figure 3 shows the water bath, 5cm radius, and central gel temperatures as a function of time for the active heating experiment. The gel was brought up to equilibrium temperature within ~1°C after approximately 3 hours, demonstrating the improved speed achieved using active heating.

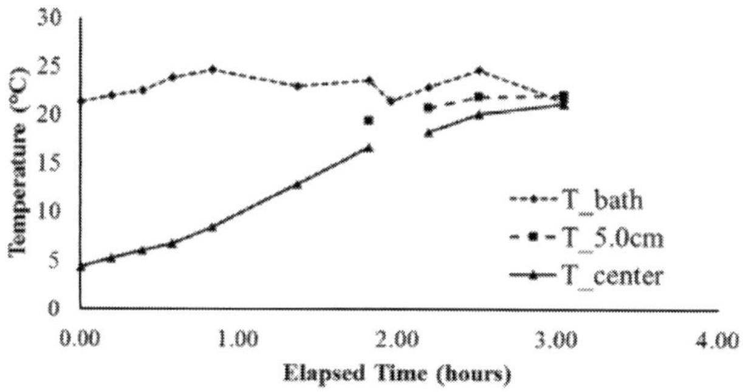

Figure 3. Hydrogel temperature versus time for active heating. Note that water in the aquarium was replaced periodically in order to maintain a bath temperature below approximately 25°C.

4. Conclusion

Active heating of hydrogels by NIR light absorption reduced the time to warm a 15 cm diameter gel dosimeter from 4°C to 20°C from more than five hours to three hours. Future work will involve increasing the number of illumination beams, increasing beam power, optimizing beam shapes and automating sample rotation to uniformly warm the gels to room temperature in one hour. Thermally uniform gel dosimeters are expected to provide more reproducible data and better agreement with independent measurements and calculations.

5. Acknowledgments

KHD acknowledges scholarship and funding support from NSERC, Western University, and the Ontario Research Fund (OCAIRO project). Disclosure: The authors have a research agreement with Modus Medical Devices Inc. concerning optical CT applications.

6. References

[1] Baldock C *et al* 2010 *Phys. Med. Biol.* **55** R1-63
[2] Babic S *et al* 2009 *Phys. Med. Biol.* **54** 6791-808
[3] Olding T and J Schreiner 2011 *Phys. Med. Biol.* **56** 1259-79
[4] Skyt P *et al* 2011 *Med. Phys.* **38** 2806-11
[5] Babic S *et al* 2008 *Phys. Med. Biol.* **53** 1637-50

Dosimetry using MRI: can it really be that difficult?

Simon J Doran[1]
[1]CRUK Cancer Imaging Centre, Institute of Cancer Research, London, UK

E-mail: Simon.Doran@icr.ac.uk

Abstract. Magnetic resonance imaging (MRI) has been used in gel dosimetry from the very first studies back in the 1980s. Almost all the imaging problems that we still encounter were known about at that time and many were described in the proceedings of the very first International Workshop on Radiation Therapy Gel Dosimetry (DOSGEL '99). The quality of MRI scanners has improved enormously in the intervening two decades, so why are there still issues? This review will draw on the wealth of previously presented material from the literature to attempt to answer this question. The reference list provides a "starter pack" that should be viewed as a "jumping off point" for further investigations by the reader, rather than an exhaustive summary of what is a large domain of work.

1. Introduction

Whilst there were a number of earlier publications that described the modification of nuclear magnetic resonance (NMR) relaxation times by absorbed radiation dose, most commentators would likely agree that the seminal work in the field of gel dosimetry is the 1984 paper by Gore and Kang [1]. It lays out a methodology for measuring a 3-D dose distribution in a tissue-equivalent phantom, by using magnetic resonance imaging (MRI). Although taken for granted nowadays, one of the novel features about this work was that it proposed to use imaging to make *quantitative* measurements of a physical property. This was one of a number of early publications that sought to move MRI on from a modality used purely for visualising the presence of pathology, to one with which accurate science could be performed.

The measurement of NMR relaxation times themselves, however, predated Gore and Kang by more than three decades. By 1948, just a few years after the discovery of the NMR phenomenon [2, 3], Bloembergen, Purcell and Pound had already produced an extensive description of NMR relaxation and had designed apparatus to measure it [4]. By 1950, ideas about radiofrequency (RF) pulses were becoming familiar [5], and in 1954, Carr and Purcell introduced their eponymous pulse sequence for measuring T_2 [6]. Arguably, by the late 1950's, most of NMR had already been invented and everything that followed was simply the out-working of these formulae. Imaging introduced an entirely new concept in the 1970's [7-9], but the factors contributing to measurement errors in spatially-resolved relaxation time mapping were easily incorporated into the framework already developed. Certainly, by the time of the 1st International Workshop on Radiation Therapy Gel Dosimetry (DOSGEL '99), the majority of the effects described later in this article had been thoroughly analysed in a variety of other contexts and are neatly summarised in Table 1 of De Deene's review article [10] from the workshop.

Much has changed since 1999: MRI scanners have improved and continue to do so; image reconstruction techniques have become more sophisticated; and successive waves of disruptive technology (parallel imaging [11, 12] simultaneous multislice imaging [13], compressed sensing [14] and MRI fingerprinting [15]) have changed the discipline to the extent that many of the pulse sequences of today would be almost unrecognisable to the early pioneers. So, with rapid, high-quality imaging at

IOP Conf. Series: Journal of Physics: Conf. Series **847** (2017) 012051 doi:10.1088/1742-6596/847/1/012051

our fingertips, why is MRI-based 3-D dosimetry not already a solved problem? In the remainder of this article, I discuss the various sources of error and try to highlight, with particular reference to previous articles in the DOSGEL and IC3DDose conference series, the reasons why 20 years of effort by the gel dosimetry community have yet to produce a turn-key solution.

2. A question of philosophy

Let me first pose a philosophical question: what we *do with* MRI-derived dose maps when we have generated them?

A fundamental problem of calibration, as has been pointed out by De Deene [16], is that we have no access to a gold standard dose distribution against which to validate our gel dosimetry. All we can do is to compare the output of one technique against another, without any guarantee that either is "the truth". For simple irradiations, we expect different techniques (e.g., gel, film, diode array, treatment planning system (TPS) and Monte Carlo) to agree closely. And so they do, as it turns out, even for some relatively complex situations [17]. However, our problem comes precisely with the extrapolation to the cases in which we are most interested, that is those where we are unsure which method to trust. It should be obvious that it is logically incoherent to argue: "My gel dosimetry system is proven to work because it matches the TPS for Irradiation A. I am therefore now going to use my gel to investigate whether the TPS produces the correct result for Irradiation B." Which modality is validating which?

Two aspects of this dilemma been addressed by Schreiner in recent IC3DDose articles: firstly, what are the *scenarios* in which gel dosimetry may be used most profitably [18] and, secondly, how best should we *compare* two measurements of a 3-D dose distribution [19]. A related question with which we, as a community, are perhaps still struggling, is "What do we *do* when our two methods give us different results?" In commissioning, how long should we spend agonising about small differences between outputs from different techniques? In clinical practice, what action levels should we set based on MR gel dosimetry and what action should we take when they are breached?

3. Chemical dosimetry

A second overarching issue is the nature of chemical dosimetry. A very large section of the gel dosimetry literature has been concerned with the ways in which the NMR properties of different types of gel are affected by radiation dose and the ways in which gel properties interact with the MRI measurement. Thus, there are studies that relate errors in MR-derived dose maps to oxygen sensitivity [20], ion diffusion [21, 22], dose rate [23], dose fractionation [24], post-irradiation evolution of the chemical system [23, 25, 26], temperature during storage and irradiation [27] and temperature during scanning [28], container size and calibration methodology [29, 30]. Ref. [31] is a topical review of the area of polymer gel dosimetry that provides an excellent starting point for further study.

The source of dose-related contrast in polymer gel [31] imaging is primarily the change in transverse relaxation resulting from radiation-induced polymerisation, whilst in Fricke gels [32-35], absorbed radiation causes the oxidation of Fe^{2+} ions to Fe^{3+}, which changes both longitudinal and transverse relaxation rates. Thus, both quantitative R_1- and R_2-mapping sequences are of interest in gel dosimetry. All of the effects described above change the *actual* R_1 ($= 1/T_1$) or R_2 ($= 1/T_2$) value of the dosimeter, which is what the quantitative MRI sequences measure. Other effects, such as water diffusion [6], have the potential to affect the *measurement process itself* and need to be accounted for.

These aspects will not be discussed further here, except to make one general point. A very significant disadvantage of gel dosimetry is that, although multi-use dosimeters have been proposed for optical CT readout [36] and at least one company has marketed Fricke gels that can be reused (TruView™, Modus Medical Devices Inc.), it is still generally true that each gel dosimeter is an individual system. By its very nature, the dosimeter cannot be characterised by repeated test irradiations, as can an ion chamber or a diode array. Absolute measurements are problematic [37] and most results are reported as "relative dosimetry". Measurements are sometimes found not to be repeatable to within the tolerances required for precision radiotherapy, and inter- and intra-batch differences are common. The most careful study of such effects thus far reported is that of Vandecasteele and De Deene [38].

Despite the speed of modern pulse sequences and steadily improving access to MR scanners through the advent of MR-based radiotherapy treatment planning, the "fabricate-irradiate-image-analyse" cycle is much more arduous on an ongoing basis than the use of a calibrated diode- or ion chamber array phantom with stable properties. At each stage of the process, errors can be introduced and these are summarised in Fig. 18 of [31].

4. Sources of error related to the MRI measurement

The high quality of images created by modern MR scanners has already been alluded to above, and this, paradoxically, leads to some concerns about their use in radiotherapy dosimetry that were less apparent in the early days of gel dosimetry. Standard clinical software can now produce, relatively quickly, quantitative maps that "look good": that is, they are free of obvious artefact and have good signal-to-noise ratio. It may thus be tempting to assume that they must also now be more *accurate* measurements than hitherto. Whilst this may be true in some cases, rigorous evaluation using well established and recently calibrated relaxation test objects is still vital. Throughout, it should be borne in mind that quantitative relaxation time measurement is a minority use case for scanner manufacturers: systems are optimised for routine clinical imaging whose requirements are different.

Useful summaries of the type of image artefact that can occur in MRI in general may be found in [39-42]. The remaining sections of this article provide an overview of the specific issues that arise in *quantitative* MRI measurements. Many of these have been discussed by previous DOSGEL and IC3DDose reviews [10, 16, 43, 44], but potential new sources of error are associated with recent developments in imaging and these are discussed briefly at the end of the article.

4.1. RF flip angle

A number of methods are available for R_1-mapping. These include, among others, inversion-recovery [42] and Look-Locker [45] techniques (both often employed in association with an echo-planar imaging readout for speed) and rapid 3-D gradient echo sequences (run several times with different flip angles [46, 47]). The method historically used to measure R_2 has generally been a multi-echo sequence, which is the imaging equivalent of the Carr-Purcell sequence [6]. Sequences based on the steady-state free precession paradigm, including DESPOT [48] and inversion-recovery trueFISP [49] allow both R_1 and R_2 to be calculated. However, in all cases, it is of great importance that the RF flip angle is well calibrated. As an example, the exponential signal decay in a Carr-Purcell sequence, relies on the accuracy of the repeated 180° pulse. Deviations in flip angle lead to incorrect echo amplitudes which do not match the mathematical model used to fit the data. Hence, *systematically* incorrect R_2 values may be generated [50], and this may occur without there being obvious image artefacts.

Why have the problems associated with RF pulses not been eradicated along with the general improvement in MR technology? The answer is that the laws of physics make it very difficult to create RF probes that give a constant flip angle over the entire volume of the sample. Similarly, slice-selective RF pulses do not impart exactly a 180° flip to magnetisation within the slice and 0° to magnetisation outside; slices do not have an ideal "square-cornered" profile, but instead a transition region at the edge, in which the flip angle varies between the two extremes [51]. At high field (e.g., 3T clinical scanners), a further complication arises in the form of incomplete RF penetration into the sample, caused by so-called standing wave effects [52, 53]. Such variations of RF flip angle across the imaging volume means that a homogeneous sample can give rise to R_1 and R_2 maps that vary systematically. Furthermore, sequences that are nominally of the same type may be implemented differently on two different MR platforms, potentially leading to differing results for the same physical sample.

4.2. Ghosting artefacts and eddy currents

Ghosting artefacts are regions of MRI signal that appear systematically "in the wrong place" in images. Typically, coherent ghosts are entire copies of the correct image that appear with a lower intensity and are shifted in the phase-encoding direction. Ghosting artefacts can have many sources, but often occur for techniques whose pulse sequences include gradient reversals and zig-zag trajectories in k-space.

IOP Conf. Series: Journal of Physics: Conf. Series **847** (2017) 012051 doi:10.1088/1742-6596/847/1/012051

Implementations of the echo-planar sequence commonly suffer from the so-called "Nyquist N/2 ghost" [54]. Simple Fourier transform theory shows that if the forward and backward echoes are not perfectly aligned and there is an alternating pattern, then the reconstructed image consists of the superposition of two separate copies of the desired image, separated by half the field-of-view.

Eddy currents are one cause of such echo misalignment in k-space. Eddy currents are a consequence of Lenz's Law: when an MRI gradient is switched rapidly, by changing the current through the gradient coil, unwanted electric currents are induced in conducting surfaces in the magnet assembly in such a way as to set up magnetic fields that oppose the original change [55, 56]. This is an area that has received much attention from equipment manufacturers and gradient performance has improved markedly since the early years of gel dosimetry. Nevertheless, ghost artefacts still have the potential to reduce the accuracy of quantitative R_1 and R_2 sequences by *mixing* signals from two separate regions of the irradiated sample, which might have different values of relaxation time (arising from different doses). The effect may be particularly pernicious for a sample that fills the field-of-view with a smoothly varying dose distribution (i.e., no obvious aliased boundaries between regions), and for ghosts of relatively low intensity that influence results subtly.

4.3. Spatial distortion

The basis of MR imaging is the observation that application of a magnetic field gradient across the sample maps the frequency of NMR spin precession onto a spatial location. Data are acquired in the temporal domain and converted (traditionally by Fourier transformation, but nowadays by a number of more complex post-processing signal models) into the *spatial frequency* domain. To make the jump to real space, we need to know the *mapping* between space and frequency. If this mapping function is incorrect, then the images created will exhibit spatial distortions. It is normally assumed that the mapping is *linear*, but this is not always the case. A brief summary follows, but for more detail, see [57, 58].

Non-linearities in magnetic field gradient are an inevitable result of the need to build gradient coils of a limited size to fit within the bore of the main magnet. The gradient system of a modern magnet is a sophisticated compromise between a number of competing requirements: linearity is traded off against eddy current performance, maximum gradient strength, self-inductance, gradient slew rate, performance of the gradient amplifiers, and bore length [59]. In practice, both the manufacturer and those performing local quality assurance go to considerable lengths to try and map and gradient-induced distortions and correct them.

Inhomogeneities in the static magnetic field (B_0) lead to local frequency perturbations and, hence, localised shifts in the apparent positions of features in images. These effects may be caused by poor shimming, but are often the result of the composition of the sample, which may contain regions of different *magnetic susceptibility*. One example of great relevance to 3-D dosimetry is the study of dose effects at air-tissue interfaces and in low density tissues such as the lungs [60]. We typically wish to measure accurate values of dose in tumours near such an interface, yet it is precisely in this region that the spatial accuracy and quantification of the MR measurement are called into question. A related effect is the so-called *chemical shift artefact*, whereby a change in chemical composition is associated with Larmor precession at a different frequency, and hence an apparent spatial position that is shifted from the true position. The effects in this paragraph can often be corrected using the "forward and reverse gradient" technique introduced by Chang and Fitzpatrick [61].

4.4. Experimental design: SNR, resolution and sampling

Compared with other imaging modalities (e.g., x-ray computed tomography), the range of options available in the design of an MRI experiment is very large. As detailed in the next section, a number of new methods are being investigated. However, for quantitative R_2 mapping, the majority of genuinely quantitative work requiring high accuracy is probably still performed using pulse sequences consisting of "trains" of spin echoes. In [16], De Deene summarises strategies for optimising this type of experiment from a signal-to-noise ratio (SNR) perspective, and a more detailed analysis is presented in

[62]. As in other applications of MRI, there is a trade-off between acquisition time, SNR and spatial resolution. If the image resolution is too poor, then partial volume effects may lead to inappropriate dose values being measured. On the other hand, if fine detail is required, then either the pulse sequence takes longer to run, or the images are more noisy, resulting in lower dosimetric precision. It is still a challenge to obtain good R_2 data for a large number of slices or an isotropic 3-D volume and this may be limiting for some polymer gel applications. 3-D quantitative R_1 measurements, which can be used for the readout of Fricke gel dosimeters, are somewhat easier to perform, and have been widely employed in dynamic contrast-enhanced MRI [63]).

4.5. Potential issues related to advances in MR image acquisition and reconstruction

The acquisition time of an MRI sequence is largely governed by the number of times one has to repeat the basic unit of (RF excitation – spin evolution – readout – recovery). Arguably, the important paradigm shift in MRI during the early years of the 21st century has been the realisation that one can acquire images of similar quality with fewer such loops.

Parallel reception (e.g., SMASH [11], SENSE [12]) makes use of the properties of multi-coil receiver arrays. In Section 2.1, I described the problem of spatial non-uniformity of transmission by an RF probe. The flip-side of this is that, in receive mode, the probe *sensitivity* varies spatially, too. Since each coil of a receiver array is located in a slightly different position in relation to the sample, the different spatial sensitivity patterns encode the NMR signals differently. Incorporated correctly into the reconstruction algorithm this spatial information can replace the data from a fraction of the phase-encoding steps, without the need to develop any new imaging sequences. Significant acceleration factors can be obtained from large arrays of coils. However, possibly the biggest drawback of the technique from a quantitative imaging point of view is the issue of spatially inhomogeneous noise amplification (so-called g-noise) [12].

Although the ideas underlying simultaneous multislice (SMS) imaging have been in development since the 1980s [64-67] it is only recently that technology has been available for the widespread deployment of the method. Instead of increasing speed by reducing the number of phase-encoding steps, one effectively acquires two or more images at once, by exciting several slices with the same, specially designed, RF pulse. Image reconstruction is achieved via one of a number of different techniques (some related to SENSE), as described in the review by Barth *et al.* [13]. This family of techniques potentially allows a signal-to-noise gain over parallel imaging, but its use in quantitative imaging may be limited by issues such as "leakage" of signal between slices.

Compressed sensing [14] is a class of method that seeks to reduce MR acquisition times by exploiting the sparsity that is "implicit" in MR images. By this, what is meant is that for some mathematical transforms, the transformed MR signal will be close to zero for certain parts of the domain, and thus signals corresponding to these parts of the domain "do not need to be acquired". This topic is very strongly related to our ability to compress such images and, for exactly this reason, the types of issues that may arise are similar to the image degradations seen when performing JPEG (or similar) compressions on image data. Whilst these methods may be widely acceptable for clinical imaging, care is needed when applying them to quantitative measurements where a high degree of accuracy is required. Furthermore, Lustig *et al.* [14] note that "the most distinct artifacts in [compressed sensing] are not the usual loss of resolution or increase in aliasing interference, but rather loss of low-contrast features in the image" and this may be a problem for dose-mapping.

Magnetic resonance fingerprinting (MRF) was first introduced in 2013 [15] and uses an entirely new image reconstruction approach. Its key assumption is that, by simulating the Bloch equations, unique patterns of evolution of the MR signal may be designed for each combination of proton density, T_1 and T_2. These patterns are stored in a *data dictionary* and the process of reconstruction involves searching this dictionary for the closest match to the observed signals. As this is a relatively new technique, relatively few multi-centre trials have yet been undertaken to standardise acquisitions or determine the accuracy of the parametric maps generated.

5. Conclusion

There is no doubt that the capabilities of modern MRI scanners are significantly greater than they were 20 years ago, and this includes the ability to make better imaging measurements of relaxation times. As a community, we are also much more aware of the pitfalls in using MRI for 3-D dosimetry and how to avoid them. Nevertheless, the demands of radiotherapy are exacting: our quantitative measurements are not likely to be useful if we cannot guarantee an accuracy of 5% or better and even greater precision.

Obtaining high-quality and consistent results is possible, but it is not a turn-key operation and is not likely to be so in the near future. It requires not just experience but also a high level of understanding of the physics and chemistry underlying the measurements. For the moment, expertise in all of these aspects together remains localised within a small number of research groups and even fewer commercial companies. To the knowledge of this author, no large-scale, multicentre trials have been funded in order to standardise gel measurements. Results are thus potentially dependent on individual scanners within an institution, and these tend to be replaced on timescales of 5 – 10 years, after which protocols need to be re-established.

To date, measurement standardisation in MRI has failed to catch the imagination of funders. However, there are indications that this state of affairs may be beginning to change, at least as far as mainstream radiology is concerned. Two organisations, the Quantitative Imaging Biomarkers Alliance (QIBA), established by the Radiological Society of North America (RSNA), and the Quantitative Imaging Network (QIN), which has grown out of the National Cancer Institute (NCI) cancer imaging programme, are making slow but steady progress.

Thus, the answer to the question posed in the title - "Can it really can be that difficult?" - is "yes." But we are getting there!

6. References

[1] Gore J *et al* 1984 *Phys. Med. Biol.* **29** 1189
[2] Bloch F 1946 *Phys. Rev.* **70** 460
[3] Purcell E M *et al* 1946 *Phys. Rev.* **69** 37
[4] Bloembergen N *et al* 1948 *Phys. Rev.* **73** 679
[5] Hahn E L 1950 *Phys. Rev.* **80** 580
[6] Carr H Y *et al* 1954 *Phys. Rev.* **94** 630
[7] Lauterbur P C 1973 *Nature* **242** 190-191
[8] Mansfield P *et al* 1973 *Journal of Physics C: Solid State Physics* **6** L422
[9] Garroway A N *et al* 1974 *J. Phys. C: Solid State Phys.* **7** 457-462
[10] De Deene Y *et al* 1999 *1st International Workshop on Radiation Therapy Gel Dosimetry* 75-89
[11] Sodickson D K *et al* 1997 *Magn. Reson. Med.* **38** 591-603
[12] Pruessmann K P *et al* 1999 *Knowledge Creation Diffusion Utilization* **962** 952-962
[13] Barth M *et al* 2016 *Magn. Reson. Med.* **75** 63-81
[14] Lustig M *et al* 2007 *Magn. Reson. Med.* **58** 1182-1195
[15] Ma D *et al* 2013 *Nature* **495** 187-192
[16] De Deene Y 2006 *J. Phys.: Conf. Ser.* **56** 72-85
[17] Ceberg S *et al* 2010 *Phys. Med. Biol.* **55** 4885
[18] Schreiner L 2008 *J. Phys.: Conf. Ser.* **164** 012001
[19] Schreiner L *et al* 2013 *J. Phys.: Conf. Ser.* **444** 012016
[20] Hepworth S *et al* 1999 *Phys. Med. Biol.* **44** 1875
[21] Baldock C *et al* 2001 *Aust. Phys. Eng. Sci. Med.* **24** 19-30
[22] Schulz R *et al* 1990 *Phys. Med. Biol.* **35** 1611
[23] De Deene Y *et al* 2006 *Phys. Med. Biol.* **51** 653
[24] Karlsson A *et al* 2007 *Phys. Med. Biol.* **52** 4697
[25] De Deene Y *et al* 2000 *Phys. Med. Biol.* **45** 859
[26] McJury M *et al* 1999 *Phys. Med. Biol.* **44** 1863
[27] Vandecasteele J *et al* 2012 *Phys. Med. Biol.* **58** 43
[28] Vandecasteele J *et al* 2012 *Phys. Med. Biol.* **58** 63
[29] Dumas E-M *et al* 2006 *J. Phys.: Conf. Ser.* **56** 239-41
[30] Oldham M *et al* 1998 *Phys. Med. Biol.* **43** 2709

[31] Baldock C *et al* 2010 *Phys. Med. Biol.* **55** R1-63
[32] Audet C 1999 1st International Workshop on Radiation Therapy Gel Dosimetry (DOSGEL) 31-45
[33] Båck S Å J *et al* 1999 1st International Workshop on Radiation Therapy Gel Dosimetry (DOSGEL) 47-61
[34] Schreiner L J 2001 2nd International Conference on Radiotherapy Gel Dosimetry (DOSGEL)
[35] Schreiner L J 2004 3rd International Conference on Radiotherapy Gel Dosimetry (DOSGEL) 10-22
[36] Juang T *et al* 2015 *J. Phys.: Conf. Ser.* **573** 012039
[37] Cosgrove V *et al* 2000 *Phys. Med. Biol.* **45** 1195
[38] Vandecasteele J *et al* 2012 *Phys. Med. Biol.* **58** 19
[39] Gaillard F *et al* https://radiopaedia.org/articles/mri-artifacts access date 6 October 2016
[40] Paudyal S http://www.slideshare.net/SudilPaudyal/mri-artifacts access date 6 October 2016
[41] Morelli J N *et al* 2011 *Radiographics* **31** 849-866
[42] Doran S J *et al* 2012 Ch. 7 in Webb's Physics of Medical Imaging 487-621
[43] De Deene Y 2009 *J. Phys.: Conf. Ser.* 012033
[44] De Deene Y 2010 *J. Phys.: Conf. Ser.* 012015
[45] Look D C *et al* 1970 *Rev. Sci. Instrum.* **41** 250-251
[46] Wang H Z *et al* 1987 *Magn. Reson. Med.* **5** 399-416
[47] Wang J *et al* 2006 *J. Magn. Reson.* **182** 283-292
[48] Deoni S C *et al* 2005 *Magn. Reson. Med.* **53** 237-241
[49] Schmitt P *et al* 2004 *Magn. Reson. Med.* **51** 661-667
[50] Majumdar S *et al* 1986 *Magn. Reson. Med.* **3** 397-417
[52] Elster A http://mri-q.com/dielectric-effect.html access date 7 October 2016
[53] Webb A G *et al* 2010 *Int. J. Imag. Syst. Tech.* **20** 2
[54] Buonocore M H *et al* 1997 *Magn. Reson. Med.* **38** 89-100
[55] Elster A http://mriquestions.com/eddy-current-problems.html access date 7 October 2016
[56] Le Bihan D *et al* 2006 *J. Magn. Reson. Imag.* **24** 478-488
[57] Sumanaweera T S *et al* 1994 *Neurosurg.* **35** 696-704
[58] Doran S J *et al* 2005 *Phys. Med. Biol.* **50** 1343-1361
[59] Turner R 1993 *Magn. Reson. Imag.* **11** 903-920
[60] De Deene Y *et al* 2006 *Med. Phys.* **33** 2586-2597
[61] Chang H *et al* 1992 *IEEE Trans. Med. Im.* **11** 319-329
[62] De Deene Y *et al* 2002 *Phys. Med. Biol.* **47** 3117
[63] Leach M *et al* 2012 *Eur. Radiol.* **22** 1451-1464
[64] Souza S *et al* 1988 *J. Comp. Asst. Tomog.* **12** 1026-1030
[65] Larkman D J *et al* 2001 *J. Magn. Reson. Imag.* **13** 313-317
[66] Maudsley A 1980 *J. Magn. Reson.* **41** 112-126
[67] Müller S 1988 *Magn. Reson. Med.* **6** 364-371

IOP Conf. Series: Journal of Physics: Conf. Series **847** (2017) 012052 doi:10.1088/1742-6596/847/1/012052

Characterization of a radiochromic silicone dosimeter

E E Carpentier[1], K M Alexander[1], S Todd[1] and L J Schreiner[1,2]
[1] Department of Physics, Queen's University, Kingston, Ontario, Canada K7L 3N6
[2] Department of Medical Physics, Cancer Centre of Southeastern Ontario at Kingston General Hospital, Kingston, Ontario, Canada, K7L 5P9

E-mail: 11eec6@queensu.ca

Abstract. A radiochromic silicone based dosimeter has been fabricated that allows for three-dimensional (3D) dose verification. This paper characterizes the silicone dosimeter and verifies previous results found by other research groups. Using a recipe found to be dose rate independent, we identify the spectra of the dosimeter irradiated to different doses and verify that the response from different dose rate deliveries is constant within uncertainty. The reproducibility of a response from identically irradiated and stored cuvettes is also investigated. A noticeable difference in response indicates that there is an inhomogeneous distribution of active ingredients due to the viscous nature of the dosimeter during fabrication. Irradiating multiple cuvettes to the same dose in different fractions shows that the response decreases as the number of fractions increases.

1. Introduction

3D dosimetry has been of much interest to the radiation oncology community [1]. Deformable 3D radiochromic dosimeters have recently been developed that use silicone to spatially fixate dose information [2-5]. Silicone dosimeters are deformable and may be formed to any clinically relevant shape, allowing for assessment of new features, such as deformable image registration algorithms in radiation delivery planning [6]. DeDeene *et al* [2, 3] and Høye *et al* [4] have collectively identified many properties of silicone dosimeters containing leucomalachite green (LMG) as a dose indicator and chloroform as a sensitizer, such as the absorption spectra when exposed to UV light [3], dose rate dependence [2-4], effects due to fractionation of dose delivery [4], short term and long term sensitivity stability [3, 4], independence of response from photon energy [2], and the effect cuvette temperature during irradiation has on the final sensitivity [3].

Recently, Høye *et al.* identified an optimal concentration of ingredients that eliminates dose rate dependence using 0.26% w/w LMG and 1% w/w chloroform, creating a dose rate independent dosimeter [5]. In this paper, the results of further investigations into these silicone radiochromic systems are reported to reproduce and confirm the results found in these previous studies, as well as further characterize the dosimetric properties of silicone dosimeters.

2. Materials and Methods

2.1. Experimental setup

The dosimeters were fabricated by dissolving 0.26% w/w LMG in 1% w/w chloroform. After the LMG was dissolved, this solution was added to the Sylgard 184 elastomer and was thoroughly mixed. The Sylgard 184 curing agent was added in a 1:10 ratio w/w with the elastomer. This mixture was then vigorously stirred for several minutes. After thoroughly combining the ingredients, the solution was desiccated until all bubbles disappeared. The dosimeter solution was separated into disposable 4.5 mL cuvettes and left to solidify for at least 72 hours in the dark at room temperature.

Irradiations were performed using 6 MV photons and delivered using various Varian linacs in our centre. Cuvettes were irradiated at a SAD = 100 cm, using 10 x 10 cm^2 fields at 1.5 cm depth in plastic water with 10 cm backscatter. Cuvettes were optically imaged using the SpectroVis Spectrophotometer (Vernier, Beaverton, OR, USA) with white diffuse light, contributing an uncertainty of ±0.002 to optical attenuation data. All attenuation information was taken at the peak wavelength of 626.6 nm.

2.2. *Irradiation*

Seven cuvettes were irradiated to 0, 5, 10, 15, 20, 25, and 30 Gy at 2 Gy/min to examine the dose dependence of the absorption spectra over wavelengths ranging from 380-700 nm. The dose rate dependence of this recipe was verified by irradiating three sets of six cuvettes to various doses at different dose rates: 2, 4, and 6 Gy/min. Each cuvette was scanned thirty minutes after irradiation in the spectrophotometer to determine the change in optical attenuation. This experiment also monitored six cuvettes, three irradiated to 10 Gy and three irradiated to 20 Gy to determine whether identical cuvettes behaved consistently after irradiation. The cuvettes were scanned every 5 minutes after irradiation for the first hour and then every few days to determine the reproducibility of the temporal changes. These cuvettes were stored in the dark at room temperature between scans. Finally, the effects of dose fractionation were investigated by irradiating four cuvettes to 20 Gy in 1, 2, 5, and 10 fractions at 6 Gy/min. After delivering each fraction, we allowed a 30 second pause before delivering the next fraction.

3. Results

The spectra of silicone cuvettes irradiated to various doses are shown in figure 1. Figure 2 shows the change in optical attenuation for cuvettes irradiated with different dose rates. The sensitivity of the silicone dosimeter is 0.0232 ± 0.0005 $cm^{-1}Gy^{-1}$ at 2 Gy/min, 0.0229 ± 0.0004 $cm^{-1}Gy^{-1}$ at 4 Gy/min, and 0.0220 ± 0.0009 $cm^{-1}Gy^{-1}$ at 6 Gy/min. The sensitivities at different dose rates vary only within measurement error thus no significant dose rate dependence could be demonstrated.

Reproducibility of the response among identical cuvettes shows that the identical silicone dosimeters can vary significantly. Figures 3 and 4 shows cuvettes irradiated to 10 Gy and 20 Gy are distinguishable despite being irradiated and stored identically. Multiple batches of silicone produced similar results (not shown). The silicone elastomer's viscosity requires vigorous stirring for several minutes during fabrication to evenly distribute active ingredients. These differing results may result from manual stirring being insufficient to create a homogeneous distribution of active ingredients, creating different concentrations of LMG in each cuvette and producing different responses upon irradiation.

Figure 5 shows the response when cuvettes are irradiated to 20 Gy with a different number of fractions. From this plot it appears that increasing the number of fractions will decrease the response. This is in agreement with results found by Høye *et al*. However, the large variability in response for cuvettes irradiated to the same dose requires allocating inherent uncertainty to individual measurements and drawing conclusions based on that level of confidence.

4. Conclusions

The measurements summarized in this work support previous reports of dose dependence [3], dose rate dependence [5], and fractionated dose delivery [5], demonstrating the consistency of radiation dosimetric properties in individual samples of dosimeters. However the observed problem of reproducibility in the temporal measurements, not reported previously, indicates that cross calibration between and within batches will need careful attention, and readout will need to be performed soon after irradiation. Further work is underway to establish how the silicone dosimeter fabrication process affects performance, in particular a new method for mixing the silicone dosimeter's ingredients is being considered to reduce response variations.

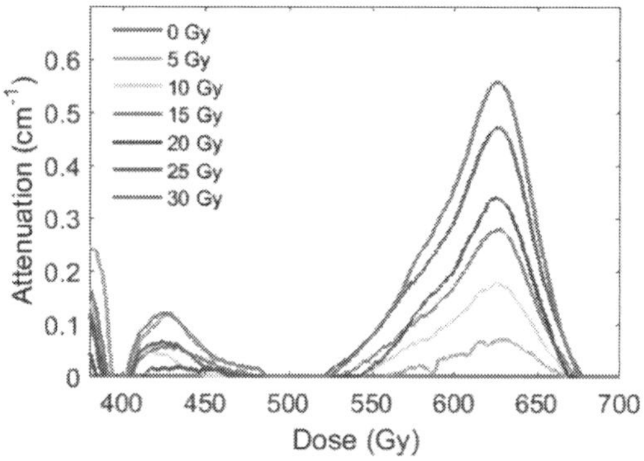

Figure 1. Plot of silicone dosimeter spectra irradiated to different doses.

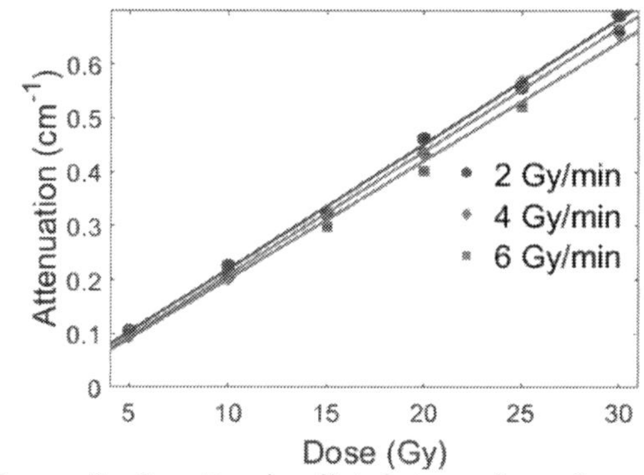

Figure 2. Cuvettes irradiated to various doses at different dose rates. Error bars are smaller than the symbol size.

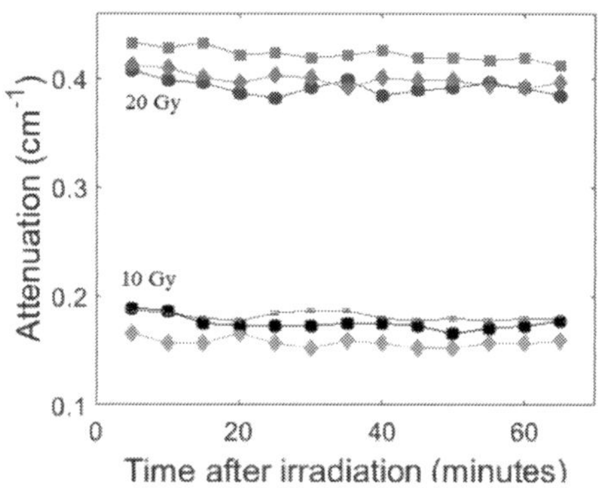

Figure 3. Reproducibility of identical cuvettes in the first hour after irradiation.

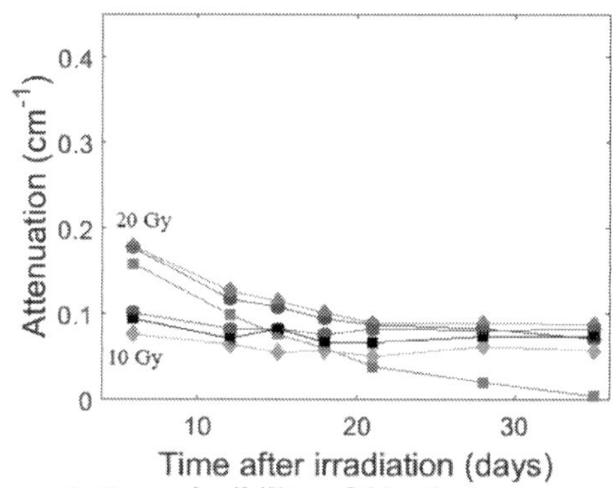

Figure 4. Reproducibility of identical cuvettes over several days after irradiation.

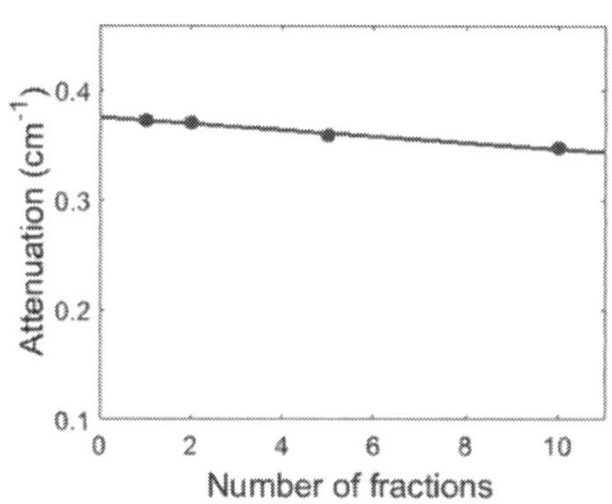

Figure 5. Results from irradiating cuvettes to 20 Gy with a different number of fractions.

5. References

[1] Baldock C *et al* 2010 *Phys. Med. Biol.* **55** R1-63
[2] De Deene Y *et al* 2015 *J. Phys.: Conf. Ser.* **573** 012025
[3] De Deene Y *et al* 2015 *Phys. Med. Biol.* **60** 1543-63
[4] Høye E *et al* 2015 *J. Phys.: Conf. Ser.* **573** 012067
[5] Høye E *et al* 2015 *Phys. Med. Biol.* **60** 5557-70
[6] Juang T *et al* 2013 *Int. J. Radiat. Oncol.* **87** 414-21

IOP Conf. Series: Journal of Physics: Conf. Series **847** (2017) 012053 doi:10.1088/1742-6596/847/1/012053

Cerium nanoparticle effect on sensitivity of Fricke gel dosimeter: Initial investigation

Ebenezer Suman Babu S, Timothy Peace Balasingh S, Benedicta Pearlin R, Rabi Raja Singh I and Paul Ravindran B
Department of Radiotherapy, Christian Medical College, Vellore-632 004, India

Email: stoebi@gmail.com

Abstract. Fricke gel dosimeters (FXGs) have been the preferred dosimeters because of its ease in preparation and water and tissue equivalency. Visible changes happen three dimensionally in the dosimeter as the ferrous (Fe^{2+}) ions change into ferric (Fe^{3+}) ions upon irradiation and the measure of this change can be correlated to the dose absorbed. Nanoparticles are promising entities that can improve the sensitivity of the gel dosimeter. Cerium Oxide nanoparticle was investigated for possible enhancement of absorbed dose in the FXG. Various concentrations of the nanoparticle based gel dosimeters were prepared and irradiated for a clinical dose range of 0-3 Gy in a telegamma unit. The optimal concentration of 0.1 mM nanoparticle incorporated in the FXG enhances the radiation sensitivity of the unmodified FXG taken as reference without modifying the background absorbance prior to irradiation. The gel recipe consisted of 5% (wt) gelatin, 50 mM Sulphuric acid, 0.05 mM Xylenol Orange, 0.5 mM Ferrous Ammonium Sulphate and 0.1 mM Cerium (IV) Oxide nanoparticle (< 25 nm particle size) and triple distilled water. The FXGs with nanoparticle showed linear dose response in the dose range tested.

1. Introduction

Gel dosimetry is gaining popularity in the field of medical physics that applies it as a dose measuring tool for quality assurance in radiotherapy treatment plans [1]. Since advances in technology have escalated the complexity in treatment delivery, the radiation delivered by the treatment unit requires verification before it can be administered to patients [2-9]. Gel dosimeters provide three dimensional (3D) information of the radiation dose absorbed, when irradiated [10-12]. However the sensitivity of the gel dosimeters to radiation dose is also important in this scenario.

Fricke chemical dosimeters have been widely used as reliable radiation dosimeters in the past decades [13, 14]. The idea of incorporating the same into a gelatin matrix to obtain the dose information in 3D was proposed by Gore in 1984 [15]. FXGs are the most preferred among the various gel dosimeters found in the literature, due to its water and tissue equivalence [16-18], ease in preparation and the availability of spatial dose information very soon after irradiation [19]. The absorbed dose in the FXG can be determined by the post irradiation measurement of the ferric (Fe^{3+}) ions [20]

Radiation sensitization effects on tumor tissues have created huge interest aiming at the therapeutic enhancement of radiation [21]. This can be achieved by the incorporation of nanoparticles (NPs) that plays a pivotal role in enhancing the absorbed dose through increased photon absorption [22]. Gold based NPs have been mostly studied and reported because of their favorable characteristics such as high biocompatibility and inertness to tissue interaction that are ideal for radiosensitization [23-26].

The effect of gold NPs in enhancing the dose response in MAGICA polymer gel has been reported by Khadem [27]. Marques reported that the absorptions interactions enhanced the local dose due to photoelectric absorption [28]. Though gold NPs have shown dose enhancement in the gel dosimeter, the high cost of the material is a disadvantage and this limits its practice. Various authors have reported the addition of additives to get better radiation sensitivity in the FXG. Recently, Maeyema reported on the clay nanoparticle for FXGs with Argon gas bubbling [29].

Nanotechnology has invaded every field from electronics to medicine in recent days. Metal oxide nanoparticles such as TiO, ZnO and CeO have been widely incorporated as catalysts [30]. Since nanoparticles are a promising entity in enhancing the response of gel dosimeters to radiation, we have explored the feasibility of incorporating Cerium nanoparticle into the FXG as a cost effective alternative for gold NPs. The dose response characteristic of the nanoparticle based Fricke gel dosimeter is reported in this study.

2. Materials and Methods

Cerium being a widely used catalyst, it was chosen for our experiments. Cerium (IV) oxide nanoparticle (<25 nm particle size, CAS No. 1306-38-3) was procured from Sigma Aldrich.

2.1. Gel dosimeter preparation

Five batches of the FXG recipe with 5% (wt) gelatin (300 bloom gelatin from porcine skin, (Sigma Aldrich) and each containing different concentrations of the nanoparticle solutions were prepared. Initially, gelatin was added to triple-distilled water and stirred using a magnetic stirrer while maintaining a temperature of 40 °C for 1 hour. The solution was then cooled to 25 °C and required quantities of Sulfuric acid (Qualigens), Xylenol Orange (Sigma), Ammonium Ferrous sulfate (SD-fine) were added. Finally the nanoparticle solution was added to each of the four batches of FXG. The final solution of the gel dosimeter consisted of 50 mM H_2SO_4, 0.05 mM Xylenol Orange and 0.5 mM Ferrous Ammonium Sulphate and the nanoparticle solution (0.05, 0.1, 0.2, 0.8 mM). The FXG without the nanoparticle solution added to it was taken as the reference. The gel solution was then transferred to cuvettes of 4 ml capacity and 1 cm path length. The samples as gel solutions in cuvettes were segregated into different boxes with labels for each concentration of the nanoparticle and kept in the refrigerator overnight and maintained at 4 °C for gelation.

2.2. Irradiation and measurement of gel dosimeters

The irradiation was performed on a telecobalt unit (Theratron 780C) with special arrangement in a water bath for the gel cuvette positioning. The gels were irradiated for doses ranging from 0-3 Gy with parallel-opposed lateral fields with a field size of 35 x 35 cm^2. The optical absorbance and transmittance of all the FXG cuvettes were measured with a UV-1800 double beam spectrophotometer (Shimadzu, Japan) at a wavelength of 585 nm.

3. Results and Discussion

3.1. Dose response

Five gel samples were irradiated for each dose in each concentration of the nanoparticle based FXG as well as the reference gel. Dose response curves were generated from absorbance in cuvettes and arithmetic mean value of the samples as shown in figure 1. The nanoparticle added FXGs exhibited linearity in dose response with R^2 values 0.9941, 0.9951, 0.996 for 0.05 mM, 0.1 mM and 0.2 mM concentration of the nanoparticle solution respectively compared to 0.9968 as the R2 value of the unmodified gel. Since p-value for the analysis of variance of the residuals was 0.3148 with 95% confidence interval, the quadratic term becomes non-significant. Hence we conclude that our data is consistent with the linear first-order model.

The gels prepared with 0.8 mM nanoparticle solution also showed improved dose response but we were not able to firmly conclude the amount of nanoparticle concentration as Cerium particle precipitates were found in it. Hence further increase in the concentration of nanoparticle in the FXG was not studied.

3.2. *Background absorbance*
Though all the nanoparticle FXG dosimeters showed improved dose response, the pre-irradiation measurements with the spectrophotometer showed higher background absorbance in the FXGs prepared with 0.2 and 0.8 mM concentrations of the nanoparticle solution suggesting darker gels if prepared in bulk quantities. Contrary to the former, the FXGs with 0.1 mM and 0.05 mM Cerium nanoparticle based FXGs showed negligible change in the background absorbance measurements.

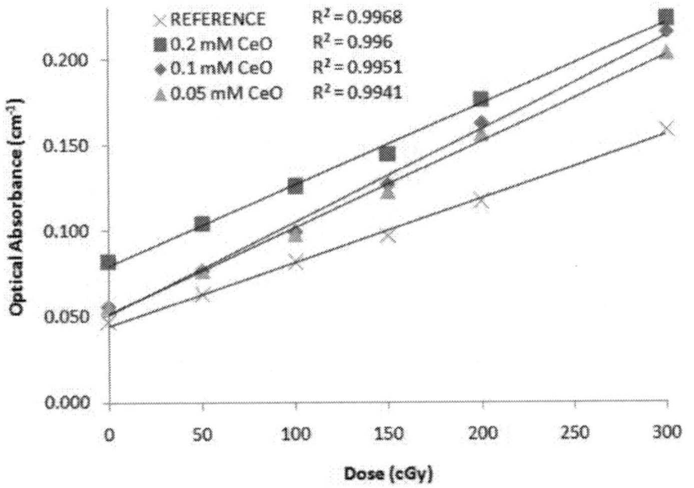

Figure 1. Comparison of dose response of 0.05 mM, 0.1 mM and 0.2 mM CeO_2 nanoparticle FXG with unmodified FXG reference)

4. Conclusion
The feasibility of incorporating Cerium oxide nanoparticle has been studied successfully. The optimal concentration of 0.1 mM nanoparticle incorporated FXGs show improved dose response making this gel dosimeter a promising tool for radiotherapy quality assurance. Further studies will be carried out to define the other radiological properties of this gel dosimeter and the possibility of using the findings in lower bloom gelatin commercially available in India.

5. Acknowledgements
Authors are grateful to the Institutional Review Board (IRB), Christian Medical College, Vellore, India for funding this project through the Fluid Research Grant.

6. References
[1] Baldock C *et al* 2010 *Phys. Med. Biol.* **55** R1-63
[2] Solberg T D *et al* 2012 *Pract. Radiat. Oncol.* **2** 2-9
[3] Ravichandran R *et al* 2011 *J. Med. Phys.* **36** 181-3
[4] Hardcastle N *et al* 2014 *Austral. Phys. Eng. Sci. Med.* **37** 45-52
[5] Mazur L M *et al* 2014 *Pract. Radiat. Oncol.* **4** 71-5
[6] Hill R F *et al* 2008 *Radiat. Meas.* **43** 1258-64
[7] Hill R *et al* 2014 *Phys. Med. Biol.* **59** R183-231
[8] Hill B *et al* 2005 *Med. Phys.* **32** 1589-97

IOP Conf. Series: Journal of Physics: Conf. Series **847** (2017) 012053 doi:10.1088/1742-6596/847/1/012053

[9] Hill R *et al* 2009 *Med. Phys.* **36** 3971-81
[10] Low D 2015 *J. Phys.: Conf. Ser.* **573** 12009
[11] Mather M L *et al* 2003 *Ultrasonics* **41** 551-9
[12] Davies J B *et al* 2008 *Radiat. Phys. Chem.* **77** 690-96
[13] Fricke H and Morse S 1927 *Am. J. Roentgenol Radium Ther. Nucl. Med.* **18** 420
[14] Fricke H and Hart E J 1966 *Radiat. Dosim.* **2** 167-239
[15] Gore J C *et al* 1984 *Phys. Med. Biol.* **29** 1189-97
[16] Kron T *et al* 1993 *Phys. Med. Biol.* **38** 139-50
[17] Chan M F and Ayyangar K 1995 *Med. Phys.* **22** 475-8
[18] Keall P and Baldock C 1999 *Austral. Phys. Eng. Sci. Med.* **22** 85-91
[19] Schreiner L J 2004 Review of Fricke gel dosimeters *J. Phys.: Conf. Ser.* **3** 9
[20] Baldock C *et al* 2001 *Austral. Phys. Eng. Sci. Med.* **24** 19-30
[21] Kwatra D *et al* 2013 *Transl. Cancer Res.* **2** 330-42
[22] Retif P *et al* 2015 *Theranostics* **5** 1030-44
[23] McMahon S J *et al* 2008 *Phys. Med. Biol.* **53** 5635-51
[24] Hainfeld J F *et al* 2004 *Phys. Med. Biol.* **49** N309-315
[25] Zhang X *et al* 2008 *Clin. Investig. Med. Médecine Clin. Exp.* **31** E160-167
[26] Zhang S X *et al* 2009 *Biomed. Microdevices* **11** 925-33
[27] Khadem-Abolfazli M *et al* 2013 *Int. J. Radiat. Res.* **11** 55-61
[28] Marques T *et al* 2010 *J. Phys.: Conf. Ser.* **250** 12084
[29] Maeyama T *et al* 2016 *J. Radiat. Res. (Tokyo)* **57** 318-24
[30] Keller A A *et al* 2010 *Environ. Sci. Technol.* **44** 1962-7

IOP Conf. Series: Journal of Physics: Conf. Series **847** (2017) 012054 doi:10.1088/1742-6596/847/1/012054

Characterization of novel preclinical dose distributions for micro irradiator

J Kodra[1], D Miles[1], S W Yoon[1], D G Kirsch[1] and M Oldham[1]
[1]Dept. of Radiation Oncology, Duke University Medical Center, Durham, NC, USA

Email: jacob.kodra@duke.edu

Abstract. This work explores and demonstrates the feasibility of utilizing new 3D printing techniques to implement advanced micro radiation therapy for pre-clinical small animal studies. 3D printed blocks and compensators were designed and printed from a strong x-ray attenuating material at sub-millimeter resolution. These techniques enable a powerful range of new preclinical treatment capabilities including grid therapy, lattice therapy, and IMRT treatment. At small scales, verification of these treatments is exceptionally challenging, and high resolution 3D dosimetry (0.5mm^3) is an essential capability to characterize and verify these capabilities, Here, investigate the 2D and 3D dosimetry of several novel pre-clinical treatments using a combination of EBT film and Presage/optical-CT 3D dosimetry in rodent-morphic dosimeters.

1. Introduction

There is significant current interest in exploring the therapeutic, radiobiological and radio-immunological effects of non-standard radiation dose distributions incorporating high spatial modulation such as GRID and LATTICE therapies [1-3]. Instead of aiming for uniform cell kill throughout the PTV, these treatments aim to reduce toxicity to normal tissue while stimulating by-stander effects, endothelial cell death and immunogenic abscopal effects [4-5] for tumor. The potential for using these treatments to 'prime' the immune response prior to more conventional radiation therapy has also been suggested. In this work we explore and demonstrate the feasibility of implementing IMRT and various highly spatially modulated radiation therapy treatments for a small-animal irradiator using 3D printing techniques to create regular modulated micro-blocks and compensators.

2. Methods

2.1 3D printing of micro-blocks and compensators

A range of 3D printers and printing materials were evaluated. Work presented here used a tungsten-filled polylactic acid (PLA) printing filament (Spool3D, Calgary, Alberta, CA) which was selected for its ease of use and ability to be printed with sub-millimeter resolution, allowing for better quality models. The tungsten within the filament acts as an efficient attenuator at orthovoltage treatment energies. Grid and lattice collimators were created using AutoCad (Autodesk, Inc., San Rafael, CA, USA). The 3D printed collimators used in this study were 5mm thick. IMRT compensators and block shapes were designed from in-house treatment planning software and small animal x-ray CT scans. All collimators, compensators and blocks were printed using an Ultimaker 2+ (Ultimaker,

Geldermalsen, Netherlands) 3D printer. A range of blocks and compensators were created to evaluate the capabilities of the 3D printing of these materials. 3D printer settings depended on the model and the amount of time needed for each print ranged from 15 to 90 minutes.

2.2 Irradiations

All irradiations were performed using the X-RAD 225Cx micro-irradiator (Precision X-ray Inc., North Branford, CT, USA) which has previously been commissioned at our institution [6-7]. Micro-compensators and blocks were mounted onto the end of a standard collimator using a novel in-house adapter (Figure 1). The adaptor enabled precise alignment and convenient replacement of the micro-blocks in a slot tray. All irradiations were performed with the default energy of this device, 225kVp.

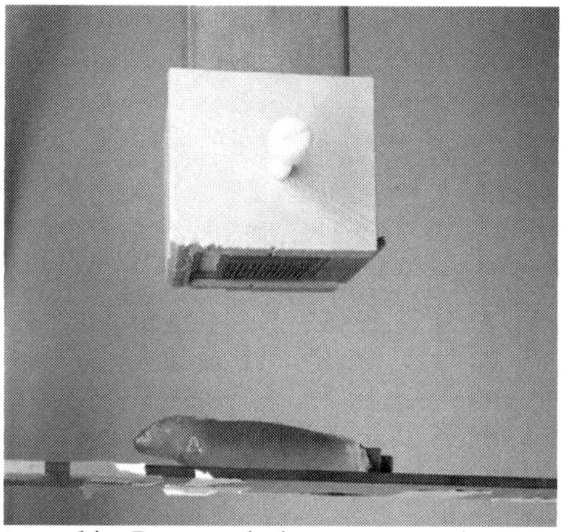

Figure 1. Rodent-morphic Presage dosimeter set-up in X-Rad irradiator with radiation grid block mounted to collimator.

2.3 Dosimetry

Dosimetric evaluation of preliminary micro-blocks and compensators was performed through a combination of simple 2D EBT Gafchromic film (Ashland Inc., Covington, KY, USA) irradiations, and high resolution 3D dosimetry in rodent-morphic phantoms using the Presage/optical-CT system [8]. Initial assessment of the attenuation capabilities of the tungsten filament and resolution of the 3D printer, was achieved by placing blocks directly onto a section of EBT2 film and irradiated with an AP field of 2 Gy. EBT2 film was analyzed using an Epson flatbed scanner (Epson America Inc., Long Beach, CA, USA) and an open source image processing software ImageJ (National Institutes of Health, Bethesda, MD, USA).

As a proof of concept, a 3D GRID treatment was designed to deliver 5 Gy to a mouse in 1mm wide bands with 1mm spacing. A 3D printed line grid was mounted to a 40x40mm collimator in the X-Rad, and used to irradiate a custom mouse-shaped Presage dosimeter via a 180-degree arc treatment (Figure 1). High resolution 3D dosimetry [9] was performed using optical-CT readout using established procedures [8, 10, 11]. The optical-CT was set-up to take 1 projection image per degree, for 360 degrees. Using an in-house reconstruction software, the data was reconstructed with 1 mm resolution.

2.4 IMRT

Preliminary testing of implementing IMRT capabilities to the X-Rad have been performed using the IMRTp platform built into CERR (National Institutes of Health, Bethesda, MD, USA) in conjunction with an in-house optimization program. Through this method, fluence maps are generated for the IMRT treatment plan and converted into 3D printed compensators.

3. Results

Radiochromic film readout shows the 3D printed grid and lattice collimators have both high resolution and efficient attenuating capabilities (Figure 2). The films show sharp edges with distinctive alternating regions of low and high dose with average peak-to-peak distance of 2.01 mm, corresponding well with the 1 mm spacing design of the 3D printed grid (Figure 3).

Figure 2. Top row-3D printed collimators of different grid and lattice spacing. Bottom row – corresponding EBT film measurements.

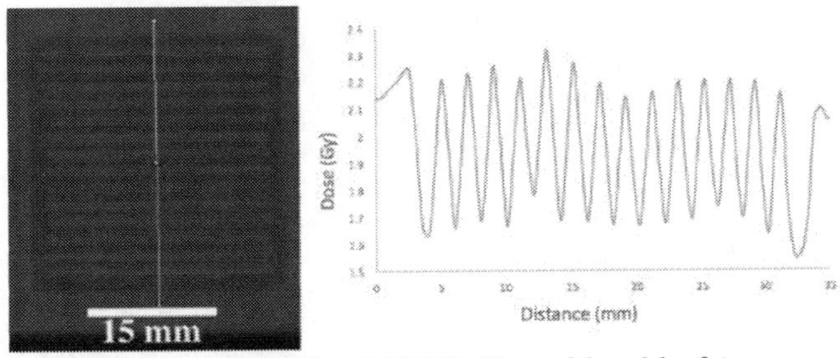

Figure 3. Left-Image an irradiated EBT2 film with grid of 1mm gaps and 1mm spacing. Right-image - Corresponding line profile showing change in optical density (proportional to dose).

Optical-CT readout of the Presage dosimeter shows high change in optical density (OD) for regions unblocked from the radiation beam (Figure 4). Another visualization of this peak-to-valley distribution can be seen in the line profile of the reconstruction (Figure 5).

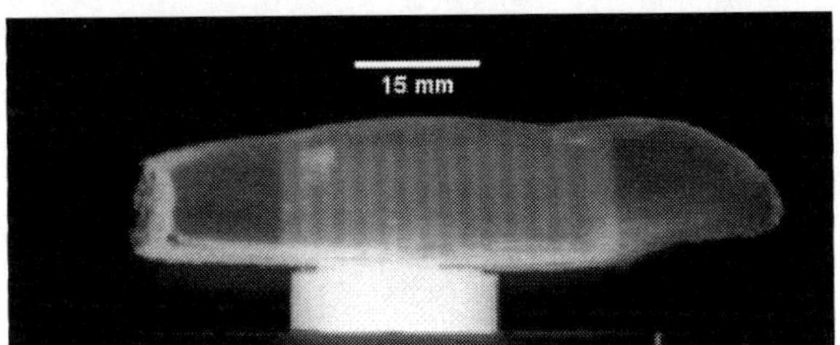

Figure 4. Projection image of PRESAGE rodent-morphic dosimeter treated with 3D printed radiation grid block in place on collimator.

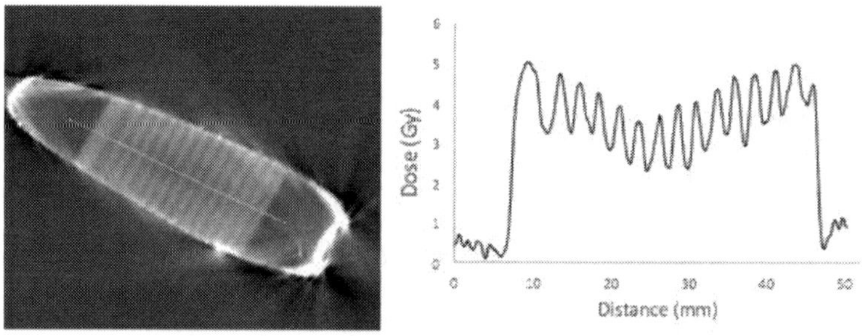

Figure 5. Left image -Reconstruction image of PRESAGE rodent-morphic dosimeter treated with 180-degree grid arc therapy. Right image -Corresponding line profile displaying the peak to valley dose distribution.

Preliminary testing of IMRT compensators show agreement between the generated fluence maps and the irradiated EBT2 film with regions of higher and lower change in OD (Figure 6).

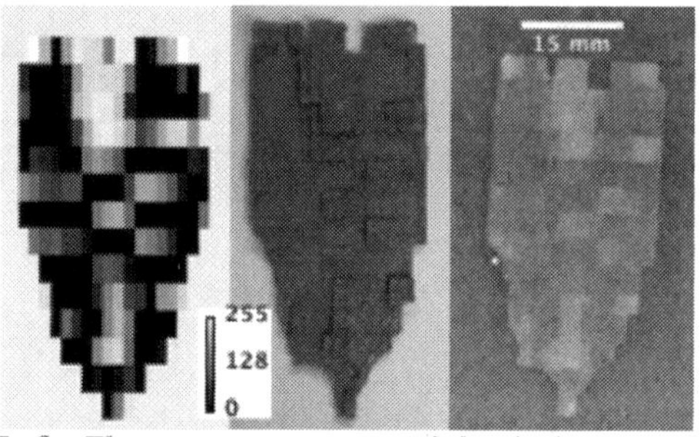

Figure 6. Left- Fluence map generated by in-house program with lighter regions corresponding to lower dose. Center-3D printed compensator of fluence map. Right- scan of an irradiated EBT2 film showing dose distribution with lower dose represented by lighter pixels.

4. Conclusions

Initial testing of 3D printed radiation blocks and compensators for use in small animal radiation therapy studies has shown promising results. A wide variety of spatially fractionated treatments are possible, including micro-IMRT for field sizes up to 4x4cm. Sub-millimeter spatial modulation of dose distribution is feasible with the techniques shown here. Further development work is required to optimize the peak-to-valley differential for GRID and LATTICE treatments. Combinatorial dosimetry approaches (e.g. both film and 3D) are considered essential to cross-validate findings in these extremely challenging dosimetric scenarios.

5. References

[1] Wu X *et al* 2010 *Cureus* **2 e9**
[2] Suarez B *et al* 2015 *Cureus* **7 e389**
[3] Kanagavelu S *et al* 2014 *Radiation Research* **182** 149-62
[4] Formenti S C *et al* 2012 *Int. J. Radiat. Oncol. Biol. Phys.* **84** 879-80
[5] Formenti S C *et al* 2013 *Int. J. Radiat. Oncol. Biol. Phys.* **105** 256-65
[6] Newton J *et al* 2011 *Med. Phys.* **38** 6754-62
[7] Thomas A *et al Med. Phys.* **38** 4846-57
[8] Rankine L J *et al* 2013 *Phys. Med. Biol.* **58** 7791-801
[9] Baldock C *et al* 2010 *Phys. Med. Biol.* **55** R1-63
[10] Bache S T *et al* 2015 *Med. Phys.* **42** 846-55
[11] Cramer C K *et al* 2015 *PloS One* **4**

IOP Conf. Series: Journal of Physics: Conf. Series **847** (2017) 012055 doi:10.1088/1742-6596/847/1/012055

Dosimetry in the presence of strong magnetic fields

D J O'Brien[1], N Schupp[2], S Pencea[2], J Dolan[2] and G O Sawakuchi[1]
[1]Department of Radiation Physics, University of Texas MD Anderson Cancer Center, Houston, TX 77030, USA
[2]Elekta Software, Elekta A. B., Maryland Heights, MO 63043, USA

E-mail: dobrien@mdanderson.org; gsawakuchi@mdanderson.org

Abstract. Magnetic resonance imaging–guided radiotherapy (MRIgRT) is an emerging technology that requires the use of radiation fields in the presence of magnetic (B) fields. In the presence of B-fields the Lorentz force influences the trajectories of the secondary electrons, which in turn affects both the dose distribution in water and the dose-response of ionization chambers and several other detectors. Thus, dosimetry in the presence of a B-field requires understanding both the B-field effects on the dose distribution and the response of detectors. In this paper we present measured data to show effects of the B-field on the dose distributions, response of ionization chambers, and presence of air-gaps surrounding the sensitive volume of the detector.

1. Introduction

Magnetic resonance imaging–guided radiotherapy (MRIgRT) integrates an MRI scanner with a radiotherapy source such as a linear accelerator or ^{60}Co for volumetric imaging guidance of the radiation beam. MRIgRT provides real-time imaging during patient treatment. MRIgRT also provides much higher soft-tissue contrast than that of conventional computed tomography (CT)-based image guided radiotherapy (IGRT) techniques such as cone-beam CT or CT on rails [1, 2]. As a result, there is great interest in developing new external-beam radiotherapy machines with integrated MRI [3-7]. However, since MRI requires magnetic (B) fields, the Lorentz force influences the trajectories of the secondary electrons, which in turn affects both the dose distribution in water [1, 8, 9] and the dose-response of ionization chambers [6, 10-12] and several other detectors [13, 14]. Thus, dosimetry in the presence of a B-field requires understanding both the B-field effects on the dose distribution and the response of detectors.

Further, how the detector is positioned relative to the B-field orientation and the beam direction also influences its response. The detector orientation relative to the Lorentz force deserves attention because of variations in the design of new MRIgRT units. The Cross Cancer Institute in Canada is developing a linac-based MRIgRT unit in which B-field orientation can be switched either longitudinally or transversely relative to the beam direction [5]. Elekta is designing a linac-based MRIgRT unit with a 1.5-T B-field fixed in a transverse configuration relative to the beam direction [15]. The Ingham Research Institute is developing an open-bore linac-based MRIgRT [7]. ViewRay has a ^{60}Co-based MRIgRT unit with low B-field strength (0.35 T) transverse to the plane of three ^{60}Co sources [3, 4] and is now developing a linac-based unit.

In this paper we present measured data to show the effects of the B-field on the dose distributions and on the response of ionization chambers, including the effect of air-gaps surrounding the sensitive volume of the detector.

IOP Conf. Series: Journal of Physics: Conf. Series **847** (2017) 012055 doi:10.1088/1742-6596/847/1/012055

2. Methods

Measurements were performed at The University of Texas MD Anderson Cancer Center using the Elekta MR-linac system. The Elekta MR-linac has a 7-MV flattening filter free beam and closed-bore design with a 1.5-T B-field. The isocenter is at 143.5 cm from the source. We performed measurements using either an in-house Elekta MRI-compatible 3D water phantom (63 × 43 × 24 cm^3) or a mixture of 30 × 30 cm^2 water-equivalent plastic slabs (Solid Water, Gammex RMI, Middleton, WI; and Virtual Water, Standard Imaging, Middleton, WI). We performed measurements with the magnet both ramped up at 1.5 T and ramped down at 0 T to compare the dose distributions in the presence and absence of a 1.5-T B-field.

All measurements performed in this work were done at a source-to-surface distance (SSD) of 133.5 cm. The isocenter was located at a water equivalent depth of 10 cm at a source-to-axis distance (SAD) of 143.5 cm. A 10 × 10 cm^2 field size defined at isocenter was used for all measurements. At least 10 cm of backscatter material was also present. Detectors were aligned by using an on-board MV imaging system.

Percentage depth dose (PDDs) and lateral profiles were measured in water using a PTW 60019 microDiamond detector and a PTW TANDEM electrometer. PDDs were scanned from the water surface to just beyond a depth of 10 cm. Profiles were measured at a depth of 10 cm.

The effect of air-gaps in plastic phantoms on the signal from three Farmer-type ionization chambers (PTW 30013, PTW 30010 and NE2571) was assessed with water-equivalent plastic slabs with the chamber placed inside a machined hole that conformed to the shape of the chamber. Each ionization chamber was rotated about its axis in increments of 45° and the response as a function of rotation angle was recorded both with and without the magnetic field.

3. Results and Discussions

3.1. PDD and Dose Profiles

Figure 1 shows PDDs for a 10 × 10 cm^2 field measured with and without a 1.5-T B-field. The depth of maximum dose d_{max} is 1.3 cm and 1.7 cm, respectively. The B-field shifts d_{max} 4-5 mm towards the water surface. The trajectories of the secondary electrons are curved by the Lorentz force, which reduces their overall range and leads to compression of the build-up region. Beyond the build-up region, the reduced electron range has shifted the depth dose curve closer to the depth kerma curve. This shift results in a 0.5% lower dose at these depths for the same photon fluence [12]. More significantly, the compression of the build-up region enhances the dose at d_{max} by 1.9%. Both of these effects have a corresponding effect on the PDD (Figure 1). As a result, the use of the PDD as a beam quality specifier must be re-examined in the presence of a B-field in terms of its relationship to the stopping power ratios. In contrast, alternative beam quality specifiers, such as the TPR_{10}^{20}, that depend on dose ratios in the transient equilibrium region are relatively unaffected by the B-field [12].

Note that in closed bore systems, such as the Elekta MR-linac, the cryostat acts as an additional source of contaminant electrons which also have the effect of shifting d_{max} closer to the surface [12]. These electrons are swept away when a transverse B-field is present and do not significantly influence the surface dose. Since electron contamination increases with field size, the relative shift in d_{max} between with and without a B-field is greater for field sizes smaller than 10 × 10 cm^2 and less for larger fields. The opposite may be true with inline MR-linac configurations where the beam and B-field are parallel, as the confining effect of the B-field on contaminant electrons in this configuration enhances the surface dose and changes the PDD relative to the case without a B-field [16].

Figure 2 shows lateral dose profiles with and without the B-field at a depth of 10 cm for a 10 × 10 cm^2 field size. In the presence of the B-field, the lateral dose profile is shifted by approximately 1 mm in the direction of the Lorentz force and the penumbra becomes asymmetric. Going from left to right on Figure 2, the penumbra starts sharp but becomes broader than the 0 T case as it enters the beam, whereas on the other side of the profile the opposite is true. Profiles measured in the direction parallel to the B-field are not significantly affected by the B-field.

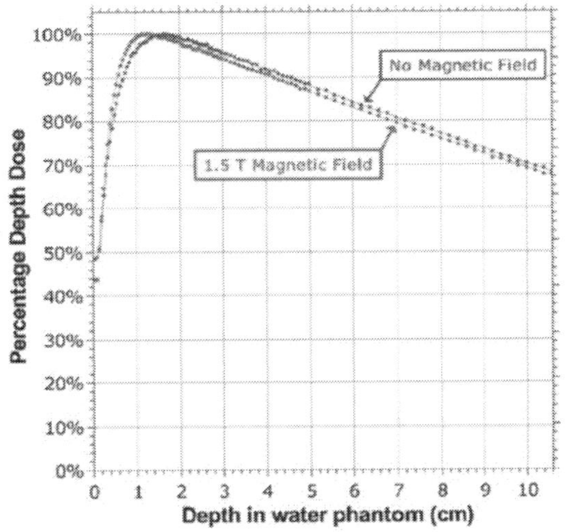

Figure 1. Percentage depth dose curves measured for a 10×10 cm^2 field with and without a 1.5 T magnetic field. These measurements were taken with a PTW 60019 microDiamond detector in a water phantom.

3.2. Ionization chamber response

Meijsing et al. [10] showed that the presence of a B-field can change the response of a Farmer-type ionization chamber by up to 11% depending on the B-field strength and on the orientation of the B-field with respect to the beam. The magnitude of the effect reaches a maximum at approximately 1 T over the range of B-fields that has been studied (≤2 T). The reason for this change in response was explained by Meijsing et al. as the result of the curving of the electron trajectories inside the air cavity of the ionization chamber, which resulted in the average path lengths of electrons traversing the cavity changing as a function of B-field strength and cavity geometry.

Reynolds et al. [11] found results similar to those of Meijsing et al. and showed that the orientation of the B-field with respect to the chamber itself also has an effect. They also found that the response of ionization chambers was relatively stable for inline MR-linacs where the B-field is parallel to the beam. In that case, they found that the chamber response did not require correction for B-fields up to 1 T regardless of the orientation of the chamber. O'Brien et al. [12] proposed a general formalism for reference dosimetry in the presence of a B-field, and they also calculated correction factors for a range of Farmer chambers for the transverse 1.5-T Elekta MR-linac system, where the beam is always perpendicular to the magnetic field. In this configuration, they showed that the correction for chambers that are aligned parallel to the B-field (1.5 T) (Figure 3a) is less than 1%. However, they also found a distinction between the two perpendicular orientations depicted in Figure 3b and Figure 3c. In Figure 3b, the chamber is rotated 90° clockwise from the parallel orientation and in this configuration the correction for the chamber response is about 4% for each of the chambers that were examined. However, in Figure 3c, the chamber is rotated 90° counter-clockwise from the parallel orientation, and in this configuration the chamber response correction was found to vary between 2.4% and 4.4% depending on the chamber model.

The reason for this discrepancy is because, in the perpendicular orientations, the chamber geometry is asymmetric with respect to the direction of the Lorentz force (which is perpendicular to the B-field). In the clockwise orientation, the Lorentz force is directed towards the stem of the chamber, whereas in the counter-clockwise orientation the Lorentz force is directed away from the chamber stem. Consequently, the geometry and materials used in the electrode and stem (which vary between different chamber models) become more important in the counter-clockwise orientation, as electrons generated in the stem are directed towards the sensitive volume of the chamber by the Lorentz force. Therefore, it is important to be aware of which perpendicular orientation is being used, and it is advisable to use the parallel orientation instead wherever possible.

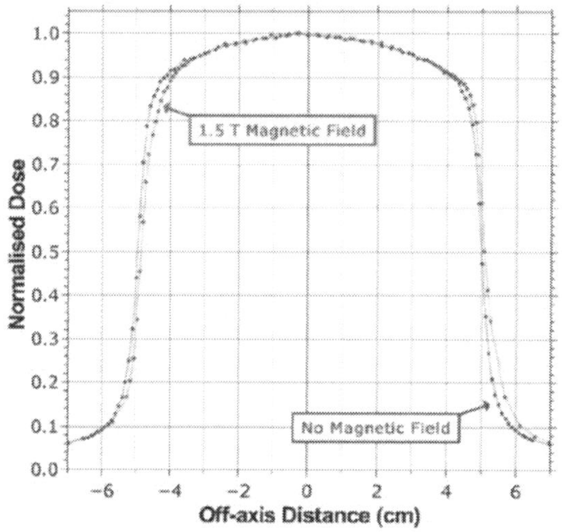

Figure 2. Lateral dose profiles measured for a 10 × 10 cm^2 field at 10 cm depth with and without a 1.5-T B-field. These measurements were taken with a PTW 60019 microDiamond detector in a water phantom.

3.3. Air gaps

Plastic phantoms are a convenient tool for routine quality assurance (QA) owing to their portability and ease of setup. In the case of MR-linacs, they are particularly attractive because of the limited space available inside the MRI bore. However, Hackett et al. [17] recently showed that ionization chambers exhibit a significantly increased rotational dependence in the presence of a B-field when used in plastic phantoms. This effect was attributed to the presence of sub-millimeter air gaps surrounding the chamber inside the phantom insert cavity. This was established by virtue of the fact that this effect was not observed when measured in water phantoms or when the insert cavity was filled with water.

Figure 4 shows the variation of three different Farmer chambers (PTW 30013, PTW 30010, and NE2571) as they were rotated about their long axis inside a plastic phantom when no B-field was present. In almost every case, the variation from the mean value was less than 0.1%. However, when a 1.5-T B-field is applied (Figure 5), some of the variations are greater than ±1% of the mean value depending on the detector and phantom used.

This is a problem for performing reference dosimetry with non-waterproof ionization chambers, in which either plastic sleeves or water phantoms with plastic inserts are used, because this variation exceeds the recommended tolerance of 0.5% [18]. It also presents a problem for performing routine QA in plastic phantoms, as consistent readings cannot be guaranteed. The air gaps responsible for these effects are very small, with Hackett et al. demonstrating an effect with a gap on the order of 0.1 mm thick.

Figure 5 shows that the effect can vary between two similarly constructed ionization chambers in the same phantom (the outer geometry of the PTW 30010 and PTW 30013 Farmer chambers are almost identical). Figure 5 also suggests that phantom/chamber combinations can be designed that reduce this effect to less than the 0.5% tolerance value, as was the case for the NE2571 chamber in phantom B. However, this would probably be difficult to achieve in general given the detector-to-detector variations.

IOP Conf. Series: Journal of Physics: Conf. Series **847** (2017) 012055 doi:10.1088/1742-6596/847/1/012055

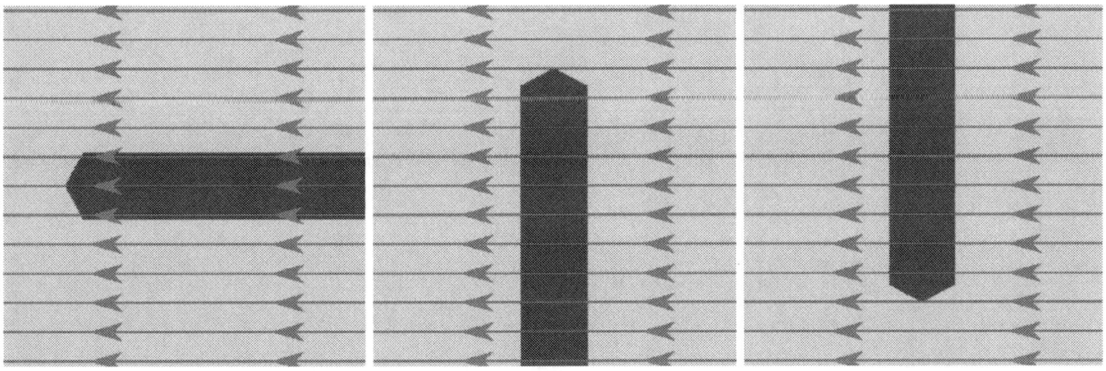

(a) (b) (c)

Figure 3. Three distinct ionization chamber orientations with respect to the magnetic field: (a) chamber parallel with magnetic field, (b) chamber rotated 90° clockwise from the magnetic field and (c) chamber rotated 90° counter-clockwise from the magnetic field.

Consequently, for any work involving the use of plastic phantoms in a B-field, the presence and effect of any sub-millimeter air gaps near the dosimeter must be carefully assessed and characterized.

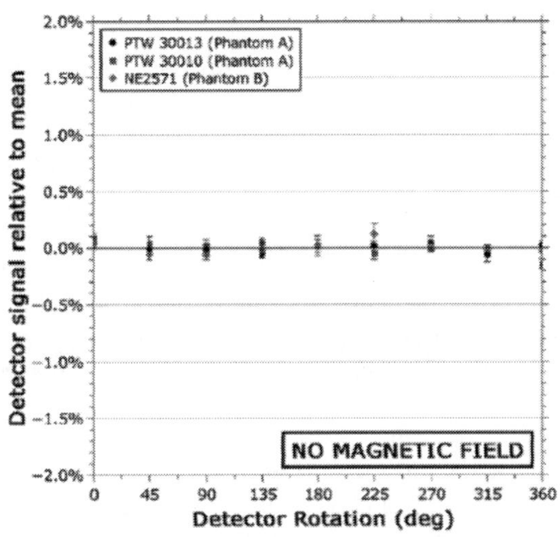

Figure 4. Variation of Farmer chamber response relative to the mean as a function of rotation about the detectors' long axis inside a plastic phantom with no magnetic field. Three detectors in two phantoms are shown.

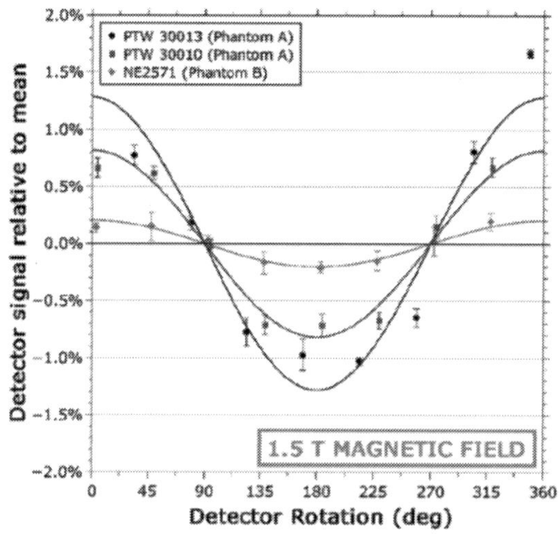

Figure 5. Variation of Farmer chamber response relative to the mean as a function of rotation about the detectors' long axis inside a plastic phantom in a 1.5 T magnetic field. Three detectors in two phantoms are shown. For clarity of presentation, the phase of oscillation of each data set has been aligned.

4. Conclusions

In the presence of B-fields, the Lorentz force perturbs the dose distribution by altering the trajectory of charged particles, which in turn affects the response of detectors. The results presented here are meant to highlight several basic effects that must be accounted for when performing dosimetry measurements in B-fields. The effects presented here are not comprehensive, and more research is needed to understand whether the standard dosimetric methods can be used in the same way in the presence of B-fields. Specific topics to be investigated include the calculation and validation of B-field correction factors for commercially available ionization chambers, investigation of the definition of the effective point of measurement in B-fields, and the study of polarity and ion recombination effects. B-field effects similar to those observed with ionization chambers may also need to be considered when using other detectors. Thus, more research is needed to develop and implement detectors to facilitate machine and patient treatment QA. Detailed investigations of these effects are important to understand the limitations of standard radiation detectors when used in B-fields as well as to develop new procedures to calibrate and perform dosimetric QA of new MRIgRT units. Particularly for QA of the MR-linac beam and patient plans, accurate dose distribution measurements are required. However, to date, only a few two-dimensional (2D) dosimetry systems have been investigated for use in the presence of B-fields [19, 20], and not many detection systems are commercially available. Therefore, development of new detectors that could be used in the presence of B-fields is essential to advance the field. Van Zijp et al. [21] have shown that many of the B-field effects on QA devices can be removed for the purposes of QA by incorporating electron-dense materials such as copper into their design. Because the MR-linac will require adaptive treatments with every fraction, QA of the adapted plans will be important. Thus, dosimetric systems that could provide accurate and fast 3D dose distribution measurements would be important for plan verification. Understanding the limitations of current dosimetric techniques in the presence of B-fields and developing new techniques to address current limitations are necessary for the safe implementation and full potential use of MRIgRT.

5. Acknowledgments

This work was partially supported by Elekta.

IOP Conf. Series: Journal of Physics: Conf. Series **847** (2017) 012055 doi:10.1088/1742-6596/847/1/012055

6. References

[1] Raaymakers B W *et al* 2004 *Phys. Med. Biol.* **49** 4109-18
[2] Lagendijk J J W *et al* 2008 *Radiother. Oncol.* 8**86** 25-9
[3] Dempsey J F *et al* 2005 *Int. J. Radiat. Oncol.* **63** S202
[4] Dempsey J *et al* 2006 *Med. Phys.* **33** 2254
[5] Fallone B G *et al* 2009 *Med. Phys.* **36** 2084-8
[6] Smit K *et al* 2013 *Phys. Med. Biol.* **58** 5945-57
[7] Oborn B M *et al* 2014 *Med. Phys.* **41** 051708
[8] Raaijmakers A J E *et al* 2005 *Phys. Med. Biol.* **50** 1363-76
[9] Raaijmakers A J E *et al* 2007 *Phys. Med. Biol.* **52** 929-39
[10] Meijsing I *et al* 2009 *Phys. Med. Biol.* **54** 2993-3002
[11] Reynolds M *et al* 2013 *Med. Phys.* **40** 042102
[12] O'Brien D J *et al* 2016 *Med. Phys.* **43** 4915-27
[13] Reynolds M *et al* 2014 *Med. Phys.* **41** 092103
[14] Stefanowicz S *et al* 2013 *Radiat. Meas.* **56** 357-60
[15] Raaymakers B W *et al* 2009 *Phys. Med. Biol.* **54** N229-37
[16] Liney G P *et al* 2016 *Med. Phys.* **43** 5188-94
[17] Hackett S L *et al* 2016 *Med. Phys.* **43** 3961-8
[18] International Electrotechnical Commission 1997 *Medical electrical equipment – dosimeters with ionization chambers as used in radiotherapy* (Genève) IEC 60731
[19] Smit K *et al* 2014 *Phys. Med. Biol.* **59** 1845-55
[20] Houweling A C *et al* 2016 *Phys. Med. Biol.* **80** N80
[21] van Zijp H M *et al* 2016 *Phys. Med. Biol.* **50** N50

IOP Conf. Series: Journal of Physics: Conf. Series **847** (2017) 012056 doi:10.1088/1742-6596/847/1/012056

How feasible is remote 3D dosimetry for MR guided Radiation Therapy (MRgRT)?

S Mein[1], L Rankine[2,3], D Miles[1], T Juang[1], B Cai[3], A Curcuru[3], S Mutic[3], J Fenoli[1], J Adamovics[4], H Li[3] and M Oldham[1]

[1]Department of Radiation Oncology, Duke University Medical Center, Durham, USA
[2]Department of Radiation Oncology, University of North Caroline at Chapel Hill, Chapel Hill, North Carolina, USA
[3]Department of Radiation Oncology, Washington University in St. Louis, St. Louis, USA
[4]Department of Chemistry and Biology, Rider University, Lawrenceville, USA

E-mail: mark.oldham@duke.edu, hli@radonc.wustl.edu

Abstract. To develop and apply a remote dosimetry protocol with PRESAGE® radiochromic plastic and optical-CT readout in the validation of MRI guided radiation therapy (MRgRT) treatments (MRIdian® by ViewRay®). Through multi-institutional collaboration we performed PRESAGE® dosimetry studies in 4ml cuvettes to investigate dose-response linearity, MR-compatibility, and energy-independence. An open calibration field and symmetrical 3-field plans were delivered to 10cm diameter PRESAGE® to examine percent depth dose and response uniformity under a magnetic field. Evidence of non-linear dose response led to a large volume PRESAGE® study where small corrections were developed for temporally- and spatially-dependent behaviors observed between irradiation and delayed readout. TG-119 plans were created in the MRIdian® TPS and then delivered to 14.5cm 2kg PRESAGE® dosimeters. Through the domestic investigation of an off-site MRgRT system, a refined 3D remote dosimetry protocol is presented capable of validation of advanced MRgRT radiation treatments.

1. Introduction

^{60}Co IMRT subject to a 0.35T lateral magnetic field was recently commissioned at Siteman Cancer Center at Barnes-Jewish Hospital and Washington University School of Medicine in St. Louis (Missouri, USA) following AAPM's TG-119 recommendations [1]. Dosimetric challenges arise with the introduction of an MR field into radiotherapy i.e. electron return effect (ERE), ferromagnetic incompatibility during quality assurance (QA), etc. [2]. Our unique approach to MRgRT dosimetry affords high-resolution 3D measurements [3] using PRESAGE® radiochromic plastics and optical-CT. Previous works qualitatively demonstrated remote dosimetry feasibility and dose-response linearity in 1kg volume PRESAGE® over 48hrs [4, 5]. Until now, remote dosimetry studies have been limited to this small field-of-view (FOV) dosimeter (8x10cm), insufficient in size for routine clinical plans. Here, we investigate the large-FOV (2kg) PRESAGE® (10x14.5cm), dose-response stability in remote dosimetry settings with a quantitative approach, to identify and model temporal and spatial behaviors over time.

IOP Conf. Series: Journal of Physics: Conf. Series **847** (2017) 012056 doi:10.1088/1742-6596/847/1/012056

2. Materials and Methods

2.1 The PRESAGE®/Optical-CT System

High resolution 3D Dosimetry was performed with the PRESAGE® dosimeter (Heuris Inc.), a radiochromic plastic, and the Duke Large Optical-CT Scanner (DLOS), an in-house bi-telecentric optical-CT scanner [6]. The linear dose-response using PRESAGE is well documented in the literature [7-9]. Formulation details are displayed in Table I. DLOS data acquisition was performed in two phases, pre- and post-irradiation, to yield change in optical density (ΔOD) post-reconstruction of 720 dosimeter projections at 0.5° increments (360° rotation). Reconstruction at 1mm isotropic resolution was performed in a MATLAB-based GUI developed in-house using filtered back-projection and a RamLak filter. ΔOD reconstructions were imported into A Computational Environment for Radiotherapy Research (CERR), a MATLAB software developed at Washington University, registered to the corresponding treatment plan, and scaled to the point of prescription dose (relative dosimetry).

Table 1. PRESAGE® formulation (DEA-1.5) and composition. Percentages are given by weight.

Component	% Make-up
Smooth-On Crystal Clear® 206	91.50%
Ethyl Acetate	5.00%
DMSO	2.00%
2-methoxy-DEA LMG	1.50%
Carbon Tetrabromide	0.50%

2.2 Small Volume Characterization

Prior to the 3D TG-119 MRgRT study, PRESAGE® sensitivity (dose-response) was tested in 4ml cuvettes to verify three key factors: linearity, MR-compatibility, and energy-independence in remote dosimetry use. Eight cuvettes from a single batch of PRESAGE® material were separated into two irradiation groups: ^{60}Co with 0.35 T MRgRT (200cGy/min), and 6 MV linac RT (600cGy/min). Four cuvettes were uniformly irradiated to 0, 5, 10, and 15Gy on both systems under the following dosimetric conditions: 10×10 cm field, 5 cm buildup, 10cm backscatter. Absorption measurements were taken and optical density was calculated prior to and after (t=48hr) irradiation using a Spectronic Genesys 20 spectrophotometer (Thermo Electron Scientific Instruments Corp., Madison, WI).

2.3 Large Volume Characterization

A detailed study of PRESAGE® dosimeters (2kg) was conducted to investigate the temporal and spatial stability of radiation induced optical density change (ΔOD) over one week. Dose-response linearity was investigated on three dosimeters irradiated with four equally-spaced square (4cm2) 6MV fields delivering doses between 10cGy and 300cGy. Spatial uniformity of ΔOD response was investigated on three separate dosimeters irradiated uniformly with 15MV extended-SSD fields with doses of 15cGy, 30cGy and 60cGy. Data from temporal and spatial stability studies were compiled, analyzed, and modeled using CERR and MATLAB's Curve Fitting tool-box. Derivation of the temporal correction for scan-time t'=48hrs (C_T) involved normalizing ΔOD reconstructions from t=48hr (delayed) by ΔOD reconstructions at t=1hr (immediate). Derivation of the spatial correction for scan-time t'=48hrs (C_R) involved the subtraction of registered ΔOD reconstructions for time-stamp t=1hr from t'=48hrs as a function of radial distance from dosimeter core. The final corrected optical density measurement $\Delta OD_{t=1hr}$ is displayed in the equation 1 below:

$$\Delta OD_{t=1hr} = \Delta OD_{t=48hrs}\left(C_T - \frac{C_R}{100\%}\right) \quad (1)$$

IOP Conf. Series: Journal of Physics: Conf. Series **847** (2017) 012056 doi:10.1088/1742-6596/847/1/012056

Application of equation 1 involves correcting each element within a scan taken at t'=48hrs, the typical post-irradiation scan-time in a remote dosimetry study, back to t=1hr.

2.4 MRgRT Study

TG-119 plans were created in the MRIdian®'s (by ViewRay®) TPS located at Washington University in preparation for an IMRT evaluation study [10]. 2x1kg and 4x2kg PRESAGE® dosimeters were scanned, pre-irradiation, at Duke and shipped. For preliminary large volume testing, we delivered an open calibration (AP) beam and a symmetrical 3-field plan (BANDS) with three adjacent regions of varying dose (3Gy, 6Gy, 9Gy) using 10cm diameter PRESAGE®. Percent depth dose and dose-response uniformity within the dosimeter under a magnetic field were analyzed. After initial testing, the remainder of TG-119 plans were delivered to 14cm 2kg PRESAGE® dosimeters. Dosimeters were returned to Duke within t=48hrs post-irradiation for dose readout. All treatment deliveries were validated with absolute dose measurements using an IBA CC01 ion chamber and Washington University's standard patient-specific QA methods for MRgRT systems.

3. Results

3.1 Formulation Characterization

In small volume (4ml), PRESAGE® responds linearly to ^{60}Co irradiation, in the presence of a 0.35T magnetic field. The sensitivity was measured at (0.0305(±0.003)cm^{-1}Gy^{-1}, R^2=0.9996) This value is within 1% of a 6MV non-MR linac irradiation (0.0302(±0.003)cm^{-1}Gy^{-1}, R^2=0.9991). In the large volume study (2kg), consistent intra-batch sensitivity (0.0930±0.002 ΔODcm^{-1}Gy^{-1}) and linearity (R2=0.9996) was observed at t=1hr. However, all dosimeters were observed to gradually darken with time (<5% per day globally) and exhibited a non-linear dose-response at later scan times (>1hr). Relative depth dose distributions are displayed in Figure 1.a for scans with time stamps t=1hr and t=4hrs for the four 4cm^2 fields: A, B, C, and D with Rx doses of 300cGy, 130cGy, 75cGy and 30cGy, respectively. A spatially uniform dose-response was observed at scan-time t=1hr. In delayed scans (>1hr), a small radially-dependent dose-response (<3% globally) was observed Figure 1.b. These results are currently under further scrutiny, but initial hypothesis attributes the phenomena to curing thermodynamics during manufacture. The refined remote dosimetry protocol (including polynomial correction terms for temporal and spatial effects, C_T and C_R) was then applied to independent dosimeters irradiated with MRgRT treatments (Figure 1.c-e).

3.2 TG-119 for MRgRT in 3D

Analysis of TG-119 clinical plans using 3D-gamma (3%/3mm, 10% threshold) give passing rates of: AP 99.3%, bands 99.6%, H&N 99.1%, prostate 98.0%, C-shape 90.8%, and multi-target 98.5%. The average 3D-gamma passing rate for the same criteria was 97.5%± 3.37%.

4. Conclusion

The presence of strong magnetic fields in MRgRT introduces new challenges and constraints to the already difficult task of complex treatment verification. The PRESAGE®/Optical-CT dosimetry system has some useful features in this context, remaining functionally unaffected by magnetic fields, and able to yield dosimetry information in high resolution in 3D. Here we demonstrate a novel 3D remote dosimetry protocol which can provide uniquely comprehensive verification data to enhance commissioning and treatment plan verification for MRgRT. The protocol uses 2kg radiochromic plastic dosimeters read-out by optical-CT within a week of treatment. Small correction factors (~5%) are required to account for spatial and temporal changes that occur in the dosimeter during the delayed readout period of one-week post-irradiation.

IOP Conf. Series: Journal of Physics: Conf. Series **847** (2017) 012056 doi:10.1088/1742-6596/847/1/012056

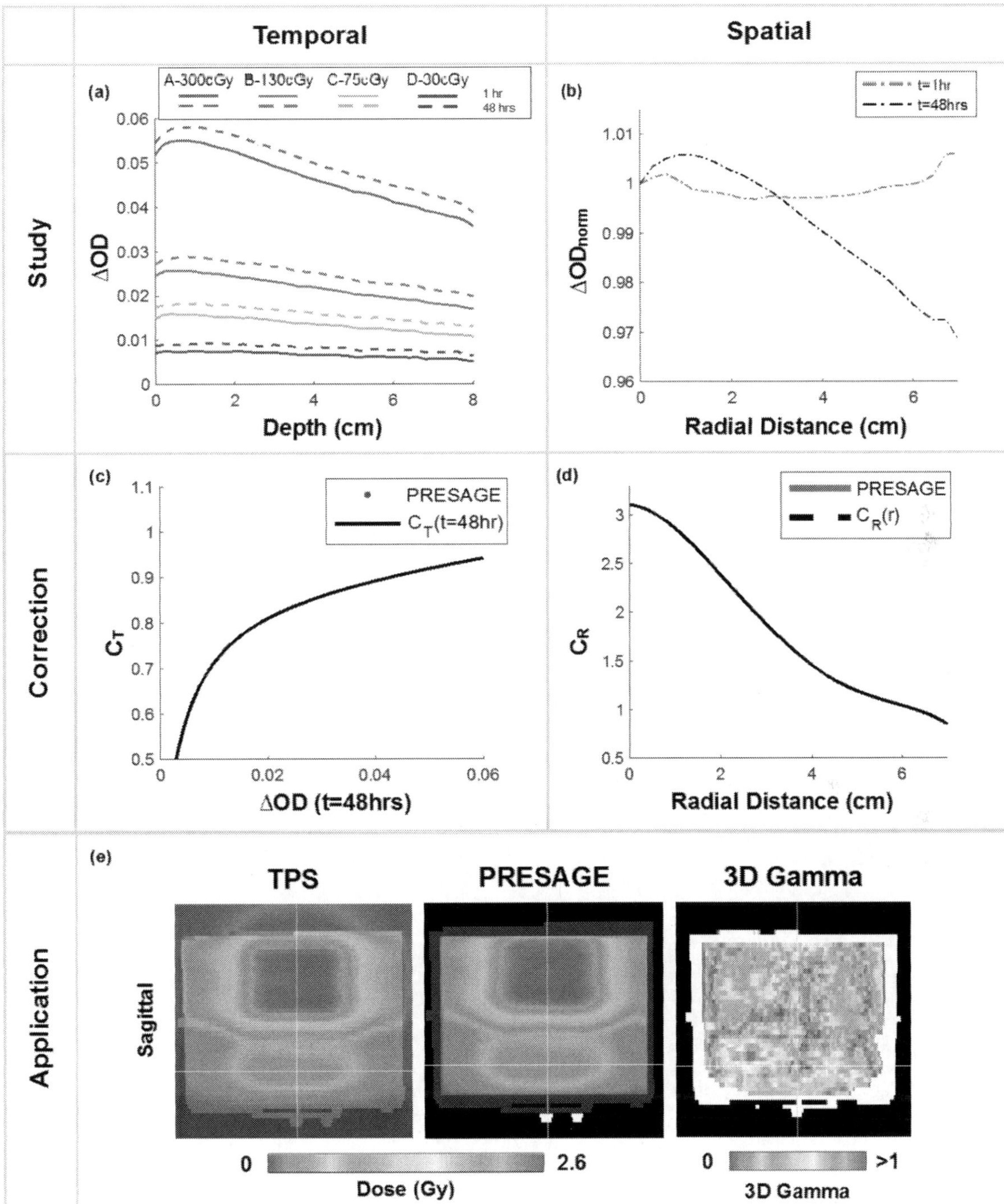

Figure 1. **(a)** Relative depth dose distribution measurements for four fields (A-300cGy, B-130cGy, C-75cGy and D-30cGy) at time-stamp t=1hr and t=48hrs, revealing non-linear increase in OD in delayed read-out. **(b)** Normalized radial dose-response between t=1hr and t=48hrs of uniformly irradiated dosimeters at 60cGy). **(c)** Temporal (relative ratio) and (**d**) spatial (%Δ) corrections. **(e)** 3D TG-119 remote dosimetry using PRESAGE® for ViewRay®. Left to right: TPS calculated dose, measured dose, and 3D-gamma (3%/3mm, 10% threshold) for Multi-target (98.5% passing rate for 3%/3mm, 10% threshold criteria).

5. References

[1] Wooten H et al 2015 *Radiother. Oncol.* **114** 402-05
[2] Li H et al 2014 *Int. J. Radiat. Onco.* **91** 65-72
[3] Baldock C *et al* 2010 *Phys. Med. Biol.* **55** R1-63
[4] Juang T et al 2014 *Med. Phys.* **41** 071706
[5] Jackson J et al 2015 *Phys. Med. Biol.* **60** 2217-30
[6] Thomas A et al 2011 *Med. Phys.***38** 4846-57
[7] Sakhalkar H et al 2009 *Med. Phys.* **36** 71-82
[8] Oldham M et al 2008 *Med. Phys.* **35** 2072-9
[9] Adamovics J et al 2006 *Radiat. Prot. Dosimetry* **120** 107-12
[10] Ezzell G et al 2009 *Med. Phys.* **36** 5359-73

IOP Conf. Series: Journal of Physics: Conf. Series **847** (2017) 012057 doi:10.1088/1742-6596/847/1/012057

Using 3D dosimetry to quantify the Electron Return Effect (ERE) for MR-image-guided radiation therapy (MR-IGRT) applications

Hannah J Lee[1,2], Gye Won Choi[1,2], Mamdooh Alqathami[1], Mo Kadbi[3], Geoffrey Ibbott[1]
[1]Department of Radiation Physics, UT MD Anderson Cancer Center, Houston, TX
[2]UT at Houston Graduate School of Biomedical Sciences, Houston, TX
[3]MR Therapy, Philips healthTech, Cleveland, OH

E-mail: gibbott@mdanderson.org

Abstract. Image-guided radiation therapy (IGRT) using computed tomography (CT), cone-beam CT, MV on-board imager (OBI), and kV OBI systems have allowed for more accurate patient positioning prior to each treatment fraction. While these imaging modalities provide excellent bony anatomy image quality, MRI surpasses them in soft tissue image contrast for better visualization and tracking of soft tissue tumors with no additional radiation dose to the patient. A pre-clinical integrated 1.5 T magnetic resonance imaging and 7 MV linear accelerator system (MR-linac) allows for real-time tracking of soft tissues and adaptive treatment planning prior to each treatment fraction. However, due to the presence of a strong magnetic field from the MR component, there is a three dimensional (3D) change in dose deposited by the secondary electrons. Especially at nonhomogeneous anatomical sites with tissues of very different densities, dose enhancements and reductions can occur due to the Lorentz force influencing the trajectories of secondary electrons. These dose changes at tissue interfaces are called the electron return effect or ERE. This study investigated the ERE using 3D dosimeters.

1. Introduction

An integrated 1.5 T magnetic resonance imaging and 7 MV linear accelerator system (MR-linac, Elekta AB, Stockholm, Sweden) is under development as an image-guided radiation therapy system. Designed with the strong magnetic field (B-field) of the MR component perpendicular to the linac beam, the MR-linac allows for superior soft tissue image contrast compared to CT and MV or kV on-board imagers. However, the design of the MR-linac system presents some dosimetric concerns on secondary particle trajectories due to the Lorentz force. This is especially true for any interfaces with organs that contain air such as soft tissue-lung interfaces [1, 2]. Hot spots occur at transitions from high to low density structures, and cold spots occur at transitions from low to high density structures. These hot and cold spots are due to the Lorentz force directing the secondary electrons back into the higher density medium and is commonly called the electron return effect (ERE). The ERE has been simulated using various Monte Carlo codes, suggesting that the dose enhancement at water-air interfaces can be up to a factor of 1.6 and at soft tissue-lung interfaces up to a factor of 1.5 [1-4]. The ERE depends on several factors, such as machine-defined B-field strength and linac beam energy and patient anatomy-defined tissue interfaces and interface shapes. These factors result in a 3D change in dose distribution where sparse sampling of points using conventional point and 2D array dosimeters may no longer be sufficient to demonstrate agreement with the treatment plan over the irradiated volume. For these reasons, 3D validation of dose delivery in the MR-linac may be necessary due to the various tissue interfaces within a human body, such as at the lung or trachea, that may result in out-of-plane dose perturbations.

As modern radiation therapy has become more complex with IMRT, VMAT, IGRT, and other treatment techniques, many forms of 3D dosimetry have been presented as tools for acquiring true volumetric dose distributions with continuous integrated dose measurements [5-9]. In this study, we utilized two types of 3D dosimeters to assess the acquisition of volumetric dose distributions in the MR-linac. Each dosimeter contained a cylindrical air cavity to induce the ERE.

2. Materials and Methods

The two types of dosimeters used for this study were radiochromic plastic PRESAGE® (Heuris Pharma, LLC, Skillman, NJ) and radiochromic gel FOX. PRESAGE® was first introduced in 2003 and has already been used for assessing IMRT and other complex treatment techniques [5-9]. The in-house made PRESAGE® dosimeters underwent an optically-visible change (clear to dark green color) upon irradiation due to the formation of malachite green from leuco malachite green (LMG). Each PRESAGE® dosimeter was prepared with a concentric cylindrical air cavity (1.5 cm diameter air cavity, 7.2 cm diameter dosimeter).

FOX was formulated in-house and had a Fricke-like composition but with greater MR contrast [10]. Similar to Fricke dosimeters, our FOX formulation underwent iron (II) oxidation upon absorbing radiation dose, resulting in optical (yellow to purple color) and MR-visible changes. Each FOX dosimeter was prepared in an 8.8 cm diameter petri dish, and a concentric cylindrical glass vial was initially inserted into the gel to mold a 2.5 cm diameter air cavity. The glass vial was removed before the gel was completely set to reduce disruption of the gel matrix.

The FOX dosimeter was primarily composed of water and was nearly equivalent to water and soft tissues (elemental compositions and effective atomic number (Z_{eff}) are listed below in Table 1). PRESAGE® had a greater Z_{eff} with its primarily polyurethane plastic composition [11].

Table 1. Elemental compositions of materials (weight fractions denoted as w_k) and Z_{eff}

Material	w_H	w_N	w_O	w_C	w_S	w_{Cl}	w_{Br}	w_{Fe}	w_{Na}	Z_{eff}
PRESAGE® [11]	0.0892	0.0446	0.2172	0.6074	-	0.0333	0.0083	-	-	8.65
FOX	0.1098	2.66E-6	0.8819	0.0081	1.12E-7	-	-	1.10E-4	1.56E-10	7.42
Water	0.1119	-	0.8881	-	-	-	-	-	-	7.42
Muscle	0.1042	0.0350	0.7289	0.1230	0.0050	-	-	-	-	7.46

2.1. *Irradiation set-up with MR-linac*

Both types of dosimeters were irradiated in the same set-up with the MR-linac. The direction of the B_0-field was along the axis of the cylindrical hole, which was centered within each cylindrical dosimeter. The linac beam (15 x 15 cm^2 field) entered perpendicularly to the B_0-field and cylindrical axis of the dosimeters (Figure 1).

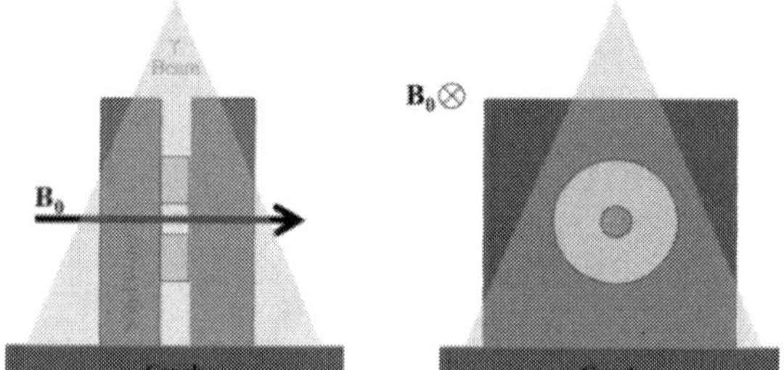

Figure 1. Irradiation set-up with MR-linac.

2.2. *Dosimeter readout techniques*

Two hours after irradiation, PRESAGE® dosimeters were readout using an optical-CT technique with the Duke Midsized Optical-CT System (DMOS) [12]. The dosimeter was placed on a rotating pedestal inside a refractive-index-matched fluid. Data were recorded for each angle at 1° intervals and reconstructed into a 3D data set.

FOX dosimeters were analyzed during and immediately after irradiation using the MR component of the MR-linac with balanced-Fast Field Echo (b-FFE) and T1-weighted sequences. Slices 5-mm thick were acquired for each dosimeter and combined into a 3D data set.

3. Results and Discussions

3.1. Experimental PRESAGE® results

Cross-sectional slices of the dose distributions for the PRESAGE® dosimeter showed a clear demonstration of the dose enhancement (red area) due to the ERE above the cavity (high to low density interface) and a region of reduced dose (blue area) below the cavity (low to high density interface) (Figure 2a).

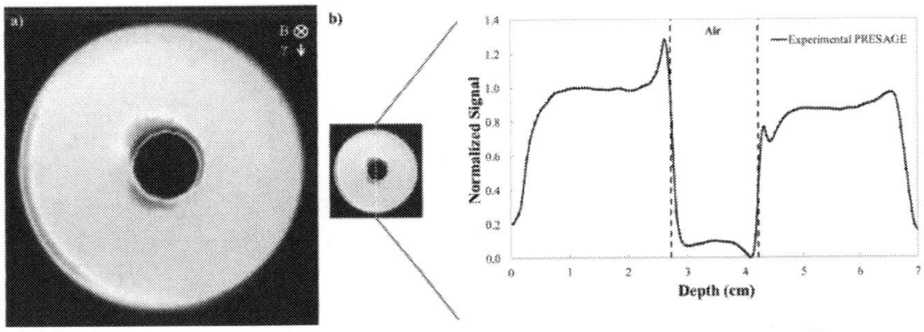

Figure 2. a) Cross-sectional slice of PRESAGE® and b) line profile.

The regions of increased and reduced doses were slightly rotated with respect to the radiation beam axis due to the average direction change of the electrons [2, 3]. Measurements from line profiles showed the dose enhanced to about 0.5 cm around the cavity for the PRESAGE® dosimeter by up to 30% (Figure 2b). The relative signal decreased 20% at cold spots for PRESAGE®.

3.2. Experimental FOX results

After irradiation, cross-sectional slices of the dose distributions for FOX dosimeters similarly showed a clear demonstration of the dose enhancement due to the ERE above the cavity and a region of reduced dose below the cavity (Figure 3a). Measurements from line profiles showed the dose enhanced to about 0.5 cm around the cavity by up to approximately 40% and reduced to about the same distance by 20% for FOX (Figure 3b).

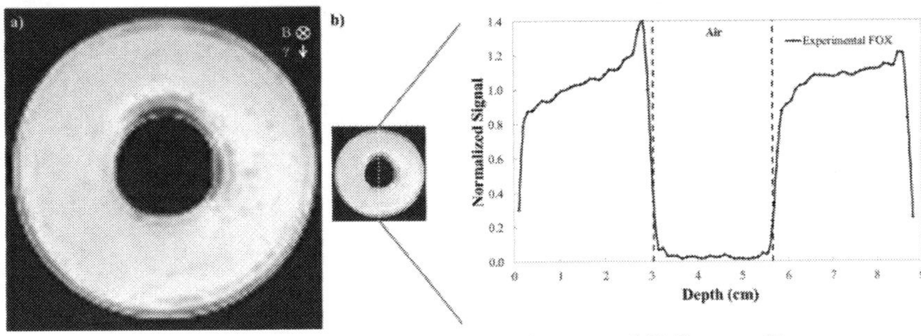

Figure 3. a) Cross-sectional slice of FOX and b) line profile.

Because the FOX dosimeter could be analyzed using MR measurements, near real-time cross-sectional slices were also acquired during irradiation. Snap shots of 6 dose points are shown in Figure 4a where the overall MR signal changes can be seen in the cross-sectional slices. However, the spatial resolution of these real-time images were not suitable for measuring the ERE in real-time (Figure 4b).

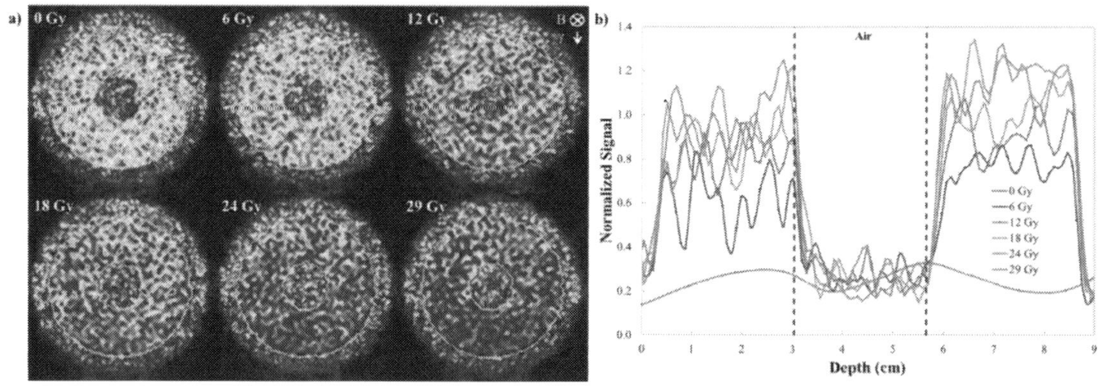

Figure 4. a) Cross-sectional slice snap shots of FOX during irradiation (approximate dose values to the center of the dosimeter) and b) line profiles.

4. Conclusions

The results of this study confirmed the feasibility of using radiochromic plastic PRESAGE® and radiochromic gel FOX for MR-linac and other MR-IGRT applications. Our results agreed well with Monte Carlo simulation results in the literature (assuming a water medium) with dose enhancement by up to 30-40% in a 1.5 T B-field for similar sized air cavities [2]. This study encourages the use of 3D dosimetry for rigorous QA of MR-IGRT in a volumetric acquisition and will be particularly advantageous at heterogeneous sites. The 3D dose readout for PRESAGE® has the advantage of greater spatial resolution with optical measurements whereas the FOX dosimeter has the advantage of readout in the same position as irradiation within the MR-linac using MR measurements, as well as the capability for real-time measurements.

5. References

[1] Lee H *et al* 2015 *Med. Phys.* **42** 3312
[2] Raaijmakers A J E *et al* 2005 *Phys. Med. Biol.* **50** 1363-76
[3] Raaijmakers A J E *et al* 2008 *Phys. Med. Biol.* **53** 909-23
[4] Rubinstein A E *et al* 2015 *Med. Phys.* **42** 5510-6
[5] Juang T *et al* 2013 *J. Phys.: Conf. Ser.* **444** 12080
[6] Lafratta R J *et al* 2015 *Phys. Conf. Ser.* **573** 012055
[7] Jackson J *et al* 2015 *Phys. Med. Biol.* **60** 2217–2230
[8] Schreiner L J 2015 *J. Phys.: Conf. Ser.* **573** 012003
[9] Baldock C *et al* 2010 *Phys. Med. Biol.* **55** R1-63
[10] Lee H *et al* 2016 *Med. Phys.* **43** 3660
[11] Brown S *et al* 2008 *Appl. Radiat. Isot.* **66** 1970-4
[12] Newton J *et al* 2010 *J. Phys.: Conf. Ser.* **250** 12078

IOP Conf. Series: Journal of Physics: Conf. Series **847** (2017) 012058 doi:10.1088/1742-6596/847/1/012058

Development of a methodology to study the effect of magnetic field on dose distributions in an MR-linac, using PRESAGE® and Monte Carlo calculations

F Costa[1], S Doran[2], S Nill[1], S Duane[3], D Shipley[3], I Billas[3], J Adamovics[4] and U Oelfke[1]

[1]Joint Department of Physics, The Institute of Cancer Research and The Royal Marsden NHS Foundation Trust, London SM2 5NG, UK

[2]CRUK and EPSRC Cancer Imaging Centre, Institute of Cancer Research, London SM2 5NG, UK

[3]Acoustics and Ionising Radiation Team, National Physical Laboratory, Hampton Road, Teddington TW11 0LW, UK

[4]Department of Chemistry and Biology, Rider University, Lawrenceville, NJ 08648, USA

Email: filipa.costa@icr.ac.uk

Abstract. At the Royal Marsden Hospital (RMH) and the Institute of Cancer Research (ICR) a new MR-linac is being installed and commissioned. Modifications to absorbed dose patterns will occur, because the paths of secondary electrons are deflected by Lorentz forces. In this paper, we describe a methodology to measure the effects of magnetic field on dose distributions, using the PRESAGE® 3D dosimeter and Monte Carlo (MC) simulations. A poly(methyl methacrylate) (PMMA) phantom has been developed to be positioned between the poles of an electromagnet and accommodate cylindrical samples of PRESAGE®. This phantom will be used to study the influence of different magnetic field strengths on radiation deposition from a Cobalt-60 (^{60}Co) beam. Both the orientation of the samples with respect to the magnetic field and radiation beam, and the size of the air gap can be changed. Preliminary MC simulations with two PRESAGE® cylinders separated by an air gap, gave a good insight about the dosimetric effects that can be obtained with our newly-developed phantom.

1. Introduction

One of the world's first combined MRI and linear accelerator (MR-linac) systems is currently being installed and commissioned at the Royal Marsden Hospital (RMH) and the Institute of Cancer Research (ICR). As part of an international consortium to develop the technology, our aim is to provide image guidance during treatment by acquiring MR images of patients, with high soft-tissue contrast.

The MRI system (Philips, Hamburg, Germany) is similar to a standard clinical scanner, but with the main magnetic field generated by a specially-designed, "split" 1.5 T magnet, with no gradient or static magnetic field (B0) coils in a central region through which radiation passes. A 7 MV linear accelerator (Elekta, Crawley, UK), with multileaf collimator, will rotate in a ring around the outside of the magnet.

IOP Conf. Series: Journal of Physics: Conf. Series **847** (2017) 012058 doi:10.1088/1742-6596/847/1/012058

A significant challenge in the delivery of radiotherapy by the MR-linac is the altered behaviour of secondary electrons. As the main magnetic field is always on, Lorentz forces will act perpendicularly to the direction of the electron travel and thus deflect the path of the cascade of secondary electrons [1]. This is particularly relevant at tissue-air boundaries where an electron return effect (ERE) is visible as electrons are forced back to the tissue, creating an increase in dose near the boundary. The effect is also dependent on the orientation of the entrance and exit surfaces with respect to the MRI static magnetic field [2]. Build-up distance is reduced and this could also create high surface doses.

Treatment planning systems have been specially developed to account for these effects [3]. However, accurate dosimetric characterization is essential to validate and commission this new technique before clinical implementation. Performing 2D data measurements of individual beams to validate IMRT treatment plans has been shown to be prone to inaccuracies [4]. Effects of magnetic field are likely to contribute to the increased complexity of dose distributions on an irradiated volume, emphasizing the need for accurate high resolution 3D dosimetric evaluation [5].

The aim of this paper is to define a methodology to study the effect of magnetic field on dose distributions, using the PRESAGE® 3D dosimeter [6] when air gaps are present. Preliminary Monte Carlo (MC) simulations are reported to test this experimental set-up.

2. Material and Methods

2.1. Phantom development and experimental set-up

A PMMA phantom was developed to accommodate cylindrical samples of PRESAGE® (Heuris Pharma, Skillman, NJ) of 2 cm diameter and a maximum of 7 cm length (figure 2B). Two pieces of PRESAGE® can be placed inside the phantom, separated by an air gap of variable size. The phantom can also be oriented in 3 different positions in relation to the beam (figure 1). The PMMA phantom consists of a number of separate ($5 \times 5 \times 1$ cm^3) slabs that can be assembled together and provide electronic equilibrium, with a density very similar to that of PRESAGE®. The black acetal base was created to fit a sample holder developed at the UK's National Physical Laboratory (NPL, Teddington, UK) for positioning ionization chambers in the magnetic field.

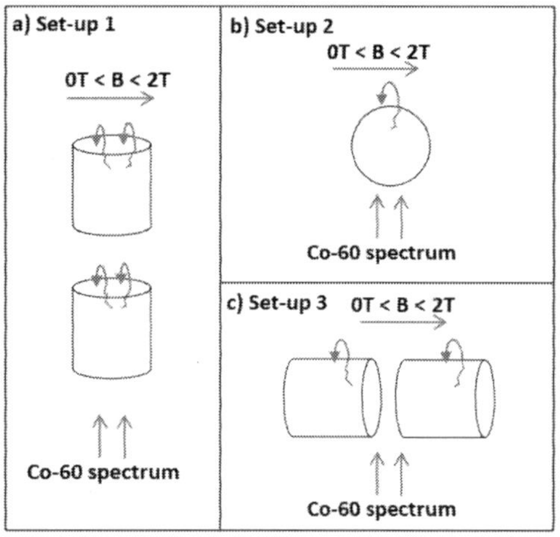

Figure 1. Three different PRESAGE® orientations that can be set-up using the PMMA phantom between the pole pieces of the electromagnet, to study the effect of an air-gap size, orientation and consequences of changing the magnetic field strength. The curved blue arrows represent schematically the electron paths under the influence of the Lorentz force.

Prior to the MR-linac itself coming into service, the phantom is placed in an electromagnet which allows us to study the effects on dose distribution under different magnetic field (B) strengths. At NPL, an electromagnet (250MM Electromagnet, GMW, San Carlos, USA) has been installed next to a ^{60}Co unit (Theratron 780C, Theratronics, Ontario, Canada). The distance between the poles of the electromagnet can be varied, but for this study, the configuration used had a 5 cm gap, giving an accessible range of magnetic field strength of 0 T – 2 T by varying the current intensity. The distance from the source to the centre of the magnetic poles was set to be 162 cm.

A schematic of three possible PRESAGE® set-up orientations is shown in figure 1. In the first figure1 (a), the long axis of the PRESAGE® is parallel to the beam direction, with an air gap in the middle. At the entrance of the first PRESAGE®, a shifted depth of dose maximum (dmax) is expected. A 4 mm shift to the left of the profile was reported for a 6 MV photon beam [2]. An increase on the dose is also expected on the distal side of both PRESAGE® cylinders due to the ERE. For this reason, electrons coming from the first PRESAGE® do not cross the air cavity, and, a new build up curve should appear at the beginning of the second PRESAGE®, as electron equilibrium has to be established again. When the magnetic field is off, a slight rebuild up is still expected due to the lack of lateral electron equilibrium. In the other configurations (figures 1(b) and 1(c)), the dose increase is expected on the exiting lateral side of both PRESAGE® cylinders. The experimental arrangement at NPL is shown in figure 2A.

2.2. *Optical CT*

Samples will be imaged using an upgraded version of the optical CT microscopy scanner described in [7], which has a maximum field-of-view of 2.5 cm.

2.3 *Monte Carlo (MC) Simulations*

EGSnrc/Cavity, a C++ user-code advanced EGSnrc application, was used to simulate the experimental set-up. A phase-space file of the Theratron numerical model has been previously implemented in BEAM, an EGSnrc-based MC code, at NPL for reference conditions (SSD=95cm, 10 × 10 cm^2 field). This phase-space file was used as the source on the EGSnrc/Cavity simulation, after modifying the jaws and the distance from the source to the centre of the poles, to be representative of the new set-up.

Simulations of two PRESAGE® cylinders with their long axes parallel to the beam and a 1.95 cm air gap in between them were performed with and without 1.5 T magnetic field [8]. Percentage depth dose (PDD) profiles were obtained for the first of the set-up orientations shown in figure 1(a).

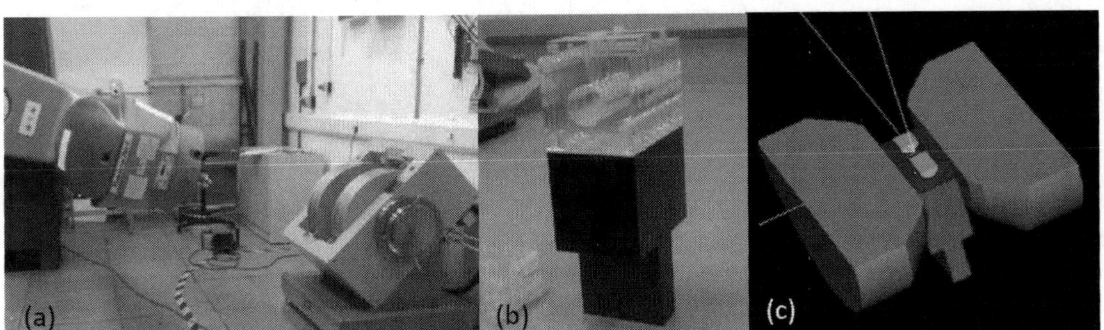

Figure 2. a) Experimental set-up at NPL. b) Phantom developed to position the PRESAGE® samples on the electromagnet. c) Axial cut of the simulated geometry displayed with C++ geometry viewer, EGS view.

3. Results and discussion

Preliminary simulations with and without the 1.5 T magnetic field are shown in figure 3 and demonstrate a dose increase of more than 40% at the beam exit of both samples due to the ERE. However, the expected reduction of dmax depth is not present. This effect may not be noticeable for the Co-60 as the beam has lower energy than the 6MV beam. Additionally, the build-up region may be modified by the increased dose effect due to the pole pieces of the electromagnet. More simulations need to be performed to answer these questions.

The pole pieces were found to contribute to an increase of approximately 8% on the PRESAGE® dose. This is likely to be due to the lateral scatter radiation coming from the pole pieces. In addition to the dose increase, when the pole pieces are simulated, a higher dose is observed at the end of the first PRESAGE® and beginning of the second PRESAGE®. This is not fully understood and further investigation is still required.

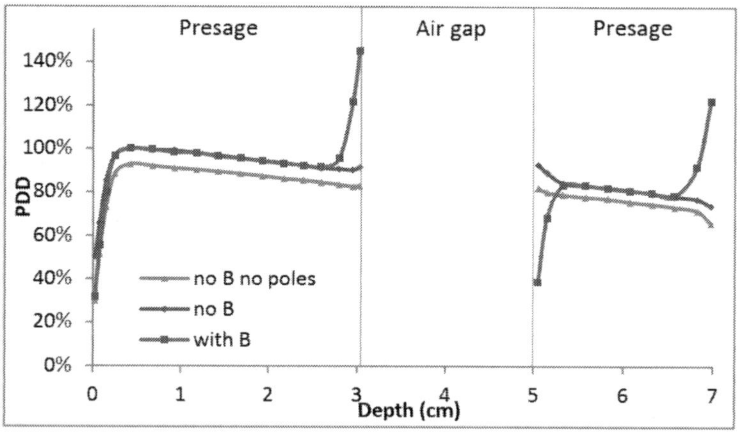

Figure 3. Preliminary results obtained by taking a central profile along the length of the two PRESAGE®s (figure 1(a), figure 2 (c), which have a 1.95cm air gap between them. Simulations were performed with and without magnetic field (B).

Work published very recently in thesis form [9] has addressed a number of issues related to this work. Its author encountered a number of problems associated with the optical-CT readout of a PRESAGE® sample with a hole in the middle. With our newly-developed phantom, it should be possible to mimic the simulated arrangement and study the effect of ERE in the presence of air gaps by scanning independently two PRESAGE® cylinders without such severe refraction-related artefacts. Furthermore, the ability to vary the air gap, the magnetic field strength and the interface orientation with respect to the magnetic field and beam entry will allow us to compare our experimental data with a wider variety of simulation results, thus resulting in a better understanding of the ERE.

It is likely, however, that the effect of an interface between PRESAGE® and the matching liquid (used to reduce the refractive index mismatch between the sample and its surroundings) will still need to be corrected in order to achieve accurate quantitative results in the build-up regions. This will be an important area of future study.

4. Conclusion

This preliminary study provided good insight about the dose distributions we expect to obtain with PRESAGE® cylinders, separated by an air gap, and in the presence of strong magnetic fields. However, some dose effects still need further investigation. Future 2D and 3D simulations of the set-up orientations described above will be used to compare with the optical-CT reconstructed images of the samples.

5. Acknowledgements

SJD acknowledges receipt of CRUK and EPSRC support to the Cancer Imaging Centre at ICR and RMH in association with MRC and Department of Health C1060/A10334, C1060/A16464 and the NHS funding to the NIHR Biomedical Research Centre and the Clinical Research Facility in Imaging. The Institute of Cancer Research is supported by Cancer Research UK under Programme C33589/A19727.

6. References

[1] Raaijmakers A J E *et al* 2004 *Phys. Med. Biol.* **49** 4109-18
[2] Raaijmakers A J E *et al* 2005 *Phys. Med. Biol.* **50** 1363-76
[3] Bol G H *et al* 2012 *Phys. Med. Biol.* **57** 1375-85
[4] Ibbott G S 2010 *J. Phys.: Conf. Ser.* **250** 012001
[5] Schreiner L J 2015 *J. Phys.: Conf. Ser.* **573** 012003
[6] Guo P Y *et al* 2006 *Med. Phys.* **33** 1338
[7] Doran S J *et al* 2013 *Phys. Med. Biol.* **58** 6279-97
[8] Kirkby C *et al* 2008 *Med. Phys.* **35** 1019
[9] Choi G W 2016 *Measurement of the Electron Return Effect Using PRESAGE Dosimeter* (Thesis)

University of Texas

IOP Conf. Series: Journal of Physics: Conf. Series **847** (2017) 012059 doi:10.1088/1742-6596/847/1/012059

The potential of polymer gel dosimeters for 3D MR-IGRT quality assurance

Y Roed[1,2], Y Ding[3], Z Wen[2], J Wang[2], L Pinsky[1] and G Ibbott[2]
[1]Department of Physics, University of Houston, Science and Research Building 1, 3507 Cullen Blvd. Houston, TX 77204 USA
[2]Department of Radiation Physics, University of Texas MD Anderson Cancer Center, 1400 Pressler St., Houston, TX 77030 USA
[3]Department of Imaging Physics, University of Texas MD Anderson Cancer Center, 1400 Pressler St., Houston, TX 77030 USA

Email: gibbott@mdanderson.org

Abstract. Advances in radiotherapy technology have enabled more accurate delivery of radiation doses to anatomically complex tumor volumes, while sparing surrounding tissues. The most recent advanced treatment modality combines a radiation delivery system (either Cobalt-60 therapy heads or linear accelerator) with a diagnostic magnetic resonance (MR) scanner to perform MR-image guided radiotherapy (MR-IGRT). For a radiation treatment plan to be delivered successfully with MR-IGRT the compliance with previously established criteria to validate the passing of such plans has to be confirmed. Due to the added strong magnetic field a new set of quality assurance standards has to be developed. Ideal detectors are MR-compatible, can capture complex dose distributions and can be read out with MRI. Polymer gels were investigated as potential three dimensional MR-IGRT quality assurance detectors.

1. Background

Image-guided radiotherapy (IGRT) is currently considered the standard in radiotherapy for treating cancer patients and is achieved by integrating on-board imaging devices with linear accelerators [1]. Patients are positioned for each treatment fraction with on-board imagers using bony anatomy and corrections may be made during the treatment itself to allow for more accurate dose delivery to the tumor while limiting the dose to the surrounding normal tissue. Quality assurance (QA) standards have been developed to ensure the precision of prescribed radiation treatment plans. Only a few millimeters deviation between the treatment plan and the actual delivered treatment is accepted for the plan to pass. These rigorous passing criteria lead to more accurate treatment delivery [2].

Magnetic resonance-image guided radiotherapy (MR-IGRT) has been proposed as an advanced treatment modality. Delineation of the tumor and adjacent organs is improved by the superior soft tissue contrast available from MR images. With the presence of a strong magnetic field, Lorenz forces act on secondary electrons influencing the shape of the delivered dose distribution in three dimensions. It remains essential to confirm that the delivered dose complies with the treatment plan within strict passing criteria. QA standards for treatment plan verification for MR-IGRT delivery systems are presently being developed. The strong magnetic field adds new challenges in the selection of QA dosimeters as electronic detectors need to be MR-compatible and their response might be influenced under such conditions.

Polymer gel dosimeters have previously been shown to be a useful tool to capture complex 3D dose distributions with steep dose gradients [3, 4]. The polymerization of the gel in response to radiation dose can be measured with an MR scanner because spin-spin relaxation rates (R2) increase in proportion to radiation dose [3, 5]. These properties make them favorable devices for QA standards in MR-IGRT.

The influence of strong magnetic fields on the steep dose gradients of radiation fields was investigated. Polymer gel dosimeters were irradiated with two different delivery systems for which different magnetic field strengths and radiation field sizes were used. This choice resulted from limited availability of one of the delivery systems. Results from exposures of the polymer gels are presented in this study.

2. Methods

Custom-designed glass cylinders filled with a proprietary polymer gel mixture were provided by the manufacturer (MGS Research, Madison, Connecticut). Each dosimeter consisted of a cylindrical sensitive volume (4 cm height, 5 cm diameter) that was attached to a 10 cm long filling port intended to minimize oxygen perfusion into the sensitive volume.

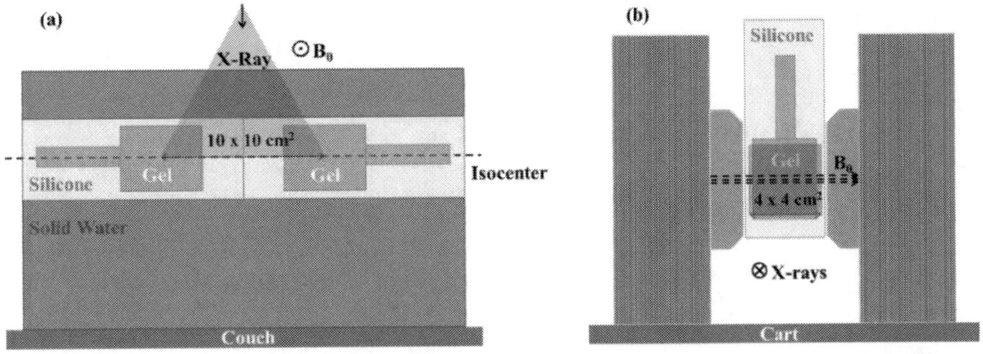

Figure 1. Experimental setup for (a) MR-Linac and (b) electromagnet combined with Versa HD™

A set of dosimeters was irradiated with a non-clinical MR-Linac pilot system (Elekta AB, Stockholm, Sweden) that combined a 1.5 T Philips Marlin MR scanner with a 7 MV linear accelerator (MR-Linac). The dosimeters were placed separately in a phantom to assure full scatter conditions and irradiated to 15 Gy. The center of each dosimeter was positioned at isocenter distance in the penumbra region of a 10x10 cm² radiation field to capture the field edges parallel to the magnetic field lines (Figure 1 a).

The experiment was repeated with the dosimeters placed in a 1.04 T magnetic field generated by an electromagnet (GMW Associates, San Carlos, California) and irradiated with a 6 MV linear accelerator (Elekta Versa HD™). A dose of 10 Gy was delivered to the sensitive volume. A source-to-axis-distance of 300 cm was used to avoid interference with the linear accelerator. At this distance the radiation field was collimated to 4x4 cm² to shield the electromagnet (Figure 1 b). The setup in the MR-Linac was reproduced, in that the X-ray beam, the magnetic field lines and the profiles were all perpendicular to each other.

All dosimeters were imaged with a 3T scanner (GE Discovery™ MR750, Waukesha, Wisconsin) 24h after irradiation using a 2D spin echo sequence with a repetition time of TR = 1000 ms and four variable echo times TE = 10, 20, 60, and 100 ms.

R2 maps were generated for all dosimeters and 18 line profiles across the sensitive volume were averaged. A Savitzky-Golay filter with window length M = 11 and polynomial order N = 3 was applied to the averaged R2 values to smooth the profiles. After smoothing, the data were normalized and the inflection point was determined for either side of the radiation field from which 80/20 penumbra widths were obtained. Data obtained with the MR-Linac were compared with measurements with

Gafchromic™ EBT3 film of the cross-plane profile of the 10x10 cm^2 field from the MR-Linac, under the same conditions.

The directions of the X-ray beam, the magnetic field and the line profile are described in the International Electrotechnical Commission gantry coordinate system, which rotates with the gantry. In this case, the beam is always directed toward $-Z_g$, the magnetic field lines appear to be pointing parallel to the $+Y_g$ axis while the MR scans have been performed parallel to the X_g axis.

3. Results

The penumbra regions of the radiation field were clearly visible in all dosimeters. The application of the Savitzky-Golay filter to the averaged line profiles smoothed the data. The line profiles across the penumbra regions showed the distinct shape of the beam profile arising from the influence of the Lorentz force on secondary electrons in the presence of the magnetic field. The Lorentz force preferentially directs secondary electrons in the $+X_g$ direction when the magnetic field lines point in the $+Y_g$ direction and the gantry is at 0 degree with the X-ray beam in the $-Z_g$ direction. The field edge corresponding to the $-X_g$ of the radiation field appeared more rounded. The absence of quasi-electronic equilibrium in the $-X_g$ penumbra of the radiation field prevented replacement of the electrons that were swept from the penumbra region into the field. On the $+X_g$ side of the field, however, a sharper penumbra appeared as secondary electrons were deflected out of the radiation field while a quasi-electronic equilibrium present inside the field replaced electrons generated in the penumbra region.

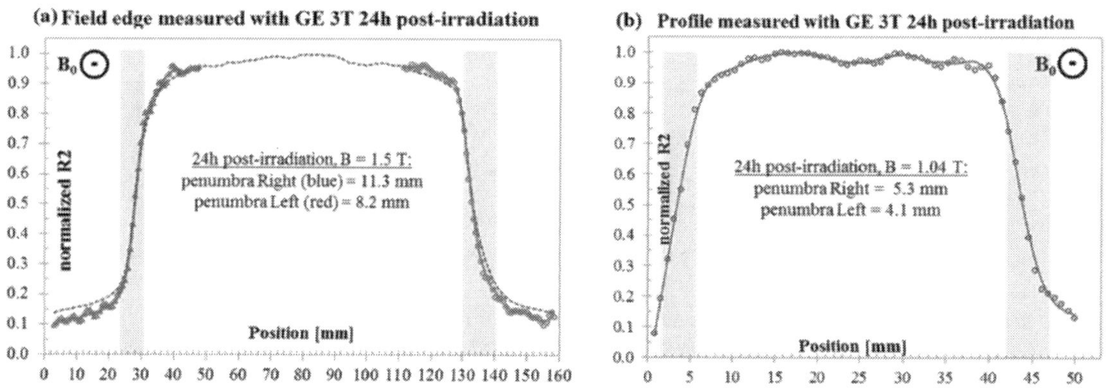

Figure 2. Averaged (circles) and smoothed (solid line) line profiles measured when irradiated with the MR-Linac (a) and with the Versa HD™ and electromagnet (b); film measurements of the cross-plane profile (dashed line) from the MR-Linac are shown for comparison.

The averaged (circles) and smoothed (solid line) line profiles across the penumbra region measured with the MR-Linac are shown in Figure 2 a. Due to their size the dosimeters did not cover the full radiation field and the field size (the distance between inflection points) was not demonstrated with this setup. The two transition regions from outside to fully inside the radiation field were visualized on R2 gray scale maps Line profiles across the regions indicated the distinct shape of the field edges in the presence of the magnetic field and agreed well with the cross-plane film profile (dashed line). The fall-off of the right field edge ($+X_g$) appeared steeper than the fall-off on the left ($-X_g$). The 80/20 penumbra regions are indicated by shading in Figure 2 a. The penumbra width on the right was determined to be 11.3 mm on the right and 8.2 mm on the left.

The collimated 4x4 cm^2 radiation field was fully captured inside the sensitive volume of the dosimeters when using the electromagnet in combination with the Versa HD™ linear accelerator as seen in Figure 2 b. However, the tails of both field edges fell outside the sensitive volume. Although the Savitzky-Golay filter smoothed the data, some fluctuations remained. The distinct shape of the beam profile due to the presence of the magnetic field was evident. The transition regions on R2 gray scale maps appeared to be very similar in width. Calculations confirmed this observation: the penumbra on

IOP Conf. Series: Journal of Physics: Conf. Series **847** (2017) 012059 doi:10.1088/1742-6596/847/1/012059

the left ($-X_g$) measured 4.1 mm while the penumbra on the right ($+X_g$) was 5.3 mm wide. The distance between the inflection points was 40.6 mm. The shading in Figure 2 b indicates the penumbra regions.

4. Discussion and Conclusions

The results showed that the polymer gel dosimeters could capture and resolve the steep dose gradients of penumbra regions of radiation fields in the presence of magnetic fields of different strengths. Smoothing the data with a Savitzky-Golay filter decreased the noise across the transition region so that the inflection point could be determined and the widths of the penumbrae could be calculated.

Investigations of field edges parallel to the magnetic field lines have been performed previously by irradiating polymer gels with an MR-Linac [6] and with Monte Carlo simulations [7]. Similar differences in penumbra width for the $-X_g$ and $+X_g$ field edges were obtained when applying a sigmoidal fit to measured data to calculate penumbra widths. The Monte Carlo study showed the same distinct shape of the penumbrae demonstrated by profiles drawn perpendicularly to the magnetic field lines.

The number of R2 values measured was defined by the MR scanner resolution and the selection of scanning parameters (matrix size, field of view). A high spatial resolution introduced noticeable noise in the MR images resulting in noisy profiles. To adequately determine penumbra widths the noisy data were smoothed. With a pixel size of 0.8 mm the differences in calculated penumbra widths fell within the dimensions of a few pixels.

Considering the difficulty of set up in the electromagnet at the extended distances and the MR scanner resolution the measured field size agreed well with the collimated field size. Using different or more optimized parameters for the Savitzky-Golay filter might improve the noise reduction after applying the filter.

This study demonstrated great promise of using polymer gels as relative 3D quality assurance devices for testing radiation treatment plans delivered by MR-IGRT delivery systems. More studies are needed to extend the set of measurements to include more geometrically complex dose distributions through delivery of intensity modulated radiotherapy treatment plans.

5. References

[1] Greer P B *et al* 2007 *Med. Phys.* **34** 4389-98
[2] Ezzell G A *et al* 2009 *Med. Phys* **36** 5359-73
[3] Baldock C *et al* 2010 *Phys. Med. Biol.* **55** R1-63
[4] Ibbott G S *et al* 1997 *Int. J. Radiation Oncology Biol. Phys.* **38** 1097-103
[5] De Deene Y *et al* 1998 *Radiotherapy and Oncology* **48** 283-91
[6] Roed Y *et al* 2016 *Med. Phys.* **43** 3661
[7] Raaymakers B W *et al* 2004 *Phys. Med. Biol.* **40** 4109-18

IOP Conf. Series: Journal of Physics: Conf. Series **847** (2017) 012060 doi:10.1088/1742-6596/847/1/012060

Clinical experience with Mobius FX software for 3D dose verification for prostate VMAT plans and comparison with physical measurements

L A Vazquez-Quino, C I Huerta-Hernandez and D Rangaraj
[1]Radiation Oncology Department, Baylor Scott & White Health, Texas A&M Health Science Center College of Medicine, Temple, TX

Email: Luis.Vazquez@BSWHealth.org

Abstract. MobiusFX, an add-on software module from Mobius Medical Systems for IMRT and VMAT QA, uses measurements in linac treatment logs to calculate and verify the 3D dose delivered to patients. In this study, 10 volumetric-modulated arc therapy (VMAT) prostate plans were planned and delivered in a Varian TrueBeam linac. The plans consisted of beams with 6 and 10 MV energy and 2 to 3 arcs per plan. The average gamma value with criterion of 3% and 3mm MobiusFX and TPS: 99.96%, 2% and 2mm MobiusFX and TPS: 98.70 %. Further comparison with ArcCheck measurements was conducted.

1. Introduction

The MobiusFX solution from Mobius Medical Systems (Mobius Medical Systems, Houston, TX) for IMRT QA uses linac treatment logs to calculate and verify the 3D dose delivered. One of the most attractive reasons to use this approach is that there is no need for ion-chamber, diode array, EPID, film, or any external measurements as historically used in the field [1-7]. Due to the nature of the Mobius FX method for IMRT QA, this in principle implies a faster method since no device needs to be positioned.

2. Methods and materials

In this study, 10 volumetric-modulated arc therapy (VMAT) prostate plans were planned and delivered in a Varian TrueBeam linac (Varian Medical Systems, Palo Alto, CA). All VMAT plans were delivered and measured at the same time with ArcCheck and Mobius means to ensure consistency in the data as shown in figure 1.

3. Results

Comparisons of profiles and gamma analyses were conducted on all VMAT QA plans (Figure 2). In order to ensure impartiality for the study the 3D dose [8] obtained by Mobius FX method was exported and introduced into the SNCPA software to analyse the comparison against ArcCheck measurement and TPS. The average gamma value with criterion of 3% and 3mm of Mobius FX vs the Treatment Planning System (TPS) was 99.96%, and 2% and 2mm of Mobius FX vs TPS was 98.70%. Similar results were observed for ArcCheck and Mobius FX gamma analysis calculation and dose profiles (Figure 3).

IOP Conf. Series: Journal of Physics: Conf. Series **847** (2017) 012060 doi:10.1088/1742-6596/847/1/012060

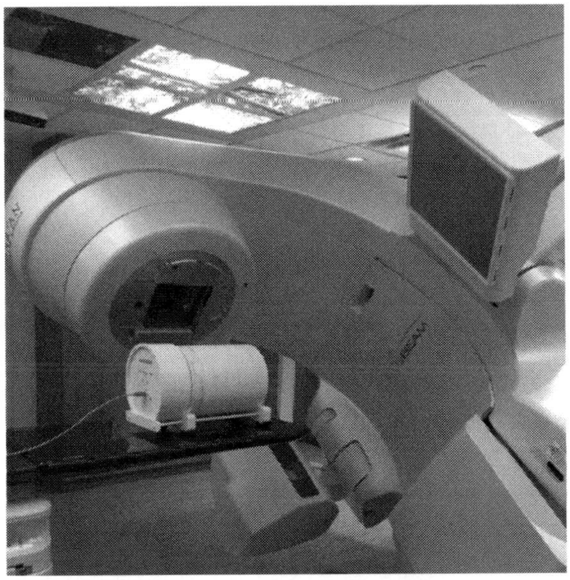

Figure 1. ArcCheck set up for VAMT delivery in a Truebeam linac. Mobius FX and ArcCheck measurements acquired at the same time.

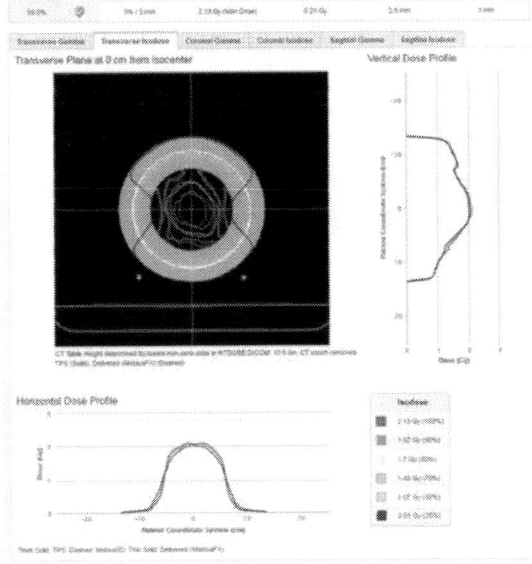

Figure 2. Mobius FX analysis of linac log files recorded at same time of ArcCheck measurements.

4. Conclusions

MobiusFX is capable of producing similar results to those produced by traditional methods for patient specific pre-treatment verification VMAT QA. Even the gamma results compared with the TPS are similar.

IOP Conf. Series: Journal of Physics: Conf. Series **847** (2017) 012060 doi:10.1088/1742-6596/847/1/012060

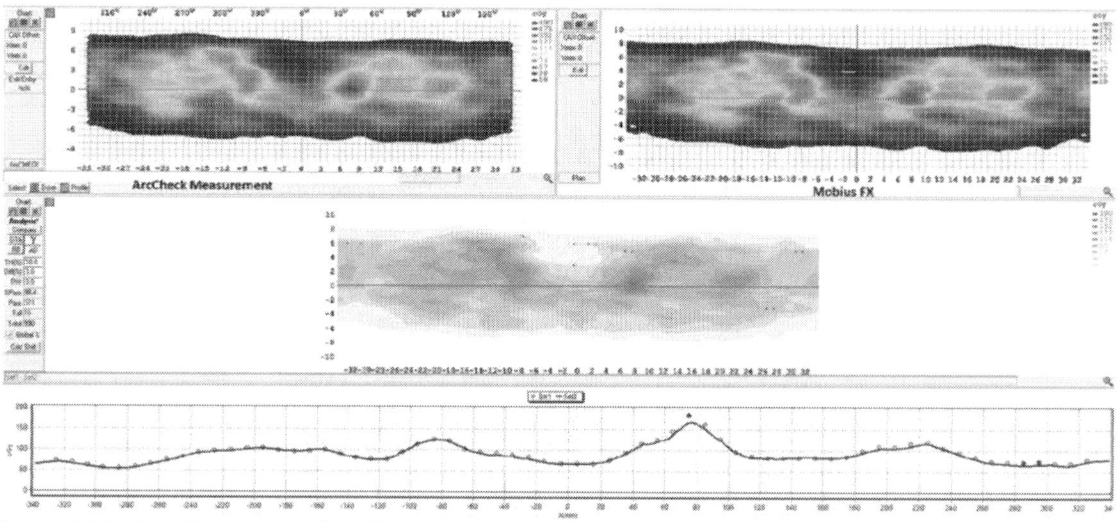

Figure 3. Mobius FX gamma and profile analysis comparison with ArcCheck measurement for a prostate VMAT plan with a gamma passing of 99.6% for 3%, 3mm criterion.

5. References

[1] Buonamici F B *et al* 2007 *Med. Phys.* **34** 1372-9
[2] Chandraraj V *et al* 2011 *J. Appl. Clin. Med. Phys.* **12** 338-49
[3] Li J G 2009 *J. Appl. Clin. Med. Phys.* **10** 62-74
[4] Vazquez Quino L A *et al* 2014 **13** 1-10
[5] Borca C 2013 *J. Appl. Clin. Med. Phys.* **14** 158-171
[6] Hill R F *et al* 2008 *Radiat. Meas.* **43** 1258-64
[7] Hill R *et al* 2009 *Med. Phys.* **36** 3971-81
[8] Baldock C *et al* 2010 *Phys. Med. Biol.* **55** R1-63

IOP Conf. Series: Journal of Physics: Conf. Series **847** (2017) 012061 doi:10.1088/1742-6596/847/1/012061

Development of 3D Slicer based film dosimetry analysis

K M Alexander[1], A Robinson[2], C Pinter[2], G Fichtinger[2] and L J Schreiner[1,3]
[1]Department of Physics, Queen's University, Kingston, Ontario, Canada K7L 3N6
[2]School of Computing, Queen's University, Kingston, Ontario, Canada, K7L 3N6
[3]Department of Medical Physics, Cancer Centre of Southeastern Ontario at Kingston General Hospital, Kingston, Ontario, Canada, K7L 5P9

E-mail: kevin.alexander@krcc.on.ca

Abstract. Radiochromic film dosimetry has been widely adopted in the clinic as it is a convenient option for dose measurement and verification. Film dosimetry analysis is typically performed using expensive commercial software, or custom made scripts in Matlab. However, common clinical film analysis software is not transparent regarding what corrections/optimizations are running behind the scenes. In this work, an extension to the open-source medical imaging platform *3D Slicer* was developed and implemented in our centre for film dosimetry analysis. This extension streamlines importing treatment planning system dose and film imaging data, film calibration, registration, and comparison of 2D dose distributions, enabling greater accessibility to film analysis and higher reliability.

1. Introduction

Dose delivery validation in modern clinical radiation therapy (RT) is difficult, as treatment techniques have become increasingly complex. This has led to the need for high-resolution detectors to accurately measure radiation dose distributions [1]. One convenient clinical dosimetry option is radiochromic film as it is near tissue equivalent, energy and dose rate independent, and is an integrating dosimeter. Film dosimeter use is widely published and has been used for quality assurance, small field dose measurements, and as part of the commissioning of new machines and treatment techniques. At the last IC3DDose conference, we presented new, open-source 3D gel dosimetry analysis software, which streamlined gel dosimeter analysis from hours to 5-10 minutes [2]. Since then, we have recognized that an open-source film dosimetry analysis software would also be a useful clinical tool for film dosimetry. This work presents the initial development and implementation of such software.

2. Materials and Methods

In our clinic, film dosimetry analysis was previously performed using commercial software which was expensive and which performed corrections and optimizations (data smoothing, non-uniformity corrections), sometimes without the user's knowledge. To overcome these challenges and software license costs, a streamlined user interface for a 2D film dosimetry workflow has been implemented in 3D Slicer, as an extension called a *slicelet*. 3D Slicer is a medical software platform with advanced graphics rendering, segmentation, and registration capabilities. 3D Slicer is a good fit for film dosimetry analysis, as it is a free, open-source and customizable computational tool used for image analysis and visualization. The film dosimetry analysis slicelet draws on tools developed for radio-

IOP Conf. Series: Journal of Physics: Conf. Series **847** (2017) 012061 doi:10.1088/1742-6596/847/1/012061

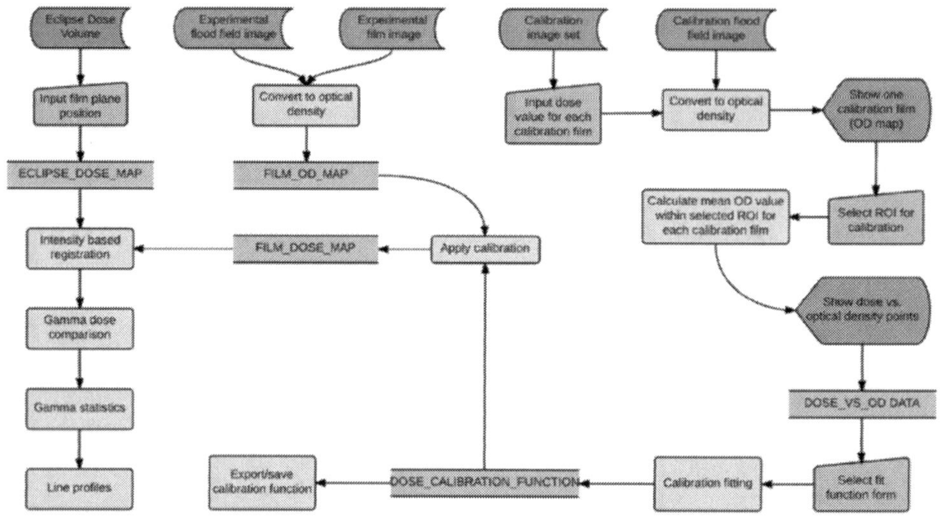

Figure 1. The workflow of the Film Dosimetry Analysis slicelet showing the planning data (upper left), film imaging data (top), film calibration (right), and dose comparison (bottom left).

therapy called *SlicerRT* [3] (www.slicerrt.org), which contains a multitude of RT-specific features, such as loading of DICOM-RT data, computation and display of dose-volume histograms, manipulation of structures, dose comparisons, and dose distribution visualization.

In this work, we present the initial development, implementation, and testing of the Film Dosimetry Analysis slicelet, which follows the general workflow shown in figure 1. The testing was performed using Gafchromic EBT3 film (International Specialty Products (ISP) Corporation, Wayne, NJ) coupled with our custom built, in-house CCD lightbox film dosimetry imaging system, using single red channel dosimetry methods. Using the CCD lightbox film imaging system with a diffuse red light field, flood field images are acquired without any film placed on the light field surface to provide a reference against which to normalize the image of the irradiated film. The irradiated film images are acquired with the film of interest placed flat on the light field surface. Using these images, the optical density for each pixel of each film is calculated. A sample film dosimetry case taken from our clinic's commissioning work on a new fractionated stereotactic radiation therapy (FSRT) brain metastases treatment is presented to demonstrate the functionality of the slicelet.

2.1. Calibration

In order to calibrate the film for dosimetry, a relationship between the optical density and the absorbed dose during irradiation is determined. This is performed by irradiating several films to known doses under well-defined conditions (e.g. 10 x 10 cm^2 fields) in order to establish the film response model. These calibration films are imaged and a flood field image is acquired. The images of these films and the flood field are then imported into the slicelet and assigned their appropriate dose value, or assigned as the flood field image. For example, in this case calibration doses of 0, 50, 100, 200, 300, 400, and 500 cGy were loaded along with a flood field image (figure 2a). Next, a region of interest is chosen, which defines a uniform region of the calibration films (figure 2b) and an average optical density value is calculated within that region for each calibration film. These corresponding optical density and dose values are then plotted and fit using the following function [4] (figure 2c):

$$\text{Dose} = a + b{\cdot}OD + c{\cdot}OD^n,$$

where fit parameters a, b, c are constants determined by a least squares fitting function, and n is limited to the range of [1.00, 4.00]. When this calibration routine is complete, the fit parameters automatically populate the fields of the slicelet. Fit parameters can be saved to a text file for future use.

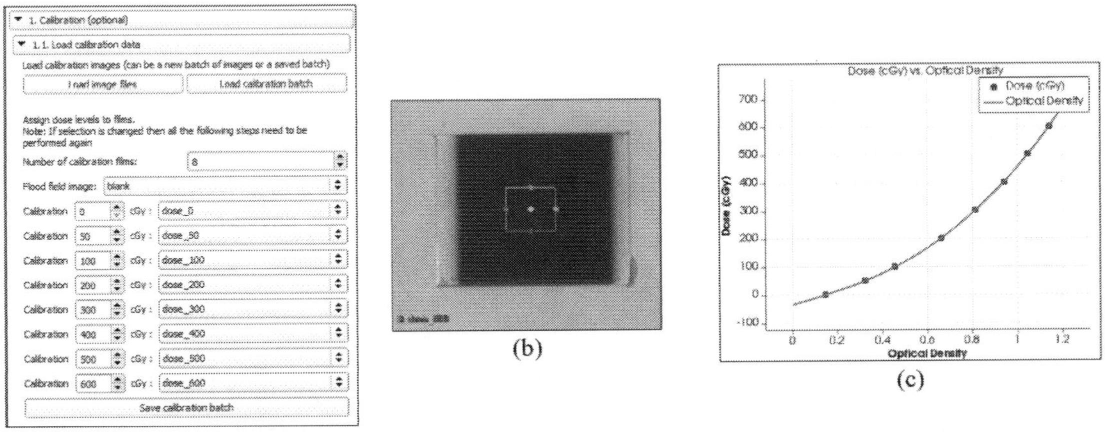

(a)

Figure 2. a) A screenshot of the graphical user interface of the calibration tab in the 3D Slicer Film Dosimetry Analysis slicelet. b) A region of interest drawn about the central, uniform region of a 600 cGy irradiated film (10 x 10 cm^2 field size). c) Calibration curve generated using the slicelet and fitted using the n-power polynomial function.

2.2. *Load experimental data and apply calibration*

The experimental film for this test case is a VMAT FSRT brain metastases treatment with 4 Gy prescribed to each of the four metastases (varying in size from 8-20 mm in diameter), and a 2 Gy whole brain dose. In this step of the slicelet, the DICOM dose and structures, and the experimental film image is also imported. The imported experimental film image, flood field image, and the DICOM dose volume are then assigned their respective roles, and the film resolution is input, in units of mm/pixel (figure 3a). The calibration function is then applied to the experimental film to convert to dose (figure 3b).

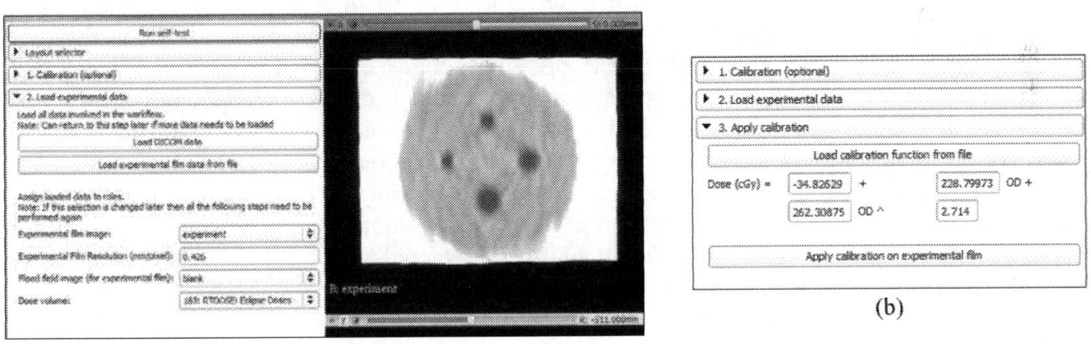

(a)

Figure 3. a) Screenshot showing the 'load experimental data' options and an image of the experimental film. b) Screenshot of the 'apply calibration' step, where fit parameters appear after performing the calibration step (step 1). These values can also be manually input.

2.3. *Film registration*

In order to register the film measurement to the calculated dose plane, the coronal plane at isocenter is extracted from the DICOM dose volume. The experimental film is also resized from pixels to distance in millimetres. The experimental film is then registered using a rigid registration built-in to the BRAINSfit registration module in 3D Slicer (figure 4a).

IOP Conf. Series: Journal of Physics: Conf. Series **847** (2017) 012061 doi:10.1088/1742-6596/847/1/012061

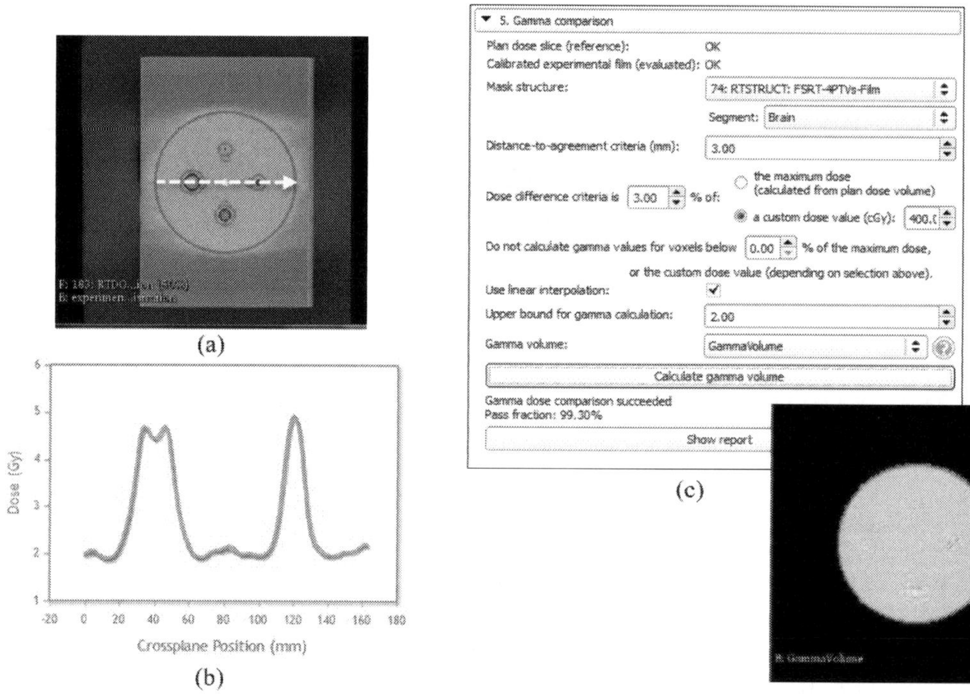

Figure 4. a) Film (green) after completing a rigid registration to the planned dose (grey) and structure set. b) Dose profiles about the registered experimental film (orange) and Eclipse (blue) dose distributions, as indicated by the line drawn in figure 4a. c) Screenshot of the user interface for the gamma comparison tool, and plane of gamma value calculated within the brain structure.

3. Results and Conclusions

After calibrating and registering the film measurement, the final step in the film dosimetry analysis slicelet is to compare that measurement to the planned dose distribution by using a 2D gamma dose comparison (figure 4c). The gamma dose comparison tool allows the user to set the distance to agreement and dose difference, as well as the option to only calculate the gamma values above a minimum dose threshold, or within a particular structure. In this example, excellent agreement between the measured and planned dose distributions was shown, resulting in a 3%/3mm gamma pass rate within the brain structure of 99.3%. Dose profiles also show excellent agreement (figure 4b).

Overall, the film dosimetry analysis slicelet streamlines complex film analysis; another example of its use can be found in a clinical example reported in this conference's proceedings [5]. This open-source analysis tool helps make film dosimetry more accessible and transparent. It also enables convenient film dosimetry, free from the constraint of commercial analysis or read out systems. Future work will look at implementing multichannel dosimetry methods in the slicelet. This poster, along with a workshop during the meeting, will show how 3D Slicer can facilitate efficient 2D and 3D dosimetry.

4. Acknowledgements

Research funding for this work has been provided by the Canadian Institutes of Health Research (CIHR) and the Ontario Consortium for Adaptive Interventions in Radiation Oncology (OCAIRO).

5. References

[1] Baldock C *et al* 2010 *Phys. Med. Biol.* **55** R1-63
[2] Alexander KM *et al.* 2015 *J. Phys.: Conf. Ser.* **573** 012042.
[3] Pinter C *et al.* 2012 *Med. Phys.* **39** 6332-7
[4] Alexander KM *et al.* 2016 *J. Phys.: Conf. Ser.* (presented at IC3DDose 2016)
[5] Devic S *et al.* 2004 *Med. Phys.* **31** 2392-401

IOP Conf. Series: Journal of Physics: Conf. Series **847** (2017) 012062 doi:10.1088/1742-6596/847/1/012062

The effect of THPS concentration on nPAG microstructure

C J Watson[1], J R Supple[1], A U Yeo[1,2], M Geso[3] and R D Franich[1]
[1]School of Science, RMIT University, Melbourne, VIC 3001, Australia
[2]Physics Department, Radiation Oncology Victoria, GenesisCare , East Melbourne, VIC 3002, Australia
[3]School of Health and Biomedical Sciences, RMIT University, Bundoora, VIC 3083, Australia

E-mail: callum.watson@rmit.edu.au

Abstract. The microstructures formed within a normoxic polyacrylamide gel (nPAG) dosimeter and their correlation to amount of dose received is investigated through the use of two scanning electron microscope modalities. The effect of gold sputtering onto the samples was examined to determine if it impacted the detectable structures. The evolution of polymer structure with increasing dose is shown to progress from long shard like polymer chains to smooth polymer structures coating the gel surface. It is demonstrated that without gold coating the samples, which is a necessity in high vacuum scanning electron microscopy, that an environmental scanning electron microscope can be used for analysis of polymerized structures. The addition of THPS as an antioxidant is shown to not alter the formation of polymer microstructures when compared to other studies involving THPC. It is also proposed that knowledge of these features could be used to manipulate the deformability, stability and polymerized opacity of an nPAG dosimeter.

1. Introduction

The polymerized microstructure of normoxic polyacrylamide gel (nPAG) must be well understood to have certainty in a measured dose from a dosimeter. The pre and post irradiation stability, in both chemical and physical terms, can have significant effects on dose response and sensitivity [1, 2]. The molecular structure of the dosimeter volume is quite significant when quantifying these effects and could be exploited to alter and improve the dosimeter utility [3], specifically with the intent of tracking and verifying dose deformation [4].

The introduction of antioxidants such as Bis [tetrakis (hydroxymethyl) phosphonium] sulphate solution (THPS) and Tetrakis (hydroxymethyl) phosphonium chloride solution (THPC) has made it easier to fabricate dosimeters that are less susceptible to negative effects due to atmospheric oxygen interference [5-8]. With more rigorous investigation of the molecular structure of the polymer gel using techniques such as scanning electron microscopy (SEM), the physico-chemical effects of individual chemicals can be better understood and manipulated to alter or tune the dose sensitivity, temporal stability and reproducibility of nPAG dosimeters [9]. The difference between the resulting microstructures from polymerization of polyacrylamide gels containing THPS and THPC could be quantified, and SEM analysis may yield quantitative metrics for the measurement of dose response and sensitivity [10], and a means of manipulating physical properties of dosimeters or adding imaging features for tracking dose warping or measuring dose deformation [11]. The separation of gelatin

IOP Conf. Series: Journal of Physics: Conf. Series **847** (2017) 012062 doi:10.1088/1742-6596/847/1/012062

structures in the hydrogel relative to the size of Bis and AAm is quite large [12], and it has very little influence over the diffusion rate and distance of these small molecules [13].

The effect that THPS has on polymer chain separation and resultant pore size is of interest for expanding nPAG dosimeter applications. The molecular structure variances caused by THPS when compared to THPC using environmental and high vacuum SEM have been investigated, while also describing the effect that SEM sample preparation has on polymer gel samples, namely any structural damage caused during deposition of a gold coating.

2. Materials and methods

The hydrogel examined in this study was similar that used by Yeo *et al* [14], where 2 wt.% gelatin was used as Sedaghat *et al* [9] had indicated that a gel concentration of 5 wt.% made it difficult to image fine molecular structure (<30μm). Six samples were prepared and stored in 15 ml glass vials, in which they were irradiated using a Clinac iX model. The samples were delivered maximum doses of 0, 4, 8, 12, 16 and 20 Gy from a 10x10 cm^2 field at 10MV while submerged in a water bath.

The samples were allowed to polymerize for 48 hours before being placed into a freezer at -80°C for 24 hours. After this period the samples were moved to an OPERON freeze dryer for 24 hours which operated at -60°C and a pressure of 0.007 mtorr. The samples were sliced thinly and mounted on a standard SEM imaging stub. A 25 nm gold layer was sputter coated onto 0, 8 and 20 Gy samples. The hydrogels were then imaged using a FEI Quanta200 SEM, the uncoated samples in low vacuum, and the gold coated samples in high vacuum modes.

3. Results and discussion

The images in Figure 1 are representative of the general structures present in each of the samples imaged in low magnification (80x). It can be seen most primarily in the 8 Gy sample (pane c) that there is large variation in the polymer chain separation across the imaged surface, indicating inhomogeneities present during the fabrication process, arising during polymerization and possible damage to the gel structure during the freeze drying process. The variance in polymer chain density is visible in Figure 1, and further magnified in Figure 2, is consistent with results determined using Raman spectroscopy [15], and the size of these structures also increases in size and length with dose until 16 Gy where the polymerization has progressed to a point that the monomer cross linking structure is no longer distinct, instead a rough surface coats much of the gelatin structure.

The surface features present after receiving radiation doses above 12 Gy are difficult to view through SEM in low vacuum conditions, and so applying a gold coating of approximately 25 nm in thickness to the gel sample allows for much higher resolution investigation of the surface topography created by polymerization. The risk in engaging a gold coating process, and indeed freeze drying, is that some of the gel and polymer structures can possibly be damaged or altered. Figure 3 evinces the large increase in resolution and sharpness of the captured images availed by the gold coating, several of the polymer structures are much more well defined even at low magnifications. The cross-linked polymer chains appear to be stable enough to sustain the gold coating without altering conformation. The images of the gold coated samples are of a quality that could be useful for image texture analysis to quantify dose, however the need to find a uniform region of depth and polymerization degree within samples makes the accuracy of dose measurements highly subject to user discretion in image capture.

4. Conclusion

This investigation describes the preparation and imaging via SEM of gold coated and uncoated nPAG samples. The images of polymer structures present in nPAG material is much sharper when gold coated and imaged on the SEM's used in this study, and does not appear to result in any damage to the polymerized structures present in the sample. The addition of THPS has no visual influence on the formation of polymer structures within the nPAG while it does influence the size of the gel pores in the sample.

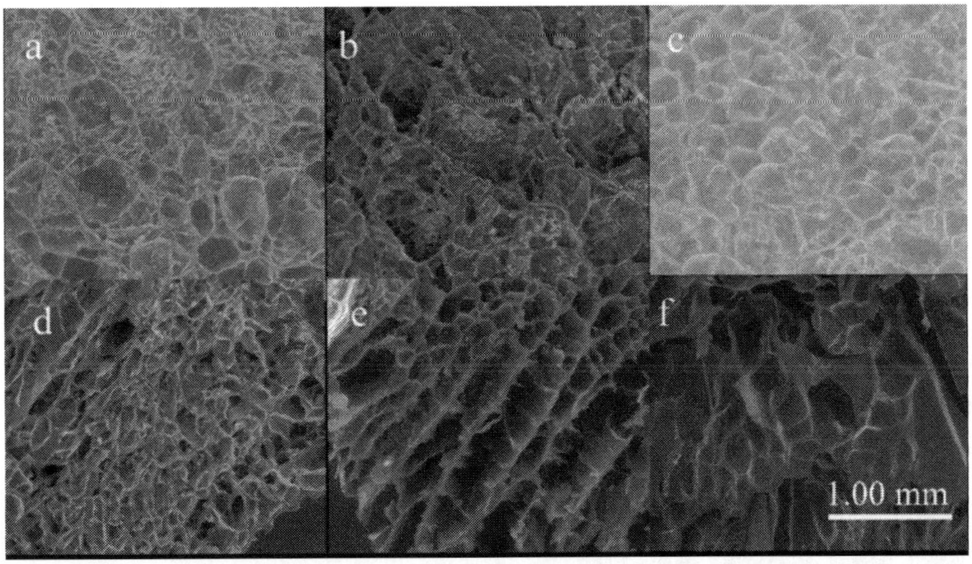

Figure 1. Low magnification SEM images of nPAG structures that have received 0 Gy(a), 4 Gy(b), 8 Gy(c), 12 Gy(d), 16 Gy(e) and 20 Gy(f). Captured on a FEI Quanta200 SEM in low vacuum mode, accelerating voltage of 15kV.

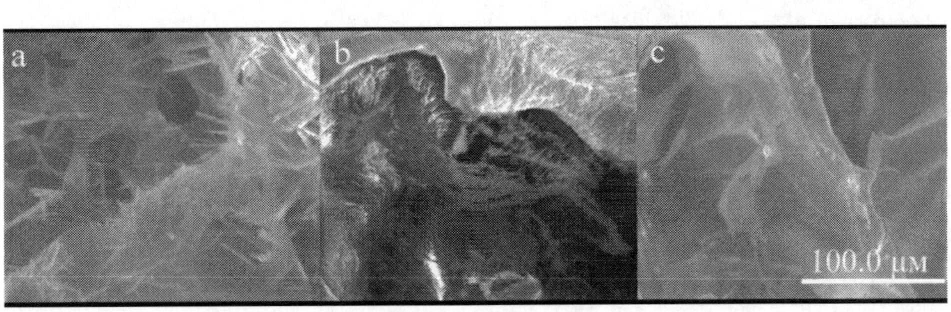

Figure 2. Environmental SEM images of nPAG samples that have received 0 Gy(a), 8 Gy(b) and 16 Gy(c) radiation doses. The samples are subject to small amounts of charge being deposited by the electron beam, resulting in saturation of some image areas.

5. Acknowledgements
The authors would like to thank Dr. Edwin Mayes and the Australian Microscopy & Microanalysis Research Facility at RMIT University for their help in sample preparation and technical assistance in capturing the above images.

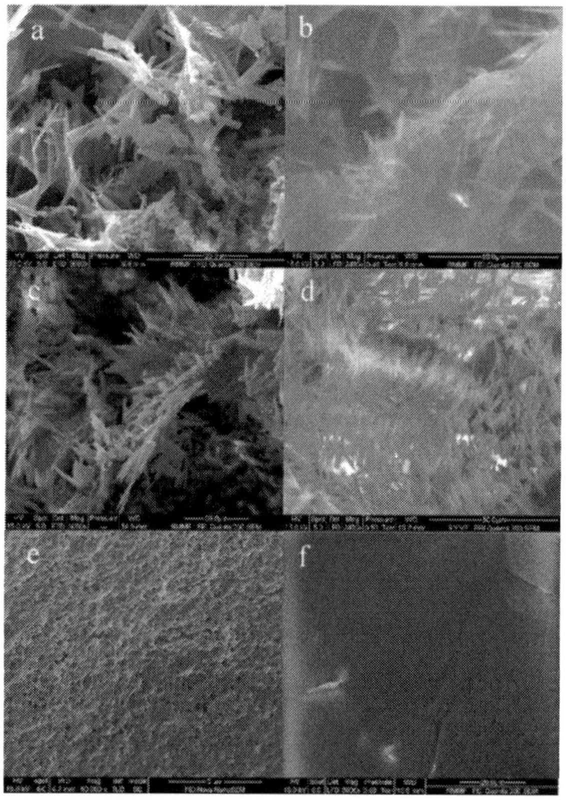

Figure 3. Images (a) and (c) are gold coated samples that have received 0 and 8 Gy respectively, with predominant feature sizes of approximately 20.0 μm, compared to images (b) and (d) which have not been gold coated and possess identical structures but are imaged in the 50.0 μm region. Image (e) is a gold coated sample that has received 20 Gy, the sample stage was tilted slightly to provide extra resolution in the depth changes present on the sample surface, the visible features in this image are less than 2 μm in size. (f) is an uncoated sample at twice the image size of (e), some of the same surface topography is visible but any higher magnification viewing of the sample resulted in charging which created artefacts that limited sharpness and the ability to capture an image displaying features that can be directly related to dose.

6. References

[1] De Deene Y *et al* 2000 *Phys. Med. Biol.* **45** 859
[2] Lepage M *et al* 2002 *Phys. Med. Biol.* **47** 1881-90
[3] Vandecasteele J, De Deene Y 2012 *Phys. Med. Biol.* **58** 43
[4] Yeo U *et al* 2013 *J. Phys.: Conf. Ser.* **444** 012107
[5] Spěváček V *et al* 2014 *Phys. Med. Biol.* **59** 5141
[6] Baldock C *et al* 1998 *Phys. Med. Biol.* **43** 695-702
[7] Baldock C 2006 *J. Phys.: Conf. Ser.* **56** 14-22
[8] Mather M L *et al* 2003 *Ultrasonics* **41** 551-9
[9] Sedaghat M *et al* 2012 *Phys. Med. Biol.* **57** 5981
[10] Shih C-T *et al* 2013 *IEEE T NUCL SCI.* **60** 2155-60
[11] Yeo U J *et al* 2012 *Med. Phys.* **39** 5065-72
[12] Baldock C *et al* 2010 *Phys. Med. Biol.* **55** R1-63
[13] Meesat R *et al* 2009 *Phys. Med. Biol.* **54** 5909
[14] Yeo U J *et al* 2012 *Med. Phys.* **39** 2203-13
[15] Jirasek A I, Duzenli C 2001 *Phys. Med. Biol.* **46** 1949

IOP Conf. Series: Journal of Physics: Conf. Series **847** (2017) 012063 doi:10.1088/1742-6596/847/1/012063

Dose integration and dose rate characteristics of a NiPAM polymer gel MRI dosimeter system

C Waldenberg[1], A Karlsson Hauer[2,3], C Gustafsson[3,4] and S Ceberg[3]
[1]Department of Radiation Physics, University of Gothenburg, Göteborg, Sweden
[2]Department of Therapeutic Radiation Physics, Sahlgrenska University Hospital, Göteborg, Sweden
[3]Department of Hematology, Oncology and Radiation Physics, Skåne University Hospital, Lund Sweden
[4]Department of Medical Physics, Lund University, Malmö, Sweden

E-mail: sofie.ceberg@med.lu.se.

Abstract. The normoxic polymer gel dosimeter based on N-isopropyl acrylamide (NiPAM) is a promising full 3D-dosimeter with high spatial resolution and near tissue equivalency. NiPAM gel samples were irradiated to different doses using a linear accelerator. The absorbed dose was evaluated using MRI and statistical significance of the analysed data was calculated. The analysis was carried out using an in-house developed software. It was found that the gel dosimeter responded linearly to the absorbed dose. The gel exhibited a dose rate dependence, as well as a dependence on the sequential beam irradiation scheme. A higher dose rate, as well as a higher dose per sequential beam, resulted in a lower dose response.

1. Introduction

Commonly today, dosimeters used for absorbed dose distribution measurements contain diodes or ionization chambers positioned in an array [1, 2]. These dosimeters have a limitation in spatial resolution as the diodes or chambers integrates the ionization over an area or a volume and the full 3D dose is estimated from a limited number of measurement points [3, 4]. Furthermore, the diodes or chambers are placed millimeters or centimeters apart and the dose between the measurement points have to be reconstructed and calculated. This makes the uncertainties of 3D quality assurance (QA) depend on both accurate measurements and calculation algorithms provided by the vendor.

Gamma pass rate evaluation is commonly used to compare two dose distributions. The concept combines the percentage dose difference between measured and planned dose as well as the distance to agreement (DTA) [5]. However, there is research that points out that gamma pass rate for virtual 3D dosimeters that utilize arrays of detectors arranged in common 3D geometries does not capture clinically relevant dosimetric differences [6]. One suggestion is to implement dose evaluation trough dose volume histogram (DVH) analysis. This requires a true 3D dosimeter with high spatial resolution such as for example the polymer gel dosimeters.

Polymer gel dosimeters have high spatial accuracy and are near tissue-equivalent, i.e, the absorbing and scattering properties for a given radiation match those of a certain biological tissue [7, 8]. However despite the promising aspects of polymer gel dosimetry, it have yet to be a common dosimeter in clinical use [8]. There could be several reasons for this. A few of them discussed by Ibbott [8] are imaging artifacts emerging during evaluation of the gel, linear energy transfer dependence and the fact that there has been a limited effort to create low and high-density gels to simulate different tissues e.g. lung tissue. Another major reason to why this type of dosimeter has low acceptance in clinics could be that it is time consuming to manufacture and analyse, and, additionally, the monomer toxicity of these dosimeters makes them hazardous and inconvenient to use and manufacture [9]. As discussed by Senden [10], there are other polymer gel formulations which are less toxic. In particular N-Isopropylacrylamide (NiPAM) gel dosimeters holds great promise [10, 11].

The aim of this study was to investigate the dose response of a NiPAM gel dosimeter. The dose rate and the sequential beam irradiation dependence were investigated.

2. Materials and methods

2.1. NiPAM gel preparation

The gel was prepared by heating deionized water together with gelatin (Sigma Aldrich, U.S.A., 5% w/w) to 45°C. When the gelatin had completely dissolved, first N-isopropyl acrylamide, NiPAM, (97% Sigma Aldrich, U.S.A., 3% w/w), and then N,N'-methylene-bis-acrylamide (98% Sigma Aldrich, U.S.A., 3% w/w) was added. Subsequently, the temperature was allowed to drop to 38 °C and the antioxidant Tetrakis hydroxymethyl phosphonium chloride, THPC, (80% solution in water, Sigma Aldrich, U.S.A.) was added to a concentration of 5 mM. The gel was manufactured under normal atmospheric conditions and constantly mixed with a magnetic stirrer. The mixture was poured into glass vials equipped with plastic screw-tops and were left in dark to form a gel over night at room temperature, approximately 21°C. Under the whole process, care was taken to minimize the gels exposure to ambient light. The gel formula was inspired by Senden [10].

2.2. Irradiation, readout and data processing

All gel vials were irradiated with a 10 MV beam, placed in a water filled PVC phantom. The gel samples were irradiated approximately 24 hours after the manufacturing. To characterize the dose rate dependence of the gel, a total of 33 vials were irradiated to 11 different absorbed doses between 0.50 and 10 Gy). Three vials were irradiated to each of the predetermined doses, each with a different dose rate, 100, 300 and 600 MU/min. To characterize the sequential irradiation dependence of the NiPAM gel, a total number of 42 vials were irradiated. Doses for each beam sequence were 0.25, 0.50, 1.0, 2.0 and 4.0 Gy. During irradiation, a dose rate of 600 MU/min was used. The beam-off time between two subsequent beams was approximately 60 seconds. The gel vials were evaluated approximately 24 hours after irradiation using a 3.0T MRI scanner (Discovery 750W General Electric Medical Systems, U.S.A.). The scanning sequence used for all scans was a T2-mapping sequence, named CartiGram, supplied by GE. The gels were scanned and data from 16 echo times were aquired with an inter echo time, ΔTE, of 80 ms (80 - 1280 ms) using a repetition time of 4000 ms.

The images generated by the MRI were processed in an in-house developed software written in Matlab (Mathworks®, Massachusetts U.S.A.). To evaluate the dose rate and sequential beam irradiation dependence, the calculated R2, and the corresponding standard deviation, were plotted as a function of the absorbed dose and linear regressions were fitted to the data.

3. Results

Linear regressions for three different dose rates (100, 200 and 600 MU min^{-1}) were carried out. It was found that the dose response of the NiPAM gel was dependent on the dose rate (Figure 1). The slope of the 100 and 300 MU min^{-1} regressions were 11.8 % and 6.5 % steeper respectively compared to the 600 MU min^{-1} regression. All three regressions were statistically significantly different from each other ($p<0.05$).

The dose response of the gel dosimeter was also dependent on sequential beam irradiation (Figure 2). The response for the single, continuous beam, compared to the response for the sequential beams were all statistically significantly different from each other with the exception of the single beam versus the 4 Gy per sequence-beam.

IOP Conf. Series: Journal of Physics: Conf. Series **847** (2017) 012063 doi:10.1088/1742-6596/847/1/012063

4. Discussion

An F-test evaluation of the regressions for the dose responses with different dose rate (Figure 1) shows that there is a difference between the regressions ($p<0.05$), i.e. the NiPAM gel is dependent on the dose rate. This is an unwanted characteristic as it complicates or limits the area of application of the gel. This result contradicts with the result in the study by Farajollahi *et al* [11] but agrees with the study by Hsieh *et al* [12].

It was found that the dose response of the gel is dependent on sequential beam irradiation (Figure 2). Although the study by Karlsson *et al* [13] investigated different types of polymer gel dosimeters (nMAG and nPAG), their findings coincides with the result in this study.

The rate of production of free radicals, and thereby the concentration of radicals, is directly linked to the dose rate of the beam which the gel is being irradiated with. Unlike polymer growth termination reactions, the polymer chain growth process involves one polymer radical. As a result, the rate of reactions is proportional to the concentration of polymer radicals. However, termination reactions involve two radicals, consequently the incidence of termination reactions is proportional to the square of the radical concentration which makes termination reactions the dominant reaction when high concentration of radicals are present [13]. This might serve as an explanation to why the gel demonstrates dose rate dependence where a higher dose rate gives a lower dose response and vice versa. The sequential beam irradiation dependence of the NiPAM gel can be explained similarly; when the radiation beam is delivered in shorter bursts with no active beam in between, the radicals will approach high concentrations only for short periods and the concentration will drop in between beams. Compared to a single continuous beam, the average lifetime of the radicals generated by the sequential beam is longer, yielding more polymer formulation and increased dose response. It is likely that the sequential irradiation dependence of the gel would be different if time between the sequential beams had been altered. If the time between the sequential beams had been reduced, the sequential irradiation dependence would probably be less pronounced.

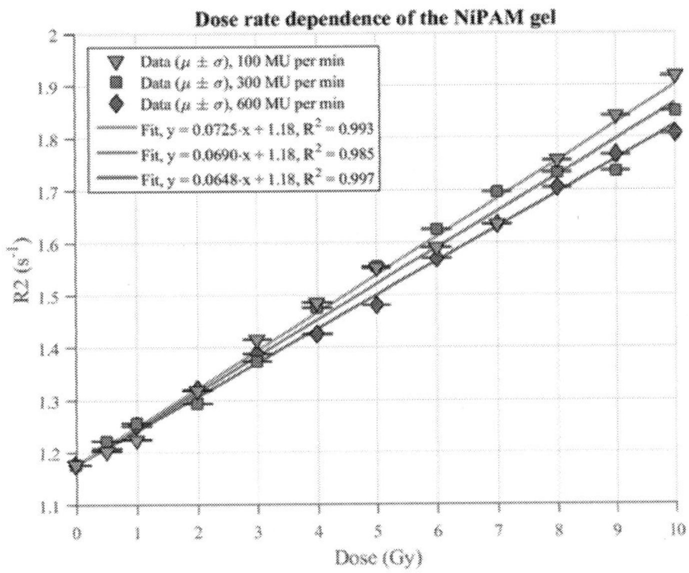

Figure 1. R2 versus absorbed dose for the dose rate dependence study of the NiPAM gel. Samples irradiated with dose rate 100, 300 and 600 MU min^{-1}.

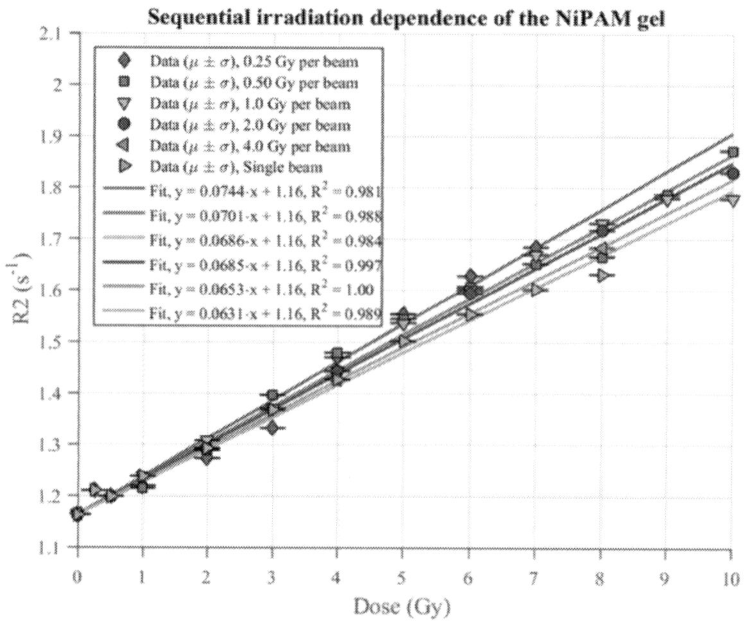

Figure 2. R2 versus absorbed dose for the sequential irradiation dependence study of the NiPAM gel. Samples were irradiated with 600 MU min^{-1}.

5. Conclusion

The dosimeter was found to respond linearly to the absorbed dose within the investigated range of doses and all regressions had a R^2-value greater than 0.98. The NiPAM polymer gel dosimeter shows both a dose rate and a sequential beam dependence.

6. References

[1] Hill R *et al* 2014 *Phys. Med. Biol.* **59** R183-231
[2] Hill R *et al* 2009 *Med. Phys.* **36** 3971-81
[3] Low D 2015 *J. Phys.: Conf. Ser.* **573** 012009
[4] Bäck A 2015 *J. Phys.: Conf. Ser.* **573** 012012
[5] Low D A *et al* 1998 *Med. Phys.* **25** 656-661
[6] Zhen H *et al* 2011 *Med. Phys.* **38** 5477- 89
[7] Baldock C *et al* 2010 *Phys. Med. Biol.* **55** R1-63
[8] Ibbott G S 2004 *J. Phys.: Conf. Ser.* **3** 58
[9] Hill B *et al* 2005 *Med. Phys.* **32** 1589-97
[10] Senden R J *et al* 2006 *J. Phys.: Conf. Ser.* **56** 156
[11] Farajollahi A R *et al* 2014 *International J. of Radiat. Res.* **12** 347-54
[12] Hsieh C-M *et al* 2015 *Radiat. Phys. Chem.* **116** 142-146
[13] Karlsson A *et al* 2007 *Phys. Med. Biol.* **52** 4697-706

IOP Conf. Series: Journal of Physics: Conf. Series **847** (2017) 012064 doi:10.1088/1742-6596/847/1/012064

Radiological properties of plastics and TLD materials its application in radiation dosimetry

E James Jabaseelan Samuel[1], K Srinivasan[1], V Poopathi[2]
[1]Department of photonics, Nuclear and medical physics division, SAS, VIT University, Vellore, India
[2] Department of Medical Physics, Apollo Gleneagles Hospitals, Kolkata, India

E-mail: ejames@vit.ac.in

Abstract. In the current study, we evaluated the tissue equivalency of nine different commonly used thermoluminescence compounds and six plastic materials over the photon energy range of 15 KeV to 20 MeV. Our result confirmed that the ratio of number of electrons per gram, electron density of the entire TLD compounds and plastic materials to ICRU-44 soft tissue was lesser than unity, except in the case of polypropylene plastics. The effective atomic number ratio of all the plastic materials was also <1 excluding Poly-vinyl-chloride, and for TLD lithium borate material, it was <1 others which showed the deviation with respect to photon energy. Mass attenuation coefficient (μ/ρ), mass absorption coefficient (μ_{en}/ρ) was calculated and the results are discussed in this paper.

1. Introduction

Tissue equivalent materials have enormous applications in a wide range of fields such as space research, phantom, radiology, nuclear engineering, health physics, radiation physics and radiation dosimetry etc [1-5]. In radiotherapy, tissue substitute materials are employed to extract the dose, similar to the human tissue organs. It is made up of low Z material such as (H,C,N,O etc) [6,16]. Radiation absorption properties of the particular material are similar to human tissue. This material is called tissue substitute or tissue equivalent material. International commission on radiation units and measurement (ICRU-44) has published various tissue substitute materials [7]. Knowledge of gamma ray interaction is important for the calculation of absorbed dose to predict the biological effect of the radiation. Photo electric effect, Compton effect and Pair production are the three major processes for photon attenuation in the energy range of 15KeV to 15MeV. Hence, mass attenuation coefficient and mass energy absorption coefficient depend on those processes. Photoelectric effect depends on the third power of atomic number of the material and it is predominant in the energy range of 20 to 100KeV [8]. KeV energy is mostly used in radiobiological study, therapy and diagnosis. On the other hand, MeV energy range plays an important role in cancer radiotherapy and imaging. Equivalent atomic number is one of the important parameters to know the photon scattering result that buildup of photons it deposited energy at some particular depth its leads to skin sparing effect. Polymethyl metha acrylate (PMMA) which is the most commonly used tissue substitute material due to ease of handling [9]. In medical physics, the most fundamental application of radiation and radioisotopes is radiation dosimetry [10, 11]. Dosimetry may involve in-patient (in-vivo), phantom (in-vitro), environmental monitoring in hospitals. Occupational exposure measurement with a personal detector helps monitor the dose levels. A good TL material can have properties such as tissue equivalency, low fading, good precision and accuracy, high sensitivity, better stability under the environmental condition such as temperature and humidity etc [12]. Thermoluminescence dosimeter plays a vital role in radiation detector particularly in radiotherapy. Only a few TL materials have the tissue equivalent properties among them lithium borate is the most attractive.

$Li_2B_4O_7$:Mn phosphor is the first TL based dosimetry but it has poor sensitivity. Depending on the application, it should be varied for therapy or diagnosis [13]. Many of the methods such as interpolating method, direct method, Auto Zeff, XMuDat computer program are available for calculating the effective atomic number. Researchers have studied the effective atomic number of different biological tissues, soils, amino acids, fatty acids etc [14, 15]. Ferreira *et al* studied the tissue equivalence of ten different materials. Among them, modelling clay and nylon are more tissue equivalent materials and they suggested that the total mass attenuation coefficient is a valuable parameter to prescribe the tissue equivalency of the material [16]. In this present analysis, we studied the tissue equivalent properties of nine various thermoluminescence and plastic materials.

2. Materials and Methods

We selected nine different thermoluminescence compounds namely Lithium borate, Lithium fluoride, Aluminium Oxide, Calcium carbonate, Calcium sulfate, Strontium sulfate, Potassium bromide, Cadmium sulfate (anhydrous salt), Barium acetate and the following six plastic materials - Polyethylene terephthalate, Poly-vinyl-chloride, PMMA, Polypropylene, Nylon, and Bakelite. Mass absorption and Mass attenuation coefficient of the element was generated using NIST Win XCom and mixture rule was used to calculate the coefficients of materials [17]. Effective atomic number, electron density and number of electrons per gram were calculated according to the literature [18]. Results were compared with ICRU-44 soft tissue (four compartments).

3. Result and discussion

Ratio of effective atomic number of all the plastics is almost unity excluding poly-vinyl-chloride. Which has the Z_{eff} of 13.86 due to high atomic number element such as chlorine (Z=17) presence of 0.567% & inclusion of K-edge absorption [19]. Ratio of effective atomic number of TL compounds to soft tissue is varied according to the materials. For Lithium borate, Lithium fluoride, Calcium sulfate, Calcium carbonate and Aluminium oxide less than 2.5 and others have the ratio of within 5.

Table 1. Electron density (ρ_e), Number of electrons per gram (n_e) and Effective atomic number (Zeff) of TL compounds and plastics

S.No	Materials name	ρ_e $\times 10^{23}$e/cm^3	n_e $\times 10^{23}$ e/g	Zeff
1	Lithium borate	1.21618	2.91883	7.248515
2	Lithium fluoride	1.05484	2.78478	8.197445
3	Aluminium Oxide	0.747376	2.95214	11.14045
4	Calcium carbonate	1.10929	3.00729	15.07936
5	calcium sulfate	1.26565	2.9363	16.69195
6	Strontium sulfate	0.711768	2.8186	29.18566
7	Potassium bromide	0.996992	2.73176	31.09977
8	Cadmium sulphate	0.590988	2.77232	38.13204
9	Barium acetate	1.27007	2.78146	43.49785
10	Polyethylene terephthalate	3.13273	2.27009	6.637982
11	Poly-vinyl-chloride	3.08252	2.51634	13.86187
12	PMMA	3.247	2.75177	6.467169
13	Bakelite	3.17485	2.44219	6.404653
14	Polypropylene	3.37635	3.56908	5.474942
15	Nylon	3.25937	2.8098	6.144411
16	ICRU-44 Soft tissue	3.30937	3.30937	7.26272

IOP Conf. Series: Journal of Physics: Conf. Series **847** (2017) 012064 doi:10.1088/1742-6596/847/1/012064

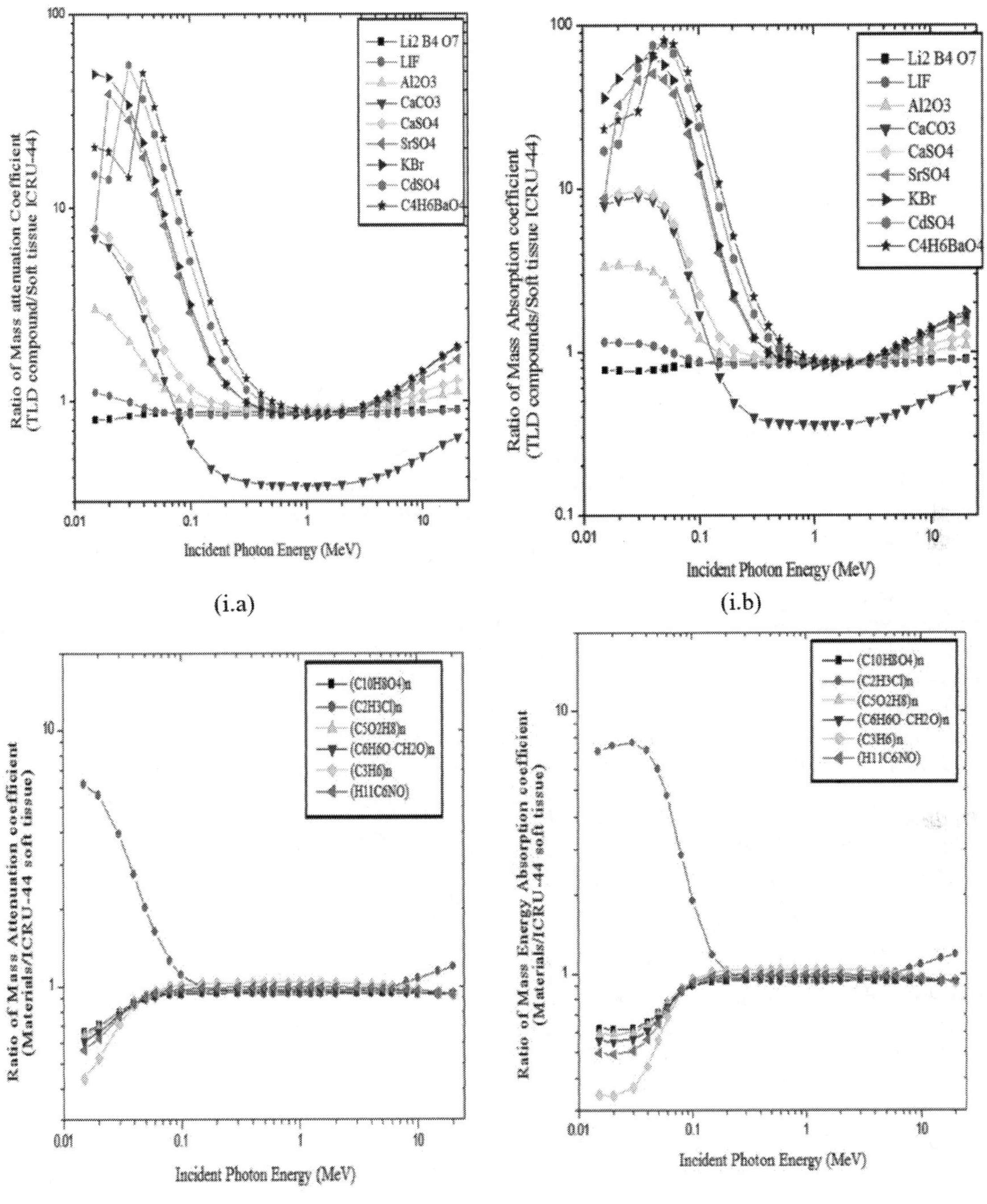

Figure 1. Ratio of Photon mass attenuation coefficient and mass energy absorption coefficient of selected materials are shown figure i a), b) is for TL compounds and figure ii a) & b) for plastics.

Ratio of mass absorption and mass attenuation coefficient of lithium borate thermoluminescence compound to ICRU-44 soft tissue is almost unity in the entire energy range of 0.015MeV to 20MeV. For other TL compounds shows the deviation in lower and higher energy range due to photoelectric and pair production process respectively, it depends on the chemical composition [16, 19]. These results can

IOP Conf. Series: Journal of Physics: Conf. Series **847** (2017) 012064 doi:10.1088/1742-6596/847/1/012064

be shown in the figure i).a & b. In medium energy range all the TL compound consider as tissue equivalent approximately 0.105MeV to 4MeV. In this energy range most probable interaction is Compton and pair production it depends on the number of electrons per gram and square of the atomic number respectively. Lithium borate and lithium acetate not contain the higher atomic number material such as S, Cl, Ba, Br, Sr hence interaction probabilities is all most close to soft tissue. Photoelectric effect is strongly depends on the cube of atomic number, these materials (lithium borate & lithium acetate) exhibits almost close to the effective atomic number of the ICRU-44 tissue. Ratio of electron density, number of electrons per gram of thermoluminescence, plastics materials to ICRU-44 tissue is unity except cadmium sulfate plastics. This can be shown in the table 1. Ratio of mass attenuation, mass absorption coefficient of Polyethylene terephthalate, PMMA, Bakelite and Nylon plastics to soft tissue are unity in the energy range of 0.15MeV to 20 MeV except polyvinyl chloride & Polypropylene. These can be shown in the figure ii a) & b). In case of polyethylene we observed the deviation in the medium energy ranging from 0.1MeV to 6 MeV it may due to compton scattering (i,e) contain high number of electrons per gram (3.56908×10^{23} e/g). B.F Wall *et al* studied the different preparation methods for lithium borate compound and analysed use of medical dosimetry. He found that according to the doping material it changes the properties (i,e) doping with Cu, Ag least attractive compare to Cu and Mn doping. It is incapable for below 500 μGy dose measurement. Li2B4O7:Cu offers more advantages below 500 μGy compare to conventional manganese doping one but suffer from the fading over an hour. His result suggests that degree of light induced fading does not affect the much more in medical dosimetry [20]. Light tight sachets may used to control the fading.

4. Conclusion

Tissue equivalency of various thermoluminacence compound and plastics was verified in terms of their radiological properties. Among them, lithium borate TL material is considered as an ICRU-44 tissue equivalent material over the entire energy range of 15KeV to 20MeV. The presence of high atomic number element destroys the properties of tissue equivalency in lower energies due to the absorption edges. Depending on the application, those tissue equivalent materials were used to construct phantom. This should be varied depending on the clinical application such as diagnosis or therapy.

5. Reference

[1] Hill B *et al* 2005 *Med. Phys.* **32** 1589-97
[2] Davies J B *et al* 2008 *Radiat. Phys. Chem.* **77** 690-96
[3] Hill R *et al* 2010 *Med. Phys.* **37** 4355-63
[4] Hill R *et al* 2014 *Phys. Med. Biol.* **59** R183-231
[5] Hill R *et al* 2009 *Med. Phys.* **36** 3971-81
[6] Kienboch R *et al* 1906 *Arch. Roentgen Ray* **11** 17
[7] ICRU 1989 Report No.44 Bethesda MD
[8] Mann K S *et al* 2012 *Appl. Radiat. Isot.***70** 681-691
[9] Manohara *et al* 2011 *J. Appl. Clin. Med. Phys.* **12** 4
[10] Juan Azorín Nieto *et al* 2004 *AIP. Con. Proc.* **724** 20
[11] Furetta C *et al* 2001 *Nucl. Instr. and Meth A.* **456** 411-417
[12] Singh V P *et al* 2014 *J. Nucl. Chem.* 7
[13] Gill H *et al* 1998 *Radiat. Phys. Chem.* **51** 671-672
[14] Baldock C *et al* 2010 *Phys. Med. Biol.* **55** R1-63
[15] Baldock C *et al* 1998 *Phys. Med. Biol.* **43** 695-702
[16] Ferreira C *et al* 2010 *J. Phys.: Conf. Ser.* **249** 012029
[17] Hubbell J H *et al* 1996 *NIST Standard Reference Database* 26
[18] Sellakumar P *et al* 2007 *Radiat. Phys. Chem.* **76** 1108-15
[19] Singh V P *et al* 2014 *Radiat. Prot.* **49** 115-21
[20] Wall B F *et al* 1982 *Phys. Med. Biol.* **27** 1023-34

IOP Conf. Series: Journal of Physics: Conf. Series **847** (2017) 012065 doi:10.1088/1742-6596/847/1/012065

Radiation dose enhancement of gold nanoparticle on different polymer gel dosimeters

E James Jabaseelan Samuel[1], K Srinivasan[1], V Poopathi[2]
[1]Department of photonics, Nuclear and Medical physics division, SAS, VIT University, Vellore, India
[2] Department of Medical Physics, Apollo Gleneagles Hospitals, Kolkata, India

E-mail: ejames@vit.ac.in

Abstract. In this work, we evaluated the dose enhancement produced by gold nanoparticle on ten different polymer gel dosimeters with a concentration of 7mgAu /g over a wide photon energy range of 15KeV to 20MeV and the results were compared with Soft tissue ICRU-44 produced. Our result showed that maximum DEF was observed at 40KeV, while it was almost negligible at higher energy range. Dose enhancement produced by AuNP on the gel dosimeter medium was varied compared to the reference ICRU-44 tissue, it was ± <1% for PAGAT, NIPAM, nPAG and ± <5% for PABIG, VIPAR, HEAG, BANG1, nMAG & ± <10% for MAGIC, ABAGIC gel dosimeters. Hence, we conclude that choosing the proper gel dosimeter is essential in dose enhancement study.

1. Introduction

Polymer gel consists of monomers [1-4]. Upon the irradiation, free radicals which initiate polymerization, the degree of polymerization are created, which is directly proportional to the radiation absorbed dose [5-7]. It was first proposed by Maryanski in 1993 [8]. Polymer gel systems are useful for complicated radiotherapy treatment plan verification and machine quality assurance [9] in IMRT, SRS, SRT, Proton therapy, Neutron boron capture therapy. The following imaging modalities can be used to extract the dose information within the exposed polymer gels such as OCT, MRI, X-ray CT etc. Gel dosimeters have a number of advantages, including true three-dimensional, soft tissue/water equivalence, directional independence, energy independence and it can be modified depending on the application [10-13]. Nanoparticle size ranges from 1 to 100 nm but definition includes particles up to 1μm which have also been reported. Due to their unique properties, nanoparticles can be used in biomedical application such as drug delivery, biosensors, imaging, and therapeutic agents [14-17]. Owing to its high atomic number, metal nanoparticle absorbs the low energy X-rays, which enhances the absorbed dose when it is loaded within the cancer tissue. Nanoparticles are well accumulated in the tumour via two mechanisms such as enhanced permeability and retention effect. Alternatively, nanoparticle binds with folate targeting material. Roeske *et al* theoretically analyzed the dose enhancement factor over the atomic number (Z) ranging from 25 to 90 at 5mg/ml concentration using 13 different energies from the radioactive source and external beam linac spectrum. They found that DEF linearly increased to Z>70, and his result suggests that high Z material joined the low energy photon, which can produce significant therapeutic advantages (dose enhancement) [17]. Marques *et al* analyzed the dose enhancement of gold nanoparticle on MAGIC-f polymer gel dosimeter by using 250KV X-rays. He found that DEF was 106%, 90% and 77% over the concentration of 0.10mM,

IOP Conf. Series: Journal of Physics: Conf. Series **847** (2017) 012065 doi:10.1088/1742-6596/847/1/012065

0.05mM and 0.02mM [18]. Also, many researchers have studied the dose enhancement effect of the gold nanoparticle using polymer gel dosimeter [19-22]. In the current study, we theoretically analyzed the dose enhancement factor over ten different polymer gel dosimeters. We also discuss the standard reference ICRU-44 soft tissue results.

2. Materials and Methods

Polymer gel elemental composition was taken from the published article [10]. Theoretical dose enhancement factor was calculated using the following formula proposed by Corde *et al* [23]. The mass absorption coefficient was generated by using NIST WinXCOM [24]. A major part of the dose enhancement study was performed at 7mg/g concentration of gold. Hence, we used the same concentration for easy comparison [25].

$$DEF = \frac{\left(\frac{\mu en}{\rho}\right)^{soft\ tissue+AuNp}}{\left(\frac{\mu en}{\rho}\right)^{soft\ tissue}}$$

Where: $(\mu_{en}/\rho) = \sum_i W_i (\mu_{en}/\rho)_i$

(μ_{en}/ρ) is the mass energy absorption coefficient.

W_i & $(\mu_{en}/\rho)_i$ are the atomic weight fraction, mass energy absorption coefficient of the ith element

3. Results and discussion

Radiation dose enhancement is defined as the ratio of the absorbed dose in the tissue in the presence and absence of any contrast agent (gold, iodine, bismuth etc). According to the published data, the photoelectric effect is the major contribution for dose enhancement. DEF was considerably varied with the function of incident photon energy, nanoparticles concentration and size; hence, finding the optimal energy was an important task to get significant clinical benefits. First, we checked the reliability of our methods. For this, we considered DEF produced in the ICRU-44 tissue medium, which was 1.45307, 1.9762 at 80KeV and 50 KeV respectively. These values were in good agreement with the previously published monte-carlo method for 100nm size gold nanoparticle at same 7mg/ml concentration is within ±<4% [23].

These variations might be due to the following two reasons. On the one hand, we studied the different target medium soft tissues /water and on the other, the exclusion of secondary photons. In our method, maximum DEF was 2.0688 at 40KeV which was consistent with the previous Brun *et al* experimental method [24]. This dose enhancement could be due to the L-shell (11.9-14.4KeV) photoelectric effect. Incident photon energy was just above the respective shell binding energy. This produced significant DE effect and the energy was equal to the corresponding electron shell which is clinically not advantageous. Because it just knocks out the electron, this photoelectron does not have sufficient kinetic energy for biological damage. For dose enhancement study, K-shell binding energy is also important but it is not necessary to remain equal to incident photon energy. The probability of photoelectric effect decreased when incident photon energy increased. Due to this, there was no considerable enhancement at higher energies.

The Chemical composition of the target medium is also one of the factors which affect the dose enhancement factor produced by the gold nanoparticle. Low energy photons readily interact with high atomic number material via photoelectric effect (Z3/E3). In this analysis, we considered ten different polymer gel dosimeters. Each gel dosimeter consisted of its own chemical composition [10]. DEF was considerably varied according to polymer gel dosimeter medium. Between the ABAGIC and MAGIC gel dosimeters, we found maximum DEF which consisted of material with the medium atomic numbers such as sulphur (Z=16) and copper (Z=29) which may contribute to the dose enhancement while with other gel dosimeters, we did not observe many variation. PAGAT, NIPAM and nPAG gel dosimeters may have the similar chemical composition of ICRU-44 soft tissue so DEF was within <1% level.

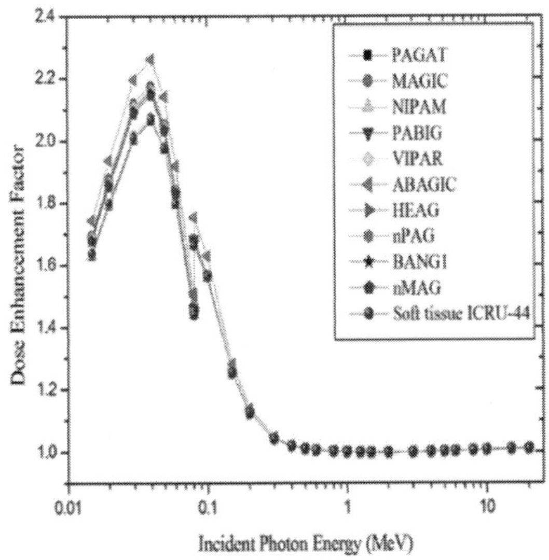

Figure 1. Dose enhancement of gold nanoparticle with 7mgAu/g

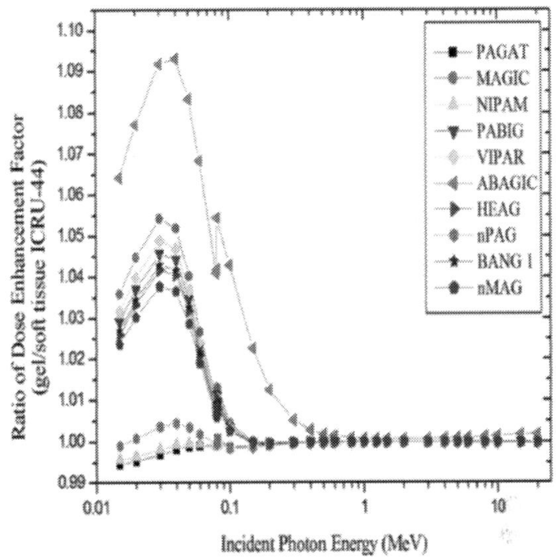

Figure 2. Ratio of dose enhancement factor of polymer gel and ICRU-44 soft tissue

4. Conclusions

Administration of high atomic number metal nanoparticle into the tumour leads to enhanced radiotherapy outcome. This type of treatment is known as Nanoparticle enhanced x-ray therapy (NEXT). Polymer gel dosimeter is one of the media for practically evaluating the DEF. Before the clinical use of metal nanoparticle in cancer treatment, many of the parameters such as biocompatibility, tumour targeting, incident photon energy, concentration and size of NP etc have to be considered. In addition, one more parameter, namely, the chemical composition of the targeting tissue is also important for better therapy dose calculation. PAGAT, NIPAM and nPAG polymer gel is the proper medium for dose enhancement study and similar DEF was absorbed as ICRU-44 soft tissue.

5. References

[1[Baldock C *et al* 1998 *Phys. Med. Biol.* **43** 695-702
[2] Baldock C *et al* 2010 *Phys. Med. Biol.* **55** R1-63
[3] Hill B *et al* 2005 *Med. Phys.* **32** 1589-97
[4] Mather M L *et al* 2003 *Ultrasonics* **41** 551-9
[5] Lepage M *et al* 2002 *Phys. Med. Biol.* **47** 1881-90
[6] Davies J B *et al* 2008 *Radiat. Phys. Chem.* **77** 690-96
[7] Murry P *et al* 2000 *Austral. Phys. Eng. Sci. Med.* **23** 44-51
[8] Maryanski M J *et al* 1993 *Magn. Reson. Imag.* **11** 253-8
[9] Vial P *et al* 2008 *Med. Phys.* **35** 4362-74
[10] Singh V P *et al* 2014 *Radiat. Phys. Chem.* **05** 033
[11] Maryanski M J *et al* 1996 *Med. Phys.* **23** 699-705
[12] Keall P *et al* 1999 *Austral. Phys. Eng. Sci. Med.* **22** 85-91
[13] Venning A J *et al* 2005 *Med. Phys.* **32** 1047-53
[14] Rippel R A *et al* 2011 *J. Nanosci. Nanotechnol.* **11** 3740-8
[15] Kim C K *et al* 2009 *Nanoscale* **1** 61-7
[16] Mahmoudi M *et al* 2011 *Nanoscale* **3** 3007-26
[17] Roeske *et al* 2007 *Tech. Can. Res. Ther.* **6** 395-401
[18] Marques T *et al* 2010 *J. Phys.: Conf. Ser* **250** 012084

[19] Khadem-Abolfazli M *et al* 2013 *Int. J. Radiat. Res.* **11** 55-61
[20] Rahman W N *et al* 2010 *AIP. Conf. Proc.* **1266** 107-10
[21] Rahman W N *et al* 2012 *Australas. Phys. Eng. Sci. Med.* **35** 301-9
[22] Kakade and Sharma *et al* 2015 *J. Can. Res. Ther.* **11** 94-7
[23] Mesbahi *et al* 2013 *Bio Impacts.* **3** 29-35
[24] Brun E *et al* 2009 *Colloids and Surfaces B: Biointerfaces* **72** 128-134
[25] Hubbell J H *et al* 1996 *NIST Standard Reference Database* 26

IOP Conf. Series: Journal of Physics: Conf. Series **847** (2017) 012066 doi:10.1088/1742-6596/847/1/012066

Silver nitrate based gel dosimeter

D Titus[1], E J J Samuel[1], K Srinivasan[1], S M Roopan[2] and C S Madhu[3]
[1]Department of Physics, School of Advanced Sciences, VIT University, Vellore, India.
[2]Department of Chemistry, School of Advanced Sciences, VIT University, Vellore, India.
[3]Department of Oncology, Prathyasa MJKM Cancer Center, Cherthala, Kerala, India

Email: ejames@vit.ac.in

Abstract. A new radiochromic gel dosimeter based on silver nitrate and a normoxic gel dosimeter was investigated using UV-Visible spectrophotometry in the clinical dose range. Gamma radiation induced the synthesis of silver nanoparticles in the gel and is confirmed from the UV-Visible spectrum which shows an absorbance peak at around 450 nm. The dose response function of the dosimeter is found to be linear upto12Gy. In addition, the gel samples were found to be stable which were kept under refrigeration.

1. Introduction

In radiation oncology dosimetry is very important [1, 2]. Gel dosimetry is the only three dimensional dosimeter for obtaining the dose distribution [8-10]. With added nanoparticles gel dosimeters have got great scope in radiotherapy as they enhance the dose and have advantages in imaging too. A review on the development of polymer gel has been reported by Baldock *et al* [11]. The different modalities of readout like Magnetic Resonance Imaging (MRI) [12, 13], Optical Computed Tomography (Optical CT) [14], X-ray CT [15, 16], Ultrasound CT (UCT) [17, 18] and vibrational spectroscopy [19, 20] are also discussed. Keeping all this is mind, in this study we have come up with a new dosimeter which works on the effect of radiation on silver nitrate trapped in a normoxic gel. Soliman [21] reported a dosimeter based on gelatin and silver nitrate in which the gelatin acts as the stabilising agent and the proposed gel was studied for its stability post radiation. Silver nanoparticles exhibited dose enhancement effect in MAGIC gel dosimeter [22]. It has got anticancer properties [23] and is less expensive compared to the high atomic number nanoparticles (Gold, Platinum, Bismuth, etc.) which are commonly used for dose enhancement [24]. Hence, the present study is based on radiation sensitive silver nitrate and PAGAT gel dosimetry irradiated with Cobalt-60 source in the clinical range. In this study we chose a normoxic PAGAT gel [25-27], as it is easy to prepare and is tissue equivalent and stable. PAGAT gel along with the silver nitrate samples were irradiated with different doses. The aim of the study was to evaluate the silver nitrate doped PAGAT gel using UV-visible spectrophotometer.

2. Materials and Method

2.1. Gel Preparation

PAGAT gel chosen for our study was prepared at normal atmospheric conditions as reported by [28]. The ingredients consisted of 89 % of distilled water, 3.5% (w/w) N'N- methylene bisacrylamide, 3.5% (w/w) acrylamide, 5% (w/w) of gelatin and 10mM of Tetrakis (hydroxymethyl) phospohonium chloride.

IOP Conf. Series: Journal of Physics: Conf. Series **847** (2017) 012066 doi:10.1088/1742-6596/847/1/012066

Silver nitrate was added at the end of PAGAT preparation at a concentration of 250 mM [21]. After preparation, the gels were filled in plastic cuvettes of 4.5 ml volume and the cuvette openings were sealed with transparent adhesive tapes to prevent oxygen entry. The samples were covered with silver foil to avoid auto reduction by light. Plain PAGAT and PAGAT with silver nitrate were prepared. The samples of gelatin and silver nitrate were also made as reported [21]. All the samples were kept in refrigerator overnight for proper gelation.

2.2. Gel Irradiation
Gel samples were irradiated with Cobalt-60 (Theratron 780c). Source to surface distance (SSD) was maintained at 80 cm with the field size of 10 x 10 cm^2. Samples were irradiated from 3 Gy to 12 Gy in the range of 3 Gy intervals. Before and after irradiation, the samples were maintained at 4 degree Celsius for 24 hours to attain complete reduction.

2.3. UV Spectrophotometry for PAGAT
U2800-Hitachi was used for UV Spectrophotometry of the gel samples. The samples were read after 24 hours prior to radiation. The 0 Gy sample, which is the unirradiated gel, was taken as the control for the irradiated gel and the same was used for the baseline correction for UV-visible spectrophotometric analysis. 350-800 nm was selected as the range of absorption wavelength with 1nm interval.

2.4. Linearity and Stability
The dose response function of the prepared gel dosimeter was plotted against the different dose levels. And UV- Visible spectrum was taken for the gel samples every alternate day for two weeks

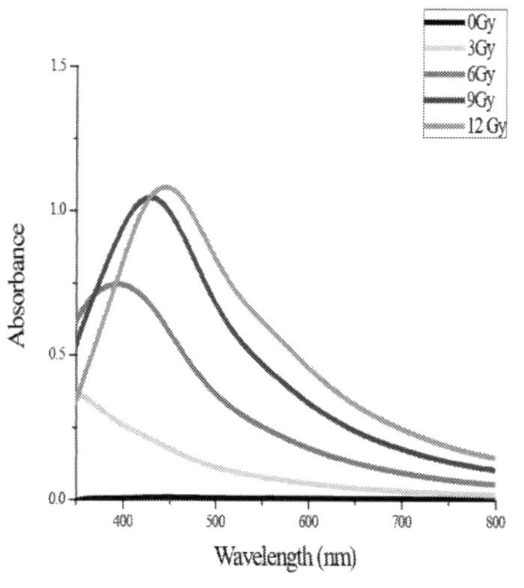

Figure 1. UV- vis spectrum of PAGAT gel.

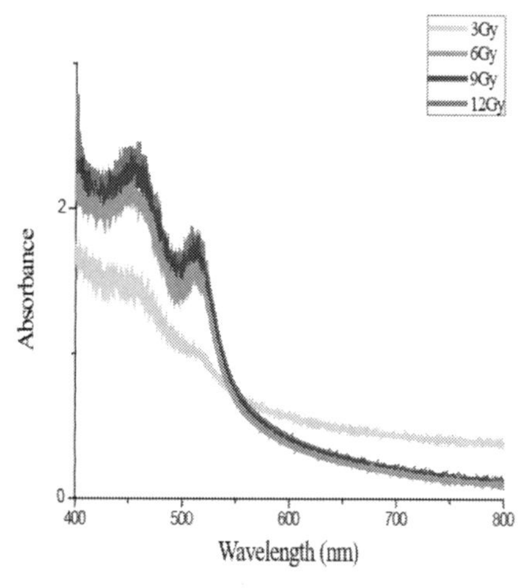

Figure 2. UV-visible spectrum of Gelatin+ $AgNO_3$.

3. Result and Discussion

3.1. UV-visible Spectrophotometry Analysis
The absorption spectrum of PAGAT gel for different dose range can be clearly seen in figure 1. No other peaks were seen for doses up to 12 Gy. Figure 2 illustrates a spectrum similar to as reported [21]. It gives the spectrum for gel samples made with gelatin and $AgNO_3$ with a concentration of 250mM which is considered as the optimum concentration. The irradiated PAGAT and PAGAT+ $AgNO_3$ can be seen in figure 3a and 3b a spectrum analogous to figure 2 is seen in figure 5 which depicts PAGAT + $AgNO_3$ gel. A peak at around 450 nm indicates silver nanoparticle presence, the surface plasmon resonance peak of silver. A second peak at around 510 nm can also be seen which may account for the increase in size

of the nanoparticle which is due to agglomeration [21]. The absorption intensity increases with the radiation dose without much shift in the position of the peak

Visually, there has been a color change from light yellow to dark brown as the dose increases as seen in Figure 3a and 3b. Soon after the preparation the gel sample turned light yellowish indicating the formation of silver nanoparticles because of the heat during the gel preparation.

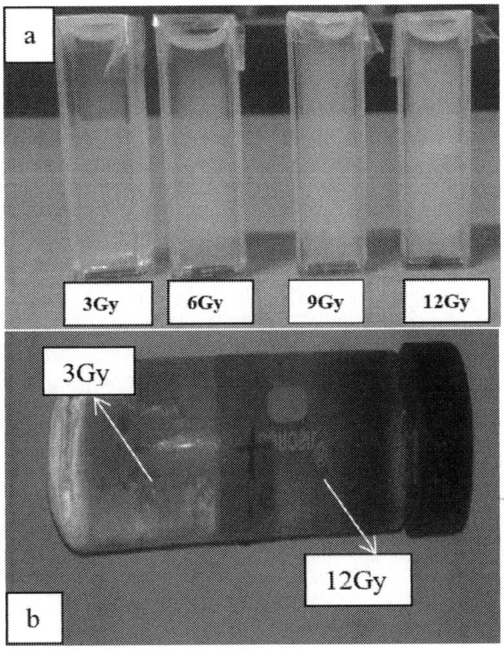

Figure 3. (a) Irradiated PAGAT gel.
(b) Irradiated PAGAT+$AgNO_3$ gel.

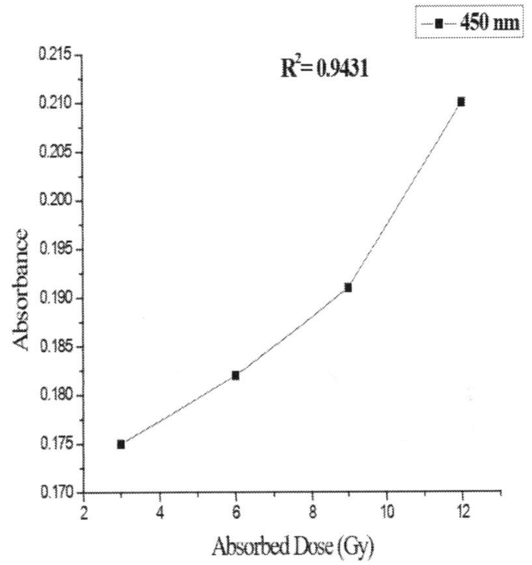

Figure 4. Dose response function at 450nm.

3.2. Linearity and stability

Figure 4 shows the linearity of PAGAT with $AgNO_3$ for 450nm wavelengths. It shows good correlation ($R^2 = 0.9431$). The gel samples with PAGAT and silver nitrate and gelatin with silver nitrate was found to be stable for two weeks when kept in fridge.

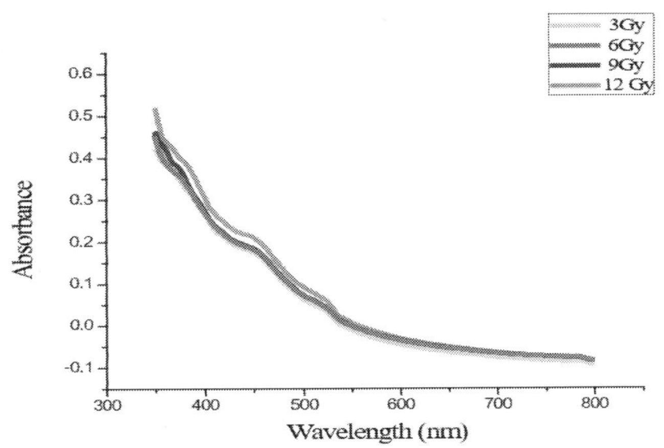

Figure 5. UV-vis spectrum of PAGAT+ $AgNO_3$

4. Conclusion

A new radiochromic gel dosimeter based on PAGAT and silver nitrate was investigated using spectrophotometry in the clinical dose range of 0-12 Gy and the Gamma radiation induces synthesis of Ag nanoparticles in PAGAT gel. The dose response of the dosimeter is linear from 3 to 12 Gy. Future work will be focused on the effects of energy, dose rate and ageing on the response of dosimeter. It also includes working on new additives to improve the radiation sensitivity and stability of the gel.

5. Acknowledgements

The authors are grateful to the DST-SERB, Government of India for their financial aid and also Dr. B Paul Ravindran, Christian Medical College, Vellore, India for the constant support.

6. References

[1] Hill R *et al* 2009 *Med. Phys.* **36** 3971-81
[2] Hill R *et al* 2014 *Phys. Med. Biol.* **59** R183-231
[3] Hurley C *et al* 2006 *Nucl. Instrum. Meth. A* **565** 801-11
[4] Baldock C 2009 *J. Phys.: Conf. Ser.* **164** 012002
[5] Hill B *et al* 2002 *Phys. Med. Biol.* **47** 4233-46
[6] Hill B *et al* 2008 *Phys. Med. Biol.* **24** 149-58
[7] Healy B J *et al* 2003 *Med. Phys.* **30** 2282-91
[8] Baldock C *et al* 1998 *Phys. Med. Biol.* **43** 695-702
[9] Baldock C 2006 *J. Phys.: Conf. Ser.* **56** 14-22
[10] Davies J B *et al* 2008 *Radiat. Phys. Chem.* **77** 690-96
[11] Baldock C *et al* 2010 *Phys. Med. Biol.* **55** R1-63
[12] Venning A J *et al* 2004 *J. Phys.: Conf. Ser.* **3** 155-8
[13] Murry P *et al* C 2000 *Austral. Phys. Eng. Sci. Med.* **23** 44-51
[14] Bosi S G *et al* 2008 *Phys. Med. Biol.* **54** 275-83
[15] Hill B *et al* 2005 *Med. Phys.* **32** 1589-97
[16] Baldock C *et al* 1994 *J. R. Soc. Med.* **87** 806-8
[17] Mather M L *et al* 2003 *Ultrasonics* **41** 551-9
[18] Mather M L *et al* 2003 *Phys. Med. Biol.* **48** N269-75
[19] Baldock C *et al* 1998 *Phys. Med. Biol.* **43** 3617-27
[20] Rintoul L *et al* 2003 *Appl. Spect.* **57** 51-57
[21] Soliman Y S *et al* 2014 *Radiat. Phy. Chem.* **102** 60-7
[22] Hassan M 2016 *J. Pak. Med. Assoc.* **66** 45-48
[23] Firdhouse M J and Lalitha P 2015 *Prog. Biomater.* **4** 113-21
[24] Alqathami M *et al* 2013 *J. Phys.: Conf. Ser.* **444** 0102025
[25] Venning A J *et al* 2005 *Phys. Med. Biol.* **50** 3875-88
[26] Brown S *et al* 2008 *Appl. Rad. Isotop.* **66** 1970-1974
[27] Braun K *et al* 2009 *J. Phys.: Conf. Ser.* **164** 012050
[28] Subramanian B *et al* 2006 *Med. Phys.* **31** 72-7

IOP Conf. Series: Journal of Physics: Conf. Series **847** (2017) 012067 doi:10.1088/1742-6596/847/1/012067

Turnbull Blue Gel (TBG) evaluation as optical dosimeter

E Borguezan Neto, M R Batistuti, J F Pavoni and L Bachmann
Physics Department, University of São Paulo – USP, Av. Bandeirantes, 3900, Ribeirão Preto, SP, Brazil

Email: l.b@usp.br

Abstract. The radiochromic Turnbull Blue Gel (TBG) is sensitive to X- and γ-rays. When exposed to high-energy radiation, TBG can simulate biological tissues, so this gel has potential use in ionizing radiation dosimetry. After irradiation, the yellow TBG changes to the blue Turnbull Blue dye. This work aims to evaluate how TBG responds to exposure to ultraviolet, visible, and infrared radiation from the sun. The gel was irradiated with a solar simulator, which sensitized TBG and transformed it into the Turnbull Blue dye; a change in the optical absorption ensued. The fluence values delivered by the source did not provide a linear response, but it was still possible to use TBG as an optical dosimeter. In conclusion, TBG can be applied to detect solar radiation because it is sensitive enough to measure sun exposure values at time intervals of few minutes.

1. Introduction

The main natural source of optical radiation is the sun. Exposure to the sun is necessary for the human body to produce the vitamins required for the organism to function properly [1,2]. However, overexposure to sun radiation can produce skin burns that may ultimately lead to skin cancer. Studying exposure to solar radiation is essential to understand biological responses. These studies are also important to agriculture because they cast light on photosynthetic processes and help to control the growth of fungi and other microorganisms [3, 4]. The World Health Organization recommends that personal exposure to solar radiation be monitored, so that the percentage of sun radiation received by the population can be established. In this context, optical dosimeters constitute an important tool to measure exposure to sunlight [5]. The dosimetric gel used in this work falls into the category of radiochromic gels because it is based on the radio-induced generation of the insoluble Turnbull Blue (TB) dye. During exposure of the gel to the sun, ferric ions (Fe^{III}) originating from the salt interact with the electrons from the ionized medium. The Fe^{III} ions are then reduced to ferrous ions (Fe^{II}), which subsequently interact with potassium ferricyanide, to form the Turnbull blue dye, $K [Fe^{II}Fe^{III} (CN)_6]$ [6].

In radiobiology, the term "absorbed dose" refers to the energy deposited by charged particles in a mass of tissue volume [J/kg] [7]. This term has been applied in different ways in photomedicine and photobiology. However, this term should not be used in the way it has been employed because it is not possible to measure the amount of energy deposited per kilogram. As the optical tissue penetration is shallow, the term ends up being understood as the energy given as fluence or energy density in an area element [J/m^2] [8].

This study aimed to measure the response of the Turnbull Blue Gel after exposure to sunlight and to evaluate its ability to function as an ultraviolet dosimeter.

IOP Conf. Series: Journal of Physics: Conf. Series **847** (2017) 012067 doi:10.1088/1742-6596/847/1/012067

2. Methodology

2.1. TBG Preparation
The dosimeter composition is Agarose (4% w/w), Potassium Ferricyanide (1.5 mM), $K_3Fe(CN)_6$ and Iron(III) Chloride (0.5mM), $FeCl_3.6H_2O$. The detail to manufacturing the TBG is better described in the reference [6]. After manufacture, the gel (liquid) was placed in quartz cuvettes with optical path of 1cm and kept in the dark at 7° C until use.

2.2. Irradiation
A solar simulator (Xe-3, Q-Lab Corporation, USA) was used to reproduce the solar irradiance in the ultraviolet, visible, and part of the infrared spectral regions. Before exposure, the spectral irradiance was measured with a calibrated spectroradiometer (USB2000, Ocean Optics, USA). The detector and the TBG samples were placed 10 cm below the source. The samples were irradiated at 25°C. Eleven samples were employed in this experiment. One sample was not irradiated and served as control of the natural formation of TB dye in the gel. The other ten samples were irradiated for different times, from 1 to 10 minutes. After irradiation, the samples were kept in the dark at 25°C until the absorbance was measured.

2.3. *Absorbance measurement*
To assess how the color of TBG changed inside the cuvettes, the absorbance of the samples was recorded between 350 and 900 nm. The spectral resolution was 1 nm. The spectra were collected on an Ultrospec 2100 Pro, UV-Vis Spectrophotometer. The blank consisted of a quartz cuvette containing Milli-Q water.

3. Results
Figure 1 shows the spectral irradiance of the solar simulator. The total irradiance was 443.8 W/m^2. The UV, visible, and infrared radiation corresponded to 11, 71, and 18% of the total irradiance, respectively.

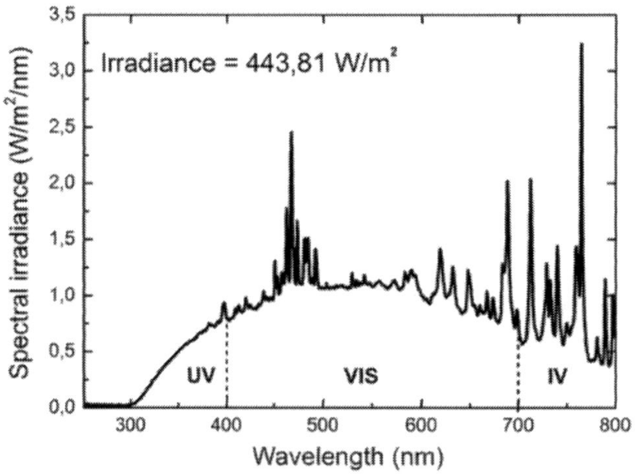

Figure 1. Spectral irradiance of the solar simulator used in this work. The data was obtained by placing the detector at the same place of the gel samples.

When the gel was exposed to high-energy radiation, the radiolysis of water took place. Consequently, free radicals emerged and reduced Fe^{III} ions to Fe^{II}, to give the TB dye. There has been much discussion about the categorization of UV radiation from the sunlight spectrum as ionizing or non-ionizing radiation with respect to water. The radiation used in this study, albeit non-ionizing, can place highly reactive ions of the solution in excited states, which can also reduce ferric ions and produce the

IOP Conf. Series: Journal of Physics: Conf. Series **847** (2017) 012067 doi:10.1088/1742-6596/847/1/012067

TB dye. This can be seen below in the comparison of the absorption spectra of the samples recorded before (Figure 2) and after (Figure 3) irradiation; Figure 4 shows the difference between them.

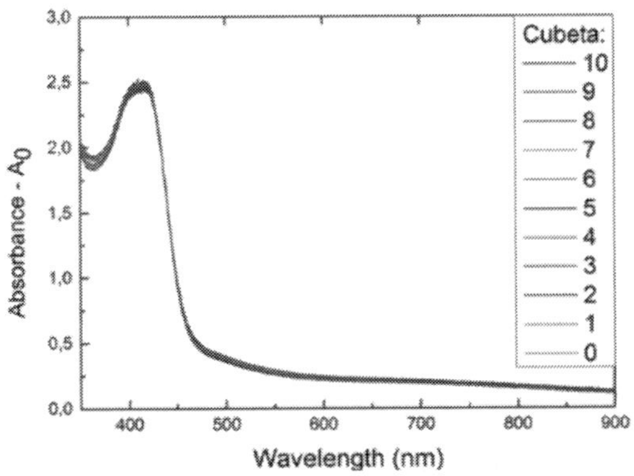

Figure 2. Absorption spectra of eleven samples before irradiation.

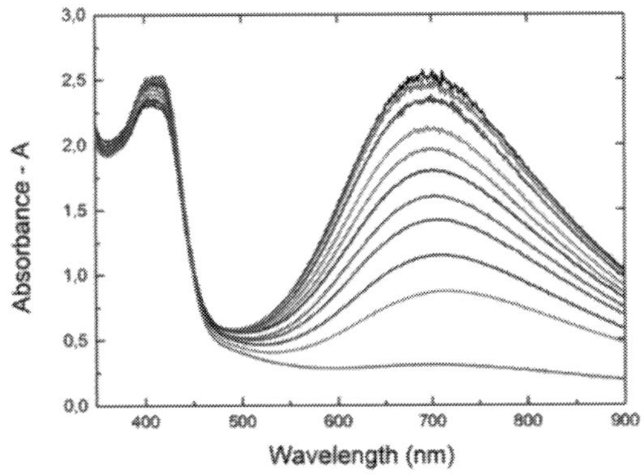

Figure 3. Absorption spectra of eleven samples after irradiation.

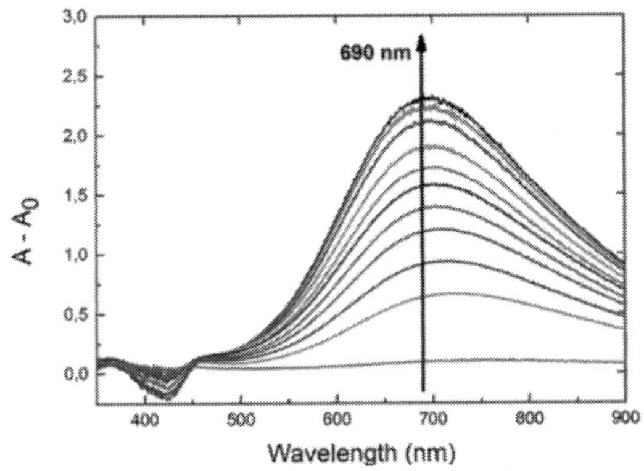

Figure 4. Absorbance difference at 690nm. The absorption peak reveals the Turnbull Blue dye formation.

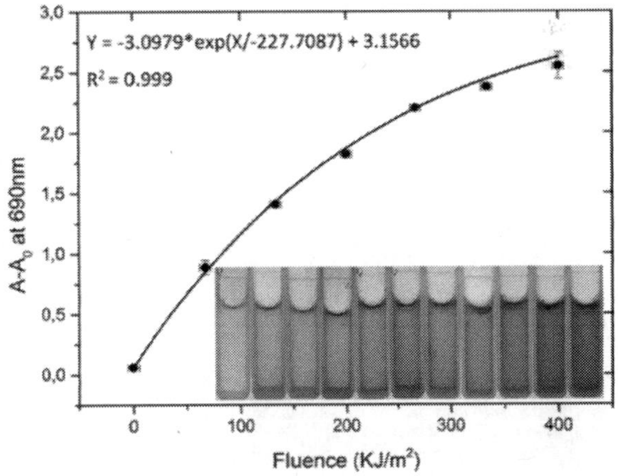

Figure 5. The gel response as a function of fluence delivered by the source. Detail: 11 samples after irradiation from non-irradiated gel to 10min.

On-going from the non-irradiated sample to the sample irradiated for one minute, the absorption changed the most significantly. Therefore, the fluence received for only one minute was enough to obtain a measurable response from the gel. If we consider 10 minutes of exposure as the maximum change in the response of the gel, the irradiation received by the sample during the first minute corresponded to 20% of the maximum response, whereas the smallest change occurred between 9 and 10 minutes (less than 5%), the period during which the curve approached saturation. Figure 5 illustrates the nonlinear relationship between the change in the absorption of the gel at 690 nm and the fluence delivered by the solar simulator. The detail in this figure shows the eleven samples used in the experiment. Considering the uncertainty of triplicate samples, it was possible to obtain a relationship expressed by the equation $Y = -3.0979 * \exp(X / -227.7087) + 3.1566$.

4. Conclusions

The dosimetric Turnbull Blue gel showed measurable and reproducible response after irradiation with a solar simulator. The spectrum issued by the solar simulator (300-800nm) consisted of UV light, the whole visible spectrum, and part of the infrared spectrum. Although it was not possible to pinpoint which

IOP Conf. Series: Journal of Physics: Conf. Series **847** (2017) 012067 doi:10.1088/1742-6596/847/1/012067

wavelength the gel was sensitive to, TBG responded to all the spectral range emitted by the solar simulator, which attested to its ability to detect and dose solar radiation.

5. References

[1] Holick M F 2004 *Am. J. Clin. Nutr.* **79** 362-71
[2] Leffell D J and Brash D E 1996 *Scientific American Magazine* **275** 52-9
[3] Braga G U L *et al* 2001 **74** 734-9
[4] Parisi A V *et al* 2010 UV *Radiation in Global Climate Change* 192-202
[5] Lucas R *et al* 1994 WHO (World Health Organization) - *Environmental Health Criteria 160: Ultraviolet Radiation* Geneva 256
[6] Solc J and Spevácek V 2009 **164** 1-5
[7] ICRU - Report 60 1998 Bethesda, Maryland, USA
[8] Sliney D H 2007 *Photochem. Photobiol.* **83** 425-32

IOP Conf. Series: Journal of Physics: Conf. Series **847** (2017) 012068 doi:10.1088/1742-6596/847/1/012068

Computational simulations of the influence of background noise removal in optical-CT imaging of gel dosimeters

Yi Du[1, 2], Gongyi Yu[1, 3], Xincheng Xiang[1], Xiangang Wang[1], Yves De Deene[2]
[1]Institute of Nuclear and New Energy Technology, Tsinghua University, Beijing, 100084, China
[2]The Biomedical Engineering Laboratory, Department of Engineering, Macquarie University, Sydney, 2109, Australia
[3]Department of Radiation Oncology, Fudan University Shanghai Cancer Center, Shanghai, 200032, China

E-mail: wangxiangang@tsinghua.edu.cn

Abstract. Background noise removal is an easy way to diminish the impact of imperfect signals in the projection blank regions in optical CT imaging of gel dosimeters. The influence of this approach to reconstruction accuracy and noise behaviour is assessed by computational simulations using synthetic phantoms. The results show that an offset is induced in the reconstructed phantom region by background noise removal, but the amplitude of the offset is negligible. Also, no apparent noise or artefact patterns are noticeable. We can conclude that the background noise removal method doesn't impact the reconstruction accuracy of filtered backprojection reconstruction algorithm for optical CT imaging.

1. Introduction

Gel dosimetry has received interest for its outstanding volumetric dose distribution measurement performance [1]. After irradiation, radiation-sensitive gel dosimeters can be read out by either MRI or optical CT [2-4]. Optical CT, utilizing the optical density changes induced by absorbed doses, has been proposed as an alternative readout method to MRI. Several optical CT configurations have been proposed [5].

In a typical optical CT scanning, light emitted from the light source is attenuated by the gel phantom, and forms traverse projections captured by the detector. After acquiring projections over 360 degrees, the tomographic images of the optical densities (ODs) of the gel phantom can be reconstructed by either filtered backprojection or iterative reconstruction algorithms. Absolute dose maps can be derived from the OD distribution by the OD-to-dose calibration. A typical gel phantom projection can be separated in two regions: the central data region containing gel phantom attenuation information, and the peripheral background region, which also contains noise and possible artefacts originating from impurities, such as scratches and stains). It's intuitive to set the background region at zero, which is called *background noise removal.* Although this approach works well for iterative reconstruction algorithms as shown in [6], for the classic filtered backprojection algorithm, currently the standard reconstruction method for optical CT, background noise removal may induce errors and impair the reconstruction accuracy, because of the filtering operation.

In this study, we investigate the influence of background noise removal to reconstruction accuracy of optical CT imaging of gel dosimeters.

2. Methods and Materials

2.1. Noise Modelling
As in [7], the system noise is assumed as stochastic white noise, which corresponds with the combined effect of thermal noise and shot noise. We derived the noise properties from measurements on our cone-beam optical CT scanner [8]. The standard deviation in four selected ROIs in background region was used as the estimate of the noise. The noise standard deviation is 1.5% normalized to the maximal signal in the CCD camera.

2.2. Simulation
Additive stochastic white noise was generated and added to simulated traverse projections as detailed in the following subsection. To assess the influence of the background noise removal method, two simulation scenarios are set with a cylindrical phantom and an elliptical phantom respectively.

2.3. Synthetic Phantoms

2.3.1. Cylindrical Phantom
Since gel dosimeters are usually prepared in a cylindrical cast, a cylindrical phantom is commonly used for gel dosimetry. A synthetic cylindrical phantom is considered to mimic a square-field dose distribution from a 6-MV linac with 100 cm SSD [7], as in Figure 1(a). The optical density (irradiation) field is 2 cm wide and 2 cm offset from the centre to avoid any bias created by cylindrical symmetry. The phantom is 10 cm in diameter, and 10 cm in height. The central axial optical density distribution is show in Figure 1(a). We used the specifications of our optical CT scanner for simulation, while stochastic noise was added to the projection as shown in Figure 1(b)-(c). To nullify the background noise, a threshold was applied to segment the background in the projection image (Figure 1(d)-1(e)).

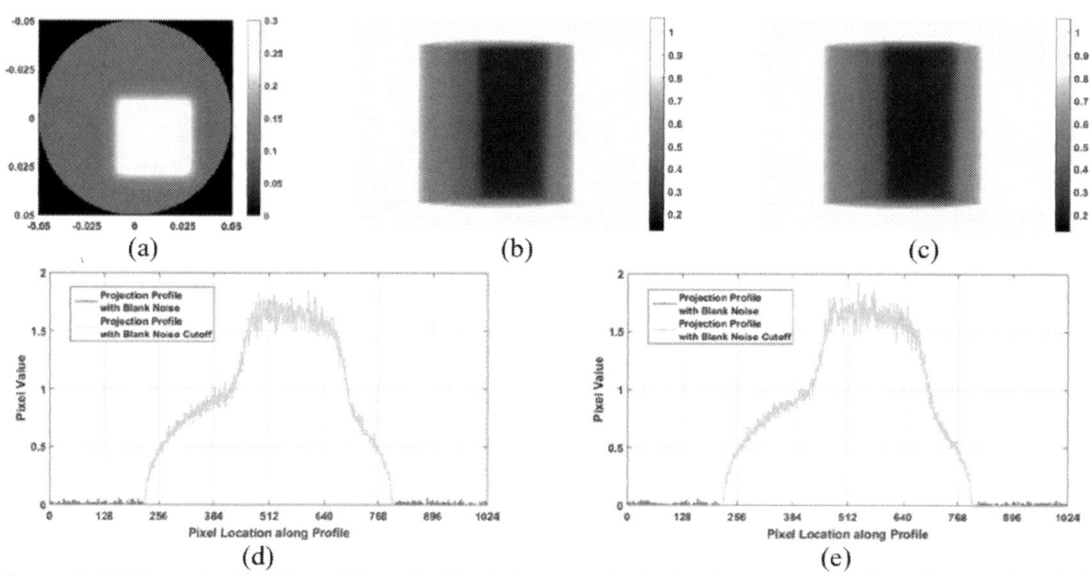

Figure 1. (a) Central axial slice of the cylindrical phantom. Projection images at 0 ° (b) and 90 ° (c), and the corresponding central row profiles shown in (d) and (e) with background noise removed.

2.3.2. Elliptical Phantom
To study the effect of background noise removal to non-cylindrical gel phantom, an elliptical phantom is used here, of which the major-axis is 10 cm and minor-axis 5 cm. The optical density (irradiation) field is 1 cm wide, and 2 cm offset from the centre in y-direction. The central axial optical density

distribution is show in Figure 2(a). The same simulation parameters as 2.3.1 are used. Projections with noise are shown in Figure 2(b)-(c), and the central row profiles in Figure 3(d)-(e) to show the noise removal effect.

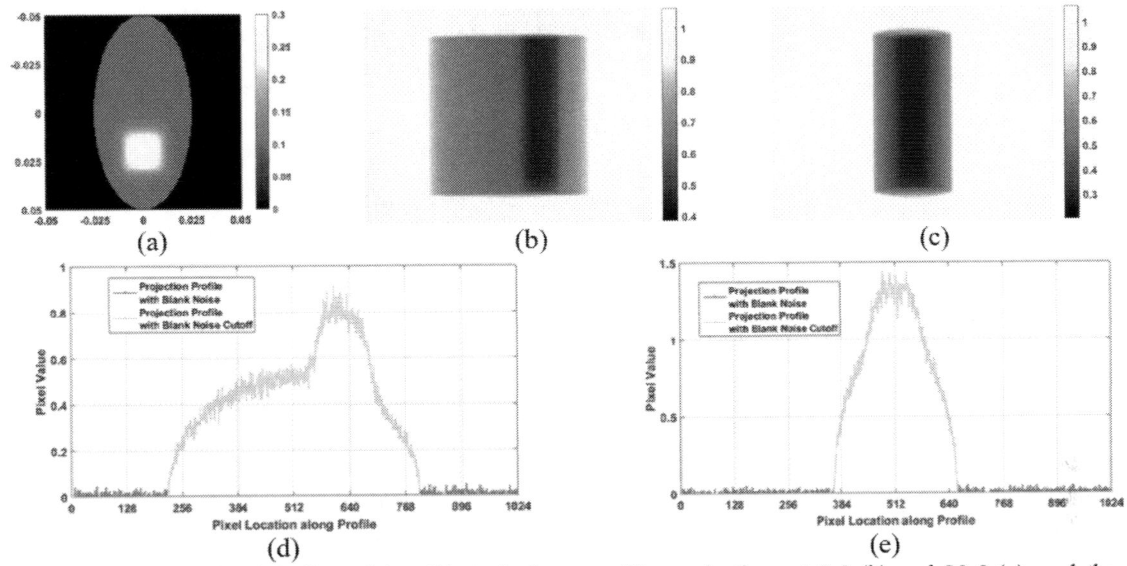

Figure 2. (a) Central axial slice of the elliptical phantom. The projections at 0 ° (b) and 90 ° (c), and the central row profiles are shown in (d) and (e), with the background noise removal effect plotted.

3. Results and Discussion

3.1. Cylindrical Phantom

Tomographic images of the cylindrical phantom were reconstructed via the cone-beam filtered backprojection algorithm (FDK) with a Hann filter for its outstanding noise performance (shown in Figure 3(a)-(b)). In the difference map (Figure 3(c)), we can observe a flat offset, but the amplitude is very small. As shown in Figure 3(d)-(e), the influence of background noise removal is no more than 0.3% normalized to the ΔOD in the square field.

3.2. Elliptical Phantom

The reconstruction results of the elliptical phantom were shown in Figure 4, from which we can draw the same conclusions as above.

Note that, except for the flat offset in Figure 3(d)-(e) and Figure 4(d)-(e), we didn't see any obvious artefact pattern caused by noise cutoff. Also, we calculated the Signal-to-Noise Ratio (SNR) and Contrast-to-Noise Ratio (CNR) inside and outside the OD field in both phantoms, and we find that the background noise removal method will increase the SNR in reconstructed images, but the improvement is also very small.

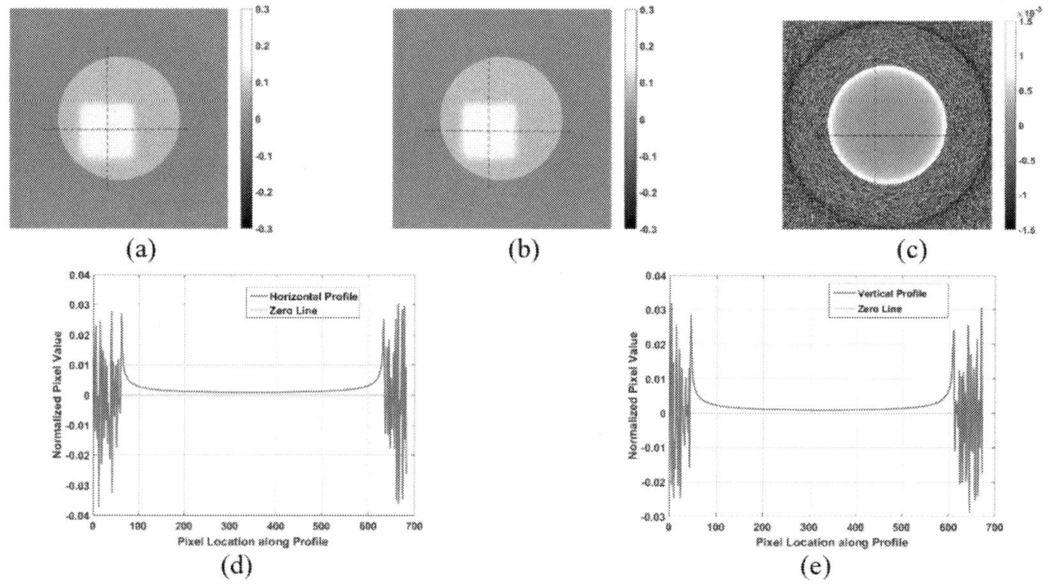

Figure 3. Central axial slices of the cylindrical phantom projections (a) without and (b) with background noise removal. The difference map between (a) and (b) is shown in (c). The horizontal and vertical profiles are plotted in (d) and (e), where the pixel value is normalized by the mean value of the ΔOD in the square filed.

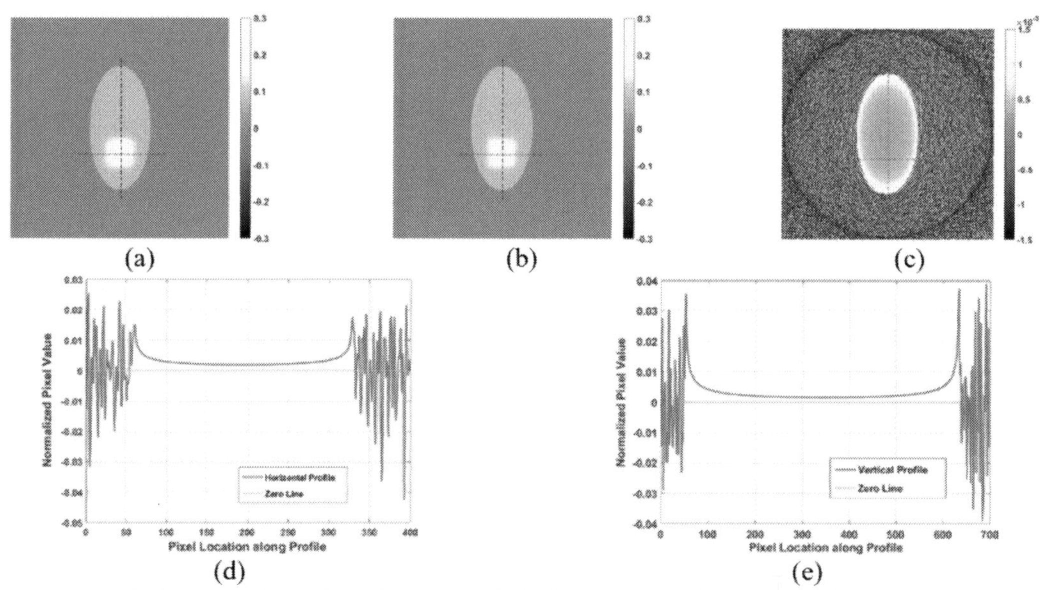

Figure 4. Central slice reconstructed results of the elliptical phantom projections (a) with blank region noise and (b) with blank region noise cutoff. The difference map between (a) and (b) is shown in (c). The horizontal and vertical profiles are plotted in (d) and (e) with normalized pixel values.

4. Conclusions

In both cylindrical and elliptical scenarios, a flat offset is observed, but the amplitude is so small that the effect can be considered negligible. Meanwhile, no apparent noise or artefact pattern is noticeable. We can conclude that impact of the background noise removal to the reconstruction accuracy of filtered backprojection algorithm is negligible in the interested phantom and dose regions. This also implies that, as an efficient trick to accelerate the reconstruction process, cropping the projections to a smaller size won't induce any noticeable loss to accuracy. Besides, the method described in this study can be easily applied to investigate the influence of artefacts removal outside the sample region, and we believe that similar conclusions could be drawn.

5. Acknowledgement

Jointly Supported by the National Science Foundation of China (No. 61571262, No.11575095, No. 61171115), and the National Key Research and Development Program (No. 2016YFC0105406).

6. References

[1] Baldock C *et al* 2010 *Phys. Med. Biol.* **55** R1-63
[2] De Deene Y *et al* 2002 *Phys. Med. Biol.* **47** 2459
[3] De Deene Y and Baldock C 2002 *Phys. Med. Biol.* **47** 3117-41
[4] De Deene Y *et al* 2002 *Phys. Med. Biol.* **47** 3441-63
[5] Doran S J 2009 *J. Phys.: Conf. Ser.* **164** 012020
[6] Yi D *et al* 2015 *J. Phys.: Conf. Ser.* **573** 012063
[7] De Deene Y 2015 *J. Phys.: Conf. Ser.* **573** 012076
[8] De Deene Y 2015 *J. Phys.: Conf. Ser.* **573** 012058

IOP Conf. Series: Journal of Physics: Conf. Series **847** (2017) 012069 doi:10.1088/1742-6596/847/1/012069

Computational simulations of the influence of angular rotation deviations in optical CT imaging of gel dosimeters

Yi Du[1,2], Gongyi Yu[1,3], Xincheng Xiang[1], Xiangang Wang[1], Yves De Deene[2]
[1]Institute of Nuclear and New Energy Technology, Tsinghua University, Beijing, 100084, China
[2]The Biomedical Engineering Laboratory, Department of Engineering, Macquarie University, Sydney, 2109, Australia
[3]Department of Radiation Oncology, Fudan University Shanghai Cancer Center, Shanghai, 200032, China

E-mail: wangxiangang@tsinghua.edu.cn

Abstract. Two types of angular rotation deviations are observed in our optical CT scanner for gel dosimetry: a large deviation at the starting position, and small deviations during the scanning. In order to investigate the influence of these angular rotation deviations on the reconstructed OD maps, we applied the angular deviations on a synthetic phantom for optical CT simulation. Meanwhile, three reconstruction schemes are proposed for compensation. The simulation results show that the influence of both types of rotation deviation is as small as less than 1% over the central reconstruction phantom region. Based on the reconstructed results via different methods, we can conclude: 1) reconstruction with the regridded projections can better decrease the errors caused by the large angle deviation; 2) for small angle deviations, reconstruction with actual-angle compensation delivers less pronounced results; 3) for real cases, the hybrid approach is shown to reduce reconstruction errors.

1. Introduction

Gel dosimetry has received increasing research interest for its outstanding volumetric dose distribution measurement performance [1-4]. Optical CT, utilizing the quantitative optical density changes induced by absorbing dose, has been proposed as an alternative readout method to MRI [5-7]. An in-house dual-wavelength optical CT scanner for radiochromic gel dosimeter imaging has been developed [8]. In a typical optical CT scanning, the gel phantom is rotated slowly step by step with projections at different acquired. When projections over 360 degrees are acquired, tomographic images of the optical densities(ODs) of the gel phantom can be reconstructed by either filtered backprojection or iterative reconstruction algorithms. In each step of the optical CT imaging chain, errors may occur that lead to overall uncertainties on the reconstructed dose maps [9]. One of these errors involves the deviations of the angular increments by the stepper motor have been observed.

In this study, we discuss the influence of these angular deviations on the accuracy of the OD maps.
in projection angles induced by the optical CT phantom rotation system.

2. Methods and Materials

2.1. Rotation Deviation Analysis
The rotation control program is in-house developed in *Matlab*. For typical optical CT scanning, projections are obtained over 360 degrees with an angular increment of 1 degree. However, in practice small deviations from its target rotation angles are observed. We measured the deviations in rotation angles over 20 scans, and Figure 1(a) shows the rotation angle deviation curves in three typical scans. We can see that in most cases the actual projection angles are equal to the respective aimed ones within 0.1 degree deviations. Only a small portion of actual angles deviate more than 0.1 degree, and all deviations are negative, i.e. the actual rotation angle lags behind the rotation angle target.

Based on different locations and amplitudes, we can classify the deviations into two categories: a) at the starting zero angle, the deviation is relatively as large as around -1.9 degree as in Figure 2(a); b) during scanning, rotation deviations always appear as -0.1 degree at random angles. For the former type a) deviation, since it always takes place at the initial state, we assume that there is kind of intrinsic starting or initialization error in the stepper motor. For the latter type b) deviation, the amplitude is constant, but the frequency and position of occurrence are random as in Figure 2(b).

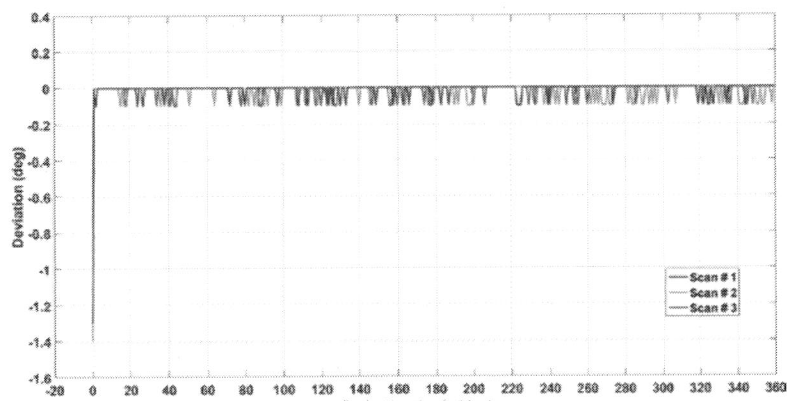

Figure 1. The rotation deviations from aimed angles in three typical experiment scans.

2.2. Deviation Modelling for Simulation
In order to assess the influence of rotation deviations to the OD maps, we used a synthetic circular phantom with square field [10]. For optical CT imaging simulation, all the geometric parameters were identical to the dual-wavelength optical CT scanner specifications [11]. In the rotation deviation simulation part, we first generated 360 exact projection angles from 0 to 359 with 1-degree interval. Then a series of numbers were generated via a deviation model, and added to the exact angle series to mimic rotation deviated angles.

We launched two simulation scenarios to assess the influence of the two types of deviations respectively, and in both scenarios we pushed the boundaries of the deviations to make them much worse than the real case. For the type a) deviation, it only happens at the initial state with a large amplitude as in Figure 2(a). The mean deviation over 20 scans is -1.5 degree, and the worst case -1.9 degree. Herein, we set the amplitude to -3 degree constantly and 6 deviations in random positions. For the type b) deviation, it occurs randomly during rotation as in Figure 2(b) but the amplitude is constant to -0.1 degree. The mean deviation count is 38.95, and the standard deviation is 11.96. To push it to the limit, we introduced a constant -0.1 degree deviation at 180 random projection angles.

2.3 Reconstruction Strategies
The classical cone-beam filtered backprojection algorithm (FDK) with a Hann filter is used for its advantageous performance in noise suppression and gradient integrity [10].

IOP Conf. Series: Journal of Physics: Conf. Series **847** (2017) 012069 doi:10.1088/1742-6596/847/1/012069

To compensate for the angular deviations, we proposed three schemes for reconstruction: 1) we reconstructed the deviated projections blindly as if there were no deviations at all; 2) for compensation, we used the actual rotation angles for backprojection in our FDK algorithm; 3) we regridded the missing projections by linear interpolation at the exact angles. As reference, exact projections with no rotation deviations were also reconstructed with FDK.

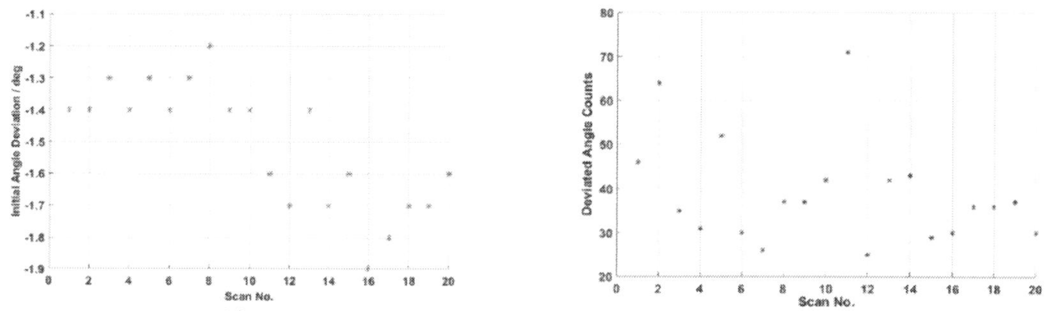

(a) (b)

Figure 2. (a) Initial angle deviations over 20 scans. (b) Counts of deviation angles over 20 scans.

3. Results and Discussion

3.1. Influence of Type a) Deviations

In this simulation, 6 angles deviated by -3 degree were generated by our Monte Carlo method, as shown in Figure 3(a). As described above, we used three reconstruction schemes: blind reconstruction, compensated reconstruction, regridded reconstruction. The comparative reconstructed results are shown in Figure 4. Note that the OD of the synthetic phantom matrix is 0.1, and the mean value OD of the central part of the square field is 0.26. The mean value of ΔOD in the central square field is about 0.16.

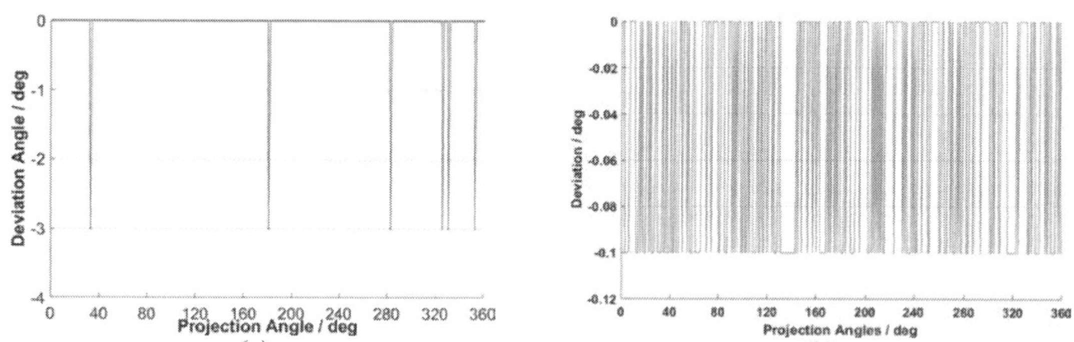

(a) (b)

Figure 3. The simulated type a) rotation deviation curve, and type b) rotation deviation curve.

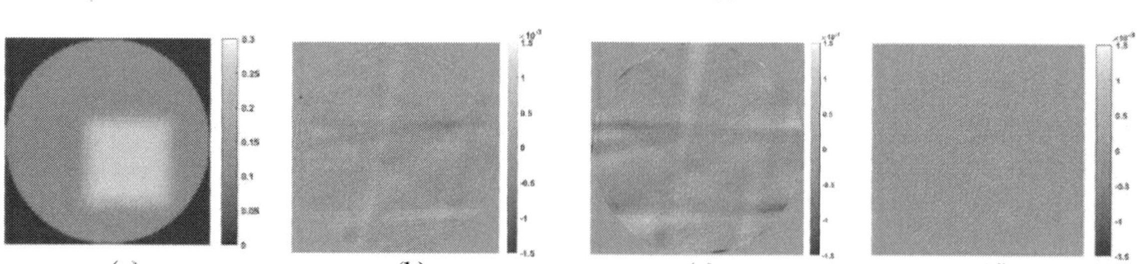

(a) (b) (c) (d)

Figure 4. (a) Reference FDK result of the middle slice with exact projections. (b) Difference map of the results between blind reconstruction and reference reconstruction. (c) Difference map of the results between compensated reconstruction and reference reconstruction. (d) Difference map of the results between regridded reconstruction and reference reconstruction.

3.2. Influence of Type b) Deviations

In this simulation, 180 angles deviated by -0.1 degree were generated (shown in Figure 3(b)). The comparative reconstructed results of the middle tomographic slice by scheme methods are demonstrated in Figure 5. The deviation induced error will corrupt the dose gradient in larger extent, but effect is small with the largest error around 0.5%.

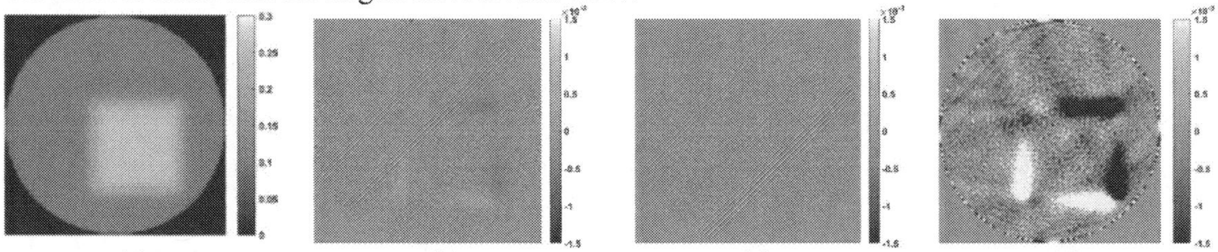

(a) (b) (c) (d)

Figure 5. (a) Reference FDK result of the middle slice with exact projections. (b) Difference map of the results between blind reconstruction and reference reconstruction. (c) Difference map of the results between compensated reconstruction and reference reconstruction. (d) Difference map of the results between regridded reconstruction and reference reconstruction.

3.3. Simulation Using Actual Angle Positions

From the comparison above, we can conclude that regriding reconstruction approach is effective for large angle deviation correction, and compensated reconstruction approach delivers better less affected results from small angle deviations. Herein, in order to assess the influence of angle deviations in a real case, we launched another simulation using actual angle positions from an experimental scanning record file, and adopted a hybrid reconstruction scheme: we first regridded the projection at the starting angle, and used compensated approach for reconstruction. As we conceive, this hybrid method gave the least influence results than the other methods, and the largest deviation in the reconstruction field is no more than 0.3% normalized by the mean value of ΔOD.

4. Conclusions

The results show that the influence of both types of rotation deviation is relatively small, less than 1% (normalized by the mean OD value of the square field) in most reconstruction area. The comparative results of different reconstruction schemes show: 1) reconstruction with regridded projections can better restore the ghost pattern caused by large angle deviation; 2) for small angle deviations, reconstruction with actual angle compensation gives less influenced results; 3) for real cases, a hybrid approach delivers results with least errors. The rationale of correction strategy suitability needs further comprehensive work. This study is based on our in-house scanner, and in the future we will use the proposed method to investigate the influence of rotation deviations in a generalized scenario. Also, we will use a hybrid iterative reconstruction algorithm, i.e. SART+OS+TV [12], which is upgraded from [13] to investigate its performance to rotation deviations.

5. Acknowledgement

Jointly Supported by the National Science Foundation of China (No. 61571262, No.11575095, No. 61171115), and the National Key Research and Development Program (No. 2016YFC0105406).

6. References

[1] Baldock C *et al* 2010 *Phys. Med. Biol.* **55** R1-63
[2] De Deene Y *et al* 2002 *Phys. Med. Biol.* **47** 2459
[3] De Deene Y and Baldock C 2002 *Phys. Med. Biol.* **47** 3117-41
[4] De Deene Y *et al* 2002 *Phys. Med. Biol.* **47** 3441-63

[5] Vandecasteele J and De Deene Y 2013 *Phys. Med. Biol.* **58** 19-42
[6] Vandecasteele J and De Deene Y 2013 *Phys. Med. Biol.* **58** 43-61
[7] Vandecasteele J and De Deene Y 2013 *Phys. Med. Biol.* **58** 63-85
[8] De Deene Y *et al* 2015 *Phys. Med. Biol.* **60** 1543-63
[9] De Deene Y and Jirasek A 2015 *J. Phys.: Conf. Ser.* **573** 012008
[10] De Deene Y 2015 *J. Phys.: Conf. Ser.* **573** 012076
[11] De Deene Y 2015 *J. Phys.: Conf. Ser.* **573** 012058
[12] Du Y *et al* 2016 *Phys. Med. Biol.* **61** 8425-8439
[13] Yi D *et al* 2015 *J. Phys.: Conf. Ser.* **573** 012063

IOP Conf. Series: Journal of Physics: Conf. Series **847** (2017) 012070 doi:10.1088/1742-6596/847/1/012070

Reduction of artefacts caused by missing ray-sum data in optical-CT imaging of implants in gel dosimeters

A Asena[1], S T Smith[1], T Kairn[1,2], S B Crowe[1,3], R D Franich[4], and J V Trapp[1]
[1]School of Chemistry, Physics, and Mechanical Engineering, Queensland University of Technology, Brisbane, Australia
[2]Genesis Cancer Care Queensland, Brisbane, Australia
[3]Royal Brisbane and Women's Hospital, Brisbane, Australia
[4]School of Science, RMIT University, Melbourne, Australia

E-mail: andre.asena3@gmail.com

Abstract. This study demonstrates the degradation in image quality, and subsequent dose evaluation inaccuracies, that are encountered when an optical-CT system reconstructs an image slice of a gel dosimeter containing an opaque implant, and evaluates the feasibility of a simple correction method to improve the accuracy of radiotherapy dose distribution measurements under these circumstances. MATLAB was used to create a number of different virtual phantoms and treatment plans along with their synthetic projections and reconstructed data sets. The results have illustrated that accurately evaluating 3D gel dose distributions in the vicinity of high-Z interfaces is not possible using the filtered back projection method, without correction, as there are serious artefacts throughout the dose volume that are induced by the missing ray-sum data. Equivalent artefacts were present in physical measurements of irradiated PAGAT gel containers when read by an optical-CT system. An interpolation correction performed prior to reconstruction via the filtered back projection algorithm has been shown to significantly improve dose evaluation accuracy to within approximately 15 mm of the opacity. With careful placement of the implant within the gel sample, and use of the linear interpolation method described in this study, there is the potential for more accurate optical CT imaging of gels containing opaque objects.

1. Introduction

Gel dosimeters show potential for measuring complex radiation dose distributions, particularly in radiotherapy [1-10]. Among the potential applications of gel dosimeters in radiotherapy are those in which the irradiated portion of the gel dosimeter surrounds an imbedded object, such as a metallic implant [11-13], head and neck phantoms [14], inserts for brachytherapy measurements [15], or implanted scintillators [16]. In these cases, options for readout of gel dosimeters are limited; there is significant potential for artefacts in MRI and x-ray CT, and missing data in optical computed tomography (optical-CT) due to opacity of the object.

In this work we investigate a methodology for reconstruction of optical-CT images of gel dosimeters which contain obstructions, thereby removing ray-sum data from projections acquired during imaging. The filtered back projection (FBP) method, which is currently employed in both x-ray CT and optical-CT systems, does not provide a mechanism to deal with incomplete sets of projection data or missing ray-sum data caused by metallic or opaque implants [17]. A number of different reconstruction methods to reduce metal artefacts in x-ray CT have been proposed [18, 19] showing mixed results.

However, these are inapplicable to OCT as ray-sum data is completely removed by opacities, rather than being significantly affected but still present in the case of x-ray CT. Here we evaluate the feasibility of a simple pre-reconstruction correction method to improve the accuracy of gel dosimetry dose distribution measurements under these circumstances.

2. Method

MATLAB (version 8.4.0.150421, R2014b) was used to generate a series of virtual phantoms and treatment plans along with their synthetic projections and reconstructed data sets. This simulated work permitted the exclusion of noise influences and light scattering effects, which allowed effects of missing ray-sum data on the image quality of an optical-CT system to be specifically evaluated. Fig. 1 illustrates how a virtual phantom was forward-projected using the RADON function to produce sinograms, which emulate the result that would be obtained via optical-CT scanning. Removing a defined range from the projection data in each sinogram models the effects of having a completely attenuating object present in the phantom. Sinograms were then reconstructed using the IRADON function for projections sampled at 0.25° intervals, using cubic interpolation, and the Hann window for frequency domain filtering. These parameters were chosen as they match the conditions used for reconstructing data obtained from the MGS Research IQScan optical-CT scanner (MGS Research, Madison, CT, USA).

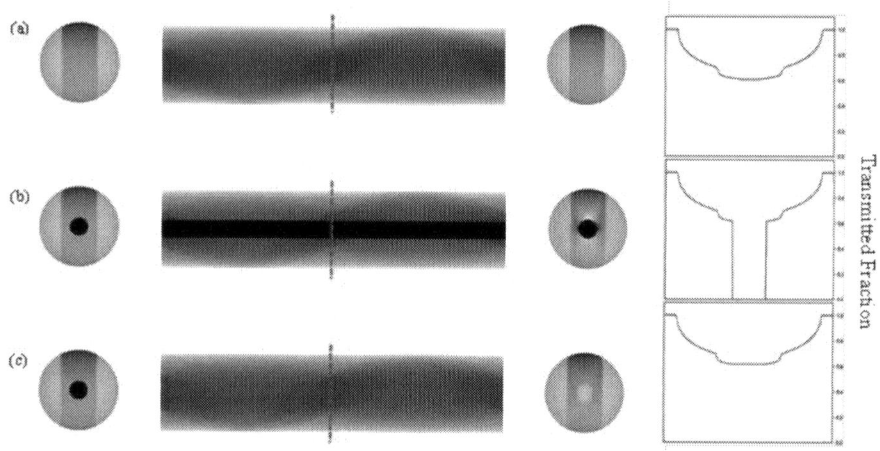

Figure 1. Simulated treatment plans of a circular phantom with a square-field irradiation. Each row of images depicts the original virtual phantom, followed by its forward projected sinogram (with a corresponding profile), and back projected reconstructions. (a) Sinogram with a complete set of ray-sum data, (b) Sinogram with missing ray-sum data, and (c) Sinogram with missing ray-sum data that has been linearly interpolated.

A variety of virtual phantoms were generated to investigate the potential for 3D gel dose distribution measurements for phantoms containing high density implants. Here we simulated a simple 6 MV photon treatment of a circular gel phantom with a high-density, opaque obstruction. The effect of the radiation beam on gel opacity was modelled as a wide, linear region of elevated pixel values passing through the simulated gel cylinder. The falloff of radiation dose (and gel opacity) with depth was modelled as a simple exponential attenuation – a decrease in the pixel values in the region from the top to the bottom of the simulated gel cylinder. Image slices were reconstructed with 1 mm, and 0.25 mm pixel size to investigate the dependence of artefacts on pixel resolution. To highlight the problems associated with dose measurements close to high-Z interfaces, the implant was modelled to reduce the downstream dose by 50%.

The missing ray-sum data was then corrected by linearly interpolating between the data points immediately adjacent to the obstruction for each angular projection. Fig. 1 (a) shows an example of a virtual gel phantom with its forward projected sinogram and back projected reconstruction. Fig. 1 (b) shows a comparable set of simulations for a similar gel phantom that contains an obstruction as

IOP Conf. Series: Journal of Physics: Conf. Series **847** (2017) 012070 doi:10.1088/1742-6596/847/1/012070

demonstrated by the depleted signal in the central axis of the sinogram. An example of the application of this method is shown in Fig. 1 (c).

A PAGAT polymer gel dosimeter was also prepared as per the recipe recommended by Venning *et al* and Khoei *et al* [20, 21]. When the preparation of the final polymer gel solution was completed, it was transferred into two 7cm diameter, cylindrical containers one of which contained a 1.5cm diameter circular obstruction. The containers were allowed to set for 24 hours by storage in a refrigerator. Irradiations were performed with a 6 MV photon beam from an Elekta linac. 3D dose distributions were read out 24 hours after irradiation, using the MGS Research IQScan optical CT scanner. The data was reconstructed to match results from simulated work using the FBP algorithm with cubic interpolation, and the Hann window for frequency domain filtering. The method of linear interpolation of missing ray-sum data was also applied to this experimental work.

3. Results

The simulated results of photon treatment plans of a circular gel phantom with a high-density obstruction are presented in Fig. 2 (a-c). A 50% downstream dose depletion caused by a high-density implant has been simulated; however, the increased doses caused by lateral-scatter and backscatter were not considered in these simulations. For high-density implant dosimetry, characterising the downstream dose depletion is of paramount importance. These specific conditions were simulated to evaluate the potential for 3D gel dose distribution measurements in the vicinity of high-density implants. Corresponding central axis profile plots in Fig. 2 (d) shows the difference between the true dose distribution and the dose distribution after reconstruction with the FBP algorithm. These profiles are not intended to present estimations of the dose inside the obstruction, but rather the doses in the region of the interface.

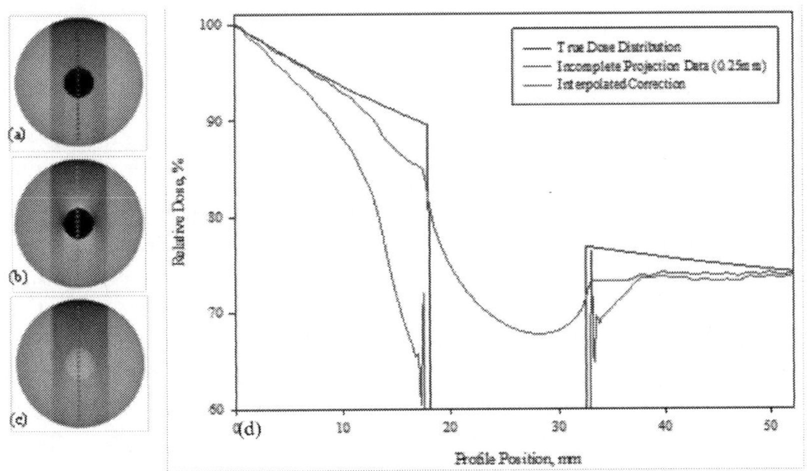

Figure 2. (a) Simulated 6 MV photon treatment plan of a circular gel phantom with opaque, high-density obstruction that reduces the dose by 50%. (b) Image slices have been reconstructed for incomplete sets of projections, (c) and for projection data that has been linearly interpolated, and (d) corresponding central axis profile plot of reconstructed data.

The uncorrected data shows significant artefacts surrounding the obstruction. Relative to the true dose, errors of 24% upstream (17 mm profile position) and 9% downstream (35 mm profile position) are present for the data sets that have been reconstructed using FBP. Applying the linear interpolation reduces the upstream error to 9% and the downstream error to 5%. This correction presents a dramatic improvement in the achievable measurement accuracy of gel dosimeters containing opaque objects.

Fig. 3 (a) displays a raw sinogram with missing ray-sum data obtained from an optical scan of a gel phantom irradiated with a simple 6 MV square photon field. Artefacts caused by reconstructing the

IOP Conf. Series: Journal of Physics: Conf. Series **847** (2017) 012070 doi:10.1088/1742-6596/847/1/012070

sinogram with incomplete ray-sum data present themselves as bright spots up- and down-stream of the obstruction. Radial streaking artefacts are also present in Fig. 3 (a).

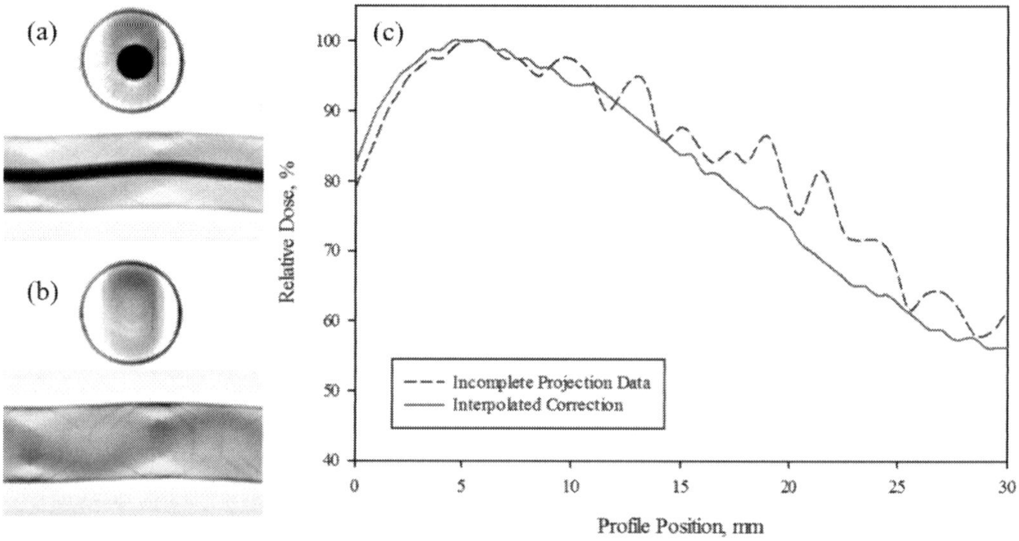

Figure 3. Sinogram and corresponding image reconstructions for a simple square 6MV photon field with an approximately water equivalent obstruction. (a) Image has been reconstructed from sinogram with incomplete ray-sum data, and (b) for data that has been linearly interpolated. (c) Uncorrected vs. interpolated off axis PDD profile plot of experimental dose distribution data.

Fig. 3 (b) shows the result of replacing missing ray-sum data through linear interpolation, with its corresponding percentage depth-dose (PDD) profile plot in Fig. 3 (c). Interpolation of the missing data has successfully minimised the radial streaking artefacts and smoothed the off-axis dose profile displayed in Fig. 3 (c).

4. Conclusions

Use of the FPB algorithm to reconstruct optical CT image slices containing opaque objects can lead to substantial inaccuracies if the effects of missing ray-sum data are not taken into account. However, the measurement uncertainties can be significantly reduced if a simple linear interpolation method is used to correct the sinograms affected by the opaque object. These results represent a dramatic improvement in the achievable measurement accuracy of optical CT readout in the presence of opaque objects. Combined with careful placement of the implant within the gel sample, the linear interpolation method described in this study has the potential to enable the broader and more accurate use of optical CT for the imaging of gels containing opaque objects.

5. References

[1] Baldock C *et al* 1998 *Phys. Med. Biol.* **43** 695-702
[2] Baldock C *et al* 2010 *Phys. Med. Biol.* **55** R1-63
[3] Hill B *et al* 2002 *Phys. Med. Biol.* **47** 4233-46
[4] Hill B *et al* 2008 *Phys. Med. Biol.* **24** 149-58
[5] Healy B J *et al* 2003 *Med. Phys.* **30** 2282-91
[6] Baldock C 2009 *J. Phys.: Conf. Ser.* **164** 012002
[7] Baldock C 2006 J. Phys.: Conf. Ser. **56** 14-22
[8] Mather M L *et al* 2003 *Ultrasonics* **41** 551-9
[9] Hill B *et al* 2005 *Med. Phys.* **32** 1589-97
[10] Davies J B *et al* 2008 *Radiat. Phys. Chem.* **77** 690-96
[11] Asena A *et al* 2015 *J. Phys.: Conf. Ser.* **573** 012061
[12] Kairn T *et al* 2013 *J. Phys.: Conf. Ser.* 444 012108.

IOP Conf. Series: Journal of Physics: Conf. Series **847** (2017) 012070 doi:10.1088/1742-6596/847/1/012070

[13] Asena A *et al* 2015 *Physica Medica* **31** 281-85.
[14] Partridge M *et al* 2006 *Physica Medica* 22 97-104.
[15] Hurley C *et al* 2006 *Nucl. Instrum. Methods A* **565** 801-11
[16] Poole C *et al* 2011 *Australas. Phys. Eng. Sci. Med.* **34** 327-32
[17] Verburg J M and Seco J 2012 *Phys. Med. Biol.* **57** 2803-18
[18] Bazalova M *et al* 2007 *Med. Phys.* **34** 2119-32
[19] Zhang Z *et al* 2011 *Med. Phys.* **38** 701-11
[20] Venning A J *et al* 2005 *Phys. Med. Biol.* **50** 3875-88
[21] Khoei S *et al* 2010 *J. Phys.: Conf. Ser.* **250** 012019

IOP Conf. Series: Journal of Physics: Conf. Series **847** (2017) 012071 doi:10.1088/1742-6596/847/1/012071

Can a commercial gel dosimetry system be used to verify stereotactic spinal radiotherapy treatment dose distributions?

T Kairn[2], A Asena[2], S B Crowe[2,3], A Livingstone[3], D Papworth[1], S Smith[2], B Sutherland[1], S Sylvander[3], R D Franich[4] and J V Trapp[2]
[1]Genesis Cancer Care Queensland, Brisbane, Australia
[2]School of Chemistry, Physics, and Mechanical Engineering, Queensland University of Technology, Brisbane, Australia
[3]Royal Brisbane and Women's Hospital, Brisbane, Australia
[4]School of Science, RMIT University, Melbourne, Australia

E-mail: t.kairn@gmail.com

Abstract. This study investigated the use of the TruView xylenol-orange-based gel and VISTA optical CT scanner (both by Modus Medical Inc, London, Canada), for use in verifying the accuracy of planned dose distributions for hypo-fractionated (stereotactic) vertebral treatments. Gel measurements were carried out using three stereotactic vertebral treatments and compared with planned doses calculated using the Eclipse treatment planning system (Varian Medical Systems, Palo Alto, USA) as well as with film measurements made using Gafchromic EBT3 film (Ashland Inc, Covington, USA), to investigate the accuracy of the gel system. The gel was calibrated with reference to a moderate-dose gradient region in one of the gel samples. Generally, the gel measurements were able to approximate the close agreement between the doses calculated by the treatment planning system and the doses measured using film (which agreed with each other within 2%), despite lower resolution and bit depth. Poorer agreement was observed when the dose delivered to the gel exceeded the range of doses delivered in the calibration region. This commercial gel dosimetry system may be used to verify hypo-fractionated treatments of vertebral targets, although separate gel calibration measurements are recommended.

1. Introduction

Radiotherapy treatments of vertebral targets are challenging to plan and deliver, due to the geometry of the anatomy involved. It is especially important to verify the dose falloff between the targeted vertebra and adjacent spinal cord, and to establish that the spinal cord is adequately spared, for hypo-fractionated (stereotactic) treatments where elevated radiation doses are delivered over few fractions [1]. While valuable dose gradient information can be derived from two-dimensional film measurements in the coronal and sagittal planes [2, 3], the three-dimensional measurement capability of dosimetry gels gives them the potential to provide information about the consistency of the dose gradient, along the whole treatment volume, as well as a more comprehensive indication of the accuracy and deliverability of the planned dose.

This study investigated the use of a commercial gel dosimetry [4] system, for use in verifying the accuracy of planned dose distributions for vertebral treatments. Gel measurements were carried out and compared with planned doses and film measurements, to investigate the accuracy of the gel system and evaluate the suitability of a simple calibration method. In addition to providing an evaluation of a specific gel dosimetry system, this study also provides a useful method for registering planned and measured 3D dose distributions, as well as an overall technique for evaluating commercial (or novel) gel dosimetry systems.

2. Method

Three different stereotactic ablative body radiotherapy (SABR) treatments were planned for vertebral targets, for delivery using a volumetric modulated arc therapy (VMAT) technique, using a Varian Truebeam linac with Millennium MLC (Varian Medical Systems, Palo Alto, USA). In all three cases, the prescribed dose to the planning target volume (PTV) was 24 Gy in 3 fractions and the maximum dose to the spinal cord was required to be below 17.5 Gy. Each of these treatment plans was copied onto a CT scan of a 20×20×10 cm^3 water-equivalent plastic phantom, for comparison with a two-dimensional dose measurement made using EBT3 radiochromic film (Ashland Inc, Covington, USA), and also copied onto CT scans of three cylinders of TruView gel (Modus Medical Inc, London, Canada) for comparison with a three-dimensional gel dosimetry measurement.

The gel measurements were made using three standard 800 ml cylinders (9.6 cm diameter, 11 cm height) of ferrous-xylenol orange based TruView gel. All three samples were individually CT scanned, with radio-opaque markers used to indicate the scanning and irradiation setup position and aid in the registration of the 3D scanned image from the irradiated gel with the 3D dose distribution from the treatment planning system.

Gel handling and irradiations followed the recommendations provided by the manufacturer [5]. Gel results were read out using optical CT, using the VISTA optical CT scanner (Modus Medical Inc, London, Canada) [6]. During scanning the gel was immersed in a refractive index matching fluid made from mixing glycerol, water and food colouring, as recommended by the manufacturer [7].

Pre-irradiation optical CT scans of all three gel samples were acquired, with radio-opaque markers in situ. Additional radio-opaque markers were added to the gel containers after irradiation, so that marker positions would remain visible even after the post-irradiation optical CT scans of the gel were background-corrected [8] via the subtraction of the pre-irradiation scan data.

For irradiation using the three different vertebral SABR treatment plans, the gel samples were positioned with their longer dimension parallel to the linac couch's longitudinal (craniocaudal) direction. The set up and delivery of each irradiation took less than four minutes. Optical CT scanning was completed 60 minutes after irradiation.

The film measurements were made with the film in the centre of the solid water, oriented in the transverse plane, and with one fraction from each of the test treatments delivered to each of three sheets of film. The handing, scanning and calibration of the EBT3 film was completed as previously described [9-11], with scans obtained before and after irradiation and results analysed using the red channel only.

3. Results

Figure 1 shows the quadratic calibration curve that was used to convert pixel values from all scans of the TruView gels into measured dose. This measurement result provides a strong indication that the dose-response of the TruView gel and VISTA readout system was non-linear. The dotted line shown in Figure 1 provides an indication of the degree to which the doses to out-of-field regions would be over-estimated if a linear dose-response was assumed (as is suggested by the manufacturer [5]) and combined with a renormalisation using the maximum dose (as is frequently applied in when gel dosimetry measurements are compared to clinical treatment plans [12]).

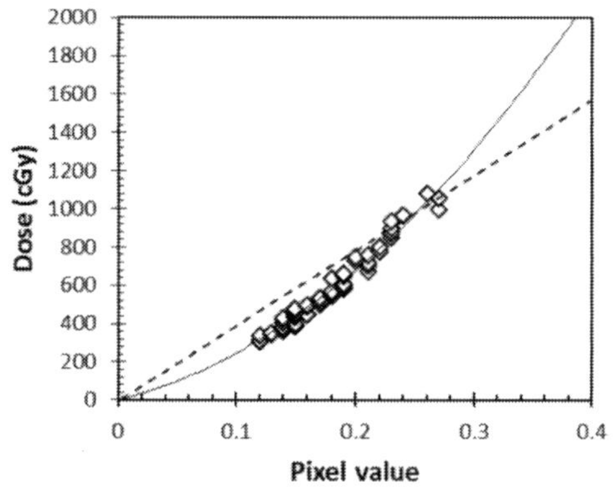

Figure 1. Calibration relationships (solid and dotted lines) derived from measurement data (diamonds) produced by irradiating a sample of TruView gel using one of the SABR test treatment plans and then assuming agreement between the planned and measured doses in a moderate-dose-gradient region (see Figure 2(a)). The solid line shows a quadratic fit to all data points. The dotted line shows a linear fit to the high-dose values only.

Figures 2(a)-(f) and 3(a)-(f) show profiles though the planned and measured doses from the three different spine SABR treatment plans. In all cases, the results of the film measurements agree closely with the planned treatment doses (mostly within 2%). The film measurement therefore confirms the accuracy of the treatment planning system's dose calculation, so that the planned dose distribution can be used as the "ground truth" against which the gel measurement can be verified. Compared to the film results, the gel profiles show the effects of increased noise, decreased scanning resolution and decreased pixel bit depth.

The gel measurement shown in Figure 2(a) was used to derive the calibration data that was used to convert pixel values from all the gel measurements into dose (shown in Figure 1) and therefore shows the closest agreement with the planned dose. Of the other two gel measurements, the closest agreement with the treatment plan (and therefore with the film measurement) was achieved in the plan that was most dosimetrically similar to the plan used for the calibration (Figures 2(b) and 3(b)), while the poorest agreement was observed in the plan involving doses that substantially exceeded the maximum calibration dose (Figures 2(c) and 3(c)).

It is possible that the level of agreement between the gel measurements and the planned doses shown in Figures 2(a)-(f) and 3(a)-(f) may have been over-estimated, due to the calibration method chosen for use in this study. This calibration method lacks the independence of a separate calibration, and produces measurement results which are, by definition, unable to identify any inaccuracies in the dose measurement within the calibration region. Given the apparent non-linear dose response produced in the TruView gel by the hypo-fractionated treatments investigated in this study, it may be advisable to perform an independent calibration measurement (using small vials of gel irradiated to different doses or using a single gel sample irradiated using a known dose profile such as a percentage depth-dose or a wedged profile) whenever TruView gels are used to examine high-dose (stereotactic) treatments. (Note that the film was independently calibrated using reference films irradiated up to 3000 cGy.)

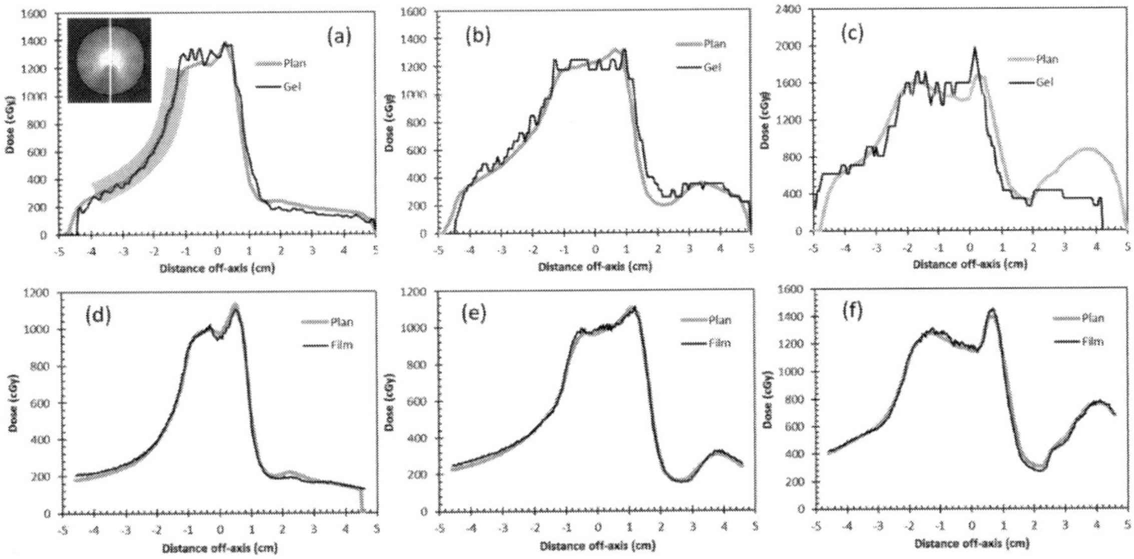

Figure 2. Anterior-posterior dose profiles through the planned target region, showing comparisons between the planned and measured doses in the gel samples (panels (a), (b) and (c)) and comparisons between the planned and measured doses in the film (panels (d), (e) and (f)), for the three different spine SABR treatments. The calibration region (where the dose measured in the gel was assumed equal to the planned dose) is highlighted in panel (a).

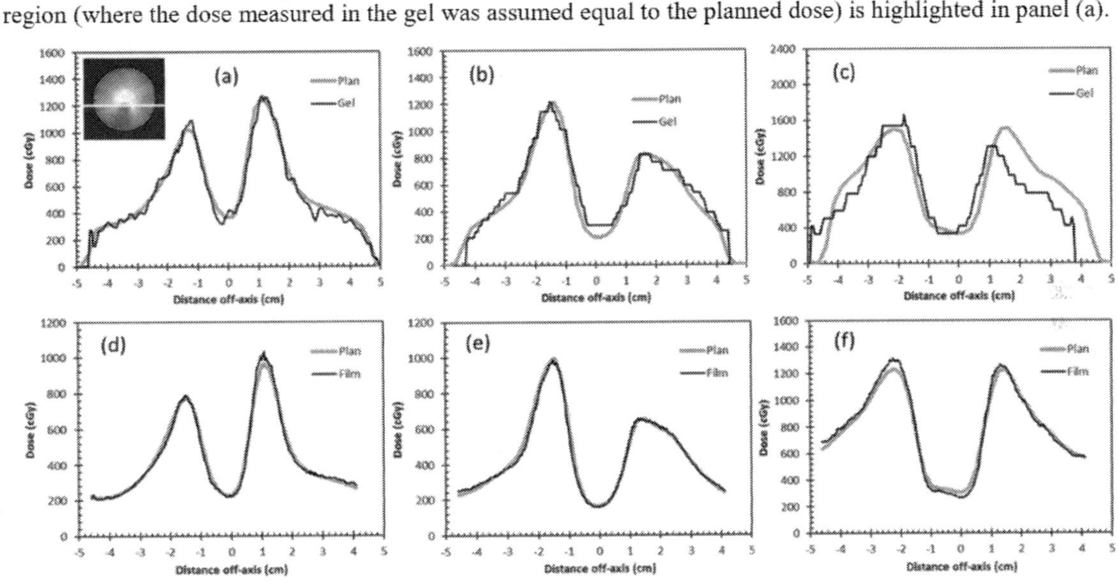

Figure 3. Lateral dose profiles through the planned spinal cord avoidance region, showing comparisons between the planned and measured doses in the gel samples (panels (a), (b) and (c)) and comparisons between the planned and measured doses in the film (panels (d), (e) and (f)), for the three different spine SABR treatments.

4. Conclusion

The results of this study suggest that for the relatively small target volumes involved in stereotactic spinal radiotherapy, TruView gel and VISTA optical CT scanner are able to approximate the doses measured using film, despite lower resolution and bit depth. Poorer agreement is observed when the dose delivered to the gel exceeds the range of doses used in the calibration. This commercial gel dosimetry system may be used to verify hypo-fractionated treatments of vertebral targets, although separate gel calibration measurements are recommended.

IOP Conf. Series: Journal of Physics: Conf. Series **847** (2017) 012071 doi:10.1088/1742-6596/847/1/012071

All doses reported in this study were calculated and measured within media with homogeneous densities. Routine quality assurance of vertebral SABR treatment plans should not use homogeneous media unless the treatment planning system has been thoroughly commissioned and shown to calculate dose accurately when highly modulated, high-dose treatments are planned for bony anatomy.

5. References

[1] Sahgal A *et al* 2008 *Int. J. Radiat. Oncol. Biol. Phys.* **71** 652-65
[2] Kairn T *et al* 2016 *Australas. Phys. Eng. Sci. Med.* **39** 325
[3] Kairn T *et al* 2016 *Med. Dosim.* **41** 258-66
[4] Baldock C *et al* 2010 *Phys. Med. Biol.* **55** R1-63
[5] Modus Medical Inc, TruView Product Data Sheet, http://modusqa.com/images/resources/MMDI_TruView_PDS.pdf
[6] Olding T and Schreiner L J 2011 *Phys. Med. Biol.* 56 1259-79
[7] Modus Medical Inc, VISTA Product Data Sheet, http://modusqa.com/images/resources/Modus_Vista_PDS.pdf
[8] Trapp J V *et al* 2001 *Phys. Med. Biol.* **46** 2939-51
[9] Aland T *et al* 2011 *Australas. Phys. Eng. Sci. Med.* **34** 251-60
[10] Kairn T *et al* 2010 *Phys. Med. Biol.* **55** L37-42
[11] Kairn T *et al* 2011 *Australas. Phys. Eng. Sci. Med.* **34** 333-43
[12] Kairn T *et al* 2012 *Phys. Med. Biol.* **57** 3359-69

IOP Conf. Series: Journal of Physics: Conf. Series **847** (2017) 012072 doi:10.1088/1742-6596/847/1/012072

Trends in gel dosimetry: Preliminary bibliometric overview of active growth areas, research trends and hot topics from Gore's 1984 paper onwards

C Baldock
Research Division, University of Tasmania, Sandy Bay, Tasmania 7005, Australia

Email: clive.Baldock@utas.edu.au

Abstract. John Gore's seminal 1984 paper on gel dosimetry spawned a vibrant research field ranging from fundamental science through to clinical applications. A preliminary bibliometric study was undertaken of the gel dosimetry family of publications inspired by, and resulting from, Gore's original 1984 paper to determine active growth areas, research trends and hot topics from Gore's paper up to and including 2016. Themes and trends of the gel dosimetry research field were bibliometrically explored by way of co-occurrence term maps using the titles and abstracts text corpora from the Web of Science database for all relevant papers from 1984 to 2016. Visualisation of similarities was used by way of the *VOSviewer* visualisation tool to generate cluster maps of gel dosimetry knowledge domains and the associated citation impact of topics within the domains. Heat maps were then generated to assist in the understanding of active growth areas, research trends, and emerging and hot topics in gel dosimetry.

1. Introduction

The use of radiation sensitive gels for the purposes of radiation dosimetry is predominantly as a result of the work originally undertaken by John Gore *et al* in 1984 [1] who showed the ferrous sulfate chemical dosimeter, initially developed by Fricke and Morse [2], could be probed by nuclear magnetic relaxometry and hence by magnetic resonance imaging (MRI). Gore's seminal 1984 paper on gel dosimetry spawned a vibrant research field ranging from fundamental science through to clinical applications [3-5].

Undertaking bibliometric studies of published research papers in a specific field allows the deduction of trends over a specific period and in so doing enable conclusions to be drawn regarding different subfields or research areas [6]. A recent study was undertaken by the author to investigate the radiation dosimetry research field with a view to determining active growth areas, research trends and hot topics from 2011-2015 [7].

A preliminary bibliometric study was undertaken of the gel dosimetry family of publications inspired by, and resulting from, Gore's original 1984 paper [1]. The Web of Science database was used to investigate the themes and trends of the gel dosimetry research field to generate cluster maps of knowledge domains and the associated citation impact of topics within the domains. The methodology used so-called term maps to visualize the totality of the gel dosimetry research field from 1984 up to 2016. A term map is a two-dimensional representation of a research field in which strongly related terms are located close to each other and less strongly related terms are located further away from each other with the term map providing an overview of the structure of a field [8]. Different areas in a map correspond with different subfields. The term maps were used to assist in the understanding of active growth areas of the gel dosimetry research field and associated research trends and emerging hot topics.

2. Methods

The titles and abstracts text corpora, corresponding to 1578 publications, was downloaded from the Web of Science database for the period from 1984 to 2016. Title and abstract information for each publication was merged into a single text corpus file for the period under consideration and analysed using the *VOSviewer* visualization tool (www.vosviewer.com) [8-11]. With an emphasis on visualization, the *VOSviewer* computer program employs a text mining function and associated natural language processing to identify relevant noun phrases in combination with a unified mapping and clustering approach to examine network co-citation data and the co-occurrence of scientific terms. The interactive functionality of the program provides an accessible and hands-on way to explore networks of bibliometric data such as citation counts and/or the co-occurrence relationships among key terms and concepts.

3. Results

3.1 Co-occurrence of terms

Figure 1 shows a co-occurrence term map, figure 2 shows a co-occurrence term density map and figure 3 shows a co-occurrence term cluster map. In figure 1, each term is represented by a circle, where the diameter of the circle and size of its label represent the frequency of the term, its proximity to another term indicates the degree of relatedness of the two terms, and its colour represents the cluster to which it conceptually belongs. The terms are spatially interrelated in multidimensional space and the figure is limited to a 2-D representation with some relationships between terms not be readily apparent. The map can be seen to contain 5 clusters of co-occurring terms. The red cluster is predominately associated with the formulation, composition and properties of gel dosimeters, the green cluster is predominately associated with dosimetry, the blue cluster is predominately associated with MRI and associated topics, the yellow cluster is predominately associated with tissue and radiological properties, and the purple cluster is predominately associated with optical properties and imaging.

3.2 Highly cited terms

Figure 4 is a co-occurrence citation impact heat map indicating the relative citation impact of the terms relative to an average citation impact for all publications in the map. Presented data is normalised to take into account the year of publication as older publications have more opportunity for being cited. Red indicates an above average citation impact, blue a below average citation impact and green an average citation impact. Figure 4 indicates that more highly cited publications occur generally on the right of the figure.

4. Discussion and Conclusions

A number of specific research topics show significant active growth and may be considered to be emerging or hot topics.

The variation in the citation impact across Figure 4 is noticeably with topics on the right of the figure having greater impact in the green (dosimtry) domain and decreasing through the red (formulation, composition and properties of gel dosimeters) and blue (MRI and associated topics) domains. There are significant variations in citation practices with, for instance, much larger reference lists in molecular biology than in mathematics resulting in publications in molecular biology on average being cited more frequently than publications in mathematics. This variation in citation practice may account for aspects of that observed in the purple domain.

Further work is continuing beyond this preliminary investigation of active growth areas and research trends to consider the past development of research areas in gel dosimetry with the aim of further identifying current and potential future so-called 'hot research topics'.

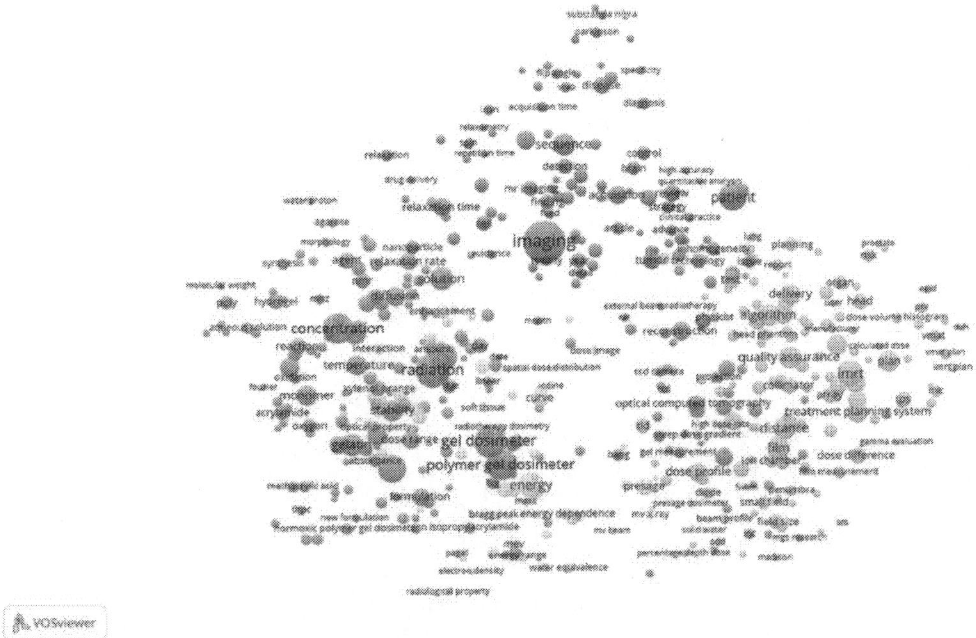

Figure 1. Co-occurrence term map.

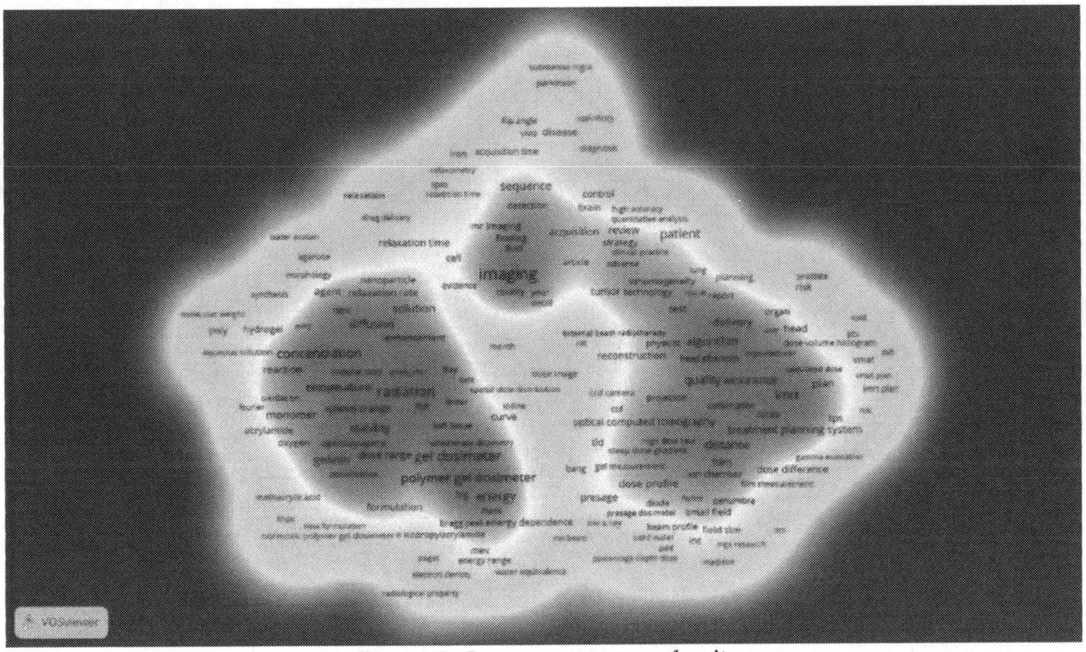

Figure 2. Co-occurrence term density map.

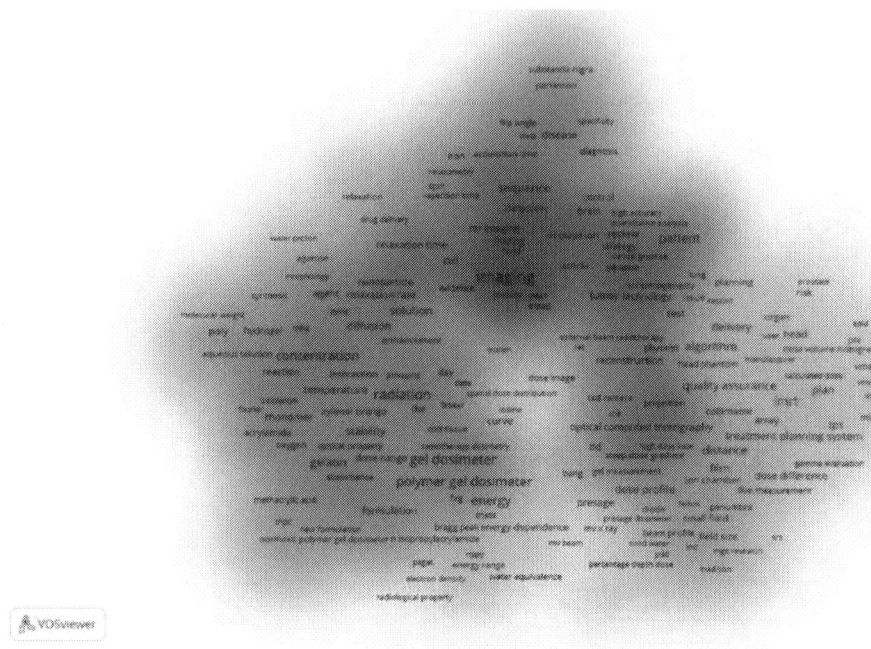

Figure 3. Co-occurrence term cluster map.

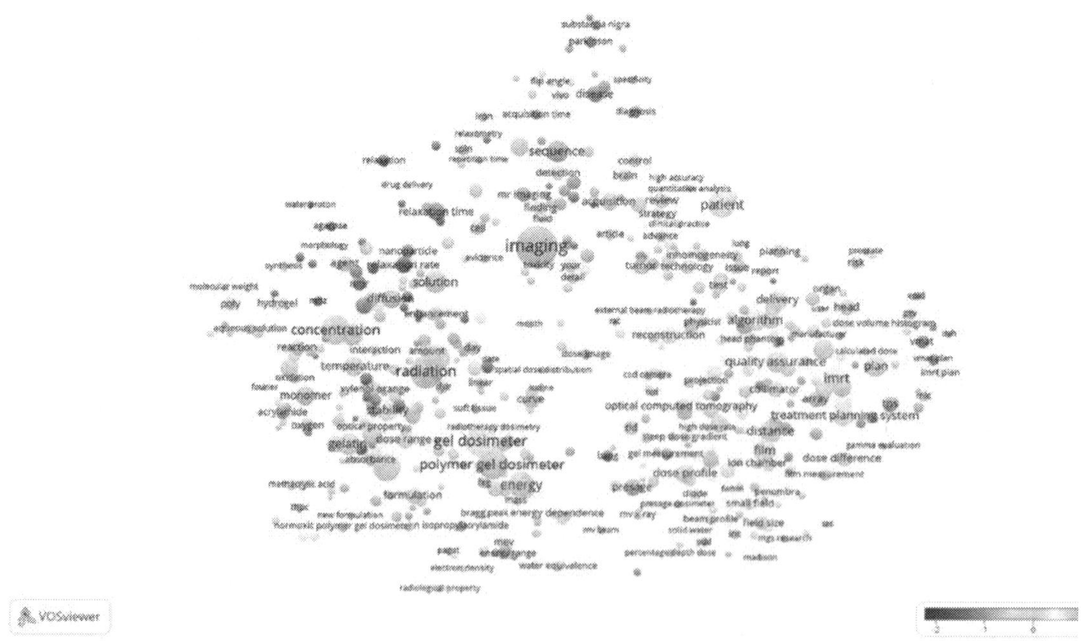

Figure 4. Co-occurrence citation impact heat map

IOP Conf. Series: Journal of Physics: Conf. Series **847** (2017) 012072 doi:10.1088/1742-6596/847/1/012072

5. References

[1] Gore J C *et al* 1984 *Phys. Med. Biol.* **29** 1189-97
[2] Fricke H and Morse S 1927 *Am. J. Roentgenol. Radium Ther. Nucl. Med.* **18** 430-2
[3] Baldock C 2006 *J. Phys.: Conf. Ser.* **56** 14-22
[4] Baldock C 2009 *J. Phys.: Conf. Ser.* **164** 01200
[5] Baldock C *et al* 2010 *Phys. Med. Biol.* **55** R1-63
[6] Gorjiara T and Baldock C 2014 *Scientometrics* **100** 121-48
[7] Baldock 2017 *J. Phys.: Conf. Ser.* **777** 012030
[8] van Eck N J and Waltman L 2010 *Scientometrics* **84** 523-38
[9] van Eck N J et al 2010 *J. American Society for Information Science and Technology* **61** 2405-16
[10] van Eck N J and Waltman L 2011 *ISSI Newsletter* **7** 50-4
[11] van Eck N J *et al* 2013 *PLoS ONE* **8** e62395